T · H · E   N · E · W

PHYSICAL OPTICS NOTEBOOK:
TUTORIALS IN FOURIER OPTICS

T · H · E   N · E · W

PHYSICAL OPTICS NOTEBOOK:
TUTORIALS IN FOURIER OPTICS

George O. Reynolds
Honeywell Electro-Optics Division

John B. DeVelis
Merrimack College

George B. Parrent, Jr.
Innovative Imaging Systems

Brian J. Thompson
University of Rochester

Copublished by
SPIE—The International Society for Optical Engineering
and
American Institute of Physics

SPIE OPTICAL ENGINEERING PRESS

Library of Congress Cataloging-in-Publication Data

The New physical optics notebook.

Rev. ed. of: Physical optics notebook / George B.
Parrent, Brian J. Thompson.
Includes bibliographies and index.
1. Optics, Physical.   I. Reynolds, George O.
II. Parrent, George B.   Physical optics notebook.
QC395.2.N48   1989   535'.2   88-34527

ISBN 0-8194-0130-7

Library of Congress Catalog Card No. 88-34527

Composition: Carrie Binschus
Design: Matt Treat
Printing: Union Printing Company

Copublished by
SPIE—The International Society for Optical Engineering
P.O. Box 10
Bellingham, Washington 98227
and
American Institute of Physics
335 East 45th Street
New York, New York 10017

Table of Contents

Preface

The three of us who author this preface do so with some considerable emotion since our colleague and co-author George Reynolds is no longer with us. George died just after the manuscript was sent to the publisher. We dedicate this volume to his memory as a colleague and a friend of many years. Those who knew George as an optical scientist will miss his continuing important contributions to our field of endeavor but will ever value his many fine original papers as well as his expository writings. Those of us who knew George as a friend will miss the warmth of his friendship and the pleasure of his quick wit even more than his science. There is a tremendous amount of George Reynolds in this book, but any errors that the reader finds must be the responsibility of his co-authors.

This volume, of course, grew out of the original *Physical Optics Notebook,* authored by George Parrent and Brian Thompson, which was first published in book form in 1969 by the Society of Photo-Optical Instrumentation Engineers (SPIE) and was subsequently reprinted. The material in the original book first appeared as a series of articles from 1964 to 1967 in SPIE's journal. The philosophy behind these articles and the resulting book was simply that some of the material we wished to teach was not readily available in a form that matched our particular pedagogical style. Furthermore, we wanted to produce a series of articles that would be helpful for self-study by the reader. To this end, we sought to keep formalism to a minimum, preferring physical implications over mathematical rigor. One of our goals was that every result developed should be accompanied by examples drawn from the physical process it was meant to describe. These goals remain a dominant feature of the present book. We have been gratified by the favorable response to this approach over the years, despite the many shortcomings in other aspects of our presentation.

The original authors resisted the pressure to revise and expand the *Physical Optics Notebook* until some years ago when the perfect co-authors, namely, our old friends and colleagues John DeVelis and George Reynolds, volunteered to undertake the task with the single caveat that they be allowed to "add a few chapters." The book has grown from 16 articles to 38 chapters! The original articles have been revised, and additional material selected from some of the tutorial writings of the authors has been revised to meet the demands of the book. Other chapters and sections of chapters have been written to complete the set of topics that we felt should be covered.

We sincerely hope that the reader, whether a student in the classroom or a self-study student, will find the mixture of fundamentals and the application of those fundamentals to real problems to be an appropriate and stimulating approach to the material.

No instructional text should be expected to stand truly alone, and this tutorial is no exception. While emphasizing interpretation and applications, we have kept the essential formulae. However, for rigor and complete derivations, we make extensive use of references. While extensive, the references in this book are by no means all-inclusive; rather, they reflect the authors' prejudices in that they are the references we have found useful for teaching. At a minimum, they will lead the reader to other related material and original papers.

The reader will find some redundancies, but we feel that repetition on occasion is of value in the learning process. Furthermore, to remove them would have interrupted the logical flow of some of the arguments.

Finally, a word about the illustrations: In our view, the many figures and photographs throughout this volume are essential to the book as we have conceived it. We have done our very best to give the proper credits for those illustrations and we apologize in advance for any errors we may have made—they were certainly not intentional.

In conclusion, we hope you will enjoy reading this tutorial, which we consider a memorial to our friend and colleague George Reynolds, without whom the project would not have been completed.

Acknowledgments

The most difficult task in writing this book is providing an appropriate expression of thanks for the many people whose efforts in one way or another contributed to its completion. For the numerous colleagues whose scientific efforts were the contribution, we have attempted to provide correct references to their work. While inadequate, it is the most direct expression of thanks available to us.

For those others whose work does not show up in other publications but rather is solely a part of this volume, we would like to express our thanks by making them a visible part of this reference. In this vein, our special thanks go to Lorretta Palagi, whose patience is surpassed only by her editorial skills, and to the SPIE publications staff.

The seemingly infinite list of references could not possibly have been correctly sorted out in time to meet the publication date without the able assistance of Barbara Granger, Chris Sohn, and the librarians at the University of Rochester.

John DeVelis wishes expressly to thank Elaine and Frank Grelle for their help with organizing and collating the many diagrams and photographs used in this book. He also wishes to thank Elaine Cherry (under whose patient tutelage he learned his editorial skills) for her help and encouragement spanning the several years during which this manuscript was written.

Special thanks go to Donald A. Servaes for his assistance in preparing many of the illustrations used here. Equally valuable were our discussions with him.

Finally, Peter F. Mueller's contributions go well beyond those that can be covered simply by reference to his published work. He contributed numerous experimental results solely for this book, as well as constructive criticism on the structure and content of several of the chapters.

October 1988   John B. DeVelis
Merrimack College

George B. Parrent, Jr.
Innovative Imaging Systems

Brian J. Thompson
University of Rochester

1 Huygens' Principle

1.1 LIGHT AS A WAVE DISTURBANCE

Fundamental to the understanding of physical optics is, of course, the concept of light as a wave disturbance. In this first chapter, we examine mathematically the simplest situations in which the wave nature of light manifests itself. Consider, for example, the passage of a light beam through a small opening in an opaque screen. The ray optics description of this situation leads to the conclusion that the size of the spot of light observed on a second screen some distance from the first will be simply proportional to the size of the hole.

This, of course, is an oversimplification. One finds, in fact, that such a proportionality law holds quite well for fairly large holes (the language is purposely vague at this point) but does not apply at all for smaller holes. In fact, if the transition from illuminated to nonilluminated areas is examined carefully, the geometric predictions do not hold even for large holes. Furthermore, as the hole is made smaller, the observed spot of light will actually increase as the diameter of the hole decreases—an attempt to illustrate this point is shown in Fig. 1.1. This figure shows a series of photographs of the intensity distribution recorded in a plane 30 cm behind a series of apertures illuminated with a collimated beam of quasimonochromatic light. The magnification is $20\times$ and the size of each aperture is indicated. In each case, the letters of the alphabet correspond to each other. Quite clearly, simple geometric predictions are inadequate. This chapter is devoted to obtaining a mathematical description of phenomena such as those illustrated here.

From the point of view of basic physics, the wave nature of light is fundamental, stemming from the consideration that light is an electromagnetic disturbance and hence is propagated by the vector wave equations, which are readily derived from Maxwell's equations. While such an approach to the problem is philosophically pleasing, it involves a level of mathematical complexity that is out of place in this book. Those interested in this aspect should find Stratton[1] a very readable treatment. For our purposes, we shall assume a much more pragmatic point of

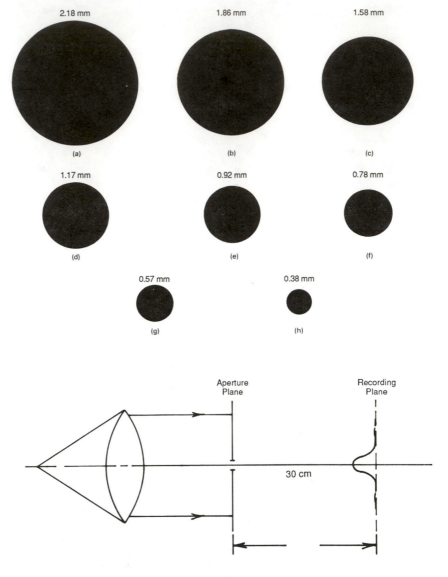

Fig. 1.1 (left) Experimental arrangement for producing diffraction patterns of circular apertures placed in the aperture plane along with the circular apertures.

view. We take it as an empirically established fact that a large class of optical phenomena may be accurately described by the hypothesis that light is a scalar, monochromatic wave. At a later point in the discussion, when we will be in a better position to do so, we will sharpen this hypothesis considerably.

1.2   WAVE PROPAGATION

The basic problem of diffraction is thus simply the determination of the manner in which a wave propagates from one surface to another. Thus, we shall be concerned with the solution of the wave equation. In its most general form, the scalar wave equation may be written as

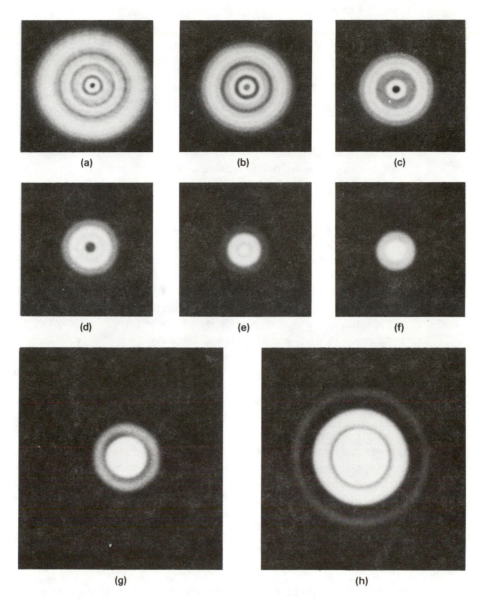

(a)　　　　　　　　(b)　　　　　　　　(c)

(d)　　　　　　　　(e)　　　　　　　　(f)

(g)　　　　　　　　　　(h)

Fig. 1.1 (right) The corresponding intensity distributions observed in the recording plane of Fig. 1.1 (left).

$$\nabla^2 V(\mathbf{x}, t) = \frac{1}{c^2} \frac{\partial^2 V(\mathbf{x}, t)}{\partial t^2} , \tag{1.1}$$

where $V(\mathbf{x}, t)$ is the optical disturbance. For monochromatic waves, $V(\mathbf{x}, t)$ separates to a form

$$V(\mathbf{x}, t) = \psi(\mathbf{x}) e^{i2\pi\nu t} , \tag{1.2}$$

where $\mathbf{x} = x\mathbf{i} + y\mathbf{j} + z\mathbf{k}$, $\boldsymbol{\xi} = \xi\mathbf{i} + \eta\mathbf{j} + z\mathbf{k}$ [see Eq. (1.6)], and $\mathbf{i}$, $\mathbf{j}$, and $\mathbf{k}$ are unit vectors along the coordinate axes; ν is the frequency of the wave; and $\psi(\mathbf{x})$

describes the spatial variation of the amplitude and phase of the disturbance. By substituting this monochromatic form into the general wave equation, the time dependence is eliminated and the spatial part of the disturbance is seen to satisfy the Helmholtz equation:

$$\nabla^2 \psi + \left(\frac{2\pi\nu}{c}\right)^2 \psi = 0 \ . \tag{1.3}$$

The early chapters, therefore, will be concerned with the solution of this equation, which can be written in a more convenient form as

$$\nabla^2 \psi + k^2 \psi = 0 \ . \tag{1.4}$$

Here ψ is the wave disturbance, ∇ is the differential operator, and k is the wave number, $2\pi/\lambda$; λ is the wavelength. Equation (1.4) may be rigorously solved using Green's theorem (see, for example, O'Neill[2]). For the present, however, we shall be content with an approximate solution which emphasizes the physical factors of the problem. The relation to the rigorous solution is discussed in Chapter 11. Here we shall be concerned with the Huygens' principle development of the solution. Thus, our solution is constructed from the following principle: A geometric point source of light will give rise to a spherical wave emanating equally in all directions. To construct a general solution from this particular one, we note that the Helmholtz equation is linear and hence a *superposition* of solutions is permitted. Next we require only the point of view that an arbitrary wave shape may be considered as a collection of point sources whose strength is given by the amplitude of the wave at that point. The field, at any point in space, is simply a sum of spherical waves. This argument, while physically pleasing, ignores the fact that the wave has a preferred direction. We may sharpen the development by the inclusion of an inclination factor to take into account this preferred direction.

The preceding argument may be expressed mathematically as follows: a spherical wave is described by the equation

$$\psi_{sp} = \frac{e^{\pm ikr}}{r} \ , \tag{1.5}$$

where r is the distance from the point source to the point of observation and the $\pm$ indicates a diverging and a converging wave, respectively. Thus, if the disturbance across a plane aperture is described by a wave function $\psi'(\boldsymbol{\xi})$ (here $\boldsymbol{\xi}$ is the position vector in the aperture plane), then the Huygens' principle development for the field at a point $\mathbf{x}$ beyond the screen leads to the expression (see Fig. 1.2)

$$\psi(\mathbf{x}) = \int_{\text{aperture}} \psi'(\boldsymbol{\xi}) \Lambda(\mathbf{x}, \boldsymbol{\xi}) \ \frac{e^{+ikr(\mathbf{x}, \boldsymbol{\xi})}}{r(\mathbf{x}, \boldsymbol{\xi})} \ d\boldsymbol{\xi} \ . \tag{1.6}$$

That is, Eq. (1.6) simply expresses the fact that a spherical wave of amplitude $\psi'(\boldsymbol{\xi})$ emanates from each point $\boldsymbol{\xi}$ in the aperture; $\Lambda(\mathbf{x}, \boldsymbol{\xi})$ is the inclination factor referred to above. The exact form of this factor is not important in this chapter and will be developed in a later chapter when it is important. For the present, we need only note that $\Lambda(\mathbf{x}, \boldsymbol{\xi})$ is essentially constant if we restrict $\mathbf{x}$ and $\boldsymbol{\xi}$ to a suitably small region of the neighborhood of the axis, i.e., a line normal to the

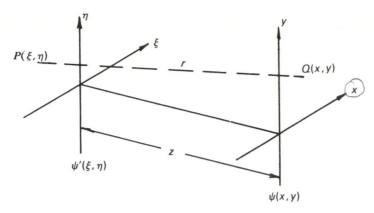

Fig. 1.2 Schematic diagram for wave equation analyses with all pertinent coordinates and distances labeled.

aperture plane and passing through the center of the aperture. Thus, with this restriction on the observation point, the solution takes the form

$$\psi(\mathbf{x}) = K \int \psi'(\boldsymbol{\xi}) \frac{e^{+ikr(\mathbf{x},\boldsymbol{\xi})}}{r(\mathbf{x},\boldsymbol{\xi})} \, d\boldsymbol{\xi} \, . \tag{1.7}$$

The remainder of this chapter is devoted to obtaining solutions of Eq. (1.7) for various aperture geometries. First we note that the function $r(\mathbf{x},\boldsymbol{\xi})$ occurs twice in Eq. (1.7). The r in the denominator affects only the amplitude of the wave and may be regarded as a slowly varying function if we restrict $\mathbf{x}$ and $\boldsymbol{\xi}$ as indicated above. The r in the exponent, however, is multiplied by $2\pi/\lambda$ and affects the phase of the radiation. Hence, while ignoring the effects on amplitude of the variations in r, we must retain the variation of r in the exponent. Accordingly, Eq. (1.7) may be further simplified to

$$\psi(\mathbf{x}) = \frac{K}{z} \int \psi'(\boldsymbol{\xi}) e^{+ikr(\mathbf{x},\boldsymbol{\xi})} \, d\boldsymbol{\xi} \, . \tag{1.8}$$

Since the amplitude varies slowly with the distance r, we may approximate r by z in the amplitude term. Thus, if we denote the Cartesian coordinates in the aperture by ξ and η, the coordinates in the observation plane by x and y, and the separation of the planes by z, we may use (see Fig. 1.2) the Pythagorean theorem and write

$$r(\mathbf{x},\boldsymbol{\xi}) = [(x-\xi)^2 + (y-\eta)^2 + z^2]^{1/2}$$

$$= R\left[1 + \frac{\xi^2+\eta^2}{R^2} \frac{-2(\xi x + \eta y)}{R^2}\right]^{1/2}, \tag{1.9}$$

where $R = (x^2 + y^2 + z^2)^{1/2}$. We now restrict our attention to relatively large distances so that Eq. (1.9) may be expanded in a binomial series and approximated by its first two terms. Thus,

$$r(\mathbf{x},\boldsymbol{\xi}) \simeq R + \frac{\xi^2+\eta^2}{2R} - \frac{\xi x + \eta y}{R} \, . \tag{1.10}$$

This approximation characterizes most of the optical phenomena in which we are interested. Exceptions [i.e., cases where Eq. (1.10) is not allowed] are discussed in a later chapter. On the basis of Eq. (1.10), diffraction problems are customarily divided into groups depending on the relative magnitude of the last two terms. Thus, for those circumstances in which the $(\xi^2 + \eta^2)/2R$ term either vanishes or may be neglected, we observe Fraunhofer or far-field diffraction, while for those circumstances in which the $(\xi^2 + \eta^2)/2R$ may not be neglected, we observe Fresnel diffraction [see, e.g., Figs. 1.1(a) to (f)]. The term $(\xi^2 + \eta^2)/2R$ may be eliminated in either of the two ways described below.

1.2.1   Far-Field Approximation

The term $(\xi^2 + \eta^2)/2R$ may be eliminated by increasing R to such a point that

$$\frac{k(\xi^2+\eta^2)_{max}}{2R} \ll 1 \ . \tag{1.11}$$

This condition is called the "far-field" or "far-zone" approximation and is of particular importance at microwave frequencies but it also merits attention in many physical optics situations including the design of a pinhole camera since it is exactly this condition that must be met. Using this condition and Eqs. (1.10) and (1.8) and noting that $z^2 \gg x^2 + y^2$, we obtain for the field

$$\psi(x,y) = \frac{Ke^{-ikz}}{z} \iint \psi'(\xi,\eta)\exp\left[\frac{-2\pi i}{\lambda z}(\xi x + \eta y)\right] d\xi \, d\eta \ , \tag{1.12}$$

where $z \cong R$. Figures 1.1(g) and (h) are examples of this condition.

1.2.2   Fraunhofer Condition

This condition is achieved by placing a lens in the (ξ, η) plane and observing the diffraction pattern formed at the focus of the lens. To examine this approximation, it is necessary to recall the function of a lens from the point of view of physical optics. Thus, by definition, a lens is a device that converts a plane wavefront into a spherical wavefront of radius f. The concept is illustrated in Fig. 1.3. Here P is a plane wave incident on the lens; S is a spherical wave emergent from the lens; ρ is the radial height to an arbitrary point on S; and x is the radial distance from the foot of the perpendicular from S to the wavefront. From the Pythagorean theorem we have $(f - x)^2 + \rho^2 = f^2$ or $2xf = \rho^2 - x^2$.

For paraxial optics, x is small and hence x^2 may be ignored by comparison. Thus,

$$x = \frac{\rho^2}{2f} \ , \tag{1.13}$$

and Eq. (1.13) is referred to as the "sagittal" approximation. The phase change introduced by a lens is therefore

$$\phi = kx = k\frac{\xi^2 + \eta^2}{2f} \ . \tag{1.14}$$

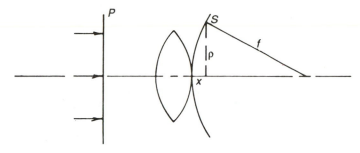

Fig. 1.3 Schematic diagram for lens analysis demonstrating the conversion of a plane wave-front to a spherical wavefront of radius f.

The lens placed in the (ξ, η) plane yields an additional term $\exp[-ik(\xi^2 + \eta^2)/2f]$ because it produces a converging spherical wave [see Eq. (1.5)]. This term cancels the term in $(\xi^2 + \eta^2)$ arising from Eq. (1.10) when $f = z$. Thus, the field at the point (x, y) in the focal plane is given by

$$\psi(x,y) = \frac{Ke^{-ikf}}{f} \iint \psi'(\xi,\eta)\exp\left[\frac{-2\pi i}{\lambda f}(\xi x + \eta y)\right] d\xi\, d\eta \ . \qquad (1.15)$$

Note that Eq. (1.15) is identical to Eq. (1.12) if f is substituted for z. These equations state that the field in the far zone or in the focal plane of the lens is the Fourier transform of the field across the diffracting aperture. They constitute the basic equations of Fraunhofer diffraction theory and are the principal results of this chapter. The use of this integral to determine diffraction patterns is discussed further in later chapters. The experimental realization of these conditions and photographic illustrations will be given.

REFERENCES

1. J. A. Stratton, *Electromagnetic Theory,* McGraw Hill Inc., New York (1941).
2. E. L. O'Neill, *Introduction to Statistical Optics,* Addison-Wesley Publishing Co., Inc., Reading, Mass. (1963).

2 Fourier Transforms

2.1 **INTRODUCTION**

The principal result of Chapter 1 was the demonstration that the Fraunhofer diffraction pattern associated with the field distribution existing across an aperture is the Fourier transform of that field. To be precise, the Fraunhofer diffraction pattern of an aperture distribution is obtained when the point of observation is infinitely distant from a coherently illuminated aperture. In practice, of course, this condition never describes a physical situation. However, the Fraunhofer theory provides an adequate approximation in many physically significant experiments. These experiments are characterized by one of the following conditions:

1. If the plane containing the point source and the plane of observation are parallel conjugate planes of a well-corrected optical system and both source and point of observation lie near the optical axis, then the Fraunhofer diffraction pattern of the limiting aperture is observed.

2. The distances from the source to the diffracting aperture z' and from the diffracting aperture to the plane of observation z are such that

$$|z| \text{ and } |z'| \gg \frac{(\xi^2 + \eta^2)_{max}}{\lambda} , \qquad (2.1)$$

where ξ and η are coordinates of a general point in the diffracting aperture and λ is the wavelength of the incident wave. Conditions 1 and 2 are those expressed in Chapter 1 by Eqs. (1.12) and (1.15).

The Fourier transform plays an important role in modern physical optics, not only in the determination of diffraction patterns and the description of interference phenomena, but also in the description of imaging systems and in spectral analysis. Therefore, it is useful to devote an entire chapter to this mathematical topic. Rather than presenting a purely mathematical discussion, the techniques of Fourier analysis are illustrated in the context of diffraction theory. Thus, the purpose of this chapter is to illustrate the use of Fourier transforms in the determination of some simple diffraction patterns. We conclude with a summary of some Fourier transform pairs useful in diffraction calculations.

2.2   DIFFRACTION PROBLEMS

Under the conditions stated in Sec. 2.1, the amplitude and phase of the field in the focal plane are described by the Fourier transform of the aperture distribution, i.e., apart from constant factors,

$$\psi(x,y) = \exp\left(\frac{ik|\mathbf{x}|^2}{2f}\right) \iint A(\xi,\eta)\exp\left[\frac{-ik}{f}(\xi x+\eta y)\right] d\xi\, d\eta \ . \qquad (2.2)$$

Here $A(\xi,\eta)$ corresponds to $\psi'(\xi,\eta)$ used in Chapter 1 and describes the amplitude and phase distribution across the aperture and $\psi(x,y)$ describes the field in the focal plane. Since only the intensity is detected, the quadratic phase term multiplying the Fourier transform in Eq. (2.2) may usually be neglected. It is important only in the case that the diffraction pattern is allowed to interfere with a coherent background; it is this interference with a coherent background that gives rise to the formation of holograms. For the remainder of this chapter, however, the quadratic phase term will be omitted. Thus, the basic formulation of the diffraction problem will take the form

$$\psi(x,y) = \iint A(\xi,\eta)\exp\left[\frac{-ik}{f}(\xi x+\eta y)\right] d\xi\, d\eta \ . \qquad (2.3)$$

Transform equation (2.3) is indeed exact in an optical system if the diffracting aperture is in the front focal plane and the recording plane is in the back focal plane of the lens.

2.2.1   Slit Aperture

As the first example, we consider an infinitely long slit of width $2a$ centered along the η axis and uniformly illuminated. That is, we take

$$A(\xi,\eta) = \begin{pmatrix} A & |\xi| < a \\ 0 & |\xi| > a \end{pmatrix} = A\,\mathrm{rect}(\xi|a) \ . \qquad (2.4)$$

This is essentially a one-dimensional problem and the diffraction integral may be written as

$$\psi(x) = \int A(\xi)\exp\left(\frac{-ik}{f}\xi x\right)d\xi = A\int_{-a}^{a} \exp\left(\frac{-ik\xi x}{f}\right)d\xi \ . \qquad (2.5)$$

[handwritten margin note: $\int f(\eta)\,d\eta$ has been absorbed into A]

The integral in Eq. (2.5) is readily evaluated to give

$$\psi(x) = \frac{-Af}{ikx}\left[\exp\left(\frac{-ikax}{f}\right) - \exp\left(\frac{+ikax}{f}\right)\right] = 2aA\,\frac{\sin\left(\frac{kax}{f}\right)}{\left(\frac{kax}{f}\right)} \ . \qquad (2.6)$$

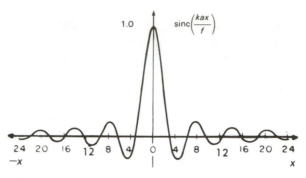

Fig. 2.1 Diffraction by a slit aperture: the main features of the function sinc (*kax/f*).

The notation sinc $\theta = \sin\theta/\theta$ is frequently used and in terms of this function $\psi(x)$ may be written as

$$\psi(x) = 2aA \, \text{sinc} \, \frac{kax}{f} \, \cdot \tag{2.7}$$

Since we have ignored the quadratic phase term in the calculation of the pattern, Eq. (2.7) must be interpreted with caution. However, since amplitude and phase are never detected directly in optical experiments, we may neglect this omission except in those cases where $\psi(x)$ is coherently added to another field. Thus, no error is involved in using Eq. (2.7) to compute the intensity being discussed. The intensity $I(x)$, defined as $\psi(x)\psi^*(x)$, is given by

$$I(x) = 4a^2 A^2 \, \text{sinc}^2 \left(\frac{kax}{f} \right) \cdot \tag{2.8}$$

Figure 2.1 shows a portion of the function sinc(kax/f) plotted as a function of x.

2.2.2   Rectangular Aperture

As our second example, we evaluate the diffraction pattern arising from a uniformly illuminated rectangular aperture of width 2*a*, length 2*b*, and amplitude A centered on the axis of the ξ and η plane. Then

$$\psi(x,y) = A \int_{-a}^{+a} \int_{-b}^{+b} \exp\left[\frac{-ik}{f} (\xi x + \eta y) \right] d\xi \, d\eta \; ; \tag{2.9}$$

performing the integration as before we obtain

$$\psi(x,y) = 4Aab \, \text{sinc} \left(\frac{kax}{f} \right) \text{sinc} \left(\frac{kby}{f} \right) \cdot \tag{2.10}$$

The intensity is

$$I(x,y) = \psi(x,y)\psi^*(x,y)$$

$$= 16 A^2 a^2 b^2 \, \text{sinc}^2 \left(\frac{kax}{f} \right) \text{sinc}^2 \left(\frac{kby}{f} \right) \cdot \tag{2.11}$$

2.2.3   Circular Aperture

It is more convenient for this particular example to work in polar rather than rectangular coordinates in Eq. (2.3). We again apply the same conditions for the amplitude and phase across the diffracting aperture. A general point in the diffracting aperture of radius a will have polar coordinates (ρ, ϕ) related to the rectangular coordinates (ξ, η) in the usual way

$$\xi = \rho\cos\phi , \quad \eta = \rho\sin\phi .$$

(2.12)

Similarly, a general point in the transform plane has polar coordinates (r, θ). Hence,

$$\psi(r, \theta) = \int_0^a \int_0^{2\pi} A\exp\left(\frac{-ik}{f} r\cos\theta \,\rho\cos\phi\right) \exp\left(\frac{-ik}{f} r\sin\theta \,\rho\sin\phi\right) \rho d\rho \, d\phi$$

$$= \int_0^a \int_0^{2\pi} A\exp\left[\frac{-ik}{f} r\rho\cos(\theta-\phi)\right] \rho d\rho \, d\phi .$$

(2.13)

This integration may be performed using the integral representation of the Bessel function, i.e.,

$$J_n(x) = \frac{i^{-n}}{2\pi} \int_0^{2\pi} e^{-ix\cos\gamma} e^{in\gamma} d\gamma$$

(2.14)

and, in particular,

$$J_0(x) = \frac{1}{2\pi} \int_0^{2\pi} e^{-ix\cos\gamma} d\gamma .$$

(2.15)

Hence,

$$\psi(r, \theta) = 2\pi A \int_0^a J_0\left(\frac{kr\rho}{f}\right) \rho d\rho .$$

(2.16)

Since

$$\int_0^x x' J_0(x') \, dx' = x J_1(x) ,$$

(2.17)

then

$$\psi(r, \theta) = A\pi a^2 \frac{2J_1\left(\frac{kar}{f}\right)}{\left(\frac{kar}{f}\right)} .$$

(2.18)

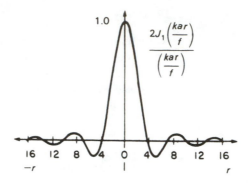

Fig. 2.2 Diffraction by a circular aperture: the main features of the function $2J_1(kar/f)/(kar/f)$.

The function $2J_1(kar/f)/(kar/f)$ (often called a "Besinc" function) is plotted in Fig. 2.2. The intensity distribution often called the "Airy" pattern, after G. B. Airy who first calculated it, is given by

$$I(r,\theta) = \psi(r,\theta)\,\psi^*(r,\theta) = \pi^2 a^4 A^2 \left| \frac{2J_1\left(\dfrac{kar}{f}\right)}{\left(\dfrac{kar}{f}\right)} \right|^2 . \tag{2.19}$$

2.2.4   Delta Function

In many diffraction and interference problems, it proves convenient to make use of an extremely irregular function, the Dirac delta function. This function is defined by the following property: let $f(\xi)$ be any function (satisfying some very weak convergence conditions which need not concern us here) and let $\delta(\xi - \xi')$ be a delta function centered at the point ξ'; then

$$\int_a^b f(\xi)\,\delta(\xi - \xi')\,d\xi = \begin{bmatrix} f(\xi') & (a < \xi' < b) \\ 0 & \text{otherwise} \end{bmatrix} . \tag{2.20}$$

We note, therefore, that

$$\int_{-\infty}^{\infty} \delta(\xi - \xi')\,d\xi = 1 .$$

The Fourier transform of the delta function is given by

$$\psi(x) = \int \delta(\xi - \xi') \exp\left(\frac{-ikx\xi}{f}\right) d\xi , \tag{2.21}$$

which by the definition of the delta function becomes

$$\psi(x) = \exp\left(\frac{-ikx\xi'}{f}\right) . \tag{2.22}$$

The amplitude is constant and the phase function $(kx\xi'/f)$ depends on the origin.

When $\xi' = 0$ then the delta function is at the origin and the transform is a constant. The converse is also true, of course, that a constant extending from $-\infty$ to $+\infty$ transforms to a delta function.

2.3   CONCLUSION

As a shorthand notation, we introduce the tilde, $\sim$, to denote the spatial Fourier transform. This notation is used throughout the remainder of this book; i.e.,

$$\tilde{\psi}(x,y) = \int \psi(\xi,\eta)\, e^{-ik(\xi x + \eta y)}\, d\xi\, d\eta \ . \tag{2.23}$$

In conclusion we should like to collect a number of Fourier transform pairs to form a table as shown in Fig. 2.3. In each case, the function $\psi(\xi,\eta)$ transforms to the function $\tilde{\psi}(x,y)$ and vice versa. The table is self explanatory and needs no further comment, except to state that the curves are diagrammatic representations only.

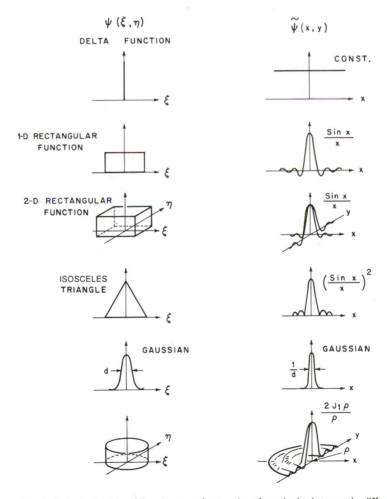

Fig. 2.3 A short table of Fourier transform pairs of particular interest in diffraction.

3   Array Theorem

3.1   INTRODUCTION

In this chapter, we continue the study of Fraunhofer or far-field diffraction. We have seen in the first chapter that such diffraction is characterized by the fact that the diffracted amplitude distribution is proportional to the Fourier transform of the amplitude distribution across the diffracting aperture. In Chapter 2, some examples of diffraction calculations were given for simple geometries. In this chapter, we examine the effect of combining apertures of similar geometry. In this class of problems, the diffraction integral assumes an interesting and characteristic form and gives rise to a subclass of diffraction effects that is important enough to receive a special nomenclature and study, namely, interference by division of wavefront.

A large class of interference effects can be treated with a single theorem, the array theorem. In the next section, this theorem is derived and discussed in terms of a simple example. Then this class of interference effects is illustrated by means of photographic examples.

3.2   THE ARRAY THEOREM

A large number of interference problems involve the mixing of similar diffraction patterns. That is, they arise in the study of the combined diffraction patterns of an array of similar diffracting apertures. This entire class of interference effects can be described by a single equation, the array theorem. This unifying theorem is easily developed as follows: Let $\psi(\xi)$ represent the amplitude and phase distribution across one aperture centered in the diffraction plane, and let the total diffracting aperture consist of a collection of these elemental apertures at different locations ξ_n. This concept is illustrated in Fig. 3.1. We require first a method of representing such an array. The appropriate representation is obtained readily by means of the delta function. Thus, if an elemental aperture is positioned such that its center is at the point ξ_n, the appropriate distribution function is $\psi(\xi - \xi_n)$.

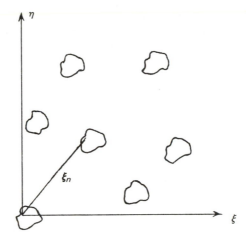

Fig. 3.1 An array of identical apertures.

The combing property of the delta function allows us to represent this distribution as follows:

$$\psi(\xi - \xi_n) = \int \psi(\xi - \alpha)\,\delta(\alpha - \xi_n)\,d\alpha \ . \tag{3.1}$$

The integral in Eq. (3.1) is termed a "convolution" integral and plays an important role in Fourier analysis. Thus, if we wish to represent a large number N of such apertures with different locations, we could write the total aperture distribution $\Psi(\xi)$ as a sum, i.e.,

$$\Psi(\xi) = \sum_{n=1}^{N} \psi(\xi - \xi_n) \ . \tag{3.2}$$

Or in terms of the delta function we could write, combining the features of Eqs. (3.2) and (3.1),

$$\Psi(\xi) = \sum_{n=1}^{N} \int \psi(\xi - \alpha)\,\delta(\alpha - \xi_n)\,d\alpha \ . \tag{3.3}$$

Equation (3.3) may be put in a more compact form by introducing the notation

$$A(\alpha) = \sum_{n=1}^{N} \delta(\alpha - \xi_n) \ , \tag{3.4}$$

whence Eq. (3.3) becomes

$$\Psi(\xi) = \int \psi(\xi - \alpha)A(\alpha)\,d\alpha \ , \tag{3.5}$$

which is physically pleasing in the sense that $A(\alpha)$ characterizes the array itself. That is, $A(\alpha)$ describes the location of the apertures and $\psi(\xi)$ describes the distribution across a single aperture. We are now in a position to calculate the Fraunhofer diffraction pattern associated with the array. From Chapter 1, we have the theorem that the Fraunhofer pattern is the Fourier transform of the aperture distribution. Thus, the Fraunhofer pattern $\tilde{\Psi}(\mathbf{x})$ of the distribution $\Psi(\xi)$ is given by

$$\tilde{\Psi}(\mathbf{x}) = \int \Psi(\xi) \exp\left(\frac{-2\pi i \xi \cdot \mathbf{x}}{\lambda f}\right) d\xi \; ; \tag{3.6}$$

substituting from Eq. (3.5) gives

POSITION VECTOR

$$\tilde{\Psi}(\mathbf{x}) = \left[\int \int \psi(\xi - \alpha) A(\alpha) \, d\alpha\right] \exp\left(\frac{-2\pi i \xi \cdot \mathbf{x}}{\lambda f}\right) d\xi \; . \tag{3.7}$$

A very important theorem of Fourier analysis (a proof is given in the Appendix, Sec. 3.5, at the end of this chapter) states that the Fourier transform of a convolution is the product of the individual Fourier transforms. Thus, Eq. (3.7) may be written as

$$\tilde{\Psi}(\mathbf{x}) = \tilde{\psi}(\mathbf{x}) \tilde{A}(\mathbf{x}) \; , \tag{3.8}$$

where $\tilde{\psi}(\mathbf{x})$ and $\tilde{A}(\mathbf{x})$ are the Fourier transforms of $\psi(\xi)$ and $A(\alpha)$. Equation (3.8) is the array theorem and states that the diffraction pattern of an array of similar apertures is given by the product of the elemental pattern $\tilde{\psi}(\mathbf{x})$ and the pattern that would be obtained by a similar array of point sources, $\tilde{A}(\mathbf{x})$. Thus, the separation that first arose in Eq. (3.5) is retained. To analyze the complicated patterns that arise in interference problems of this sort, one may analyze separately the effects of the array and the effects of the individual apertures.

3.3   APPLICATIONS OF ARRAY THEOREM

3.3.1   Two-Beam Interference

In this section, we use Eq. (3.8) to describe the simplest of interference experiments, Young's double-slit experiment in one dimension. To facilitate interpretation of the results, the transform is written in the sharpened form as given in Chapter 1. Thus, the individual aperture will be described by

$$\psi(\xi) = \begin{pmatrix} C & |\xi| \leq a \\ 0 & |\xi| > a \end{pmatrix} = \text{rect}(\xi|a) \; . \tag{3.9}$$

Here C is a constant representing the amplitude transmission of the apertures.
From Chapter 2 [Eq. (2.7)] we have the result that the elemental distribution in the Fraunhofer plane is

$$\tilde{\psi}(x) = 2aC \, \text{sinc} \, \frac{2\pi a x}{\lambda f} \; . \tag{3.10}$$

The array in this case is simply two delta functions; thus,

$$A(\xi) = \delta(\xi - b) + \delta(\xi + b) \; . \tag{3.11}$$

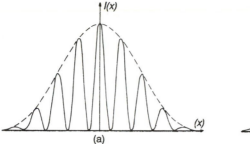
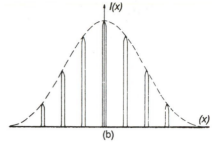

(a)   (b)

Fig. 3.2 (a) The normalized intensity distribution in the Fraunhofer pattern of two apertures having the ratio $b/a \simeq 4.5$ and (b) the normalized intensity distribution in the Fraunhofer pattern of a larger number of apertures having the same ratio as (a).

The array pattern is, therefore,

$$\tilde{A}(x) = \int [\delta(\xi - b) + \delta(\xi + b)] \exp\left(\frac{-2\pi i \xi x}{\lambda f}\right) d\xi \; ; \tag{3.12}$$

Eq. (3.12) is readily evaluated by using the combing property of the delta function. Thus,

$$\tilde{A}(x) = \exp\left(\frac{2\pi ibx}{\lambda f}\right) + \exp\left(\frac{-2\pi ibx}{\lambda f}\right) = 2\cos\left(\frac{2\pi bx}{\lambda f}\right). \tag{3.13}$$

Finally, the diffraction pattern of the array of two slits is

$$\tilde{\Psi}(x) = 4aC\text{sinc}\left(\frac{2\pi ax}{\lambda f}\right)\cos\left(\frac{2\pi bx}{\lambda f}\right). \tag{3.14}$$

The intensity is

$$I(x) = 16a^2C^2\text{sinc}^2\left(\frac{2\pi ax}{\lambda f}\right)\cos^2\left(\frac{2\pi bx}{\lambda f}\right). \tag{3.15}$$

From Eq. (3.15) it is clear that the resulting pattern has the appearance of cosine^2 fringes of period $\lambda f/b$ with an envelope sinc$^2(2\pi ax/\lambda f)$. A typical distribution is shown in Fig. 3.2(a).

In a precisely similar manner, we can use our previous results to build up the expressions for the interference observed by using two square apertures and two circular apertures. It is suggested that the reader solve these two problems as an exercise in the use of the array theorem.

3.3.2 One-Dimensional Array

For N identical apertures equally separated by a distance $2b$, the array theorem takes the general form derived below, of which the two-beam example is a special case. The array of delta functions will be represented by a sum of the form

$$A(\xi) = \sum_{n=0}^{N-1} \delta(\xi - 2nb) \; . \tag{3.16}$$

Thus, the Fourier transform of the array is given by

$$\tilde{A}(x) = \sum_{n=0}^{N-1} \exp\left(\frac{-2\pi i 2bxn}{\lambda f}\right) , \tag{3.17}$$

which may be written as

$$= \frac{1 - \exp\left(\dfrac{-2\pi i N 2bx}{\lambda f}\right)}{1 - \exp\left(\dfrac{-2\pi i 2bx}{\lambda f}\right)} . \tag{3.18}$$

Therefore,

$$\tilde{\Psi}(x) = 2aC \left[\frac{1 - \exp\left(\dfrac{-2\pi i N 2bx}{\lambda f}\right)}{1 - \exp\left(\dfrac{-2\pi i 2bx}{\lambda f}\right)}\right] \operatorname{sinc}\left(\frac{2\pi a x}{\lambda f}\right) \tag{3.19}$$

and

$$I(x) = 4a^2 C^2 \left[\frac{1 - \cos\left(\dfrac{N 2\pi 2bx}{\lambda f}\right)}{1 - \cos\left(\dfrac{2\pi 2bx}{\lambda f}\right)}\right] \operatorname{sinc}^2\left(\frac{2\pi a x}{\lambda f}\right) . \tag{3.20}$$

Hence, rewriting the term in square brackets,

$$I(x) = 4a^2 C^2 \left[\frac{\sin^2\left(\dfrac{2\pi N bx}{\lambda f}\right)}{\sin^2\left(\dfrac{2\pi bx}{\lambda f}\right)}\right] \operatorname{sinc}^2\left(\frac{2\pi a x}{\lambda f}\right) . \tag{3.21}$$

Note that Eq. (3.21) reduces to Eq. (3.15) when $N = 2$. Figure 3.2(b) shows the distribution for N large.

3.4   SOME EXAMPLES

In this section, a number of illustrations are given covering the mathematical discussions of Chapter 2 and the preceding discussion of the array theorem.

Figure 3.3 shows a series of photographs of the Fraunhofer diffraction patterns of single apertures. The apertures are shown, with the correct orientation, below the photographs.

Figure 3.4 illustrates diffraction by an array of two to six slits. The whole of these patterns is contained in the first maximum of the sinc^2 envelope function Note that, as the number of slits increases, the main maxima stay in the same position but get sharper and there are $N - 2$ subsidiary maxima.

Finally, Fig. 3.5 illustrates some other examples of the array theorem.

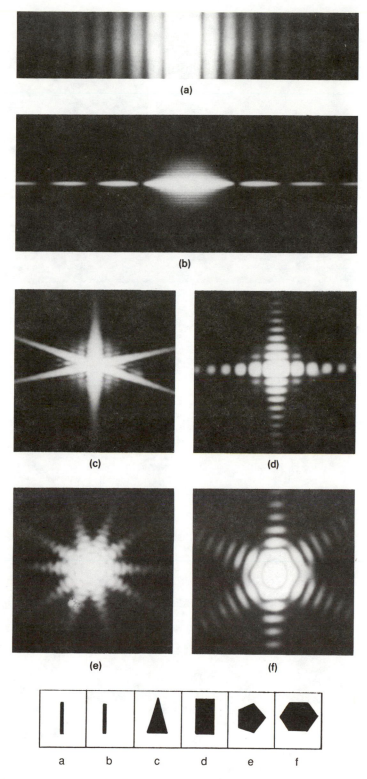

Fig. 3.3 Photographs of the Fraunhofer diffraction patterns (Fourier transforms) of various apertures: (a) slit aperture with a slit source, (b) slit aperture with point source, (c) triangular aperture, (d) rectangular aperture, (e) pentagonal aperture, and (f) hexagonal aperture.

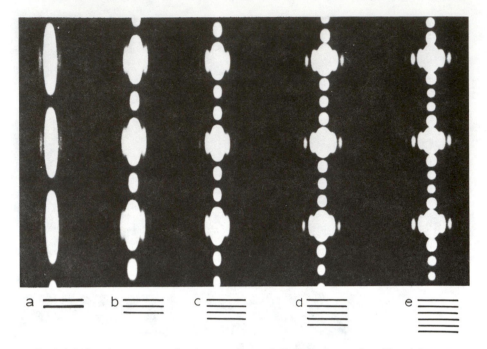

Fig. 3.4 Diffraction patterns of various numbers of slits from two to six, with point source.

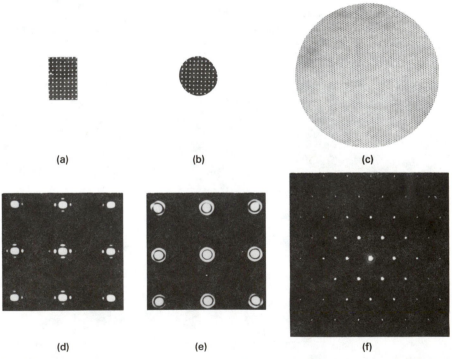

Fig. 3.5 Illustration of the array theorem showing (a) an array of circular apertures contained within a rectangular aperture, (b) an array of circular apertures contained within a circular aperture, (c) a hexagonal array of circular objects and (d), (e), and (f), respectively, the corresponding Fraunhofer diffraction patterns.

3.5   **APPENDIX: THE CONVOLUTION THEOREM**

The derivation of the convolution theorem is greatly facilitated by making use of an integral representation of the delta function. A common and useful one-dimensional representation may be obtained by taking the Fourier transform of both sides of Eq. (2.20) in the preceding chapter. It is

$$\delta(x - x') = \int e^{-2\pi i \mu (x - x')} \, d\mu \ . \tag{3.22}$$

This representation is given here without proof; the reader will do well to consult the suggested reference.

Let $f(x)$ and $g(x)$ be two functions that possess Fourier transforms, i.e.,

$$\tilde{f}(\mu) = \int f(x) \, e^{2\pi i \mu x} \, dx \ , \tag{3.23}$$

$$\tilde{g}(\mu) = \int g(x) \, e^{2\pi i \mu x} \, dx \ . \tag{3.24}$$

By the inversion theorem, we write

$$f(x) = \int \tilde{f}(\mu) \, e^{-2\pi i \mu x} \, d\mu \ , \tag{3.25}$$

$$g(x) = \int \tilde{g}(\mu) \, e^{-2\pi i \mu x} \, d\mu \ . \tag{3.26}$$

We wish to find an expression for the Fourier transform of $h(x) = f(x)g(x)$. Thus,

$$\tilde{h}(\mu) = \int h(x) \, e^{2\pi i \mu x} \, dx$$

$$= \int f(x)g(x) \, e^{2\pi i \mu x} \, dx \ . \tag{3.27}$$

Substituting from Eqs. (3.25) and (3.26) into Eq. (3.27) and introducing the dummy variables τ and η to avoid the confusion of too many μ's, we obtain

$$\tilde{h}(\mu) = \int \left[\iint f(\tau)g(\eta) \, e^{-2\pi i \tau x} \ e^{-2\pi i \eta x} \, d\tau \, d\eta \right] e^{2\pi i \mu x} \, dx \ . \tag{3.28}$$

Rearranging the integrals in Eq. (3.28), we write

$$\tilde{h}(\mu) = \iiint f(\tau)g(\eta) \, e^{2\pi i x [\mu - (\tau + \eta)]} \, dx \, d\tau \, d\eta \ . \tag{3.29}$$

The innermost integral in Eq. (3.29) is just a Dirac delta function according to Eq. (3.22); thus, Eq. (3.29) reduces to

$$\tilde{h}(\mu) = \iint f(\tau)g(\eta) \, \delta[\tau - (\mu - \eta)] \, d\tau \, d\eta \ . \tag{3.30}$$

Using the combing property of the delta function,

$$\tilde{h}(\mu) = \int f(\mu - \eta)g(\eta)\,d\eta \ . \tag{3.31}$$

Equation (3.31) is the convolution theorem and expresses the Fourier transform of $h(x)$ in terms of the convolution of the transforms of its products. The inversion theorem applied to Eq. (3.31) gives us the convolution theorem in the form required for the array theorem. Thus,

$$h(x) = f(x)g(x) = \int \tilde{h}(\mu)\, e^{-2\pi i \mu x}\, d\mu$$

$$= \int \left[\int f(\mu - \eta)g(\eta)\, d\eta \right] e^{-2\pi i \mu x}\, d\eta \ . \tag{3.32}$$

There are, of course, other ways of deriving this identity but we have purposely chosen the method using the delta function as a further illustration of its use.

REFERENCE

1. M. J. Lighthill, *Fourier Analysis and Generalized Functions,* Cambridge University Press, Mass. (1959).

4 Image Formation: The Impulse Response

4.1 INTRODUCTION

We are now in a position to begin a study of the physical optical aspects of image formation. The developments in this and the next few chapters will draw heavily on the results of the preceding chapters. All results will be derived from wave theoretical considerations. From this approach, the customary geometric laws relating to object and image distances, magnification, etc., are easily derived. More important, however, is the fact that quality limitations are immediately evident from such a development. The word "quality" is used here to describe the subjective impression. Considerable discussion will be necessary before we are able to replace the subjective quality with some physically defined, meaningful, and measurable quantity.

Let us start the problem of image formation from the physical optical point of view by considering objects that are incoherently illuminated.

4.2 IMPULSE RESPONSE

To give a completely general and consistent development of image formation with incoherent light requires several theorems from the theory of partial coherence. As we have not yet discussed this theory, and as it will prove useful to have the basic concepts of image theory well in mind before doing so, we resort to a two-step development in this chapter. This development, while correct and physically pleasing, is not detailed enough to include the subtleties that can be treated later. We are concerned with incoherent quasimonochromatic light. The precise definition of the term *quasimonochromatic* requires some concepts from coherence theory and, therefore, is rigorously defined in Chapter 11. Here, and in the intervening chapters, it is sufficient to think of a quasimonochromatic beam as one such that its bandwidth is a small fraction of its central frequency and which is used in experiments involving sufficiently small path differences. Without the rigor of coherence theory, we have to characterize such a system by a single concept, which must be accepted on faith at this point. This concept may be simply stated: An optical system employing incoherent radiation may be regarded as linear and stationary in intensity. This statement implies two properties: linearity and stationarity, which are explained in the following sections. For mathematical simplicity, the development in this chapter is in one dimension. In later chapters, the analysis is extended to two dimensions where necessary.

4.2.1   Linearity

A system is linear if the addition of inputs produces an addition of corresponding outputs. Throughout this discussion, the arrow, $\rightarrow$, should be read as "produces." If $f_1(x')$ is an input that produces an output $g_1(x)$ denoted by

$$f_1(x') \rightarrow g_1(x) \ , \tag{4.1}$$

and if

$$f_2(x') \rightarrow g_2(x) \ , \tag{4.2}$$

then

$$af_1(x') + bf_2(x') \rightarrow ag_1(x) + bg_2(x) \ , \tag{4.3}$$

where a and b are constants.

4.2.2   Stationarity

The property of stationarity implies that if the location of the input is changed, i.e., $f_1(x')$ is replaced by $f_1(x' - x_0')$, the only effect on the output is to change its location, i.e.,

$$f_1(x' - x_0') \rightarrow g_1(x - x_0) \ . \tag{4.4}$$

These properties of linearity and stationarity are fundamental to the development of image theory and considerable care should be taken to be sure that their implications are well in mind before proceeding. Under suitable conditions, optical systems possess these properties.

To develop an image theory from these concepts, it is necessary only to find an expression for the image of a point input. This conclusion may be obtained from the following argument. Consider an object consisting of two points located at x_1' and x_2' and let the intensity at these two points be given by $f(x_1')$ and $f(x_2')$, respectively. The object intensity $I_{obj}(x')$ may be expressed using delta functions as

$$I_{obj}(x') = f(x')\delta(x' - x_1') + f(x')\delta(x' - x_2') \ . \tag{4.5}$$

Here δ is the Dirac delta function discussed in Chapter 3. The intensity image of one point may be denoted by $S(x)$, i.e.,

$$\delta(x') \rightarrow S(x) \ , \tag{4.6}$$

where $S(x)$ is called the impulse response or point spread function. The images of the two points can be written down immediately using the properties of linearity and stationarity. Thus,

$$f(x')\delta(x' - x_1') \rightarrow f(x_1)S(x - x_1) \ , \tag{4.7}$$

and

$$f(x')\delta(x' - x_2') \rightarrow f(x_2)S(x - x_2) \ ; \tag{4.8}$$

or

$$f(x')\delta(x' - x'_1) + f(x')\delta(x' - x'_2)$$

$$\rightarrow f(x_1)S(x - x_1) + f(x_2)S(x - x_2) \ . \tag{4.9}$$

Similarly, if the object consisted of a large set of points, i.e.,

$$I_{obj}(x') = \sum_{n=1}^{N} f(x')\delta(x' - x'_n) \ , \tag{4.10}$$

then the output or image will be given by a sum of impulse responses, i.e.,

$$\sum_{n=1}^{N} f(x')\delta(x' - x'_n) \Rightarrow \sum_{n=1}^{N} f(x_n)S(x - x_n) \ . \tag{4.11}$$

We may define the image intensity as $I_{im}(x)$ and rewrite the above expression as an equation:

$$I_{im}(x) = \sum_{n=1}^{N} f(x_n)S(x - x_n) \ . \tag{4.12}$$

Up to this point, we have considered only objects consisting of discrete points. However, the arguments are readily generalized to continuously varying objects as follows: Let $f(x'')$ describe the intensity variation across a continuously varying scene. We may write $f(x'')$ as a "sum of delta functions" by using the "combing" property of the delta function (see Chapter 3). Thus,

$$I_{obj}(x') = f(x') = \int f(x'')\delta(x' - x'')\, dx'' \ . \tag{4.13}$$

The only difference between this representation of the object and the preceding one is that $I_{obj}(x')$ is now given by a continuous sum. Still the properties of linearity and stationarity allow us to write the output as a continous sum of impulse responses, i.e.,

$$\int f(x'')\delta(x' - x'')\, dx'' \Rightarrow \int f(x')S(x - x')\, dx' \ .$$

Hence, denoting the image by $I_{im}(x)$, we write

$$I_{im}(x) = \int f(x')S(x - x')\, dx' \ . \tag{4.14}$$

Equation (4.14) is, of course, exactly analogous to Equation (4.12) where the integral (a continuous summation) replaces the discrete summation. Equation (4.14) is, in fact, the starting point for the analysis of any linear stationary system.

It is clear from Eq. (4.14) that if we can obtain an expression for the image of a point object, i.e., the point spread function, we can determine the image I_{im} by convolving the point spread function with the object distribution $f(x')$. Throughout this book, the terms "impulse response," "point spread function," and "point diffraction pattern" are used interchangeably. We may then use this answer to describe the image of an arbitrary object by using Eq. (4.14). To obtain an expression for the impulse response, we must determine the wave theoretic solution for the image of a point object. This is accomplished in the next section.

4.3   IMAGE OF A POINT OBJECT

Let us consider an optical field $\Psi_{obj}(x')$ existing in a plane at a distance z from a lens of aperture $A(\xi)$ (see Fig. 4.1). We wish to determine the distribution $\Psi_{im}(x)$ existing in the plane a distance z' the other side of the lens. Here a one-dimensional situation is discussed but the analysis is easily extended to the two-dimensional case. Let r be the distance from the plane $\mathbf{x}$ to the aperture plane ξ and s from the plane ξ to the plane $\mathbf{x}'$. Furthermore, in Chapter 1, the effect of the lens was shown to be the introduction of a phase factor $\exp(-ik\xi^2/2f)$. Hence,

$$\Psi_{im}(x) = K_1 \iint \Psi_{obj}(x') \frac{e^{ikr}}{r} A(\xi) \exp\left(\frac{-ik\xi^2}{2f}\right) \frac{e^{iks}}{s} dx' d\xi ; \quad (4.15)$$

i.e., the first e^{ikr}/r represents a spherical wave propagating from a point in the object to the lens plane, and e^{iks}/s is the spherical wave from each point on the wavefront immediately after the lens.

In the following development, we shall, for simplicity, limit our attention to behavior in a single dimension. The extension to a full two-dimensional treatment is straightforward and is left as an exercise for the student. Hence,

$$r^2 = R'^2 - 2\mathbf{x}' \cdot \xi + |\xi|^2 , \quad (4.16)$$

where

$$R'^2 = |\mathbf{x}'|^2 . \quad (4.17)$$

Similarly,

$$\left.\begin{array}{l} |s|^2 = R^2 - 2\mathbf{x} \cdot \xi + |\xi|^2 , \\[2mm] R^2 = |\mathbf{x}|^2 . \end{array}\right\} \quad \cdots \quad (4.18)$$

Expanding Eqs. (4.17) and (4.18) binomially and neglecting terms in second order and above gives:

$$\left.\begin{array}{l} |r| \simeq R' - \dfrac{x'\xi}{R'} + \dfrac{\xi^2}{2R'} , \\[4mm] |s| \simeq R - \dfrac{x\xi}{R} + \dfrac{\xi^2}{2R} . \end{array}\right\} \quad \cdots \quad (4.19)$$

Substitution for r and s from Eq. (4.19) into Eq. (4.15) yields

$$\Psi_{im}(x) = K_1 \frac{e^{ik(R+R')}}{RR'} \iint \Psi_{obj}(x') A(\xi)$$

$$(4.20)$$

$$\times \exp\left(\frac{-ik\xi^2}{2}\right)\left(\frac{1}{R} + \frac{1}{R'} - \frac{1}{f}\right) \exp\left[ik\xi\left(\frac{x'}{R'} + \frac{x}{R}\right)\right] dx\, d\xi .$$

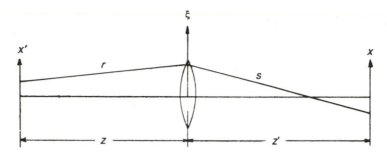

Fig. 4.1 System geometry for image formation problem.

If $x \ll f$ and $x \ll z$ with similar constraints on x, then we have, essentially, paraxial optics. Then, combining the constant factors outside the integral,

$$\Psi_{im}(x) = K \iint \Psi_{obj}(x') A(\xi)$$

$$\times \exp\left(\frac{ik\xi^2}{2}\right)\left(\frac{1}{z} + \frac{1}{z'} - \frac{1}{f}\right)\exp\left[-ik\xi\left(\frac{x'}{z'} + \frac{x}{z}\right)\right] dx'\, d\xi \ , \tag{4.21}$$

where

$$K = K_1 \frac{e^{ik(R+R')}}{RR'} \cdot \tag{4.22}$$

The first exponential term goes to unity if

$$\frac{1}{z} + \frac{1}{z'} = \frac{1}{f} \cdot \tag{4.23}$$

This will be recognized as the Gaussian focusing condition from geometrical optics (see the Appendix, Sec. 4.5). Hence, under this condition,

$$\Psi_{im}(x) = K \iint \Psi_{obj}(x') A(\xi) \exp\left[-ik\xi\left(\frac{x'}{z'} + \frac{x}{z}\right)\right] dx'\, d\xi \ . \tag{4.24}$$

Rearranging

$$\Psi_{im}(x) = K \iint \Psi_{obj}(x') \exp\left(\frac{-ik\xi x'}{z'}\right) A(\xi) \exp\left(\frac{-ik\xi x}{z}\right) \, d\xi\, dx' \ . \tag{4.25}$$

It will be noted that Eq. (4.25) contains two Fourier transform terms. Performing the x integration first results in

$$\Psi_{im}(x) = K \int \tilde{\Psi}_{obj}\left(\frac{\xi}{z\lambda}\right) A(\xi) \exp\left(\frac{-ik\xi x}{z}\right) \, d\xi \ . \tag{4.26}$$

Here $\tilde{\Psi}_{obj}(\xi/z\lambda)$ is the Fourier transform of $\Psi_{obj}(x')$.

Since we are interested in the impulse response, we may now consider the object distribution to be a delta function, i.e.,

$$\Psi_{obj}(x') = \delta(x') \ . \tag{4.27}$$

Thus,

$$\tilde{\Psi}_{obj}\left(\frac{\xi}{z\lambda}\right) = \int \delta(x') \exp\left(\frac{-ik\xi x'}{z'}\right) dx' = 1 \ . \tag{4.28}$$

Using this relation, Eq. (4.26) reduces to

$$\Psi_{im}(x) = K \int A(\xi) \exp\left(\frac{-ik\xi x}{z}\right) d\xi \ . \tag{4.29}$$

Equation (4.29) is an important result. It states that the amplitude distribution in the image of a point is given by the Fourier transform of the aperture distribution function, i.e., the function describing the amplitude and phase variation introduced by passage through the lens. After passage through the lens, any deviation of the phase front from a reference sphere is termed an *aberration*. It should be remembered that the constant K in Eq. (4.29) contains a spherical phase factor [see Eqs. (4.22), (4.17), and (4.18)]. While this factor may be safely ignored at this point, it plays a crucial role in coherent imaging and holography. Therefore, the factor will be reintroduced where appropriate in later chapters.

However, since we are considering incoherent systems here, we require an expression for the intensity impulse response $S(x)$. This is, of course, readily obtained from Eq. (4.29) by simply forming the squared modulus. Thus,

$$S(x) = \left| K \int A(\xi) \exp\left(\frac{-ik\xi x}{z}\right) d\xi \right|^2 \ . \tag{4.30}$$

Equations (4.29) and (4.30) will play a central role in the detailed study of imaging in subsequent chapters. For our present purposes, it is sufficient to have shown that the impulse response can be obtained from a knowledge of the aperture transmission function $A(\xi)$.

4.4   CONCLUSIONS

In summary the results of the present chapter are:

1. An optical imaging system employing incoherent light may be considered to be a linear and stationary system in intensity.

2. The detailed distribution of light in the image of an extended object can be calculated from a knowledge of the distribution in the image of a point.

3. The distribution in the image of a point may be determined directly from a knowledge of the aperture transmission function.

While Eqs. (4.30) and (4.14) are sufficient to describe the image forming properties of an optical system, their use in any but the simplest problems is very involved. Consequently, an alternative formulation which better lends itself to intuitive interpretation will be given later. In the next chapter, the use of the convolution integrals of Eqs. (4.14) and (4.30) to compute images will be illustrated for some simple examples in order to provide some insight into the implication of these results in the area of image quality prediction and evaluation.

4.5   APPENDIX: THE RELATIONSHIP TO GEOMETRICAL OPTICS

Equations (4.23) and (4.25) may be used to illustrate the relationship between wave optics and geometrical optics. First, it should be noted that the wave optics constraint to eliminate the quadratic phase error, e.g., Eq. (4.23), is precisely the Gaussian focusing condition from geometrical optics. Of course, in the study of aberrations, the quadratic term in the Seidel expansion is referred to as the *focusing error*.

Furthermore, interesting relationships between the two different ways of modeling optical phenomena may be obtained by imposing the basic tenet of geometrical optics on Eq. (4.25). In geometrical optics, diffraction effects are ignored. We can accomplish that in this case by assuming that the diffraction pattern has no width, i.e., it is a Dirac delta function. This condition arises by assuming uniform illumination across the aperture in Eq. (4.24) and further assuming it to be infinitely wide. Under these conditions, Eq. (4.24) may be evaluated to yield

$$\psi_{im}(x) = K \int \psi_{obj}(x')\delta\left(\frac{x'}{\lambda z'} + \frac{x}{\lambda z}\right) dx' . \qquad (4.31)$$

The integral in Eq. (4.31) may now be evaluated to yield

$$\psi_{im}(x) = K \psi_{obj}\left(-\frac{z}{z'} x\right) . \qquad (4.32)$$

Note that in this case the image is the same function as the object. That is, there has been no degradation due to diffraction. However, the argument is negative indicating that the image is inverted relative to the object. Also, the scale of the argument in the image function is changed by the magnification, $m = -z/z'$; that is, the image is magnified by the ratio of the object distance to the image distance. Both of these results are well known from geometrical optics.

After we have discussed coherence theory, it will be possible to extend this analysis to include incoherently illuminated objects. For now we include the result without proof. If $I_{im}(x)$ denotes the intensity distribution in the image of such an object and $I_{obj}(x')$ the intensity distribution of the object, then the two are related by

$$I_{im}(x) = K' \int I_{obj}(x')S\left(x' + \frac{z}{z'} x\right) dx' , \qquad (4.33)$$

where $S(x)$ is the intensity point image or intensity impulse response as calculated from Eq. (4.30) and K' is a constant.

5 Image Formation in Terms of the Impulse Response

5.1 INTRODUCTION

In the previous chapter, we showed that a stationary linear optical system using incoherent light could be completely described in terms of a convolution integral relating the object distribution $I_{obj}(x')$ to the image distribution $I_{im}(x)$. In such a description, the optical system is characterized by its impulse response, i.e., the point image. Mathematically, these statements are summarized in the following integral [see Eq. (4.14)]:

$$I_{im}(x) = \int I_{obj}(x')\, S(x - x')\, dx' \; . \tag{5.1}$$

In this chapter, several relatively simple imaging problems will be analyzed with the aid of Eq. (5.1). In general, this approach to the imaging problem is too cumbersome to be used for detailed calculations and a more generally useful technique will be introduced in another chapter. However, it is instructive to carry out the convolution process for a few simple examples in order to develop some insight into the physical limitations on image quality which are introduced by the diffraction theory. To keep the computational difficulty to a minimum, the examples in this chapter are limited to a one-dimensional optical system. The concepts are readily generalized to two dimensions; however, details of the calculations become unwieldy.

5.2 IMPULSE RESPONSE FOR A CYLINDRICAL LENS

As a first example, we consider a diffraction-limited cylindrical lens used to image simple one-dimensional objects. To proceed it is necessary to determine the impulse response of such a diffraction-limited system. We recall from the previous chapter that the amplitude impulse response is given by the Fourier transform of the aperture illumination function. For a diffraction-limited cylin-

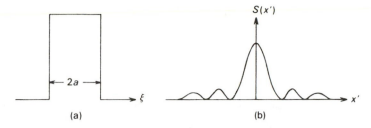

Fig. 5.1 (a) The aperture illumination function and (b) the corresponding impulse response.

drical lens, the aperture illumination function is a real constant across the aperture, here taken to be unity, and zero outside the aperture, as shown in Fig. 5.1(a). Denoting the aperture width by $2a$, the intensity impulse response $S(x')$ is given by

$$S(x) = \left| \int_{-a}^{a} \exp\left(\frac{-ik\xi x}{2f}\right) d\xi \right|^2 . \tag{5.2}$$

This expression is equivalent to Eq. (4.30) with both K and $A(\alpha)$ set equal to unity. Here k is the mean wave number of the light, f is the focal length of the lens, and ξ is the aperture coordinate. The integral in Eq. (5.2) is readily evaluated to give

$$S(x) = 4a^2 \operatorname{sinc}^2\left(\frac{kax}{2f}\right) . \tag{5.3}$$

The evaluation of Eq. (5.2) is shown in detail in Chapter 2, Eqs. (2.4) to (2.8).

The impulse response is shown in Fig. 5.1(b). Having determined the form of the impulse response, we are now in a position to apply the convolution integral of Eq. (5.1) to the determination of some simple diffraction-limited images.

5.3   IMAGE OF A BAR

Let us consider the image of a bar located many focal lengths away from the lens aperture so that the image is formed in the focal plane. That is, we consider an incoherently illuminated object of half-width b and intensity I surrounded by a dark background. The object may be described mathematically by $I_{obj}(x')$ where

$$I_{obj}(x') = \begin{pmatrix} I & |x'| \le b \\ 0 & |x'| > b \end{pmatrix} = IR(x'|b) . \tag{5.4}$$

Here the symbol $R(x'|b)$ is read "rectangular function of x' of half-width b."

Using Eqs. (5.1), (5.3), and (5.4) and changing variables, we may express the image $I_{im}(x)$ as

$$I_{im}(x) = 4a^2 I \int \operatorname{rect}(x - x'|b) \operatorname{sinc}^2\left(\frac{kax'}{2f}\right) dx' . \tag{5.5}$$

Using the more explicit form of Eq. (5.4), we may write

$$I_{im}(x) = 4a^2 I \int_{x-b}^{x+b} \text{sinc}^2 \left(\frac{kax'}{2f} \right) dx' \; . \tag{5.6}$$

Even in this elementary example we encounter difficulty in getting numerical results since the integral in Eq. (5.6) cannot be expressed in terms of simple familiar functions. Fortunately, the function

$$s(x) = \int_0^x \text{sinc}^2 \left(\frac{u}{2} \right) du \tag{5.7}$$

has been tabulated.[1] Our integral can be cast in this form by the following change of variables:

$$\frac{y}{2} = \frac{kx'a}{2f} = \frac{kx'}{4f^\#} \; , \tag{5.8}$$

where $f^\# = f/2a$. Using Eq. (5.8), the normalized image can be expressed in the form

$$\mathcal{S}(x) = \left(\frac{\pi}{2a^2 Ic} \right) I_{im}(x) = \int_{2\pi(x-b)/c}^{2\pi(x+b)/c} \text{sinc}^2 \left(\frac{y}{2} \right) dy \; . \tag{5.9}$$

In Eq. (5.9) we have introduced the spot size (beam width; in the two-dimensional case, this would be the Airy disk diameter $c = 2\lambda f^\#$). Finally, we write

$$\mathcal{S}(x) = \int_0^{2\pi(x+b)/c} \text{sinc}^2 \left(\frac{y}{2} \right) dy + \int_{2\pi(x-b)/c}^{0} \text{sinc}^2 \left(\frac{y}{2} \right) dy$$

$$= s \left[\frac{2\pi(x+b)}{c} \right] - s \left[\frac{2\pi(x-b)}{c} \right] \quad x > 0 \; . \tag{5.10}$$

Since $s(x)$ is symmetric, the expression for $x > 0$ is adequate for our purposes.

Using the tabulated values of $s(x)$, the diffraction-limited image of a bar may be readily plotted for various values of the ratio $2b/c$. In the calculations presented in Fig. (5.2), the spot size was fixed and the size of the slit $2b$ was varied to achieve the various values of α. The effects of the convolution of the object with the fixed impulse response, as described by Eq. (5.5), are most readily apparent at the edges of the slit in Fig. 5.2. This effect shows that the spread of the edge due to convolution is effectively the same for all slit widths. A variation in α, achieved by changing the spot size for a fixed slit width, would have dramatically changed the edge slope in Fig. 5.2. (The experimental data presented in Fig. 5.7 were made by varying the spot size, showing this edge sharpening effect with increasing α.) In the curve of Fig. 5.2, the abscissa is the normalized coordinate $\eta = x/c$ and α is the ratio of the slit size ($2b$) to the width of the impulse response c

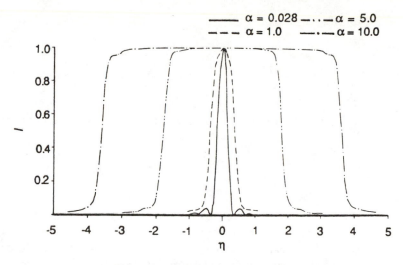

Fig. 5.2 Calculated images of bar objects with various values of α.

(spot size). With $\alpha = 0.028$, the image of the bar is not noticeably different from the impulse response of the lens itself. This situation is described by stating that the object is not "resolved" by the lens. For $\alpha = 1$, the image is significantly different from the impulse response, permitting the conclusion that the object is finite; but very little more. For the cases of $\alpha = 5$ and $\alpha = 10$, the image resembles the object very closely. These cases are said to be "well resolved." However, even in these "well-resolved" images, a definite rounding of the "shoulder" and "toe" on the bar can be noticed. This rounding or softening of the edges is characteristic of the convolution integral and is the reason for the frequently used terminology "smoothing integral" for Eq. (5.1).

5.4 IMAGE OF TWO BARS

As a second example, we consider the image of two bars of equal width and with a center-to-center spacing of two bar widths. Thus, the intensity of the object (shown in Fig. 5.3) is given by

$$I_{obj}(x') = \begin{pmatrix} 0 & |x'| < b \\ 1 & b \leq |x'| \leq 3b \\ 0 & |x'| > 3b \end{pmatrix} . \tag{5.11}$$

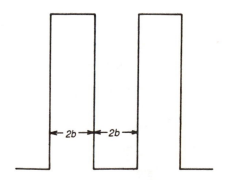

Fig. 5.3 A two-bar object.

Denoting the normalized image in this case by $\mathcal{S}_2(x)$, we write

$$\mathcal{S}_2(x) = \int_{x-3b}^{x-b} \text{sinc}^2\left(\frac{kax'}{2f}\right)dx' + \int_{x+b}^{x+3b} \text{sinc}^2\left(\frac{kax'}{2f}\right)dx' \ . \tag{5.12}$$

Equation (5.12) can be put in a more pleasing form by noting that

$$\mathcal{S}_2(x) = \int_{x-3b}^{x+3b} \text{sinc}^2\left(\frac{kax'}{2f}\right)dx' - \int_{x-b}^{x+b} \text{sinc}^2\left(\frac{kax'}{2f}\right)dx' \ . \tag{5.13}$$

Equation (5.13) will be recognized as the image of a bar of half-width $3b$ minus the image of a bar of half-width b. This representation of one image in terms of two simpler ones is a result of the linearity (see Chapter 4) of the imaging process and will play a very important role in subsequent articles. Using the same change of variables as before and introducing the notation $\mathcal{S}_1(\eta|\alpha/2)$ to denote the normalized image of a bar of width $\alpha/2$, we may write Eq. (5.13) as

$$\mathcal{S}_2(\eta) = \mathcal{S}_1(\eta|3\alpha/2) - \mathcal{S}_1(\eta|\alpha/2) \ , \tag{5.14}$$

which is plotted in Fig. 5.4 for various values of α. Again, α is varied by changing the width of the slit and the separation $2b$ for a constant spot size c, resulting in a similar edge slope in all the images because of the convolution process. Here again we find that for $\alpha = 0.028$ the object is not resolved at all because the image

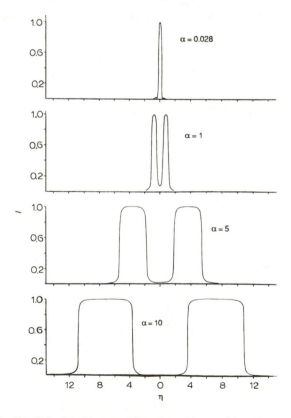

Fig. 5.4 Calculated images of two-bar objects with various values of α.

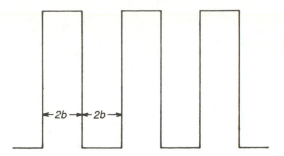

Fig. 5.5 A three-bar object.

is essentially indistinguishable from the impulse response of the system. For the case $\alpha = 1$, we note that the two bars are reasonably well resolved. That is, it is clear that there are two objects, but it is not clear that these objects are bars. Note that the intensity does not go to zero between the images. For $\alpha = 5$, the images are well separated and the central intensity is zero. For $\alpha = 10$, we find once more that the image is rather well resolved. However, the edges are still rounded because of the smoothing of the convolution integral. This problem is closely related to the two-point (or two-line) resolution limit determination, which is one of the principal methods of judging the quality of optical systems. We return to the two-point resolution problem in Chapter 6.

5.5   IMAGE OF THREE BARS

Another factor frequently used in judging optical systems is the three-bar resolution limit; and, as our final example of the convolution integral approach, we calculate the image of a three-bar target. For this case, the object intensity distribution is as shown in Fig. 5.5 and is described by

$$I_{obj}(x') = \begin{pmatrix} 1 & |x'| \leq b \\ 0 & b < |x'| < 3b \\ 1 & 3b \leq |x'| \leq 5b \\ 0 & |x'| > 5b \end{pmatrix} . \tag{5.15}$$

By arguments similar to those in the preceding example, we may express the normalized image of three bars, $\mathcal{S}_3(x)$, as the image of a bar of half-width $5b$ minus the image of a bar of half-width $3b$ plus the image of a bar of half-width b, i.e.,

$$\mathcal{S}_3(x) = \mathcal{S}_1(\eta|5\alpha/2) - \mathcal{S}_1(\eta|3\alpha/2) + \mathcal{S}_1(\eta|\alpha/2) . \tag{5.16}$$

The resultant image is plotted in Fig. 5.6 for fixed spot size c and a variable bar width and separation $2b$. (In the three-bar experimental results of Fig. 5.8, the bars of width and spacing $2b$ were fixed, and the spot size c was varied for experimental simplicity.) As before, we can see the image passes from completely unresolved for $\alpha = 0.028$, to just resolved for $\alpha = 1$, to quite well resolved for $\alpha = 5$ and $\alpha = 10$.

5.6   EXPERIMENTAL ILLUSTRATIONS

The calculated examples in the previous sections were restricted to one-dimensional optical systems. Furthermore, the lens aperture was considered fixed and

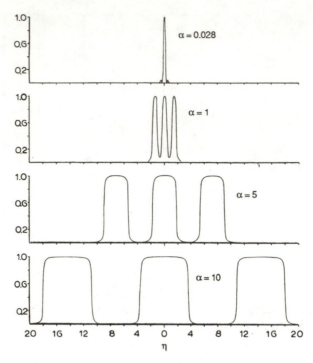

Fig. 5.6 Calculated images of three-bar objects with various values of α.

the scale of the object was varied. Naturally, if the scale of the target had been fixed and the aperture diameter of the lens changed, then similar effects would have resulted. Experimentally, it is more convenient to use a two-dimensional lens and then to change the size of the impulse response by changing the aperture diameter of the lens. The illustrations in this section were all conducted in this manner. The impulse response for the two-dimensional system is given by the square of the Fourier transform of the aperture distribution and is, therefore, a $|2J_1(\phi)/\phi|^2$ function where $\phi = kar/f$ and a is the aperture radius [see Chapter 2, Eq. (2.18)].

Figure 5.7 shows the image of a single bar with various f numbers and hence various values of α. With $f^\# = 11.7$ ($\alpha = 7.5$), the bar is well resolved, but as the $f^\#$ is increased, the resolution gets worse until by the time $f^\# = 780$ ($\alpha = 0.12$) the bar is completely unresolved and all we record is the impulse response of the lens still slightly distorted along the length of the bar.

Figure 5.8 shows another example, this time for a three-bar object. Again with $f^\# = 11.7$ ($\alpha = 7.5$), the bars are well resolved until at $f^\# = 350$ ($\alpha = 0.26$) the information that there are three bars is lost. Finally, the impulse response is all that is recorded.

As a final example, the source is a small circular object. In Fig. 5.9(a) it is well resolved. As the $f^\#$ is increased, the area occupied by the image increases until finally in (e) only the impulse response is seen.

REFERENCE

1. W. W. Gerbes, G. E. Reynolds, M. R. Hoes, and C. J. Drane, "Table of $S(x)$ and its first eleven derivatives," ASTIA No. AD146837-8-9, AFCRC-TR-58-117 (I, II, III) (1958).

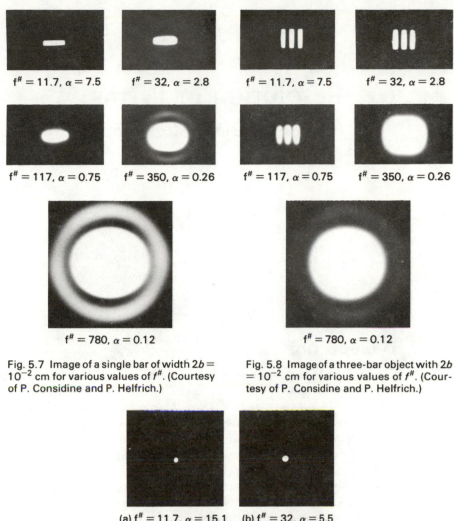

$f^{\#} = 11.7, \alpha = 7.5$ $f^{\#} = 32, \alpha = 2.8$ $f^{\#} = 11.7, \alpha = 7.5$ $f^{\#} = 32, \alpha = 2.8$

$f^{\#} = 117, \alpha = 0.75$ $f^{\#} = 350, \alpha = 0.26$ $f^{\#} = 117, \alpha = 0.75$ $f^{\#} = 350, \alpha = 0.26$

$f^{\#} = 780, \alpha = 0.12$

$f^{\#} = 780, \alpha = 0.12$

Fig. 5.7 Image of a single bar of width $2b = 10^{-2}$ cm for various values of $f^{\#}$. (Courtesy of P. Considine and P. Helfrich.)

Fig. 5.8 Image of a three-bar object with $2b = 10^{-2}$ cm for various values of $f^{\#}$. (Courtesy of P. Considine and P. Helfrich.)

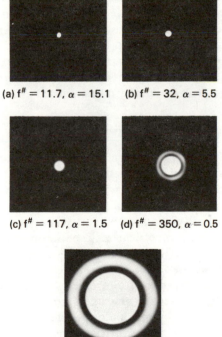

(a) $f^{\#} = 11.7, \alpha = 15.1$ (b) $f^{\#} = 32, \alpha = 5.5$

(c) $f^{\#} = 117, \alpha = 1.5$ (d) $f^{\#} = 350, \alpha = 0.5$

(e) $f^{\#} = 780, \alpha = 0.23$

Fig. 5.9 Image of a circular aperture of diameter 2×10^{-2} cm for various values of $f^{\#}$. (Courtesy of P. Considine and P. Helfrich.)

6 Resolution in Terms of the Impulse Response

6.1 INTRODUCTION

The problem of the definition and the determination of an image quality criterion has long been and still is a major one in the field of image evaluation and assessment. A large variety of measures have been used, from a two-point resolution criterion to fidelity defect, etc. None of these measures is, however, completely satisfactory. In this chapter, we wish to discuss some of the simple criteria in terms of the intensity impulse response of the optical system. This allows us a preliminary look at the image assessment problem and also allows us to follow through further examples on the use of the intensity impulse response. We concern ourselves only with the Rayleigh and Sparrow criteria for two-point resolution; more sophisticated image quality criteria will be discussed in Chapter 19.

6.2 TWO-POINT RESOLUTION

Historically, one of the first measures established for the evaluation of optical systems was to specify how well the system could resolve a two-point object. The two objects were incoherent with respect to each other. Idealized point objects are, of course, not essential; all we require is that the individual objects are not themselves resolved, i.e., the intensity impulse response of the system is much broader than the geometrical image of the object. Clearly, two objects which are themselves resolved would be observable as two individual objects for all separations. What we are trying to establish now is a criterion that will allow us to determine the presence of two objects when the objects themselves are not resolved. Under these circumstances, the image will consist of the sum of the intensity impulse response of the optical system located at each image point—in general, these two intensity distributions will overlap. What is the separation of the centers of the impulse responses and hence the separation of the objects that will allow us to distinguish them as two separate impulse responses and hence conclude that the object consisted of two distinct parts? Naturally, this type of discussion is only applicable if indeed the object to be imaged did consist of two equally bright unresolved objects. However, this problem does occur in viewing a double star with a telescope or when viewing two adjacent cells under a micro-scope. Note that using the results of this type of study does not give a useful indication of the performance of the same optical system when used to image a continuous as opposed to a two-point object.

Two criteria have been proposed and used extensively for the two-point resolution measurement: the well-known Rayleigh criterion and the Sparrow criterion. We will discuss these criteria in some detail but it must be borne in mind that they are only criteria and as such are useful only for the comparison of one system with another. They do not measure the absolute limit of resolution. At best, resolution criteria such as these are approximate treatments. A complete treatment requires an analysis of the noise characteristics, which are the ultimate resolution limiting factors.

6.3   IMAGE OF TWO POINTS: ONE DIMENSIONAL

Consider a one-dimensional optical system that is to be used to image a two-point object. The one-dimensional analogue of a point is a line; therefore, in this section, we use line objects. Let us further imagine that the separation of the two lines can be varied. From the discussion of the last two chapters, the image of a single line object is given by [see, e.g., Eq. (4.30)]

$$S(x) = 4a^2 \operatorname{sinc}^2 \left(\frac{kax}{z} \right) , \tag{6.1}$$

where a is the half-width of the aperture of the cylindrical lens, $k = 2\pi/\lambda$, and x is the position coordinate in the image plane. Hence, for two object points A and B separated by a distance $2b$ (Fig. 6.1), which are imaged at A' and B', the object intensity distribution is described by

$$I_{obj}(x') = I_0[\delta(x' - b) + \delta(x' + b)] ; \tag{6.2}$$

hence, the image intensity distribution is

$$I_{im}(x) = 4a^2 I_0 \int_{-\infty}^{\infty} [\delta(x' - b) + \delta(x' + b)] \operatorname{sinc}^2 \left[\frac{ka(x - x')}{z} \right] dx' . \tag{6.3}$$

Equation (6.3) assumes, of course, unit magnification where $z = z' = 2f$. Performing the integration of Eq. (6.3) yields

$$I_{im}(x) = 4a^2 I_0 \left[\operatorname{sinc}^2 \frac{ka}{z} (x - b) + \operatorname{sinc}^2 \frac{ka}{z} (x + b) \right] , \tag{6.4}$$

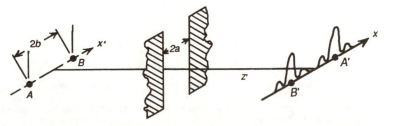

Fig. 6.1 System geometry for one-dimensional resolution analysis.

i.e., the impulse response is put down at the image of each point object individually. The form of the resultant intensity is clearly going to depend on the separation of the two sinc^2 functions in image space. Two-point resolution criteria depend on the separation of these two sinc^2 functions.

6.3.1   Rayleigh Criterion

The Rayleigh criterion chooses the separation of the sinc^2 functions such that the central maximum of one coincides with the first minimum of the other. Hence, the resultant intensity distribution has a small minimum at the center of the two-point images. From knowledge of the sinc^2 function, the separation $2b$ is then given by

$$2b = \frac{z'\lambda}{2a} . \tag{6.5}$$

Figure 6.2 illustrates this resultant intensity distribution when $2b < z'\lambda/2a$, $2b = z'\lambda/2a$, and $2b > z'\lambda/2a$.

6.3.2   Sparrow Criterion

The Sparrow criterion chooses the separation of the two sinc^2 functions such that

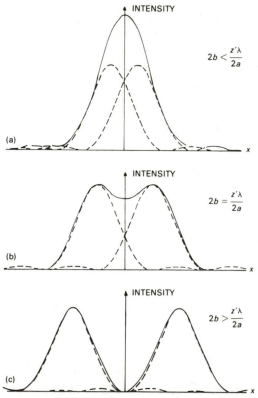

Fig. 6.2 Plots of two-point intensities in the image plane for various separations of the object points $2b$ and a fixed lens aperture $2a$ yielding a fixed impulse response size: (a) $2b < z'\lambda/2a$, unresolved points; (b) $2b = z'\lambda/2a$, two points just resolved by the Rayleigh criterion; and (c) $2b > z'\lambda/2a$, two points well resolved (——————— resultant intensity; _ _ _ _ _ _ _ sinc^2 functions).

$$\frac{\partial^2}{\partial x^2} [I(x)] \bigg|_{x=0} = 0 \ . \tag{6.6}$$

Hence, we have to evaluate

$$\frac{\partial^2}{\partial x^2} \left[\text{sinc}^2 \frac{ka}{z'} (x - b) + \text{sinc}^2 \frac{ka}{z'} (x + b) \right] = 0 \text{ at } x = 0 \ . \tag{6.7}$$

To carry out this evaluation, we note that

$$\frac{\partial}{\partial \theta} \left(\frac{\sin^2 \theta}{\theta^2} \right) = \frac{\theta^2 \sin 2\theta - 2\theta \sin^2 \theta}{\theta^4} \ , \tag{6.8}$$

and, hence,

$$\frac{\partial^2}{\partial \theta^2} \left(\frac{\sin^2 \theta}{\theta^2} \right) = \frac{2\theta^2 \cos 2\theta - 2\theta \sin^2 \theta}{\theta^4} - \frac{2\theta^3 \sin 2\theta - 6\theta^2 \sin^2 \theta}{\theta^6} \ ,$$

$$= \frac{2\cos 2\theta}{\theta^2} - \frac{4\sin 2\theta}{\theta^3} + \frac{6\sin^2 \theta}{\theta^4} \ . \tag{6.9}$$

Furthermore, from symmetry $\text{sinc}\, x = \text{sinc}(-x)$ and, therefore,

$$\frac{\partial^2}{\partial x^2} [I_{im}(x)] \bigg|_{x=0} = \frac{k^2 a^2 b^2}{(z')^2} \cos \left(\frac{2kab}{z'} \right)$$

$$- \frac{2kab}{z'} \sin \left(\frac{2kab}{z'} \right) + 3\sin^2 \left(\frac{kab}{z'} \right) = 0 \ . \tag{6.10}$$

Solving Eq. (6.10) by the usual iteration method gives

$$\frac{2kab}{z'} = 2.606 \tag{6.11}$$

and, hence,

$$2b = \frac{2.606 z' \lambda}{2\pi a} \ . \tag{6.12}$$

Figure 6.3 shows this result.

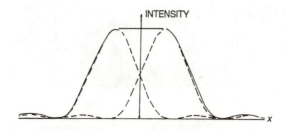

INTENSITY

Fig. 6.3 Plot of a two-point intensity in the image plane corresponding to the Sparrow criterion of Eq. (6.6) demonstrating the constant intensity in the region of the origin (_____ resultant intensity; _ _ _ _ _ _ _ _ sinc² functions).

6.4   IMAGE OF TWO POINTS: TWO DIMENSIONAL

The results obtained in the previous section are easily carried over to the two-dimensional situation. The impulse response for an optical system of radius d is now given by

$$S(x) = 4d^2 \left[\frac{2J_1\left(\frac{kdx}{z'}\right)}{\left(\frac{kdx}{z'}\right)} \right]^2 = 4d^2 \left[\Lambda_1 \left(\frac{kdx}{z'}\right) \right]^2 . \tag{6.13}$$

Hence, Eq. (6.4) becomes

$$I_{im}(x) = 4d^2 I_0 \left(\left\{ \Lambda_1 \left[\frac{kd(x-b)}{z'} \right] \right\}^2 + \left\{ \Lambda_1 \left[\frac{kd(x+b)}{z'} \right] \right\}^2 \right) . \tag{6.14}$$

The Rayleigh criterion is defined by

$$2b = \frac{1.22 z' \lambda}{2d} \tag{6.15}$$

and the Sparrow criterion by

$$2b = \frac{2.976 z' \lambda}{2\pi d} . \tag{6.16}$$

Figure 6.4 illustrates these results experimentally: In (a) the two equally bright incoherent objects are individually well resolved; as the $f^{\#}$ of the optical imaging system is increased, the individual objects cease to be resolved (g). Gradually the resolution of the two objects is lost. Figure 6.4(e) is the region of the Rayleigh criterion and (f) approximates to the position for the Sparrow criterion. These illustrations and indeed this whole analysis should be compared with the results of Chapter 5.

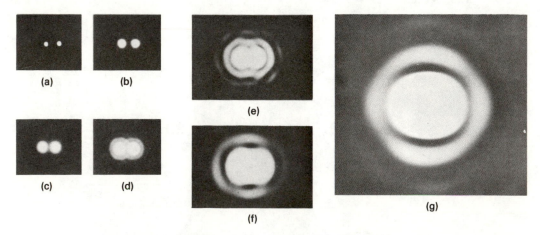

(a)   (b)   (e)   (c)   (d)   (f)   (g)

Fig. 6.4 Experimental illustrations of the variation of two-point resolution as a function of increasing the f/number of the optical system in (a) through (g); (e) illustrates the Rayleigh criterion region and (f) illustrates the Sparrow criterion region. In (g) the two points are unresolved.

6.5   CONCLUSIONS

While the two-point criteria just discussed are useful in astronomy, for more general imagery they are far too restrictive. We require a method of image evaluation that is independent of the object detail. One could, of course, consider an arbitrary object as being composed of a continuous distribution of point sources and then invoke the linearity of the imaging problem to extend the range of usefulness of these criteria. This approach, however, is quite cumbersome. A much more fruitful approach is to characterize the lens not by its impulse response but rather by its sine wave response. The linearity of the system then makes possible the decomposition of arbitrary objects into sine waves and the use of the sine wave response for general objects.

7 Image Formation: The Transfer Function

7.1   INTRODUCTION

Until now we have considered image formation in terms of convolution of the intensity impulse response (intensity diffraction pattern) with the intensity distribution of an incoherently illuminated object. This approach to image formation, while correct and frequently useful, is often difficult to apply. An alternative approach, which simplifies the solution of many problems and is well suited for complicated objects, is the transfer function analysis. We will develop this analysis by means of an example. To minimize mathematical complexities, we have limited the analysis to one dimension.

7.2   IMAGE OF A COSINUSOIDAL INTENSITY DISTRIBUTION

To relate this approach to our earlier analysis, let us consider an object intensity distribution $I_{ob}(x')$ which consists of a cosinusoidal variation, as depicted in Fig. 7.1 and given by

$$I_{ob}(x') = 1 + a \cos 2\pi \mu_0 x' \, , \tag{7.1}$$

where a is the amplitude of the cosine function, μ_0 its spatial frequency, and x' the position coordinate. The visibility of an intensity distribution is defined by

$$\text{visibility} = \frac{I_{max} - I_{min}}{I_{max} + I_{min}} \equiv V \, , \tag{7.2}$$

which in this example is a. Again, we assume that the object is illuminated with incoherent quasimonochromatic light. From Eq. (4.14), it is clear that the image intensity distribution $I_{im}(x)$ is given by

$$I_{im}(x) = \int I_{ob}(x') \, S(x - x') \, dx' \, , \tag{7.3}$$

where x is the position coordinate in image space. Here $S(x)$ is the intensity

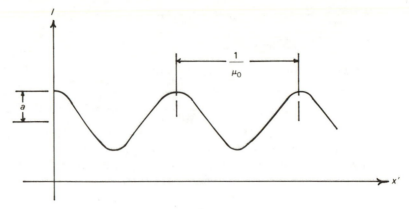

Fig. 7.1 Consinusoidal object intensity distribution.

impulse response. Substituting Eq. (7.1) into (7.3) gives the image intensity distribution as

$$I_{im}(x) = \int (1 + a \cos 2\pi\mu_0 x') \, S(x - x') \, dx' \, . \tag{7.4}$$

To evaluate Eq. (7.4), we use the convolution theorem of Chapter 3 and also recall from Eq. (2.21) that the transform of a constant is a delta function. Furthermore, the Fourier transform of a cosine function is two delta functions [see Eqs. (3.11), (3.12), and (3.13)]. Thus,

$$I_{im}(x) = \int \left\{ \delta(\mu) + \frac{a}{2} [\delta(\mu - \mu_0) + \delta(\mu + \mu_0)] \right\} \tilde{S}(\mu) e^{-2\pi i \mu x} \, d\mu \, , \tag{7.5}$$

where μ is the coordinate in frequency (transform) space.

Using the combing property of the delta function, Eq. (7.5) may be evaluated readily to give

$$I_{im}(x) = \tilde{S}(0) + \frac{a}{2} [\tilde{S}(\mu_0) e^{2\pi i \mu_0 x} + \tilde{S}(-\mu_0) e^{-2\pi i \mu_0 x}] \, . \tag{7.6}$$

To interpret Eq. (7.6), we recall from Eq. (3.23) that

$$\tilde{S}(\mu) = \int S(x) e^{2\pi i \mu x} \, dx \, . \tag{7.7}$$

Taking the complex conjugate on both sides of Eq. (7.7) and remembering that $S(x)$ is a real function, we have

$$\tilde{S}^*(\mu) = \int S(x) e^{-2\pi i \mu x} \, dx = \tilde{S}(-\mu) \, , \tag{7.8}$$

where the asterisk denotes a complex conjugate. Thus, the last two terms on the right side of Eq. (7.6) are complex conjugates of each other. Therefore,

$$I_{im}(x) = \tilde{S}(0) + \frac{a}{2} 2 \, \text{Re} \, [\tilde{S}(\mu_0) e^{2\pi i \mu_0 x}] \, , \tag{7.9}$$

where Re denotes a real part. Writing

$$\tilde{S}(\mu) = |\tilde{S}(\mu)| e^{2\pi i \phi(\mu)} , \qquad (7.10)$$

we may simplify Eq. (7.9) to give finally

$$I_{im}(x) = \tilde{S}(0) \left\{ 1 + a \; \frac{|\tilde{S}(\mu_0)|}{\tilde{S}(0)} \; \cos 2\pi [\mu_0 x + \phi(\mu_0)] \right\} . \qquad (7.11)$$

Thus, independent of the exact form of the impulse response, *the image of a cosine is a cosine of the same frequency; only the contrast and phase of the cosine can be affected by a linear system.* The visibility in the image is given by

$$V_i = a \frac{|\tilde{S}(\mu_0)|}{\tilde{S}(0)} \qquad (7.12)$$

while the visibility in the object was $V_0 = a$. The ratio of the image visibility to the object visibility is the modulus of the optical transfer function and is given by

$$|\tau(\mu_0)| = \frac{V_i}{V_0} = \frac{|\tilde{S}(\mu_0)|}{\tilde{S}(0)} . \qquad (7.13)$$

When the effect of spatial phase shifts is included, the optical transfer function is the complex quantity

$$\tau(\mu_0) = \frac{|\tilde{S}(\mu_0)|}{\tilde{S}(0)} \; e^{i\phi(\mu_0)} = \frac{\tilde{S}(\mu_0)}{\tilde{S}(0)} . \qquad (7.14)$$

The phase function $\phi(\mu_0)$ here is, of course, the phase that occurs in Eqs. (7.10) and (7.11).

7.3   PERIODIC REAL OBJECT

To illustrate the significance of the above result, let us consider two more examples before discussing general physical implications. Consider a periodic real object. It is always possible to represent such an object by a Fourier series of the form

$$I_{ob}(x') = \sum_{n=-\infty}^{\infty} a_n \cos 2\pi (n\mu_0 x' + \psi_n) . \qquad (7.15)$$

The image of such an object will be given by the convolution with the impulse response which, as in the previous example, may be expressed as

$$I_{im}(x) = \int [\tilde{I}_{ob}(\mu) \tilde{S}(\mu)] e^{-2\pi i \mu x} \, d\mu . \qquad (7.16)$$

Taking the Fourier transform on both sides of Eq. (7.15) and substituting into Eq. (7.16) yields

$$I_{im}(x) = \sum_{n=-\infty}^{\infty} \frac{a_n}{2} \int [\delta(\mu - n\mu_0) e^{i\psi_n}$$

$$+ \delta(\mu + n\mu_0) e^{-i\psi_n}] \tilde{S}(\mu) e^{-2\pi i \mu x} d\mu \ . \tag{7.17}$$

Using the symmetry of the impulse response, Eq. (7.17) reduces to

$$I_{im}(x) = \sum_{n=-\infty}^{\infty} \frac{a_n}{2} [\tilde{S}(n\mu_0) e^{i\psi_n} e^{2\pi i n\mu_0 x} + \tilde{S}(-n\mu_0) e^{-i\psi_n} e^{-2\pi i n\mu_0 x}] \ . \tag{7.18}$$

Using Eq. (7.10), Eq. (7.18) reduces quickly to

$$I_{im}(x) = \sum_{n=-\infty}^{\infty} a_n |\tilde{S}(n\mu_0)| \cos 2\pi [n\mu_0 x + \psi_n + \phi(n\mu_0)] \ . \tag{7.19}$$

Note that Eq. (7.19) is of precisely the same form as Eq. (7.15). That is, in imaging the periodic object, each cosine component is preserved with only its relative amplitude and phase changed.

To image more general objects, the same procedure is applied only the Fourier transform must be used instead of the Fourier series. Thus, the object intensity is now given by

$$I_{ob}(x') = \int \tilde{I}_{ob}(\mu) e^{-2\pi i \mu x'} d\mu \ ; \tag{7.20}$$

and the corresponding image is

$$I_{im}(x) = \int [\tilde{I}_{ob}(\mu) \tilde{S}(\mu)] e^{-2\pi i \mu x} d\mu \ . \tag{7.21}$$

From these examples, the advantage of the transfer function should be clear. Instead of evaluating a convolution of the impulse response with the object intensity to obtain the image intensity, one simply multiplies the object spectrum by the *unnormalized* transfer function $\tilde{S}(\mu)$ to obtain the image spectrum. This point is perhaps made clearer by taking the Fourier transform on both sides of Eq. (7.21), which gives

$$\tilde{I}_{im}(\mu) = \tilde{I}_{ob}(\mu) \tilde{S}(\mu) \ . \tag{7.22}$$

Of course, to obtain the detailed intensity distribution in the image, Eq. (7.21) must be evaluated, which is in general no less difficult than evaluating Eq. (7.3) directly. However, evaluating Eq. (7.22) is always a simple matter. That is, multiplication by the *unnormalized* transfer function can always be performed. Frequently, that is enough of an answer; to know which spatial frequencies are passed and to what extent they are attenuated is often all the answer that is required. This point will be illustrated by examples in the next chapter.

7.4   THE TRANSFER FUNCTION AND THE APERTURE FUNCTION

Finally, we derive an important relationship between the transfer function and the aperture illumination function.

Let $A(\xi)$ represent the complex transmission of the aperture. The intensity impulse response is then given by [see Eq. (4.30)]

$$S(x) = \left| \int A(\xi) \exp\left(-\frac{ik\xi x}{z}\right) d\xi \right|^2 . \tag{7.23}$$

The unnormalized transfer function was defined as

$$\tilde{S}(\mu) = \int S(x) e^{2\pi i \mu x} dx , \tag{7.24}$$

which using Eq. (7.23), apart from an additive constant, may be written as

$$\begin{aligned}
\tilde{S}(\mu) &= \iiint A(\xi_1) A^*(\xi_2) \exp\left[-2\pi i x \left(\frac{\xi_1 - \xi_2}{\lambda z}\right)\right] \exp(2\pi i \mu x) \, d\xi_1 \, d\xi_2 \, dx \\
&= \iint A(\xi_1) A^*(\xi_2) \int \exp\left[-2\pi i x \left(\frac{\xi_1 - \xi_2}{\lambda z} - \mu\right)\right] dx \, d\xi_1 \, d\xi_2 \\
&= \iint A(\xi_1) A^*(\xi_2) \, \delta\left(\mu - \frac{\xi_1 - \xi_2}{\lambda z}\right) d\xi_1 \, d\xi_2 \\
&= \int A^*(\xi_2) \int A(\xi_1) \, \delta[\xi_1 - (\xi_2 + \lambda z \mu)] \, d\xi_1 \, d\xi_2 \\
&= \int A^*(\xi_2) A(\xi_2 + \lambda z \mu) \, d\xi_2 .
\end{aligned} \tag{7.25}$$

Equation (7.25) is an important result which states that the unnormalized transfer function $\tilde{S}(\mu)$ is given by the unnormalized autocorrelation of the aperture function with its complex conjugate. The usefulness of this theorem will be illustrated in Chapter 8.

7.5   CONCLUSION

In this chapter, we have seen that the cumbersome convolution integral formulation of the imaging problem may be replaced by multiplication by making use of Fourier analysis. Our results are (1) the spectrum of the image is given by the spectrum of the object multiplied by the unnormalized transfer function of the lens and (2) the transfer function may be determined either by taking the Fourier transform of the impulse response or directly from the aperture illumination function itself.

8 Image Formation in Terms of the Transfer Function

8.1 INTRODUCTION

In previous chapters, we demonstrated that an imaging system may be analyzed either in the domain of spatial coordinates, i.e., in terms of its impulse response (diffraction pattern), or in the domain of spatial frequencies, i.e., in terms of its transfer function. Analysis in terms of the impulse response was illustrated in Chapter 5. Here we illustrate the transfer function analysis. To minimize mathematical complexities, we have limited the analysis to one dimension.

8.2 THE TRANSFER FUNCTION

With incoherent light the transfer function $\tilde{S}(\mu)$ is given by Eq. (7.25) as the convolution of the aperture distribution A with its complex conjugate, i.e.,

$$\tilde{S}(\mu) = \int A^{*}(\xi)\, A(\xi + \lambda z \mu)\, d\xi \; , \tag{8.1}$$

where the image distance z is left general rather than being specified as equal to z' and $2f$; here the object is at an arbitrary distance z'. For a clear aperture, i.e., no aberration and no apodization (masking of the aperture), $A(\xi)$ may be written as

$$A(\xi) = \begin{pmatrix} 1 & |\xi| \leq a \\ 0 & |\xi| > a \end{pmatrix} = \mathrm{rect}(\xi|a) \; , \tag{8.2}$$

where a is the half-width of the aperture.

The integral in Eq. (8.1) can be thought of as measuring the overlapping area as the aperture is slid across itself as illustrated in Fig. 8.1. Thus, the transfer function is given by

$$\tilde{S}(\mu) = \int_{-a}^{a-\lambda f\mu} d\xi = a - \lambda f\mu + a = 2a\left(1 - \frac{\lambda z \mu}{2a}\right) \; ; \tag{8.3}$$

and the normalized transfer function $\tau(\mu)$ is

$$\tau(\mu) = \frac{\tilde{S}(\mu)}{\tilde{S}(0)} = \left(1 - \frac{\lambda z \mu}{2a}\right); \quad \mu \geqslant 0 \; . \tag{8.4}$$

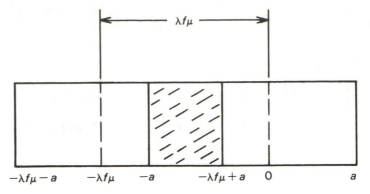

Fig. 8.1 Schematic illustration of the convolution of two rect functions.

Similarly, for negative μ,

$$\tau(\mu) = \left(1 + \frac{\lambda z \mu}{2a}\right) ; \quad \mu \leqslant 0 ; \tag{8.5}$$

or finally

$$\tau(\mu) = \left(1 - \frac{\lambda z |\mu|}{2a}\right) ; \quad |\mu| \leq \frac{2a}{\lambda f} , \tag{8.6}$$

where $2a$ is the aperture width. We define the dimensionless ratio $z/2a$ as the $z^{\#}$ and Eq. (8.6) can be put in the form

$$\tau(\mu) = (1 - \lambda f^{\#}|\mu|) ; \quad |\mu| \leq \frac{1}{\lambda f^{\#}} . \tag{8.7}$$

In this form, it is clear that the limiting spatial frequency is $1/\lambda f^{\#}$. Thus, sinusoidal intensity variations at spatial frequencies in excess of $1/\lambda f^{\#}$ will not be imaged by a system using incoherent light. In the special case that the object is at infinity, $z = f$ and $z^{\#} = f^{\#}$.

A plot of the normalized transfer function is shown in Fig. 8.2, making clear that even those spatial frequencies which are imaged by the system are imaged with a reduced contrast. This results in a photometric distortion of the image. We will illustrate this result by means of an example.

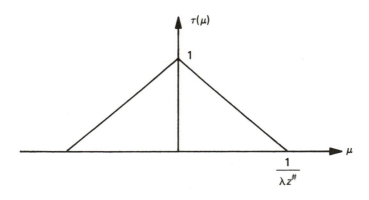

Fig. 8.2 The normalized transfer function for a one-dimensional lens.

In general, of course, the image distance is z' so that image resolution results must be scaled to the object through the system magnification, the ratio of the image distance to the object distance.

8.3   IMAGE OF A RONCHI RULING

The result derived in the previous section may be simply illustrated by considering the problem of imaging a Ronchi ruling with incoherent light. The intensity variation of such a ruling is shown graphically in Fig. 8.3, and may be represented mathematically by the expression

$$I_{obj}(x') = \begin{pmatrix} 1 & x' < \rho/4 \\ 0 & \rho/4 < x' < 3\rho/4 \\ 1 & 3\rho/4 < x' < \rho \end{pmatrix} , \tag{8.8}$$

$$I_{obj}(x' \pm \rho) = I_{obj}(x') . \tag{8.9}$$

From the impulse response analysis point of view, we would construct an impulse response around each point in $I_{obj}(x')$ and add them all to produce the image distribution. That is, we would calculate the image $I_{im}(x)$ from the convolution integral

$$I_{im}(x) = \int I_{obj}(x') S(x - x') dx' . \tag{8.10}$$

With our present point of view, we do not calculate the image $I_{im}(x)$, but rather the spatial spectrum (Fourier transform) $\tilde{I}_{im}(\mu)$ of the image by means of the equation

$$\tilde{I}_{im}(\mu) = \tilde{I}_{obj}(\mu) \tau(\mu) . \tag{8.11}$$

The spectrum of the object $\tilde{I}_{obj}(\mu)$ is obtained readily by direct integration,

$$\tilde{I}_{obj}(\mu) = \int_{-\infty}^{\infty} I_{obj}(x) e^{2\pi i \mu x'} dx'$$

$$= \sum_{n=-\infty}^{\infty} \int_{n\rho - \rho/4}^{n\rho + \rho/4} e^{2\pi i \mu x'} dx'$$

$$= \frac{\rho}{2} \operatorname{sinc} \frac{2\pi \mu \rho}{4} \sum_{n=-\infty}^{\infty} e^{2\pi i n \rho \mu} . \tag{8.12}$$

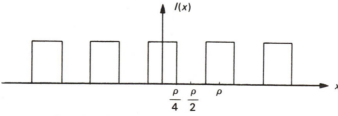

Fig. 8.3 Intensity transmittance of a Ronchi ruling.

The sum on the right side may be replaced by an equivalent sum of delta functions, i.e.,

$$\sum_{n=-\infty}^{\infty} e^{2\pi in\rho\mu} = \sum_{n=-\infty}^{\infty} \delta\left(\mu - \frac{n}{\rho}\right). \tag{8.13}$$

This theorem is demonstrated in the Appendix, Sec. 8.5, at the end of this chapter. Using this theorem reduces Eq. (8.12) to

$$\tilde{I}_{obj}(\mu) = \frac{\rho}{2} \operatorname{sinc} \frac{2\pi\mu\rho}{4} \sum_{n=-\infty}^{\infty} \delta\left(\mu - \frac{n}{\rho}\right). \tag{8.14}$$

From Eq. (8.14) it is clear that the spectrum of the Ronchi ruling is discrete, consisting only of frequencies that are multiples of $1/\rho$. Furthermore, the sinc function, which determines the magnitude of the delta functions, is zero for all the even harmonics and changes sign on every other one of the odd harmonics. A plot of the spectrum is shown in Fig. 8.4.

We may now compute the image spectrum $\tilde{I}_{im}(\mu)$ by simply multiplying $\tilde{I}_{obj}(\mu)$ by $\tau(\mu)$. Several special cases are of interest.

Case 1: $1/\rho > 1/\lambda z^{\#}$, where the product is zero; i.e., only the dc component is passed. There is no intensity variation in the image. The Ronchi ruling spacing is beyond the resolution limit of the system.

Case 2: $1/\rho < 1/\lambda z^{\#} < 3/\rho$, where the resultant image will consist only of the central order of the sinc function; i.e., the fundamental and the dc component. That is, the image will be a simple cosine wave of period ρ and with visibility reduced by $\tau(1/\rho)$. This result is illustrated in Fig. 8.5. Figure 8.5(a) is the object

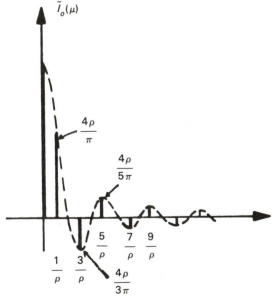

Fig. 8.4 A plot of the spectrum of a Ronchi ruling.

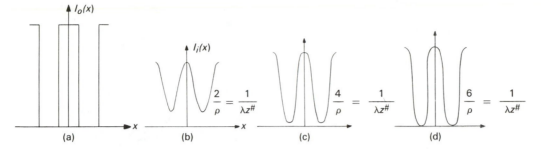

Fig. 8.5 The image intensity distributions when varying amounts of the object spectrum are allowed to contribute to the image.

square wave intensity distribution (Ronchi ruling) and (b) shows the resulting image intensity distribution for $1/\rho = \frac{1}{2}(1/\lambda z^{\#})$.

Case 3: $4/\rho = 1/\lambda z^{\#}$, where Fig. 8.5(c) illustrates the resulting image intensity distribution. The visibility of the image is now improved as compared to Fig. 8.5(b). Referencing Fig. 8.4, the part of the spectrum contributing is readily determined; in this case, the central order and the first side order of the sinc function; i.e., the dc component, the fundamental, and the first harmonic.

Case 4: $6/\rho = 1/\lambda z^{\#}$, where the resulting intensity distribution is plotted in Fig. 8.5(d). The contrast is now approaching unity but the square wave is still not well reproduced.

Figure 8.6 shows experimental results corresponding to the theoretical examples just discussed. Figure 8.6(a) shows the image of a Ronchi ruling equivalent to case 1 above while (b), (c), and (d) are experimental results equivalent to cases 2, 3, and 4, respectively. Again, the main effect is the change in the contrast of the image.

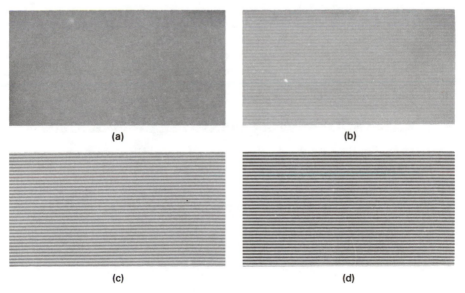

Fig. 8.6 Photographic images of a Ronchi ruling corresponding to the theoretical curves of Fig. 8.5.

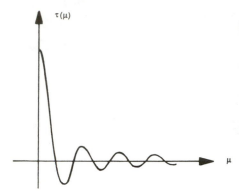

Fig. 8.7 The normalized transfer function for a badly defocused one-dimensional lens.

Fig. 8.8 Photographic image of a converging bar target recorded with a badly defocused system demonstrating spurious resolution.

8.4   DEFOCUSED LENS

Another illustrative example is the case in which the lens is badly defocused. In this case, the impulse response is approximately given by

$$S(x) \simeq \begin{pmatrix} 1 & |x| \leq b \\ 0 & |x| > b \end{pmatrix} , \tag{8.15}$$

where b is the one-dimensional analogue of the radius of the blur circle, i.e., it is the half-width of the blurred line image. The normalized transfer function for this case is

$$\tau(\mu) = \operatorname{sinc} 2\pi b \mu . \tag{8.16}$$

This is shown in Fig. 8.7. Notice that this curve becomes negative in several regions. Actually, since the transfer function is defined as the ratio of image to object contrast, it can never be negative. In the band of frequencies between $1/2b$ and $1/b$, the image Fourier components are spatially shifted by π. By convention, spatial phase shifts of π are indicated by drawing the transfer function as negative. Such negative regions produce spurious resolution, which is illustrated in Fig. 8.8.

8.5   APPENDIX: FOURIER TRANSFORM OF A DIRAC COMB

To demonstrate the theorem used in Eq. (8.13), we must recall a basic theorem from Fourier analysis and the definition of a delta function.

Theorem: If $[\phi_n(x)]$ forms a complete orthonormal set on the interval $a \leq \phi \leq b$, then an arbitrary function $f(x)$ defined on that interval can be expressed as a series of the form:

$$f(x) = \sum_{n=-\infty}^{n=\infty} c_n \phi_n(x) , \tag{8.17}$$

where

$$c_n = \int_a^b f(x)\phi_n^*(x) \, dx \tag{8.18}$$

or

$$f(x) = \sum_{n=-\infty}^{n=\infty} \left[\int_a^b f(x')\phi_n^*(x')\,dx' \right] \phi_n(x) \ . \tag{8.19}$$

In particular, if

$$\phi_n(x') = e^{-2\pi in\rho x} \ , \tag{8.20}$$

and the following changes of variables are made: $x' = \mu' + q/\rho$ and $x = \mu + q/\rho$, then

$$f\left(\mu + \frac{q}{\rho}\right) = \sum_{n=-\infty}^{n=\infty} \left[\int_{-1/(2\rho)}^{1/(2\rho)} f\left(\mu' + \frac{q}{\rho}\right) e^{2\pi in\rho\mu'}\,d\mu' \right] e^{-2\pi in\rho\mu} \tag{8.21}$$

and evaluating both sides at $\mu = 0$ we obtain

$$f\left(\frac{q}{\rho}\right) = \sum_{n=-\infty}^{n=\infty} \int_{-1/(2\rho)}^{1/(2\rho)} f\left(\mu' + \frac{q}{\rho}\right) e^{2\pi in\rho\mu'}\,d\mu' \ . \tag{8.22}$$

Next, recall that the definition of the Dirac delta function is contained in the expression of the combing property, i.e.,

$$f(\mu) = \int f(\mu')\,\delta(\mu - \mu')\,d\mu' \ . \tag{8.23}$$

With these thoughts in mind, we may now consider the theorem

$$\sum_{n=-\infty}^{n=\infty} e^{2\pi in\rho\mu} = \sum_{n=-\infty}^{n=\infty} \delta\left(\mu - \frac{n}{\rho}\right) \ . \tag{8.24}$$

To prove Eq. (8.24), we must show that

$$\int_{-\infty}^{\infty} f(\mu) \sum_{n=-\infty}^{n=\infty} e^{2\pi in\rho\mu}\,d\mu = \sum_{n=-\infty}^{n=\infty} f\left(\frac{n}{\rho}\right) \tag{8.25}$$

for any $f(\mu)$. To prove this, we proceed as follows. On the left side of Eq. (8.25) we interchange the integration and summation and, noting that the functions $e^{2\pi in\rho\mu}$ are orthonormal over a period of ρ, we write the integral as a sum of integrals. Thus,

$$\int_{-\infty}^{\infty} f(\mu) \sum_{n=-\infty}^{n=\infty} e^{2\pi in\rho\mu}\,d\mu = \sum_{n=-\infty}^{n=\infty} \sum_{q=-\infty}^{q=\infty} \int_{q/\rho-1/(2\rho)}^{q/\rho+1/(2\rho)} f(\mu)\,e^{2\pi in\rho\mu}\,d\mu$$

$$= \sum_{n=-\infty}^{n=\infty} \sum_{q=-\infty}^{q=\infty} \int_{-1/(2\rho)}^{1/(2\rho)} f\left(\mu' + \frac{q}{\rho}\right) \exp\left[2\pi in\rho\left(\mu' + \frac{q}{\rho}\right)\right] d\mu'$$

$$= \sum_{n=-\infty}^{n=\infty} \sum_{q=-\infty}^{q=\infty} \int_{-1/(2\rho)}^{1/(2\rho)} f\left(\mu' + \left(\frac{q}{\rho}\right)\right) e^{2\pi in\rho\mu'}\,d\mu'\,e^{2\pi inq} \ . \tag{8.26}$$

If we interchange the order of the summations and note that

$$e^{2\pi inq} = 1 \ , \tag{8.27}$$

Eq. (8.26) becomes

$$\int_{-\infty}^{\infty} f(\mu) \sum_{n=-\infty}^{n=\infty} e^{2\pi in\rho\mu} \, d\mu = \sum_{q=-\infty}^{q=\infty} \sum_{n=-\infty}^{n=\infty} \int_{-1/(2\rho)}^{1/(2\rho)} f\left(\mu' + \frac{q}{\rho}\right) e^{2\pi in\rho\mu'} \, d\mu \ . \tag{8.28}$$

Using Eq. (8.22) we may reduce Eq. (8.28) to

$$\int_{-\infty}^{\infty} f(\mu) \sum_{n=-\infty}^{n=\infty} e^{2\pi in\rho\mu} \, d\mu = \sum_{q=-\infty}^{q=\infty} f\left(\frac{q}{\rho}\right) \ , \tag{8.29}$$

which completes the proof.

9   Fresnel Diffraction

9.1   INTRODUCTION

In the previous chapters, we gave a good deal of attention to the phenomenon of Fraunhofer diffraction. This was necessary because of its importance to the image forming process in describing the impulse response of optical systems. However, in pursuing this aspect of diffraction, we have ignored the whole area of Fresnel diffraction. In this chapter, we discuss the general problem of Fresnel diffraction, which essentially occurs whenever results of the two conditions for Fraunhofer diffraction are not satisfied, i.e.,

1. when the plane of observation is not in the far field of the aperture or

2. when the plane of observation and the point source are not parallel conjugate planes of a well-corrected optical system.

Figure 9.1 illustrates the region of Fraunhofer and Fresnel diffraction for an aperture of diameter D illuminated with a coherent collimated beam of quasi-monochromatic light. In Fig. 9.1(a), the far field extends from $z > D^2/\lambda$ to infinity. The Fresnel region, or "near field," exists between the aperture and the start of the far field. In Fig. 9.1(b), the Fraunhofer condition exists only in a plane—the focal plane of the lens. The region to either side of the focal plane is the region of Fresnel diffraction. We will look at these conditions in detail in the following sections. While it is customary to use the terms "Fresnel region" and "near field" interchangeably, it should be noted that the Fresnel approximations are applicable from within the near field to infinity. However, since the simpler Fraunhofer approximations apply from D^2/λ onward, the use of the Fresnel representation is usually limited to the near field.

9.2   FRESNEL DIFFRACTION: NEAR FIELD

Figure 9.1(a) showed the near-field region where Fresnel diffraction phenomena can be observed. Using the coordinate system of Fig. 9.2, the condition for Fresnel diffraction becomes

$$\frac{k(\xi^2 + \eta^2)_{\max}}{z} \gg 1 \; , \tag{9.1}$$

where $k = 2\pi/\lambda$.

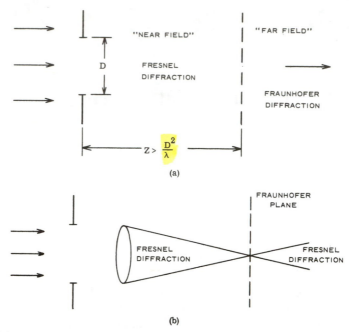

Fig. 9.1 Schematic layout illustrating the differences between Fraunhofer and Fresnel diffraction at an aperture with and without a lens.

From Eq. (1.6) we recall that the wave disturbance in the (x, y) plane is given by

$$\psi(x,y) = \iint \psi'(\xi,\eta)\,\Lambda(x,y,\xi,\eta)\,\frac{e^{ikr}}{r}\,d\xi\,d\eta \;, \tag{9.2}$$

where r is the distance from a point P in the ξ, η plane to a point Q in the x, y plane; $\Lambda(x,y,\xi,\eta)$ is the inclination factor. A more detailed solution of Eq. (1.4), the Helmholtz equation, given in Chapter 11, yields an explicit form for Λ:

$$\psi(x,y) = -\frac{i}{2\lambda} \iint \psi'(\xi,\eta)\cos\theta\,\frac{e^{ikr}}{r}\,d\xi\,d\eta \;, \tag{9.3}$$

where θ is the angle between the normal and the direction of propagation at the point P. Hence, the inclination factor is

$$\Lambda(x,y,\xi,\eta) = -\frac{i\cos\theta}{2\lambda} \;. \tag{9.4}$$

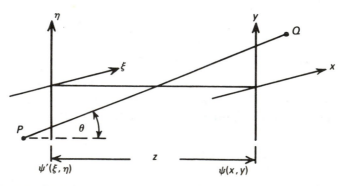

Fig. 9.2 Schematic for defining the coordinates in the diffracting plane and the observation plane.

When θ approximates to zero, then the cosine approaches unity and

$$\psi(x,y) = \frac{-i}{2\lambda} \iint \psi'(\xi,\eta) \frac{e^{ikr}}{r} d\xi\, d\eta \ . \tag{9.5}$$

Equation (9.5) is extremely useful since often in diffraction problems the intensity distribution is not required for large angular divergence. If r is now written in terms of the Cartesian coordinates, then

$$r^2 = (x - \xi)^2 + (y - \eta)^2 + z^2 \ . \tag{9.6}$$

If z is large compared to the other distance (a condition consistent with making $\theta \to 0$), Eq. (9.6) can be expanded as a binomial series, and hence

$$r \approx z + \frac{(x - \xi)^2}{2z} + \frac{(y - \eta)^2}{2z} + \dots \ . \tag{9.7}$$

Neglecting higher order terms in this expansion and noting that a change in r in the denominator is not very significant compared with a change in the value of r in the exponent, we write

$$\psi(x,y) = -\frac{i}{2\lambda}\frac{e^{ikz}}{z} \iint \psi'(\xi,\eta)\exp\left[\frac{ik(x-\xi)^2}{2z}\right]$$

$$\times \exp\left[\frac{ik(y-\eta)^2}{2z}\right] d\xi\, d\eta \ , \tag{9.8}$$

where we set $r = z$ in the denominator of the amplitude factor, which is treated here as a constant because it varies slowly compared to the phase. We can rewrite Eq. (9.8) in the following manner:

$$\psi(x,y) = K \iint \psi'(\xi,\eta)\exp\left[\frac{ik(x^2+y^2)}{2z}\right]\exp\left[\frac{ik(\xi^2+\eta^2)}{2z}\right]$$

$$\times \exp\left[\frac{-ik(x\xi+y\eta)}{z}\right] d\xi\, d\eta \ , \tag{9.9}$$

where $K = (-i/2\lambda)\exp(ikz/z)$. Clearly one way to interpret this equation is to define some hypothetical aperture distribution $\psi_h(\xi,\eta)$ such that

$$\psi_h(\xi,\eta) = \psi'(\xi,\eta)\exp\left[\frac{ik(\xi^2+\eta^2)}{2z}\right] ; \tag{9.10}$$

then

$$\psi(x,y) = K\exp\left[\frac{ik(x^2+y^2)}{2z}\right] \iint \psi'_h(\xi,\eta)$$

$$\times \exp\left[\frac{-ik(x\xi+y\eta)}{z}\right] d\xi\, d\eta \ . \tag{9.11}$$

Equation (9.11) shows that the distribution $\psi(x,y)$ is just the two-dimensional Fourier transform of this hypothetical aperture distribution. Conceptually this is important but in general it doesn't help in solving problems, although in the analysis of complex systems it can often be useful in simplifying the equations to be handled (see Chapter 32). Furthermore, since computer programs for two-dimensional Fourier analysis are now quite common, there may be a computational advantage. However, for our purposes let us return to Eq. (9.8). The solution of this equation obviously depends on evaluating integrals of the type

$$F(x) = F(\xi_1, \xi_2) = \int_{\xi_1}^{\xi_2} \exp\left[\frac{ik(x-\xi)^2}{2z}\right] d\xi \ .$$ (9.12)

This equation can be written in terms of the Fresnel integrals.

9.3   FRESNEL'S INTEGRALS

Letting

$$\alpha^2 = (x-\xi)^2 \frac{2}{\lambda z} \ ,$$ (9.13)

then

$$F(\alpha_1, \alpha_2) = K' \int_{\alpha_1}^{\alpha_2} \exp\left(i\frac{\pi}{2}\alpha^2\right) d\alpha \ ,$$ (9.14)

where K' and the limits α_1 and α_2 are obtained from Eq. (9.13). The Fresnel integrals are usually expressed in the following form

$$\left.\begin{array}{l} C(\alpha) = \displaystyle\int_0^\alpha \cos\frac{\pi}{2}\alpha^2 \, d\alpha \\[20pt] S(\alpha) = \displaystyle\int_0^\alpha \sin\frac{\pi}{2}\alpha^2 \, d\alpha \end{array}\right\} \ .$$ (9.15)

When the limits of integration are $\pm\infty$, $C = S = \pm\frac{1}{2}$. Tables of the Fresnel integrals are readily available and enable the evaluation of Eq. (9.15) to be carried out for specific values of α, i.e., α_1 and α_2. The integrals in Eq. (9.15) may be graphed in the complex plane with a real coordinate $c(\alpha)$ and an imaginary coordinate $s(\alpha)$. The resultant curve is known as "Cornu's spiral" and is illustrated in Fig. 9.3. Its use in calculating the intensity distribution for Fresnel diffraction patterns is briefly illustrated later in this chapter.

9.4   FRESNEL DIFFRACTION BY A RECTANGULAR APERTURE

For a rectangular aperture of width $2a$ and length $2b$ uniformly illuminated with a coherent collimated beam of quasimonochromatic light of unit amplitude, Eq. (9.8) becomes

$$\psi(x,y) = K \int_{-a}^{a}\int_{-b}^{b} \exp\left[\frac{ik(x-\xi)^2}{2z}\right] \exp\left[\frac{ik(y-\eta)^2}{2z}\right] d\xi \, d\eta \ ,$$ (9.16)

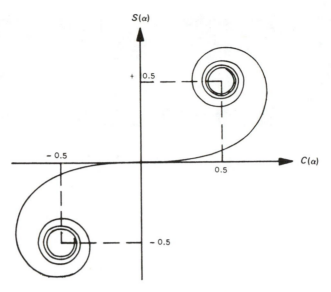

Fig. 9.3 Cornu's spiral.

using K as defined in Eq. (9.9). Note from Eq. (9.11) that in the Fresnel approximation the external spherical phase factor of Fraunhofer diffraction is replaced by a quadratic phase factor. Hence, if

$$\left.\begin{array}{l} \alpha_1 = (x+a)\sqrt{\dfrac{2}{\lambda z}} \;\; ; \;\; \alpha_2 = (x-a)\sqrt{\dfrac{2}{\lambda z}} \\[4mm] \beta_1 = (y+b)\sqrt{\dfrac{2}{\lambda z}} \;\; ; \;\; \beta_2 = (y-b)\sqrt{\dfrac{2}{\lambda z}} \end{array}\right\} , \tag{9.17}$$

then

$$\psi(x,y) = KK'K'\big(\{C(\alpha_2) - C(\alpha_1) + i[S(\alpha_2) - S(\alpha_1)]\}$$
$$\times \{C(\beta_2) - C(\beta_1) + i[S(\beta_2) - S(\beta_1)]\}\big) , \tag{9.18}$$

where K' is a constant. Clearly, Eq. (9.18) may be evaluated easily by use of the tabulated values of the Fresnel integrals. The resultant intensity distribution in the Fresnel diffraction pattern is given by

$$I(x,y) = \psi(x,y)\psi^*(x,y) . \tag{9.19}$$

The use of Cornu's spiral is readily demonstrated for this example by rewriting Eq. (9.18) in the following form:

$$\psi(x,y) = KK'K'Ue^{i\phi} Ve^{i\theta} . \tag{9.20}$$

Then the resultant intensity is simply $K'^2 K'^2 K^2 U^2 V^2$. The values of U and V may be read directly from Cornu's spiral since they are the lengths of the vectors connecting the point α_1 with α_2 and β_1 with β_2. Some typical Fresnel diffraction

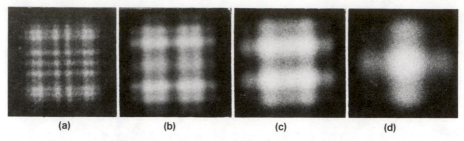

Fig. 9.4 Typical Fresnel diffraction patterns produced by a rectangular aperture: (a) is closest to the aperture and (b), (c), and (d) were taken at increasing distances from the aperture.

patterns associated with a rectangular aperture are shown in Fig. 9.4 where (a) is closest to the aperture and the remaining photographs were taken at increasing distances from the aperture.

9.5   FRESNEL DIFFRACTION BY A STRAIGHT EDGE

The resultant intensity distribution can be obtained readily from the expression obtained in Eq. (9.18) by allowing $\alpha_1 = -\infty$, $\beta_1 = \infty$, and $\beta_2 = -\infty$. Then

$$\psi(x,y) = K'KK' \sqrt{2} \, e^{ix/4} \{C(\alpha_2) - C(-\infty) + i[S(\alpha_2) - S(-\infty)]\} \qquad (9.21)$$

and the intensity is given by

$$I(x,y) = 2K^2K'^2K''^2 \left\{ \left[C(\alpha_2) + \frac{1}{2} \right]^2 + \left[S(\alpha_2) + \frac{1}{2} \right]^2 \right\}. \qquad (9.22)$$

The form of this function is shown in Fig. 9.5. The abscissa at $\alpha_2 = 0$ is the edge of the geometrical shadow. A photographic record of this type of diffraction pattern in shown in Fig. 9.6.

9.6   FRESNEL DIFFRACTION BY A CIRCULAR APERTURE

For a circular aperture of unit radius, it is much easier to express the form of Eq. (9.8) in terms of polar coordinates in the aperture (ρ, θ) and observation plane

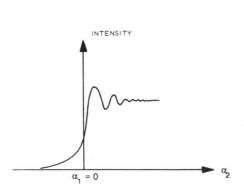

Fig. 9.5 Plot of the intensity distribution in the Fresnel diffraction pattern produced by a straight edge.

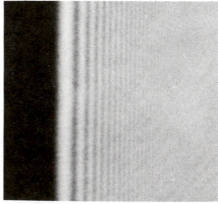

Fig. 9.6 Photograph of the Fresnel diffraction of an edge.

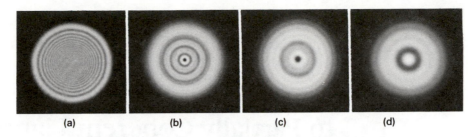

Fig. 9.7 Typical Fresnel diffraction patterns of a circular aperture: (a) is closest to the aperture and (b), (c), and (d) were taken at increasing distances from the aperture.

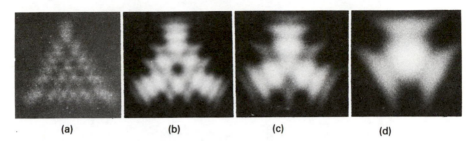

Fig. 9.8 Typical Fresnel diffraction patterns of a triangular aperture: (a) is closest to the aperture and (b), (c), and (d) were taken at increasing distances from the aperture.

(r, ϕ). In this case, we obtain the form

$$\psi(r, \phi) = K'' \int_0^1 \int_0^{2\pi} \exp\left\{ ik\left[\frac{r\rho}{z}\cos(\theta - \phi) - \frac{1}{2}\frac{\rho^2}{z} \right] \right\} \rho d\rho d\theta , \qquad (9.23)$$

where K'' represents the appropriate constants and phase factors arising from the change of variables. It is interesting to note that the integral with respect to θ is precisely the same as that for the corresponding integral for Fraunhofer diffraction by a circular aperture. The resulting ρ integral, however, must now be evaluated using special functions, e.g., Lommel functions. It is also quite apparent that, by invoking the far-field condition on the diffracting aperture, the quadratic factor in ρ may be neglected, and we obtain the complete Fraunhofer diffraction by a circular aperture. Some typical Fresnel diffraction patterns associated with a circular aperture are shown in Fig. 9.7 where (a) is closest to the aperture and the remaining photographs were taken at increasing distances from the aperture. Once again, it is clear from these photographs and those for the corresponding Fraunhofer case how the characteristic Fresnel diffraction pattern makes the transition to characteristic Fraunhofer diffraction as the observation plane in the field is moved from the near field to the far field of the aperture.

Finally, in Fig. 9.8, we see the Fresnel diffraction patterns for a triangular aperture under the same field conditions as the circular aperture example.

10 Heuristic Introduction to Partially Coherent Light

10.1 INTRODUCTION

During the course of this book, we have discussed the image forming process in some detail in Chapters 4 through 8. The discussion was predicated on the fact that the object to be imaged was illuminated incoherently so that the resultant image was formed by an intensity addition. To describe the nature of the image, we needed to know either the intensity impulse response or its Fourier transform, i.e., the transfer function of the particular imaging system. Contrary to this discussion, the first three chapters dealt with situations where quasimonochromatic fields were discussed in terms of the actual optical disturbance or, in particular, the wave disturbance (see Chapter 1). To determine the resultant intensity in a given plane, a summation of the wave disturbance was made and then that resultant squared to give the resultant intensity.

Hence, we have actually discussed two different mechanisms by which the resultant intensity is formed. Basically, we have the *incoherent addition* of Chapters 4 through 8 in which the resultant intensity is formed by taking the sum of the squares of the individual wave disturbance ψ, given by

$$I = \sum_i |\psi_i|^2 \ , \tag{10.1}$$

and the *coherent addition* of Chapters 1, 2, and 3 in which the resultant intensity is formed by taking the sum of the individual wave disturbances ψ and then performing the square, given by

$$I = \left| \sum_i \psi_i \right|^2 . \tag{10.2}$$

Clearly, Eqs. (10.1) and (10.2) represent basically different physical principles, i.e., these equations define in principle what is mathematically and physically considered to be a "linear incoherent superposition" and a "linear coherent superposition," respectively. For many years, image formation was basically considered an incoherent process and the only time that coherent addition was important was in diffraction and interference experiments. However, with the advent of the laser, the necessity for understanding image formation in coherent light arose; and the image is formed by an addition as exemplified in Eq. (10.2), a linear coherent superposition of wave amplitudes.

Unfortunately, that is not the whole story. It is not sufficient to consider only the two situations of strictly coherent and strictly incoherent light. There are at least two good reasons for this:

1. Strictly incoherent and strictly coherent fields are virtually unobtainable and are only mathematical idealizations.

2. We cannot ignore the possibility of intermediate states of coherence. What happens conceptually if we mix some strictly coherent light with some strictly incoherent light?

10.2 PARTIALLY COHERENT LIGHT

Not until the mid-1900s then has the practical necessity of studying these intermediate or partially coherent states been realized. The physical meaning and importance of the results of these studies are of paramount concern in many experimental situations. These remarks are true independent of the development of the laser, since, apart from the laser, there are a wide variety of situations in which it is necessary to consider the coherence existing in the optical field in order to successfully design the optical system, e.g., in microscopy, particularly phase contrast and dark field illumination; in "coherent" optical systems such as Schlieren, spatial filtering, optical processing, and holographic devices; in spectroscopy and interferometry; and, last but not least, in the general imaging process such as photography.

Since a number of these examples involve interference or diffraction effects, it is surprising that apart from Michelson's work these were considered strictly incoherent phenomena; especially so since in 1869 Verdet demonstrated that light from the sun will interfere in a two-beam experiment if the two beams are selected by a pair of pinholes 1/20 mm apart. Sunlight is usually considered to be incoherent yet, in looking at an object illuminated with sunlight, coherent effects will be noted if there is resolvable detail on the order of 20 lines/mm. This would, of course, require auxiliary optics since the unaided eye can't resolve this line frequency. Hence, the unaided eye is always operating as an incoherent imaging system in sunlight.

The subject of partial coherence is best introduced by means of a series of actual and gedanken experiments.

10.2.1 Experiment 1 [a]

We are given an optical field produced by an unknown distant source of finite size. In this field, a screen is erected containing two small circular apertures whose separation can be varied. Behind this screen, a converging lens of known focal length is located and a photographic or photoelectric record of the intensity distribution is obtained in its focal plane (see Fig. 10.1). At first, with one aperture, P_1, alone, the intensity distribution observed is that readily associated

---

[a]This experiment is analyzed in detail in one dimension in Chapter 22.

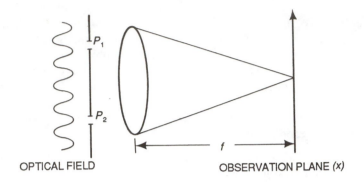

OPTICAL FIELD   OBSERVATION PLANE *(x)*

Fig. 10.1 Schematic diagram of the experimental arrangement.

with the diffraction pattern of a circular aperture. We assume that by the use of a microdensitometer we show that it has the form [see Eq. (2.19)]:

$$\left| \frac{2J_1(x)}{x} \right|^2, \tag{10.3}$$

where x is a normalized radial coordinate. We can, therefore, conclude that:

1. The amplitude distribution across the aperture P_1 is uniform.

2. The radiation across the aperture is essentially coherent.

The second aperture, P_2, alone gives a similar result. Now when the two apertures are opened together at their closest separation, two-beam interference fringes are observed that are formed by the division of the incident wavefront by the two apertures. At this closest separation, the fringes are extremely sharp [see Fig. 10.2(a)]. As the separation of the apertures increases, the photographic

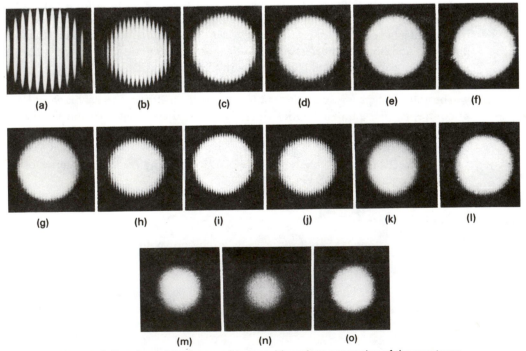

Fig. 10.2 Two-beam interference fringes with various separation of the apertures.

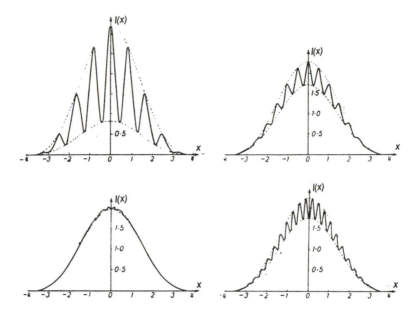

Fig. 10.3 Intensity plots of typical results of Fig. 10.2.

record looks like the results shown in Figs. 10.2(a) through (o). The fringes essentially disappear at (f) only to reappear faintly in (g) through (l), only to fade again at (m), and reappear very faintly at (n) and (o). Intensity plots corresponding to a typical sample of these photographic records are shown in Fig. 10.3. From the results of Fig. 10.2, the following facts are recorded. As the separation of P_1 and P_2 increases,

1. the fringe spacing decreases,

2. the minima are never zero,

3. the relative heights of the maxima above the minima steadily decrease until (f) where they start to increase,

4. the absolute heights of the maxima decrease and the heights of the minima increase until (f),

5. eventually the fringes disappear, at which point the resultant intensity is just twice the intensity observed with one aperture alone, and

6. the fringes reappear with increasing separation but the fringes contain a central minimum not a central maximum.

Items 1 through 5 may be summarized by defining a visibility V [first introduced by Michelson for this very purpose and previously introduced in Chapter 7 as Eq. (7.2)]:

$$V = \frac{I_{max} - I_{min}}{I_{max} + I_{min}} . \tag{10.4}$$

If this visibility function is plotted against the separation of the apertures P_1 and P_2 for the example given in Fig. 10.2, a curve similar to that shown in Fig. 10.4 results. For the closest separation, the addition is approximately coherent and hence the actual wave disturbances have to be added. Let ψ_1 and ψ_2 be the actual

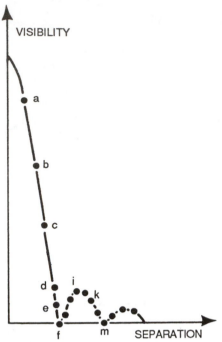

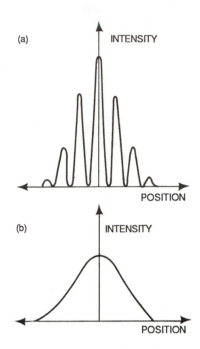

Fig. 10.4 Visibility of the fringes as a function of separation of the two apertures where the letters correspond to those used in Fig. 10.2.

Fig. 10.5 (a) Coherent addition of two beams—the resultant intensity pattern and (b) incoherent addition—resultant intensity pattern.

wave disturbances due to apertures P_1 and P_2 independently. Hence, the resultant wave disturbance is ψ_R given by

$$\psi_R = \psi_1 + \psi_2$$

and the resultant intensity by

$$I_R = \psi_R\psi_R^* = \psi_1\psi_1^* + \psi_2\psi_2^* + \psi_1\psi_2^* + \psi_2\psi_1^* \; , \tag{10.5}$$

where the asterisk denotes a complex conjugate. Noting that $\psi_1\psi_1^*$ is the intensity I_1 due to aperture 1 alone, etc., we may conclude

$$I_R = I_1 + I_2 + 2\sqrt{I_1 I_2}\cos\delta \; , \tag{10.6}$$

where δ is the path difference existing between the two beams. In our example, $I_1 = I_2$ and we may write

$$I_R = 2I(1 + \cos\delta) \tag{10.7}$$

and I has the form $|2J_1(x)/x|^2$ [see Eq. (2.19)]. The resultant intensity distribution given by Eq. (10.7) is illustrated in Fig. 10.5(a). The visibility of these fringes as computed by Eq. (10.4) is unity. Figure 10.5(a) is essentially this type of result.

For the separation resulting in Figs. 10.2(f) and (m), no resultant fringes are seen and the visibility is zero. Here the addition of the two beams of light is essentially incoherent and the resultant intensity is given by an intensity addition of the two independent intensities. Hence,

$$I_R = I_1 + I_2$$

$$= 2I \text{ when } I_1 = I_2 . \tag{10.8}$$

This result is illustrated in Fig. 10.5(b) for comparison purposes. Finally, a measurement of the spectral width of the incident light produces a value of $\Delta\nu$ corresponding to a line width of several angstroms.

10.2.2   Experiment 2

The second experiment is conducted with the same apparatus as that used for experiment 1. Now a different optical field is incident on the two apertures. The resultant fringe patterns now take a slightly different form. A typical result is shown in Fig. 10.6(a); Fig. 10.6(b) shows the resultant intensity for the same separation of apertures but with the optical field of experiment 1. If fringe visibility is computed for Fig. 10.6(a), the value of V varies with position in the field gradually decreasing as the distance from the central fringe increases. The fringe visibility—if computed for the central region of the fringe pattern as a function of the aperture separation—would follow the curve of Fig. 10.4. The final measurement in this experiment is to determine the line width of the incident light. This time it is considerably broader being about 50 Å wide.

It could be concluded correctly that the possible explanation of the different results of these two experiments is connected with the quite different spectral width of the two incident light fields. There is an obvious connection here between the result of experiment 2 and the existence of white light fringes in classical interference experiments. The white light fringes rapidly fade with path difference of the interfering beams. This conclusion is strengthened by the results of experiment 3.

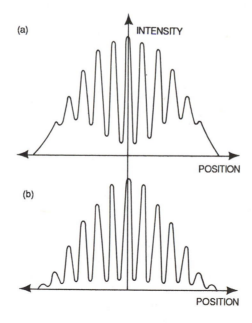

Fig. 10.6 Illustration of two different types of fringes: (a) visibility variable over fringes and (b) visibility constant over fringes.

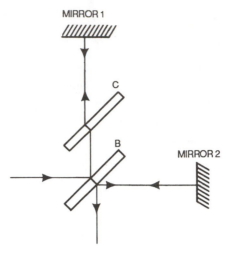

Fig. 10.7 Schematic diagram of Michelson's interferometer.

10.2.3   Experiment 3 [b]

The optical fields of experiments 1 and 2 are allowed to be incident on a Michelson interferometer as depicted diagramatically in Fig. 10.7. The incident light is divided by the beam splitter B, one beam passing through a compensating plate C to mirror 1 where it is reflected back along the same path and then passes through the beam splitter. The second beam initially goes through the beam splitter and to mirror 2 and is reflected back and off the beam splitter, emerging parallel with the first beam. If mirrors 1 and 2 are parallel, circular fringes are seen when looking in toward the beam splitter from the emergent direction of the beams. If mirror 1 is slowly moved backward, the fringe visibility as defined by Eq. (10.4) decreases as a function of the path difference introduced by moving the mirror. For the two optical fields under test, field 2 rapidly loses visibility of the fringe whereas field 1 retains visibility over much larger path differences.

10.3   CONCLUSIONS

It can be readily concluded from these experiments that it is not sufficient to merely be able to describe coherent and incoherent addition. Furthermore, it is evident that these conditions are rather special results of a more generalized phenomena. Equations (10.6) and (10.8) relate only to limiting forms of a more generalized equation. We need them to reformulate the laws of addition of optical fields to account for the phenomena described in the three experiments.

In the next chapter, we develop the basic mathematics necessary for a classical description of partially coherent fields which can then be applied to solve real problems in diffraction and image formation. The importance of the coherent and incoherent limits in image formation must be stressed in that, under appropriate circumstances, the incoherent imaging results of Chapters 4 through 8 are applicable. However, it is important to know and understand how to evaluate the appropriate circumstances. In a similar manner, the coherent limit has its place in image formation. This is particularly true of image formation using laser illumination. We develop the coherent situation in Chapter 13.

---

[b]The Michelson interferometer is analyzed in detail for measuring temporal coherence in Chapter 23.

11 Elementary Theory of Optical Coherence: Part I*

11.1 INTRODUCTION

The time period prior to the experimental discovery of the laser witnessed the theoretical development and experimental verification of classical coherence theory, as well as the application of communication and information theory concepts, in the design of optical systems. Communication techniques applied to optical systems led to the introduction of the optical transfer function as a viable tool in the design and analysis of incoherent optical imaging systems. The work in coherence theory was primarily concerned with the propagation of radiation. However, application of this theory to the imaging problem determined the parameters and conditions of coherence for which the optical system is linear.

The advent of the laser stimulated considerable research using a coherent optical source. In imaging systems equipped with laser sources, researchers observed deleterious effects in the image, such as edge ringing, edge shifting, and the presence of speckle noise. Simultaneously, the long coherence length of the laser led to a rebirth of holography, since improved three-dimensional imaging effects were observed. The subsequent use of holograms as filters also revived interest in optical data processing.

The incoherent and coherent limits of the theory of partial coherence were utilized to describe these various incoherent and coherent imaging phenomena. Techniques for linearizing the coherent imaging system and reducing the effect of speckle noise were also developed.

In addition, interest in high-resolution optical analyzing instruments led to the examination of the partially coherent imaging problem. This resulted in a generalized treatment of the problem in which it was shown that optical imaging systems become nonlinear when the degree of spatial coherence in the object plane becomes comparable in size to the resolved object. Techniques for avoiding these system nonlinearities were subsequently developed.

---

*Reprinted with permission from "Review of optical coherence effects in instrument design: part I," by G. O. Reynolds and J. B. DeVelis, *Opt. Eng.* 20(3), SR-084–SR-094 (May/June 1981). Parts II and III are reprinted, respectively, as Chapters 17 and 18 of this book.

In this chapter, we review the effects of optical coherence in instrument design by first describing the necessary elements of classical coherence theory from an experimental viewpoint, building on the heuristic discussion of Chapter 10. We treat spatial and temporal coherence effects separately, and then review those aspects of the theory of partial coherence which are important in the design of optical instruments.

For spatial coherence, this approach results in a relationship between conditions for coherence of the radiation and optical system parameters which define the operating region of linearity for any optical system. These effects are explored in a variety of instruments and result in general guidelines useful in instrument design.

Here, and throughout the remainder of this book, we treat coherent imaging problems separately and show that the introduction of partial coherence tends to degrade such systems. We also discuss design considerations for optimizing the degree of partial coherence in such systems in a series of examples.

Finally, we discuss the effects of temporal coherence and illustrate how its control optimizes system performance, using some examples. The results outlined should be useful in optimizing the coherence effects in designing optical instruments.

11.2   ELEMENTS OF CLASSICAL COHERENCE THEORY

In this section, we continue a heuristic and experimental approach to the subject of classical coherence theory.[a] The theory of partial coherence is an appropriate starting point from which the formulations of the imaging problem can be developed.[3] However, we have restricted our discussions to those elements of the theory that are essential to the investigation of coherence effects in instrument design. Examples of instruments in which coherence effects influence the instrument performance are high-resolution instruments, such as microscopes, microdensitometers, contact and projection printers, as well as holographic and other instruments that use laser sources.

The meaning of the terms *coherent radiation field* and *incoherent radiation field* is often sought. The definition usually given is that coherent light interferes or diffracts and that incoherent light does not. For this reason, we think of the sun as an incoherent source because we do not normally observe diffraction effects in our everyday observations. However, it is a simple matter to demonstrate that this naive explanation is deficient. Such an observation was first made[4] by Verdet in 1869. This experiment may be illustrated: two small pinholes, spaced a distance d apart. are placed in front of a lens, e.g., the human eye, to create a Young's two-point interferometer. The restrictions in the experiment are that the size of the pinhole must be several times larger than the average wavelength of the radiation and that the distance of the observation plane from the pinholes must be large compared to the pinhole size. When this interferometer is used to observe the sun and the separation of the pinholes is on the order of 100 μm, interference fringes appear. For pinhole separations less than 100 μm, the fringe contrast increases. Figure 11.1 is a schematic of this experimental arrangement.

---

[a] A rigorous description of this theory is available in the published literature.[1,2]

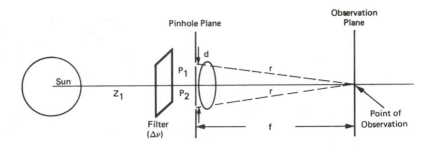

Fig. 11.1 Schematic arrangement of the two-pinhole experiment for observing interference fringes with sunlight.

Figures 11.2, 11.3, and 11.4 show the observed radiation fields and their corresponding microdensitometer traces when the pinhole separations d are 50 and 100 μm. Figure 11.2 shows the intensity distribution in the observation plane and its microdensitometer trace, with a pinhole separation d equal to 100 μm. In this case, we see the superposition of two circular diffraction patterns modified

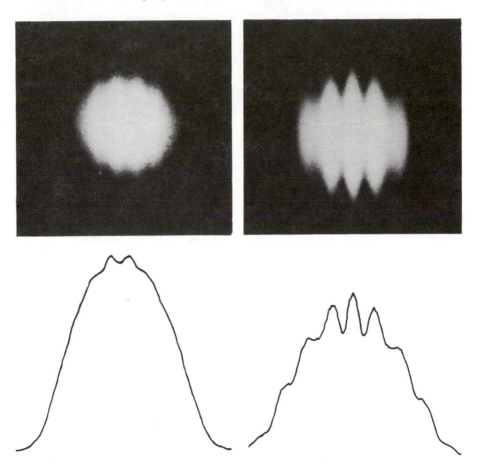

Fig. 11.2 Output of sunlight interferometer experiment and corresponding microdensitometer trace for a pinhole separation of 100 μm. No spectral filter was used in this experiment. (Courtesy of B. Justh.)

Fig. 11.3 Output of sunlight interferometer experiment and corresponding microdensitometer trace for a pinhole separation of 50 μm. No spectral filter was used in the experiment. (Courtesy of B. Justh.)

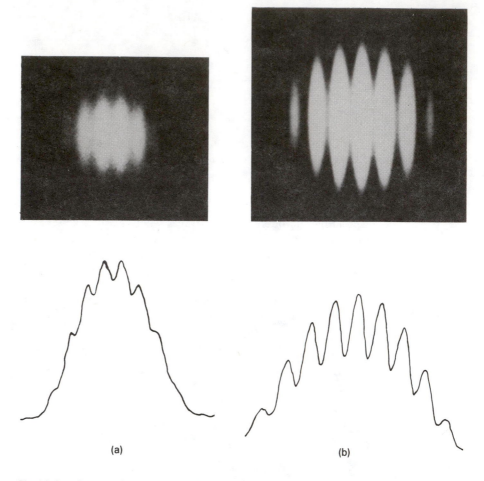

Fig. 11.4   Output of sunlight interferometer experiment and corresponding microdensitometer traces: (a) 100-μm pinhole separation with a No. 58 Wratten filter over the pinholes; the filter has a spectral width of approximately 50 Å; (b) 50-μm pinhole separation with a No. 58 Wratten filter over the pinholes. (Courtesy of B. Justh.)

by fringes occurring as a result of the interference between the radiation fields from the pinholes. Figure 11.3 shows the intensity distribution in the observation plane with its accompanying microdensitometer trace for a pinhole separation of 50 μm. Interference fringes of higher contrast are observed due to the smaller separation between pinholes. The experiment was repeated with a spectral filter placed over the pinholes to reduce the spectral bandwidth of the radiation to approximately 50 Å, compared to the spectral width of the sunlight (3000 Å in the visible). Figures 11.4(a) and (b) show the results. This reduction in bandwidth yielded a further increase in fringe contrast. From this experiment, it is deduced that radiation which emanates from a large thermal source, such as the sun, can give rise to interference fringes. This experiment demonstrated that the contrast of the fringes from the sunlight radiation and the pinhole separation had some intimate relationship with one another.

When Michelson performed similar experiments on the star Betelgeuse (α-Orionis) with his stellar interferometer,[5] the measured region of spatial coherence was on the order of 3 m, suggesting that the angular subtent of the source determines the size of the region over which interference fringes can be observed.

These results demonstrate that primary incoherent sources can give rise to radiation fields that are coherent over small regions of space as determined by their angular subtent. Furthermore reducing the spectral width of the radiation used in the experiment improves the contrast of the fringes in all cases. The increase of fringe contrast with decreasing pinhole separation and/or bandwidth indicates that the coherence properties of the radiation field are variable and cannot be described by a single number. In fact, the theory of partial coherence relates these fringe variations to different degrees of spatial and temporal coherence which leads to the important concept of a *coherence volume*. As we will show, the dimensions of this coherence volume become very important in the design of certain classes of instruments.

The full explanation of these experiments is complicated because of the high frequency of the complex optical field amplitudes (10^{14} to 10^{15} Hz). The measurement of the complex field was and remains beyond the range of existing detectors. Square law detectors, which perform a long-time average, were used to measure the square of the complex field amplitude (intensity). The intensity of the radiation field, which is the measurable quantity, does not satisfy the wave equation and, hence, the existing theories were inadequate for explaining the states of partial coherence observed in the Verdet experiment.

The development of the theory of partial coherence circumvented this problem because it was formulated in terms of an observable quantity (the mutual coherence function) which satisfies a pair of coupled-wave equations. The theory includes the effects of polychromatic sources and enables measurement of boundary conditions necessary for the solution of the wave equations.

11.3   REVIEW OF THE THEORY OF PARTIAL COHERENCE [b]

11.3.1   Introduction

In this section, we expand on the treatment of Chapter 10 by examining the two-pinhole experiment using the theory of partial coherence to relate the varying fringe contrast (observed in the sunlight experiment) to the degree of partial coherence of the radiation field. In our approach, we restricted ourselves to the scalar field and, hence, we ignored polarization effects. These polarization effects can be incorporated readily into the theory by utilization of the full vector nature of the field.[2,6]

An observable quantity that is measurable is the mutual coherence function introduced by Wolf.[1] This function (of a field assumed to be stationary in time) is defined as a mathematical complex cross-correlation of the optical disturbance at two typical field points, $\mathbf{x}_1$ and $\mathbf{x}_2$; that is,

$$\Gamma_{12}(\tau) \equiv \Gamma(\mathbf{x}_1, \mathbf{x}_2, \tau) = \langle V_1(\mathbf{x}_1, t + \tau) V_2^*(\mathbf{x}_2, t) \rangle , \qquad (11.1)$$

where V_1 is the optical disturbance at the point $\mathbf{x}_1$; V_2 is the optical disturbance at the point $\mathbf{x}_2$; $\tau = t_2 - t_1$ and is given by $\Delta \ell / c$ (where $\Delta \ell$ is the path difference between the two beams and c is the speed of light in a vacuum); $\mathbf{x}_n$ denotes a vector quantity; and the angular brackets denote a long time average, i.e.,

---

[b]This section closely follows the development in Chapter 3 of the *Theory and Application of Holography*, by J. B. DeVelis and G. O. Reynolds, Addison-Wesley Publishing Co., Inc., Reading Mass. (1967).

$$< f(t) > \, = \, \lim_{T \to \infty} \frac{1}{2T} \int_{-T}^{T} f(t) \, dt \;. \tag{11.2}$$

When the two field points coincide, the cross-correlation becomes an autocorrelation, reducing the mutual coherence function to the self-coherence function which, in turn, reduces to the intensity at the point $\mathbf{x}_1$ for zero time delay $[I(\mathbf{x}_1) = \Gamma_{11}(0)]$. The complex degree of coherence is defined as the normalized mutual coherence function:

$$\gamma_{12}(\tau) \, = \, \Gamma_{12}(\tau)/[\Gamma_{11}(0)\,\Gamma_{22}(0)]^{1/2} \;. \tag{11.3}$$

It should be emphasized that both the modulus and phase of this function are measurable quantities.

Using the definition of $\Gamma_{12}(\tau)$ and the fact that $V(\mathbf{x}, t)$ satisfies the wave equation, it may be shown[1,2] that the mutual coherence function in a vacuum satisfies a pair of coupled-wave equations for propagation; namely,

$$\nabla_n^2 \Gamma(\mathbf{x}_1, \mathbf{x}_2, \tau) \, = \, (1/c^2) \frac{\partial^2 \Gamma(\mathbf{x}_1, \mathbf{x}_2, \tau)}{\partial \tau^2} \;, \quad n = 1, 2 \;, \tag{11.4}$$

where c is the speed of light and ∇_n^2 is the Laplacian that operates on the coordinates of the point $\mathbf{x}_n$. Thus, since the mutual coherence function, in principle, is an optically observable quantity obeying a pair of wave equations, it can be measured on a boundary and specified as a boundary condition. Equation (11.4) can then be solved subject to this boundary condition.

11.3.2   Quasimonochromatic Approximation

Consider the system shown in Fig. 11.5(a). It will become clear that the intensity in the $\mathbf{x}$ plane depends on the coherence properties of the radiation diffracted at the two pinholes, S_1 and S_2, in the $\boldsymbol{\xi}$ plane. It follows from the linearity of the wave equation that the optical disturbances from S_1 and S_2 yield an intensity distribution in the $\mathbf{x}$ plane at the point P of the form

$$I_P = |A_1 e^{i\phi_1} + A_2 e^{i\phi_2}|^2 = A_1^2 + A_2^2 + 2\,\mathrm{Re}[A_1 A_2^* e^{i(\phi_1 - \phi_2)}] \;, \tag{11.5}$$

where $A_1 e^{i\phi_1}$ is the diffracted optical disturbance in the $\mathbf{x}$ plane due to S_1; $A_2 e^{i\phi_2}$ is the diffracted optical disturbance in the $\mathbf{x}$ plane due to S_2; and Re denotes the real part of a complex quantity. For the case of completely incoherent illumination, the intensity at the point P should, based on our previous observations, reduce to

$$I_P = A_1^2 + A_2^2 = I_1 + I_2 \;, \tag{11.6}$$

where I_1 and I_2 are the intensities at P due to S_1 and S_2, respectively. The cross term in Eq. (11.5) is the distinguishing and important term to be considered. Although Eqs. (11.5) and (11.6) describe the principle of the interference phenomena, we must return to the coherence theory formulation for a rigorous mathematical description of this experiment.

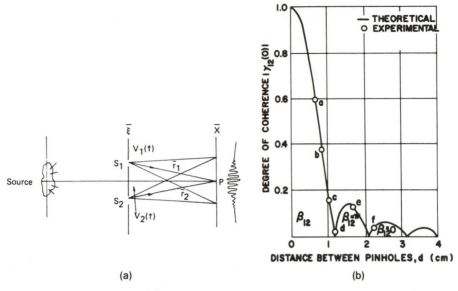

Fig. 11.5 Measurement of modulus of complex degree of coherence: (a) schematic for two-pinhole experiment where P denotes a general point of observation in the x plane (from Ref. 2); (b) modulus of complex degree of coherence as a function of pinhole separation for the experimental arrangement shown in Fig. 11.5(a) where phase change $\beta_{12} = 0$ and π (from Ref. 1).

The quasimonochromatic form of the coherence theory is characterized by the condition that the spectral width of the radiation ($\Delta\nu$) is taken to be very small compared to the mean frequency ($\bar{\nu}$):

$$\Delta\nu \ll \bar{\nu} \; . \tag{11.7}$$

If, in addition, we assume that all path differences in the experiment of interest satisfy the condition

$$\Delta\ell \ll c/\Delta\nu \; , \tag{11.8}$$

then the radiation behaves much as if it were monochromatic of frequency $\bar{\nu}$. Thus, the second condition [Eq. (11.8)] puts a strong restriction on all path differences to which this quasimonochromatic theory may be applied. Since the approximations given by Eqs. (11.7) and (11.8) are assumed here, the details of the spectral composition of the light may be ignored, and all frequency-dependent parameters may be evaluated at the mean frequency $\bar{\nu}$.

In Fig. 11.5(a), S_1 and S_2 are two sources, and $V_1(t)$ and $V_2(t)$ are stationary random functions which describe the light oscillations at S_1 and S_2. The conditions under which the results given above are valid can now be considered. Since the light disturbances are propagated by linear differential wave equations (see Chapters 1, 2, and 3), the disturbance at P may be represented as a superposition of the contributions from S_1 and S_2 in the form

$$V_P(t) = K_1 V_1(t - r_1/c) + K_2 V_2(t - r_2/c) \; , \tag{11.9}$$

where K_1 and K_2 are propagators and r_1 and r_2 are the distances from P to S_1 and S_2, respectively.

The intensity at P is then given by

$$I(P) = <V_P(t) V_P^*(t)> ,$$ (11.10)

which may be rewritten as

$$I(P) = <| K_1 V_1(t - r_1/c) + K_2 V_2(t - r_2/c)|^2>$$

$$= I_1 + I_2 + 2\,\mathrm{Re}[\,K_1 K_2^* <V_1(t - r_1/c) V_2^*(t - r_2/c)>] ,$$ (11.11)

where I_1 and I_2 are the intensities at P due to S_1 and S_2. Note the similarity between this result and that stated in Eq. (11.5). Because of the time invariance, the term inside the angular brackets depends only on $r_1/c - r_2/c = \tau$ and not on r_1/c and r_2/c explicitly. Thus, the origin of time can be changed and Eq. (11.11) can be rewritten as

$$I(P) = I_1 + I_2 + 2\,\mathrm{Re}[\,K_1 K_2^* \Gamma_{12}(\tau)] ,$$ (11.12)

where Γ_{12} is described by Eq. (11.1). Equation (11.12) could be termed the *generalized interference law* for partially coherent light. Since only quasimono-chromatic light is being considered, the experiment is restricted to small path differences defined by Eq. (11.8). For this case, we form the function $\Gamma_{12}(0)$, which is termed the *mutual intensity function*. Furthermore, because K_1 and K_2 are purely imaginary, the notation

$$\Gamma_{12}(\tau) = |\Gamma_{12}(0)|\, e^{i\Phi_{12}} = |\Gamma_{12}|\, e^{i\Phi_{12}}$$ (11.13)

is introduced, where $\Phi_{12} = 2\pi\bar{\nu}\tau + \beta_{12}$; $2\pi\bar{\nu}\tau = $ phase difference $= (2\pi/\bar{\lambda})$ $(|r_1 - r_2|)$; and $\beta_{12} = \arg[\Gamma_{12}(\tau)]$. (For a further discussion of this point, see Refs. 1 and 2.) Equation (11.12) can be rewritten as

$$I(P) = I_1 + I_2 + 2|\,K_1 K_2^*|\,|\Gamma_{12}|\cos\Phi_{12} .$$ (11.14)

Using Eq. (11.3), we can rewrite Eq. (11.14) as

$$I(P) = I_1 + I_2 + 2\sqrt{I_1 I_2}\,|\gamma_{12}|\cos\Phi_{12} .$$ (11.15)

Equation (11.15) is the generalized interference law for partially coherent quasi-monochromatic light. Setting $|\gamma_{12}| = 0$ reduces Eq. (11.15) to:

$$I(P) = I_1 + I_2 ,$$ (11.16)

which is the incoherent limit. Equation (11.6) states that there is no interference, as expected, since incoherent light was assumed to superimpose in intensity. Setting $|\gamma_{12}| = 1$, Eq. (11.15) becomes

$$I(P) = I_1 + I_2 + 2\sqrt{I_1 I_2}\,\cos\Phi_{12} ,$$ (11.17)

which is the well-known interference law for completely coherent radiation. For intermediate values of $|\gamma_{12}|$, the radiation is termed partially coherent and the resulting interference is governed by Eq. (11.15).

A qualitative description of some basic experiments that illustrate coherence effects can now be given. It should be emphasized again that the effects of coherence in a radiation field are all contained in the complex function $\gamma_{12}(\tau)$. However, as an aid in understanding coherence principles, it is useful to separate coherence effects into two categories. The first, termed *spatial coherence,* arises primarily from angular source size considerations and the second, termed *temporal coherence* or *coherence length,* arises from considerations of the finite spectral width of the radiation. Thus, $\gamma_{12}(0)$ measures spatial coherence effects and $\gamma_{11}(\tau)$ is a measure of temporal coherence. In general, the two-pinhole experiment discussed in deriving Eq. (11.15) is used for measuring spatial coherence effects. To isolate this effect, we demand $|r_1 - r_2| \ll \Delta\ell$ where $\Delta\ell$ was defined in Eq. (11.8).

To illustrate an important technique for measuring the spatial coherence between two disturbances, we consider Eq. (11.15) where $I_1 = I_2 = I$ and the equation then reduces to

$$I(P) = 2I(1 + |\gamma_{12}|\cos\Phi_{12}) ; \tag{11.18}$$

that is, an interference pattern with reduced visibility is observed.

The visibility of the fringes V defined by Eq. (7.2) is

$$V = \frac{I_{max} - I_{min}}{I_{max} + I_{min}} , \tag{11.19}$$

and if we then use this equation to compute the visibility of the fringes described by Eq. (11.18), the resulting visibility becomes

$$V = |\gamma_{12}| . \tag{11.20}$$

Equation (11.20) demonstrates the relationship between the modulus of the complex degree of coherence and the fringe visibility for this experiment. The phase of the complex degree of coherence is related to the relative position of the central fringe to the origin when the pinholes are symmetrically displaced with respect to the optical axis. To measure the degree of spatial coherence between two disturbances, their intensities are adjusted to be equal, and the visibility of the resulting fringes is measured by allowing them to interfere. A typical curve demonstrating the variation of the modulus of the complex degree of coherence as a function of pinhole separation is shown in Fig. 11.5(b). In this figure, the solid line represents the theoretical value and the designated points are the experimental values quantitatively described in Fig. 11.6. Figures 11.6 and 11.7 show a comparison of the experimental and theoretical results, demonstrating the effects of fringe visibility and the corresponding phase shifts in Φ_{12}, given by Eq. (11.18), which arise from a distant incoherent source.[7,8] The theoretical coherence function in this case is calculated by utilizing the Van Cittert-Zernike theorem,[1,2] which is derived later. In Fig. 11.6, the fringe visibility was varied by changing the separation distance between the pinholes in a two-pinhole interference experiment; in Fig. 11.7, the degree of spatial coherence was changed by varying the size of the pinhole in a collimator.

The qualitative effect of temporal coherence (coherence length) and an example demonstrating the validity of the quasimonochromatic approximations [Eqs. (11.7) and (11.8)] are shown in Fig. 11.8. In one of these experiments, Figs.

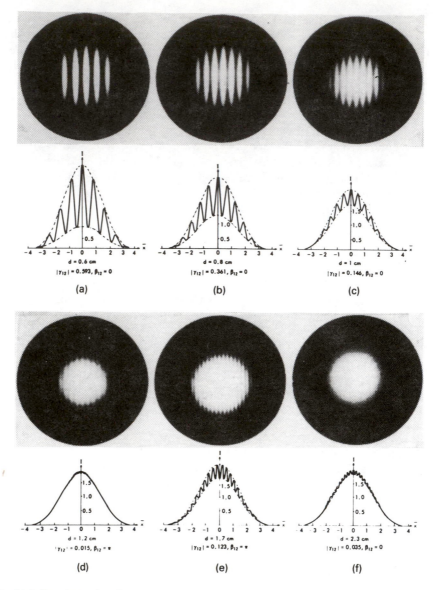

Fig. 11.6 Two-beam interference experiment utilizing partially coherent light and exhibiting spatial coherence effects. The upper portions of (a) through (f) show fringe visibility patterns; the lower portions, theoretical intensity curves where the dashed lines represent the values of maximum and minimum intensity. In these experiments, the fringe visibility and frequency were varied by changing the separation distance d between the pinholes. Figures (a) through (c) show the decrease in fringe visibility and corresponding increase in frequency as the separation distance between the two pinholes increases; (d) and (e) show a phase shift $\beta = \pi$ in the fringe pattern with respect to the origin and the resulting fringe visibility and increased frequencies for two additional separation distances. Figure (f) shows another π phase shift resulting in a maximum intensity at the origin and corresponding fringe visibility and increased frequency due to further separation of the pinholes. Figures (a) through (c) correspond to points on the first lobe of the $|\gamma_{12}|$ curve [Fig. 11.5(b)]; (d) and (e), to points on the second lobe; and (f), to a point on the third lobe in Fig. 11.5(b) (from Ref. 8).

11.8(a) through (d), path lengths were varied by placing different thicknesses of optically flat glass in front of one of the pinholes and illuminating the interferometer with both laser light and filtered collimated thermal light. The thickness of the glass was chosen to exceed the coherence length of the thermal light,

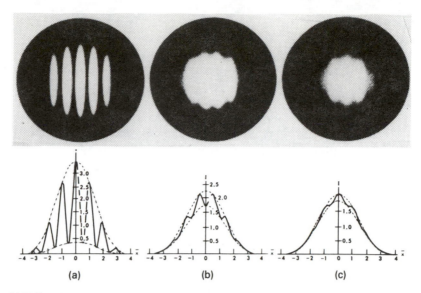

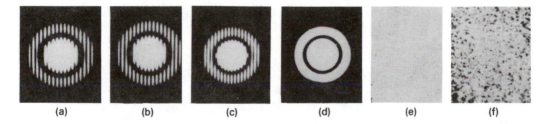

Fig. 11.7 Two-beam interference experiment utilizing partially coherent light, showing change of phase and exhibiting spatial coherence effects. The upper portions of (a) through (c) show the observed patterns; the lower portions, the theoretical intensity curves where the dashed lines represent the maximum and minimum intensities. In these experiments, the pinhole separation was fixed so that the resulting fringe frequency was constant. The degree of coherence at the two pinholes was varied by changing the size of the primary incoherent source used in the experiment and is illustrated by the different fringe visibilities in (a) through (c). The π phase shift apparent from examining the central fringe of these figures results from sampling different lobes of the coherence function at the pinhole plane (from Ref. 7).

Fig. 11.8   Two-beam interference experiment utilizing partially coherent light and exhibiting temporal coherence effects: (a) shows high-contrast fringes formed with a He-Ne gas laser illuminating a pair of small circular apertures. In (b) a piece of plain optical-quality glass, 0.5 mm thick, was introduced in front of one of the apertures to add an extra optical path. Again illuminating with light from a He-Ne gas laser, no difference in the fringe contrast is observed. Figure (c) shows high-contrast fringes when the experiment is repeated with a coherent field produced by a collimated beam from a mercury arc lamp without the glass plate; however, with the glass in place, the fringes disappear as shown in (d). (The scale change between the two pairs of illustrations is due to different wavelengths: 6328 Å for He-Ne and 5461 Å for Hg.) Figures (e) and (f) are photographs of a cement block illuminated with a collimated beam from a mercury arc lamp and a He-Ne gas laser, respectively. Both illuminating fields have approximately the same degree of spatial coherence, but the laser has a coherence length greater than the depth of roughness on the surface of the block. The absence of speckling in (e) allows one to see the surface structure (from Ref. 9).

but not that of the laser light. In the other experiment, Figs. 11.8(e) and (f), the radiations were chosen so that the surface irregularities were greater in depth than the coherence length of one source but not the other, the result being a loss of interference in the former case. The experiments presented in Fig. 11.8 demonstrate the importance of the quasimonochromatic approximation and the necessity of an experiment to measure temporal coherence effects [that is, $|\gamma_{11}(\tau)|$]. One experiment for measuring $\gamma_{11}(\tau)$ uses a Michelson interferometer with a path length compensator in one arm.[1,2] In such an experiment, a beam splitter is used to divide the quasimonochromatic radiation into two separate beams, which are recombined after one of them is passed through a compensator to adjust the optical path length. The fringe visibility, given by Eq. (11.19), at the observation point is varied by moving one of the mirrors until the fringes disappear.

11.3.3   Coherence Volume

One interpretation of the results of the two-pinhole experiment gives rise to the concept of the coherence volume. The coherence volume defines a region of space, in any given experiment, for which radiation from any pair of points within the volume will interfere. Similarly, radiation emanating from points separated by distances larger than those defined by the dimensions of the coherence volume will not interfere. The two-pinhole experiment described a lateral dimension of the incoming wave field for which interference fringes were observed. This dimension is called the spatial *coherence interval*. The spectral bandwidth experiments indicated a maximum longitudinal dimension of the incoming wave field over which interference effects could be observed. We call this maximum length $\Delta\ell$, as defined by the equality in Eq. (11.8), the *coherence length* of the radiation field.

Both spatial and temporal coherence dimensions are characterized by gradual rather than sharp changes in fringe contrast, as seen in Fig. 11.5(b) for the case of spatial coherence. These two characteristic dimensions of "spatial coherence interval" and "temporal coherence" length define a "coherence volume" for the radiation field, as shown schematically in Fig. 11.9. In Fig. 11.9, we see a

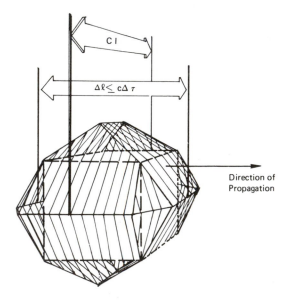

Fig. 11.9 Schematic of coherence volume whose length $\Delta\ell$ is determined by the temporal coherence and whose cross-sectional area is determined by the spatial coherence interval, CI. The transitional region (cross-hatched areas outside the cube) is not sharp because the fringe visibility gradually goes to zero as indicated in Fig. 11.5(b) for the case of spatial coherence.

TABLE 11.I
Temporal Coherence Length for Various Types of Sources

Source Type	Mean Wavelength (Å)	Bandwidth (Å)	Coherence Length
Hg arc (high pressure)	5461	100	<0.03 mm
Hg arc (low pressure)	5461	10	<0.3 mm
Ag laser	5445	$\approx 9 \times 10^{-2}$	<3 cm
He-Cd laser	4416	$\approx 7 \times 10^{-3}$	<27.5 cm
He-Ne laser	6328	$10^{-4} - 10^{-5}$	$\approx 10^4$ cm

representation of the coherence volume whose characteristic area normal to the direction of propagation is determined by the maximum value of the spatial coherence interval (CI), and whose third dimension along the direction of propagation of the wave is determined by $\Delta\ell$, the coherence length of the radiation.

Typical values of the coherence lengths for various sources are given in Table 11.I. Values for the spatial coherence intervals are dependent on the angular subtent of an incoherent source, or on the collimator diameter in the case of laser sources, as shown in Table 11.II.

Although we have treated the coherence interval and the coherence length as separate entities, the reader should not be misled into thinking of them as characteristics that can be analyzed and evaluated as independent properties. We consider them separately only because it is easier for us to do so. To exceed either the coherence length of the radiation or the spatial coherence of the radiation is sufficient to remove the interference effect as seen in Fig. 11.8 for the case of temporal coherence.

Also, we have seen from the experiments in Figs. 11.4 and 11.8 that a quasimonochromatic field can arise from a polychromatic incoherent source. The mercury arc and the sun are both polychromatic incoherent sources and both can create quasimonochromatic partially coherent wave fields as discussed earlier. The description of this process in terms of coherence theory gives rise to one of the most useful theorems of the theory, the *Van Cittert-Zernike theorem,* derived in Sec. 11.3.4.3.

TABLE 11.II
Spatial Coherence Intervals for Various Sources

Source	Source Size (μm)	Distance (mm)	λ (μm)	Angular Subtent (rad)	fℓ (cm)	Diameter (cm)	f/#	Spatial Coherence Interval (to first zero)
Hg arc	10	500	1/2	2×10^{-5}	—	—	—	≈30 mm
Hg arc	1000	500	1/2	2×10^{-3}	—	—	—	≈0.3 mm
Laser	——	——	1/2	——	24	12	f/2	12 cm
Laser	——	——	1/2	——	210	70	f/3	12 cm

11.3.4   Solution of the Coupled-Wave Equations

The second-order mutual coherence function for stationary scalar optical fields obeys a pair of coupled-wave equations [Eq. (11.4)] whose boundary conditions are measurable. A large class of interesting physical situations is described using this approach.

The second-order mutual coherence function for stationary scalar optical fields can be written in terms of its Fourier transform with respect to frequency, i.e.,

$$\Gamma(\mathbf{x}_1,\mathbf{x}_2,\tau) = \int_0^\infty \hat{\Gamma}(\mathbf{x}_1,\mathbf{x}_2,\nu)\exp(-2\pi i\nu\tau)\,d\nu \; , \tag{11.21}$$

where $\hat{\Gamma}(\mathbf{x}_1,\mathbf{x}_2,\nu)$ is the mutual power spectrum.

Equations (11.7) and (11.8) define two conditions that make up the quasi-monochromatic approximation. By dividing Eq. (11.8) by c we may write

$$\tau \ll \frac{1}{\Delta\nu} \; , \tag{11.22}$$

showing that the time delay τ is much less than the time associated with the reciprocal of the bandwidth of the radiation. Figure 11.10 illustrates the mutual power spectrum of the quasimonochromatic approximation.

From Eq. (11.7) we know that

$$\hat{\Gamma}(\mathbf{x}_1,\mathbf{x}_2,\bar{\nu}) \simeq 0 \; , \tag{11.23}$$

for $|\nu - \bar{\nu}| \gg \Delta\nu$; i.e., the function $\hat{\Gamma}(\mathbf{x}_1,\mathbf{x}_2,\bar{\nu})$ is approximately zero outside of the shaded region in Fig. 11.10. In the quasimonochromatic approximation, Eq. (11.21) can be written as

$$\Gamma(\mathbf{x}_1,\mathbf{x}_2,\tau) = \int_0^\infty \hat{\Gamma}(\mathbf{x}_1,\mathbf{x}_2,\nu)\exp(-2\pi i\nu\tau)\,d\nu$$

$$= \exp(-2\pi i\bar{\nu}\tau) \int_0^\infty \hat{\Gamma}(\mathbf{x}_1,\mathbf{x}_2,\nu)$$

$$\times \exp[-2\pi i(\nu - \bar{\nu})\tau]\,d\nu \; . \tag{11.24}$$

Consider the complex exponential function under the integral in Eq. (11.24). From the assumption given by Eq. (11.22), this exponential is approximately equal to unity, i.e.,

$$\exp(-2\pi i\Delta\nu\tau) \simeq 1 \; . \tag{11.25}$$

Thus, from our two assumptions [Eqs. (11.7) and (11.8)], the form for the second-order mutual coherence function for stationary quasimonochromatic (QM) scalar optical fields is

$$\Gamma_{QM}(\mathbf{x}_1,\mathbf{x}_2,\tau) = 0 \; , \quad \text{for } |\nu - \bar{\nu}| \gg \Delta\nu \; , \tag{11.26}$$

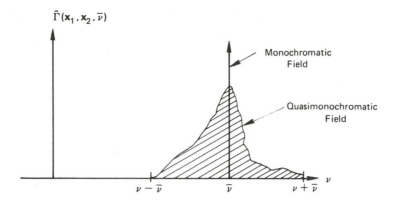

$\hat{\Gamma}(\mathbf{x}_1, \mathbf{x}_2, \overline{\nu})$

Monochromatic
Field

Quasimonochromatic
Field

$\nu - \overline{\nu}$ $\overline{\nu}$ $\nu + \overline{\nu}$ ν

Fig. 11.10 Schematic plot of the mutual power spectrum for the quasimonochromatic approximation showing the finite spectral width of the radiation.

and from Eq. (11.24), we have

$$\Gamma_{QM}(\mathbf{x}_1, \mathbf{x}_2, \tau) = \exp(-2\pi i \overline{\nu} \tau) \Gamma(\mathbf{x}_1, \mathbf{x}_2, 0) \ ,$$

for $|\nu - \overline{\nu}| \ll \Delta \nu$ and $\Delta \nu \tau \ll 1$.   (11.27)

Thus, the second-order mutual coherence function for quasimonochromatic fields reduces to a product of a spatial function and a simple oscillatory time function. The spatial function $\Gamma(\mathbf{x}_1, \mathbf{x}_2, 0)$ is called the *mutual intensity function*. Equation (11.27) describes a working range within the polychromatic field problem which retains some of the spectral properties of such fields. The coupled-wave equations [Eq. (11.4)] can be solved for quasimonochromatic partially coherent fields by using Eq. (11.27). Finally, it should be emphasized that the quasimonochromatic field can be coherent, incoherent, or partially coherent, the actual case of interest being determined by the functional form of $\Gamma(\mathbf{x}_1, \mathbf{x}_2, 0)$ in Eq. (11.27).

11.3.4.1   Coherent Limit

For completely coherent fields, $\Gamma(\mathbf{x}_1, \mathbf{x}_2, 0)$ becomes separable into a product of two functions,[2] i.e.,

$$\Gamma(\mathbf{x}_1, \mathbf{x}_2, 0) = U(\mathbf{x}_1) U^*(\mathbf{x}_2) \ .$$ (11.28)

Consistent with Eq. (11.3), the magnitude of the complex degree of coherence for this limit is $|\gamma_{12}(0)| = 1$. Substitutions of Eqs. (11.27) and (11.28) into Eq. (11.4) shows that each of the functions $U(\mathbf{x}_1)$ and $U^*(\mathbf{x}_2)$ satisfies the Helmholtz equation:

$$\nabla^2 U_s(\mathbf{x}) + k^2 U_s(\mathbf{x}) = 0 \ , \quad s = 1, 2 \ .$$ (11.29)

Thus, the quasimonochromatic coherent limit reduces to the solution of the monochromatic Helmholtz equation.

11.3.4.2   Incoherent Limit

For incoherent fields, the magnitude of the complex degree of coherence is zero, i.e., $|\gamma_{12}(0)| = 0$. This means that for all pinhole separations the fringe visibility is zero. For this case, $\Gamma(\mathbf{x}_1, \mathbf{x}_2, 0)$ is approximately described by[2,6,10]

$$\Gamma(\mathbf{x}_1, \mathbf{x}_2, 0) \cong I(\mathbf{x}_1)\delta(\mathbf{x}_1 - \mathbf{x}_2) \ . \tag{11.30}$$

In essence, this equation states that each point in the field is independent of every other point, implying a lack of correlation between all pairs of points.

11.3.4.3 Van Cittert-Zernike Theorem

The Van Cittert-Zernike theorem logically explains how the degree of spatial coherence of a radiation field from a primary incoherent source is related to the geometry of the source. The general Green's function solution of the coupled-wave equations with the quasimonochromatic approximation is

$$\Gamma(\mathbf{x}_1, \mathbf{x}_2, 0) = \iint\limits_{-\infty}^{\infty} \Gamma(\mathbf{S}_1, \mathbf{S}_2, 0)$$

$$\times \frac{\partial G_1(\mathbf{S}_1, \mathbf{x}_1)}{\partial n} \frac{\partial G_2^*(\mathbf{S}_2, \mathbf{x}_2)}{\partial n} d\mathbf{S}_1 \, d\mathbf{S}_2 \ , \tag{11.31}$$

where G is the Rayleigh-Sommerfeld Green's function[2,6]; $\partial G/\partial n$ is the normal derivative of the Green's function [see Eq. (11.33)]; $\mathbf{S}_1$ and $\mathbf{S}_2$ are the variables on the source surface; and $\mathbf{x}_1$ and $\mathbf{x}_2$ are the variables in the field.

Substituting Eq. (11.30) for an incoherent source as a boundary condition in Eq. (11.31), and assuming that the distance between the source plane and the observation plane z_1 is much greater than the characteristic diameter of the source, one can show that[1,2]

$$\Gamma(\mathbf{x}_1, \mathbf{x}_2, 0) \propto \int\limits_{-\infty}^{\infty} I(\mathbf{S}_1)\exp[ik\mathbf{S}_1(\mathbf{x}_1 - \mathbf{x}_2)/z_1] d\mathbf{S}_1 \ . \tag{11.32}$$

Equation (11.32) states that the mutual intensity function arising from a primary incoherent source is proportional to the Fourier transform of the intensity distribution of the source. The experimental results shown in Fig. 11.5(b) illustrate the close comparison between the theoretical predictions of Eq. (11.32) and two-pinhole measurements for the case of a circular incoherent source. In this case, the normalized mutual intensity function is a Besinc function arising from the Fourier transform of the uniform, circularly symmetric source intensity distribution.

11.3.5 The Imaging Problem: Coherent and Incoherent Limits

We now discuss the conditions of the scalar optical radiation field under which an optical imaging system may be treated as a linear system and indicate the parameter in terms of which the linearity of the optical imaging system exists. The relationship between the mutual intensity function of the object and image is obtained by solving the pair of coupled-wave equations given by Eq. (11.4). The mutual intensity function of the object, $\Gamma_{ob}(\boldsymbol{\xi}_1, \boldsymbol{\xi}_2)$, and the mutual intensity function of the image, $\Gamma_{im}(\mathbf{x}_1, \mathbf{x}_2)$, are related by the geometry shown in Fig. 11.11 with the lens plane denoted by $\mathbf{x}'$.

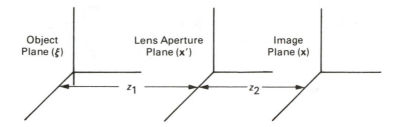

Fig. 11.11 Schematic diagram for the imaging problem.

The general solution to the pair of coupled-wave equations in the quasi-monochromatic limit is given by Eq. (11.31).

For the imaging system shown in Fig. 11.11, Eq. (11.31) gives the solution of the coupled-wave equations between the ξ and the x' planes. To avoid confusion with prime variables, we use z_1 and z_2 in place of z and z' as in previous chapters. This mutual intensity function, multiplied by the lens function, is used as a boundary condition for propagation between the x' and x planes. Use of Eq. (11.31) yields the mutual intensity function in the image plane x. In performing this calculation, the paraxial approximation on the Green's function, i.e.,

$$\frac{\partial G(\xi, x')}{\partial n} = -\frac{ik}{2\pi z_1} e^{ikz_1} \exp\left[ik\frac{(\xi - x')^2}{2z_1}\right],$$
(11.33)

and the Gaussian imaging condition [see Eq. (4.23)],

$$\frac{1}{z_1} + \frac{1}{z_2} = \frac{1}{f},$$
(11.34)

are applied. The resulting mutual intensity function in the image plane is

$$\Gamma_{im}(\mathbf{x}_1, \mathbf{x}_2, 0) = D \iint\limits_{-\infty}^{\infty} \Gamma_{ob}(\boldsymbol{\xi}_1, \boldsymbol{\xi}_2, 0) \exp\left[ik\frac{(|\boldsymbol{\xi}_1|^2 - |\boldsymbol{\xi}_2|^2)}{2z_1}\right]$$

$$\times K\left(\frac{\mathbf{x}_1}{z_2} + \frac{\boldsymbol{\xi}_1}{z_1}\right) K^*\left(\frac{\mathbf{x}_2}{z_2} + \frac{\boldsymbol{\xi}_2}{z_1}\right) d\boldsymbol{\xi}_1\, d\boldsymbol{\xi}_2\ ,$$
(11.35)

where D is a complex constant,

$$K\left(\frac{\boldsymbol{\xi}_n}{z_1} + \frac{\mathbf{x}_n}{z_2}\right) = \int \text{rect}(\mathbf{x}'_n | a)$$

$$\times \exp\left[-ik\mathbf{x}'_n \cdot \left(\frac{\boldsymbol{\xi}_n}{z_1} + \frac{\mathbf{x}_n}{z_2}\right)\right] d\mathbf{x}'_n \quad \text{for } R = 1, 2$$
(11.36)

is the amplitude impulse response of the lens, and

$$\text{rect}(\mathbf{x}'_n | a) = \begin{cases} 1 \text{ for } |\mathbf{x}'_n| < a \\ 0 \text{ for } |\mathbf{x}'_n| > a \end{cases}$$
(11.37)

is the lens transmission function.

In Eqs. (11.34) through (11.37), f is the focal length of the imaging lens and D is a complex constant containing quadratic phase factors in $\mathbf{x}_1$ and $\mathbf{x}_2$ as well as the obliquity factors. Equation (11.35) relates the mutual intensity in the object to the mutual intensity in the image, utilizing the amplitude impulse response of the imaging lens, which is spatially stationary. The system magnification is given by $m = z_2/z_1$. The image intensity is obtained from Eq. (11.35) by letting $\mathbf{x}_1 = \mathbf{x}_2$.

Quasimonochromatic coherent images are obtained from Eq. (11.35) by using Eq. (11.28) to express the mutual intensity in the object plane. This yields

$$I_{im}(\mathbf{x}_1) = \left| D \int U(\boldsymbol{\xi}) \exp \left(\frac{ik|\boldsymbol{\xi}_1|^2}{2z_1} \right) K \left(\frac{\mathbf{x}_1}{z_2} + \frac{\boldsymbol{\xi}_1}{z_1} \right) d\boldsymbol{\xi}_1 \right|^2 , \qquad (11.38)$$

which indicates that coherent imaging systems are linear in amplitude. Furthermore, for those situations where the quadratic phase factor is stationary (i.e., the object is in the near field of the lens),[11,,12] the amplitude transfer function is directly related to the lens aperture function. (This subject is covered in Chapters 12, 13, and 14.)

Quasimonochromatic incoherent images are realized from Eq. (11.35) by using the incoherent form of the mutual intensity function, Eq. (11.30), which yields

$$\Gamma_{im}(\mathbf{x}_1, \mathbf{x}_2, 0) = D \int I_{ob}(\boldsymbol{\xi}_1) K \left(\frac{\mathbf{x}_1}{z_2} + \frac{\boldsymbol{\xi}_1}{z_1} \right)$$

$$\times K^* \left(\frac{\mathbf{x}_2}{z_2} + \frac{\boldsymbol{\xi}_1}{z_1} \right) d\boldsymbol{\xi}_1 . \qquad (11.39)$$

To recover the image intensity from Eq. (11.39), we let $\mathbf{x}_1 = \mathbf{x}_2$ to obtain the following result apart from a constant factor:

$$I_{im} \left(\frac{\mathbf{x}_1}{m} \right) = \int I_{ob}(\boldsymbol{\xi}_1) \left| K \left(\frac{\mathbf{x}_1}{m} + \boldsymbol{\xi}_1 \right) \right|^2 d\boldsymbol{\xi}_1 . \qquad (11.40)$$

Equation (11.40) shows that incoherent imaging is linear in intensity and $|K|^2$ is the intensity impulse response (point spread function) of the imaging lens. The transfer function for this case is the autoconvolution of the lens aperture function, i.e., the Fourier transform of the intensity impulse response. (This subject was covered in more detail in Chapters 4 through 8.)

11.3.5.1   *Imaging with Partially Coherent Light*

In discussing the Van Cittert-Zernike theorem, it was shown that propagation of radiation in free space increased the spatial coherence of the radiation. To illustrate the presence of partially coherent radiation in an imaging system, we observe [from Eq. (11.39)] that, even though we assumed incoherent illumination in the object plane, the image is partially coherent. This again indicates that the spatial coherence of the radiation increases with propagation through the imaging system.

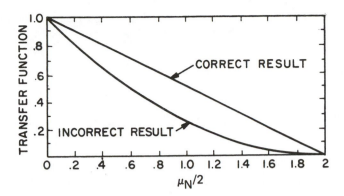

Fig. 11.12 Comparison of correct and cascaded functions for measurement of the lens system transfer function. The correct result predicted by partially coherent analyses is obtained by autoconvolving the lens system pupil function, whereas the incorrect result is obtained by multiplying the transfer functions of the individual lenses in the system. The abscissa μ_N represents a normalized spatial frequency (from Ref. 13).

An example where spatial coherence in the image plane arising from an incoherent object is important is the case of cascaded optical systems. In this situation, the coherence formulation showed that the lens aperture functions must be multiplied and then autoconvolved to yield the system transfer function as opposed to multiplying the transfer functions of the individual lens components.[13] This is illustrated in Fig. 11.12.

In general, partially coherent imaging effects for transparent objects are obtained by assuming a spatially invariant coherence function in the object plane (consistent with the Van Cittert-Zernike theorem), i.e.,

$$\Gamma_{ob}(\boldsymbol{\xi}_1, \boldsymbol{\xi}_2, 0) = \Gamma_{ob}(\boldsymbol{\xi}_1 - \boldsymbol{\xi}_2, 0) , \tag{11.41}$$

and introducing complex functions $t(\boldsymbol{\xi}_1)$ and $t(\boldsymbol{\xi}_2)$ to describe the object transmittance. The general imaging equation, Eq. (11.35), for a quasimonochromatic, spatially invariant input field becomes

$$\Gamma_{im}(\mathbf{x}_1, \mathbf{x}_2, 0) = D \int \Gamma_{ob}(\boldsymbol{\xi}_1 - \boldsymbol{\xi}_2, 0)$$

$$\times t(\boldsymbol{\xi}_1) t^*(\boldsymbol{\xi}_2) K \left(\frac{\mathbf{x}_1}{z_2} + \frac{\boldsymbol{\xi}_1}{z_1} \right)$$

$$\times K^* \left(\frac{\mathbf{x}_2}{z_2} + \frac{\boldsymbol{\xi}_2}{z_1} \right) d\boldsymbol{\xi}_1 \, d\boldsymbol{\xi}_2 . \tag{11.42}$$

The quadratic phase factors in $\boldsymbol{\xi}_1$ and $\boldsymbol{\xi}_2$ in Eq. (11.35) are negligible when the object is in the near field of the lens,[12] i.e.,

$$|\boldsymbol{\xi}|_{max} \ll \left| \frac{ma}{4\pi} \right| , \tag{11.43}$$

where m is the system magnification and a is the radius of the imaging lens. In the use of microscopes, numerical apertures of 0.9 and less have been shown to be consistent with this assumption.[14]

The image spectrum can be obtained from Eq. (11.42) by setting $\mathbf{x}_1 = \mathbf{x}_2$ and taking its spatial Fourier transform to obtain[3,11]

$$\tilde{I}_{im}(\mu) = D \int \tilde{\tau}(\rho)\tilde{\tau}^*(\mu - \rho)$$

$$\times \left[\int \tilde{\Gamma}_{ob}(\mu - \sigma - \rho)\tilde{K}(\mu - \sigma)\tilde{K}^*(\sigma)\, d\sigma \right] d\rho \, , \qquad (11.44)$$

where $\sim$ denotes the spatial Fourier transform and σ and ρ are dummy variables conjugate to ξ and μ, respectively. The integral in the square brackets is the *transmission cross coefficient,* or generalized transfer function, for an imaging system which is linear in the mutual intensity function. However, this function is nonlinear in intensity. Specific examples of instruments exhibiting these non-linearities due to coherence effects and design considerations aimed at reducing or eliminating them are discussed in the next section and in Chapters 17 and 18.

11.3.5.2   *Effects of Spatial Coherence on the Two-Point Resolution Limit*

An example that illustrates the effects of spatial coherence in an imaging system is illustrated with the determination of the two-point resolution limit.[2,3,15–17] Two-point resolution was described in Chapter 6 and these coherence effects are discussed in detail in Chapter 13. These results were obtained by using a two-point object illuminated with partially coherent light as the boundary condition in Eq. (11.42). Application of the Sparrow criterion, i.e.,

$$\left. \frac{d^2 I(x)}{dx^2} \right|_{x=0} = 0 \, , \qquad (11.45)$$

shows that the just resolvable two-point resolution limit varies as a function of the degree of spatial coherence. This variation, which is shown in Fig. 11.13, illustrates[2,17] a factor of 1.56 difference in the two-point resolution values for the coherent and incoherent limits.

These results have also been presented as image intensity distributions[16] and one typical result is shown in Fig. 11.14 where the separation of the two points just exceeds the incoherent Rayleigh limit. In the incoherent limit, the two points are just resolved (i.e., two peaks), whereas in the coherent limit the two points are not resolved. As the radiation becomes more spatially coherent, the apparent separation between the two points decreases.

The resolvability of a two-point object in the case of completely coherent illumination can also be affected by the relative phases of the object, as seen[18] in Fig. 11.15. In this example, we note that zero phase difference between the two objects renders the points unresolved, just as in the coherent limit of Fig. 11.14. A phase difference of 90 deg between the two points gives a result similar to the incoherent limit of Fig. 11.14. A phase difference of 180 deg between the two points appears to improve the resolution capability of the system while shifting the locations of the two-point sources. This phase variation illustrates the non-linear behavior of the coherently illuminated system.

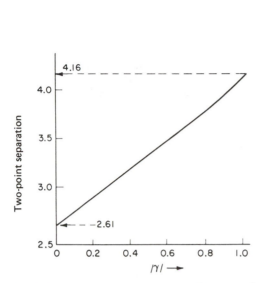

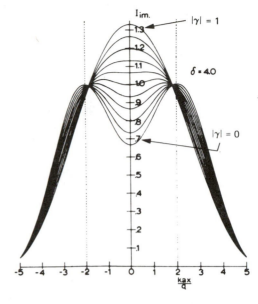

Fig. 11.13 Two-point resolution as a function of the degree of coherence. The two-point separation is normalized for a unit magnification system (from Ref. 2).

Fig. 11.14 Image intensity distribution for various values of $|\gamma|$ from $|\gamma| = 0$ to $|\gamma| = 1.0$ in steps of 0.1 for $\delta = 4.0$, where δ represents the pinhole separation normalized with the imaging parameters of the system and the abscissa is normalized by the wave number k, the lens radius a, and the image distance q (from Ref. 16).

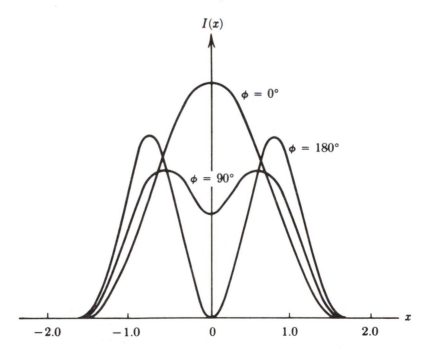

Fig. 11.15 Image intensity distribution for two mutually coherent point sources having various phase relationships and separated by the Rayleigh distance (from Ref. 18).

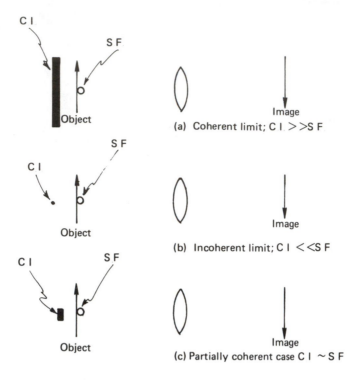

Fig. 11.16 Schematic diagram illustrating the dimension of the spatial coherence interval, CI, to the diameter of the point spread function, SF, of the lens in the object plane: (a) coherent imaging; spatial coherence interval much greater than spread function diameter; (b) incoherent imaging; spatial coherence interval much less than spread function diameter; (c) partially coherent imaging; spatial coherence interval approximately equal to spread function diameter.

11.3.5.3   Classification of Linear and Nonlinear Regions of Operation

The two-point resolution example suggests a relationship between the coherence interval of the radiation in the object plane and the resolution capability (impulse response diameter) of the imaging system as measured in the object plane. These parameters are further illustrated in Fig. 11.16, which demonstrates that the system is nonlinear in intensity, unless the coherence interval is much less than the spread function diameter of the imaging system [Fig. 11.16(b)] projected to the object plane. Since the spread function diameter is directly related to the resolution capability of the imaging system, nonlinearities in the imaging system will occur in those cases where the partial coherence of the illumination of high-resolution (small) objects causes radiation to be diffracted into angles exceeding the collection capability of the imaging lens. These nonlinear errors manifest themselves as mensurational or radiometric changes in the image intensity distribution and are different for different objects, i.e., the system is object-dependent and cannot be completely characterized by a system transfer function as in the linear case. This classification concept will prove to be extremely useful later, when different instruments are examined with respect to coherence effects in Chapters 17 and 18.

REFERENCES

1. M. Born and E. Wolf, *Principles of Optics,* Pergamon Press, New York (1964).
2. M. J. Beran and G. B. Parrent, Jr., *Theory of Partial Coherence,* Prentice Hall, Englewood Cliffs, N.J. (1964).
3. B. J. Thompson, *Image Formation with Partially Coherent Light, Progress in Optics,* Vol. 7, E. Wolf, ed., pp. 169–230, North Holland Publishing Co., Amsterdam (1969).
4. E. Verdet, *Ann. Scientif. l'Ecole Normale Superieure,* Vol. 2, p. 291 (1865); *Leçons d'Optique Physique,* Vol. 1, p. 106 (1869).
5. A. A. Michelson, "On the application of interference methods to astronomical measurements," *Astrophysics J.* 51(5), 257-262 (1920); A. A. Michelson and F. G. Pease, "Measurement of the diameter of (alpha) Orionis with the interferometer," *Astrophysics J.* 53(4), 249-259 (1921).
6. E. L. O'Neil, *Introduction to Statistical Optics,* Addison-Wesley Publishing Co., Reading, Mass. (1963).
7. B. J. Thompson, "Illustration of the phase change in two-beam interference with partially coherent light," *J. Opt. Soc. Am.* 48, 95–97 (1958).
8. B. J. Thompson and E. Wolf, "Two-beam interference with partially coherent light," *J. Opt. Soc. Am* 47, 895–902 (1957).
9. B. J. Thompson, "Advantages and problems of coherence as applied to photographic situations," *SPIE J.* 4(1), 7-11 (1965).
10. G. B. Parrent, Jr., and T. J. Skinner, *Optica Acta* 8(93), 1193 (1961).
11. G. B. Parrent, Jr., "Basic theory of partial coherence," *American Federation of Information Processing Societies Conf. Proc.* 28, 17 (1966).
12. T. J. Skinner, "Energy considerations propagation in random medium and imaging in scalar coherence theory," PhD Thesis, Boston Univ. (1965).
13. J. B. DeVelis and G. B. Parrent, Jr., "Transfer function for cascaded optical systems," *J. Opt. Soc. Am.* 57(12), 1486-1490 (1967).
14. D. Nyyssonen, "Linewidth measurement with an optical microscope: the effect of operating conditions on the image profile," *Appl. Opt.* 16(8), 2223-2230 (1977).
15. G. B. Parrent, Jr., and B. J. Thompson, *Physical Optics Notebook,* 1st ed., pp. 48-51, SPIE, Bellingham, Wash. (1969); reprinted here as Chapter 13.
16. D. N. Grimes and B. J. Thompson, "Two-point resolution with partially coherent light," *J. Opt. Soc. Am.* 57(11), 1330-1334 (1967).
17. F. Rojak, "Two-point resolution with partially coherent light," MS Thesis, Lowell Technological Institute, Lowell, Mass. (1961).
18. J. W. Goodman, *Introduction to Fourier Optics,* p. 129, McGraw-Hill Book Co., New York (1968).

12 Image Formation with Coherent Light

12.1 INTRODUCTION

In this chapter we discuss a result stated earlier, Eq. (11.28). This result was obtained from coherence theory and justifies solving coherent imaging problems using a single complex wave equation even though all optical detectors are square law devices. We first review the square law property of detectors which results in losing the optical phase. We obtain several of the same results described in Chapter 11 using this simplified approach. In the usual circumstances of image formation, the object is illuminated incoherently and the image is formed by an addition of intensity of the light reaching the image plane. The characteristics of the image can be described if the intensity impulse response of the imaging system is known (see Chapter 4). Recall that the intensity impulse response is the distribution of intensity in the image plane produced by a point object. Hence, the image of an extended object is formed by a summation of the individual impulse responses suitably weighted by the intensity at the various points. In Chapter 7, an intensity transfer function was defined as the Fourier transform of the intensity impulse response, and it was seen that image formation could be discussed equally well in terms of this intensity transfer function.

Note that above we have used the terminology *intensity* impulse response and *intensity* transfer function. This is to distinguish them from the impulse response and transfer functions required for describing image formation in coherent light, i.e., the amplitude impulse response and the amplitude transfer function. The words *incoherent* and *coherent* are sometimes used to replace intensity and amplitude in the above terms.

12.2 THE MEASUREMENT OF INTENSITY

All detectors of optical radiation such as the eye, photocells, and film are responsive to the intensity of the incident radiation. The quantity that we normally call *intensity* is the long time average of the square of the actual optical disturbance. Let the complex amplitude of an optical field be $V(\mathbf{x}, t)$. The instantaneous intensity $I(\mathbf{x}, t)$ is then given by

$$I(\mathbf{x},t) = V(\mathbf{x},t)V^*(\mathbf{x},t) \ . \tag{12.1}$$

The observable here is, of course, the quantity defined by

$$I(\mathbf{x}) = \ <I(\mathbf{x},t)> \ , \tag{12.2}$$

where the angle brackets denote a time average, i.e.,

$$<f(t)> \ = \ \lim_{T \to \infty} \ \frac{1}{2T} \int_{-T}^{T} f(t)\,dt \ . \tag{12.3}$$

Hence, the intensity $I(\mathbf{x})$ is given by

$$I(\mathbf{x}) = \ <V(\mathbf{x},t)V^*(\mathbf{x},t)> \ . \tag{12.4}$$

12.3   ADDITION OF OPTICAL FIELDS

Consider two optical fields given simultaneously by $V_1(\mathbf{x},t)$ and $V_2(\mathbf{x},t)$. If these two fields are added, the resultant field is

$$V(\mathbf{x},t) = \ V_1(\mathbf{x},t) + V_2(\mathbf{x},t) \tag{12.5}$$

and

$$\begin{aligned}
I(\mathbf{x},t) = \ &V_1(\mathbf{x},t)V_1^*(\mathbf{x},t) + V_2(\mathbf{x},t)V_2^*(\mathbf{x},t) \\
&+ V_1(\mathbf{x},t)V_2^*(\mathbf{x},t) + V_1^*(\mathbf{x},t)V_2(\mathbf{x},t) \ .
\end{aligned} \tag{12.6}$$

The time-averaged observable $I(\mathbf{x})$ is

$$\begin{aligned}
I(\mathbf{x}) = \ &<V_1(\mathbf{x},t)V_1^*(\mathbf{x},t)> + <V_2(\mathbf{x},t)V_2^*(\mathbf{x},t)> \\
&+ <V_1(\mathbf{x},t)V_2^*(\mathbf{x},t)> + <V_1^*(\mathbf{x},t)V_2(\mathbf{x},t)> \ .
\end{aligned} \tag{12.7}$$

The first two terms are the intensities associated with the two individual optical fields as defined by Eq. (11.10), i.e.,

$$\left.\begin{aligned}
<V_1(\mathbf{x},t)V_1^*(\mathbf{x},t)> \ &= \ I_1(\mathbf{x}) \ , \\
<V_2(\mathbf{x},t)V_2^*(\mathbf{x},t)> \ &= \ I_2(\mathbf{x}) \ .
\end{aligned}\right\} \tag{12.8}$$

The remaining two terms are the cross-correlation functions of the two fields at two points in space but at the same time and are illustrated in the generalized interference law of Eq. (11.15). If $V_1(\mathbf{x},t)$ and $V_2(\mathbf{x},t)$ both vary randomly with time and also randomly with respect to each other, the cross-correlation terms are zero and Eq. (12.7) becomes

$$I(\mathbf{x}) = \ I_1(\mathbf{x}) + I_2(\mathbf{x}) \ , \tag{12.9}$$

which clearly represents a linear incoherent superposition of intensities [see Eq. (11.16)].

Quasimonochromatic Light: Under the conditions of the quasimonochromatic approximation, a coherent field may be fully described by an optical disturbance $U(\mathbf{x}, \tau)$, which satisfies the scalar wave equation. The optical disturbance $U(\mathbf{x}, \tau)$ satisfies the scalar wave equation (see Chapter 1)

$$\nabla^2 U(\mathbf{x}, \tau) = \frac{1}{c^2} \frac{\partial^2 U(\mathbf{x}, \tau)}{\partial t^2} . \tag{12.10}$$

Assuming a time harmonic field

$$U(\mathbf{x}, \tau) = \psi(\mathbf{x}) e^{-i2\pi\bar{\nu}\tau} , \tag{12.11}$$

where $\bar{\nu}$ is the mean frequency of the wave and $\psi(\mathbf{x})$ describes the spatial variation of the amplitude $A(\mathbf{x})$ and phase $\phi(\mathbf{x})$ of the disturbance, i.e.,

$$\psi(\mathbf{x}) = A(\mathbf{x}) e^{i\phi(\mathbf{x})} . \tag{12.12}$$

Thus, the observable measure, the intensity, is now

$$I(\mathbf{x}) = \psi(\mathbf{x}) \psi^*(\mathbf{x}) = |A(\mathbf{x})|^2 . \tag{12.13}$$

Under these conditions, Eq. (12.7) becomes:

$$I(\mathbf{x}) = |\psi_1(\mathbf{x}) + \psi_2(\mathbf{x})|^2 . \tag{12.14}$$

Equations (12.12), (12.13), and (12.14) point out very strongly the problem associated with recording the intensity—all information about $\phi(\mathbf{x})$ is completely lost. The measurement of intensity does not then uniquely determine the optical field. This same problem exists in other areas of the electromagnetic spectrum. For example, at x-ray wavelengths, the loss of the phase information makes the determination of crystal structures from x-ray diffraction measurements a research undertaking rather than a routine calculation. If the intensity and the phase could be measured at each point in the diffraction pattern, then the diffracting structure could be calculated by direct Fourier transformation. However, when only intensity is measured, an infinite number of possible structures may exist.

The difficulties are easily seen by recourse to a very simple example. Let the complex amplitude $\psi(x) = \cos(x)$ as shown in Fig. 12.1(a). The resultant intensity is then shown in Fig. 12.1(b) as $\cos^2(x)$. The complex amplitude described by

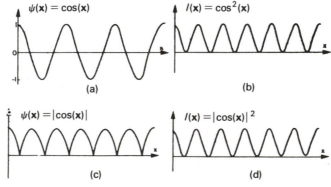

Fig. 12.1 Illustration that two distinct optical fields (a) and (c) can produce identical intensity distributions (b) and (d).

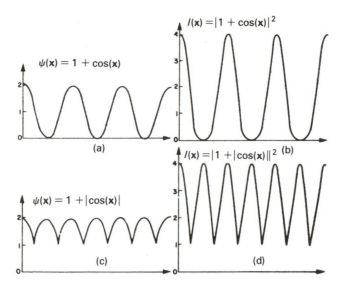

Fig. 12.2 Addition of a coherent background term to the optical fields of Figs. 12.1(a) and (c) results in distinct intensity distributions.

$\psi(x) = |\cos(x)|$ [Fig. 12.1(c)] also yields an intensity distribution given by $\cos^2(x)$. Hence, by measuring the intensity alone, the two complex amplitudes of Figs. 12.1(a) and (c) cannot be distinguished.

It is of interest to note that a technique does exist for distinguishing between these two fields. This method consists of adding a coherent background, often in the form of a constant. This idea was first introduced by Zernike[1] for the display of weak interference fringes (see Chapter 35) and subsequently used by Pinnock and Taylor[2] for determining the approximate relative phases of various portions of optical diffraction patterns. More important still is the use of this same idea as one of the primary steps in holography (see Chapters 25, 26, and 27).

Returning to the two optical fields discussed above, the addition of a coherent background yields resultant complex amplitudes $1 + \cos(x)$ and $1 + |\cos(x)|$, respectively. The intensity in each case is now $|1 + \cos(x)|^2$ and $|1 + |\cos(x)||^2$. These functions are plotted in Figs. 12.2(a) through (d). Clearly, now the two fields are distinguishable and furthermore the value of $\psi(x)$ can be recovered by taking the positive square root of the intensity.

12.4   THE IMAGING PROBLEM

If an optical field $U_{obj}(\mathbf{x}', \tau)$ is imaged to $U_{im}(\mathbf{x}, \tau)$ by an imaging system, then apart from a constant factor,

$$U_{im}(\mathbf{x}, \tau) = \int_{-\infty}^{\infty} U_{obj}(\mathbf{x}', \tau) K(\mathbf{x} - \mathbf{x}') \, d\mathbf{x}' \, , \qquad (12.15)$$

where $K(\mathbf{x} - \mathbf{x}')$ is the resulting complex amplitude in the image plane due to a point object and is called the *amplitude impulse response* of the system. Here we have assumed that $\Delta\nu \ll \bar{\nu}$. The resultant intensity in the image plane is given by

$$I_{im}(\mathbf{x}) = <U_{im}(\mathbf{x}, \tau) U_{im}^*(\mathbf{x}, \tau)> \, . \qquad (12.16)$$

Substituting from Eq. (12.15) into (12.16),

$$\therefore I_{im}(\mathbf{x}) = \left\langle \int_{-\infty}^{\infty} U_{obj}(\mathbf{x}', \tau) K(\mathbf{x} - \mathbf{x}') \, d\mathbf{x}' \int_{-\infty}^{\infty} U_{obj}^*(\mathbf{x}'', \tau) K^*(\mathbf{x} - \mathbf{x}'') \, d\mathbf{x}'' \right\rangle$$

(12.17)

and, since the impulse responses are not time dependent,

$$I_{im}(\mathbf{x}) = \int_{-\infty}^{\infty}\int_{-\infty}^{\infty} <U_{obj}(\mathbf{x}', \tau) \, U_{obj}^*(\mathbf{x}'', \tau)> K(\mathbf{x} - \mathbf{x}') K^*(\mathbf{x} - \mathbf{x}'') \, d\mathbf{x}' \, d\mathbf{x}''.$$ (12.18)

This equation is the reduced form of Eq. (11.35) where the quadratic phase factor is stationary, i.e, the object is in the near field of the lens.

12.4.1   Incoherent Fields

If the object distribution is incoherent, each point in the object is assumed to be statistically independent (as far as time variations are concerned) of every other point, as defined by Eq. (11.30). Hence, the time-averaged quantity is then written as[3]

$$<U_{obj}(\mathbf{x}', \tau) \, U_{obj}^*(\mathbf{x}'', \tau)> = I_{obj}(\mathbf{x}')\delta(\mathbf{x}' - \mathbf{x}'') \, ,$$ (12.19)

where $I_{obj}(\mathbf{x}')$ is the object intensity distribution and $\delta(\mathbf{x}' - \mathbf{x}'')$ is the Dirac delta function. Hence, Eq. (12.18) becomes

$$I_{im}(\mathbf{x}) = \int_{-\infty}^{\infty}\int_{-\infty}^{\infty} I_{obj}(\mathbf{x}')\delta(\mathbf{x}' - \mathbf{x}'') K(\mathbf{x} - \mathbf{x}') K^*(\mathbf{x} - \mathbf{x}'') \, d\mathbf{x}' \, d\mathbf{x}'' \, .$$ (12.20)

Performing the z integration yields

$$I_{im}(\mathbf{x}) = \int_{-\infty}^{\infty} I_{obj}(\mathbf{x}') K(\mathbf{x} - \mathbf{x}') K^*(\mathbf{x} - \mathbf{x}') \, d\mathbf{x}' \, .$$ (12.21)

The quantity $K(\mathbf{x} - \mathbf{x}') K^*(\mathbf{x} - \mathbf{x}')$ is the intensity impulse response.

12.4.2   Coherent Fields

When the object distribution is coherent, as in Eq. (11.28), Eq. (12.18) becomes

$$I_{im}(\mathbf{x}) = \int_{-\infty}^{\infty}\int_{-\infty}^{\infty} \psi_{obj}(\mathbf{x}') \psi_{obj}^*(\mathbf{x}'') K(\mathbf{x} - \mathbf{x}') K^*(\mathbf{x} - \mathbf{x}'') \, d\mathbf{x}' \, d\mathbf{x}'' \, .$$ (12.22)

Hence,

$$I_{im}(\mathbf{x}) = \left| \int_{-\infty}^{\infty} \psi_{obj}(\mathbf{x}') K(\mathbf{x} - \mathbf{x}') \, d\mathbf{x}' \right|^2 \, .$$ (12.23)

The complex amplitude associated with the object is convolved with the amplitude impulse response and the result is squared to determine the intensity. Quite obviously, Eqs. (12.21) and (12.23) yield different results for the image intensity distribution [see Eq. (11.38)]. In the coherent situation, the system is linear in intensity and, hence, the whole system including the detector constitutes a linear system. For coherent fields, the optical system (excluding the detection step) is linear in the complex amplitude and the entire system (including the detection step) constitutes a nonlinear system.

12.5   THE AMPLITUDE IMPULSE RESPONSE

The particular image-forming process we discuss here and in the chapters immediately following is that described as *coherent imaging* by Eq. (12.23). The quality of the image formed will be determined by the amplitude impulse response. To determine this function, we return to the analysis developed in Chapter 4. For completeness, we summarize the development here but extend it to two dimensions.

Let us consider an optical disturbance $\psi(\xi, \eta)$ existing in a plane at a distance z_1 from a lens of aperture $A(x', y')$ (see Fig. 12.3). We wish to determine the distribution $\psi(x, y)$ existing in the plane a distance z_2 on the other side of the lens. Let r be the distance from a point in the (ξ, η) plane to a point in the (x', y') plane and s the distance from the (x', y') plane to the (x, y) plane. Hence, following Eq. (4.15),

$$\psi(x,y) = K_1 \iiiint \psi(\xi,\eta) \frac{e^{ikr}}{r}$$

$$\times A(x',y') \exp\left[\frac{-ik(x'^2 + y'^2)}{2f}\right] \frac{e^{iks}}{s} dx' dy' d\xi d\eta , \quad (12.24)$$

where K_1 is a constant. Writing r and s in terms of the square root of the sum of the squares of the differences of the coordinates, expanding binomially with paraxial approximations, and adding the imaging condition $1/z_1 + 1/z_2 = 1/f$ gives

$$\psi(x,y) = K \iiiint \psi(\xi,\eta) A(x',y') \exp\left[-ikx'\left(\frac{\xi}{z_1} + \frac{x}{z_2}\right)\right]$$

$$\times \exp\left[-iky\left(\frac{\eta}{z_1} + \frac{y}{z_2}\right)\right] dx' dy' d\xi d\eta , \quad (12.25)$$

where K is a constant. Since we are interested in the amplitude impulse response,

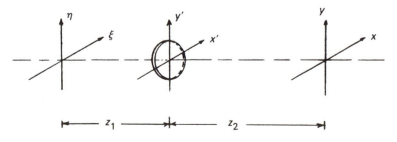

Fig. 12.3 Coordinate system for image formation.

we may consider the object distribution to be a delta function, i.e.,

$$\psi(\xi, \eta) = \delta(\xi, \eta) \ . \tag{12.26}$$

Substituting Eq. (12.26) into (12.25) and integrating, we obtain

$$\psi(x, y) = K \iint A(x', y')$$

$$\times \exp\left(\frac{-ikxx'}{z_2}\right) \exp\left(\frac{-ikyy'}{z_2}\right) dx' \, dy' \ , \tag{12.27}$$

which is the two-dimensional equivalent of Eq. (4.29) and states that the amplitude distribution in the image of a point is given by the Fourier transform of the aperture distribution function, i.e, the function describing the amplitude and phase variation introduced by passage through the lens.

For a perfect lens (diffraction limited) with a circular aperture, Eq. (12.27) can be solved by rewriting in terms of polar coordinates. A general point in the aperture of radius a will have polar coordinates (ρ, θ) associated with the rectangular coordinates (x', y'). Similarly, a general point in the image plane has polar coordinates (r', ϕ) related to the rectangular coordinates (x, y). Hence, following Chapter 2 (diffraction by a circular aperture),

$$\psi(r', \phi) = K \int_0^a \int_0^{2\pi} \exp\left[\frac{-ik}{z_2} r\rho\cos(\theta - \phi)\right] \rho d\rho d\theta \ . \tag{12.28}$$

Performing the θ integration yields

$$\psi(r', \phi) = 2\pi K \int_0^a J_0\left(\frac{kr'\rho}{z_2}\right) \rho d\rho \ , \tag{12.29}$$

and finally

$$\psi(r', \phi) = K\pi a^2 \frac{2J_1\left(\dfrac{kr'a}{z_2}\right)}{\left(\dfrac{kr'a}{z_2}\right)} \ . \tag{12.30}$$

This function is displayed in Fig. 12.4(b) and compared to the intensity impulse response of the same optical system of Fig. 12.4(a).

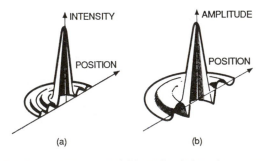

Fig. 12.4 The (a) intensity and (b) amplitude impulse responses.

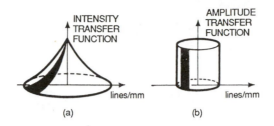

Fig. 12.5 The (a) intensity and (b) amplitude transfer functions.

12.6   THE AMPLITUDE TRANSFER FUNCTION

The transfer function is only really defined for the case of incoherent object illumination when it is given by the Fourier transform of the intensity impulse response (see Chapter 7). For an ideal spherical lens, this intensity transfer function is the Fourier transform of $|\psi(r',\phi)|^2$ squared. This is shown in Fig. 12.5(a). By comparison, we can define an amplitude transfer function as the Fourier transform of the amplitude impulse response. Clearly, this is the aperture function $A(\xi,\eta)$ of Eq. (12.27) [see Fig. 12.5(b)].

12.7   CONCLUSIONS

In this chapter, we have determined the basic building block of coherent imaging—the amplitude impulse response. We are now in a position to discuss some particular examples of coherent image formation to determine the similarities and differences with incoherent image formation.

REFERENCES

1. F. Zernike, "Diffraction and optical image formation," *Proc. Phys. Soc. London* 61, 158–164 (1948).
2. P. R. Pinnock and C. A. Taylor, "The determination of the signs of structure factors by optical methods," *Acta Cryst.* 8(11), 687–691 (1955).
3. M. J. Beran and G. B. Parrent, Jr., *Theory of Partial Coherence,* Prentice Hall, Englewood Cliffs, N.J. (1964).

13   Coherent Imaging: Resolution

13.1   INTRODUCTION

In the introduction to Chapter 6, we stated that the determination of an image quality criterion has long been and still is a major problem in the field of image evaluation and assessment. If this statement was true when only incoherent objects had to be considered, it is doubly true now that we need to allow for coherent and partially coherent objects. To attempt to obtain some insight into the problems encountered with partially coherent light, we will discuss a number of simple objects that are often used in image evaluation. Clearly, the most simple object is the one that consists of just two points which are themselves not resolved, i.e., the amplitude impulse response of the imaging system in object space is much greater than the physical dimensions of the "point." Our discussions here relate directly to Chapter 6, which discussed the two-point imaging problem for incoherent light. The analysis was applicable to the problem of two stars, which can be considered incoherent with respect to each other. However, it is unlikely to apply to the viewing of two small adjacent cells in a microscope.

13.2   IMAGE OF A TWO-POINT OBJECT

The Rayleigh criterion used in Chapter 6 and first introduced by Lord Rayleigh in 1902 states that the two-point objects are resolvable as two objects when the maximum of illuminance produced by one point coincides with the first minimum of the illuminance of the other. This criterion is based on an assumption of incoherently illuminated objects. The second criterion used in Chapter 6, the Sparrow criterion, states that the two points are just resolved if the second derivative of the resultant image illuminance vanishes at the point midway between the respective Gaussian image points. This latter criterion may be used independently of the coherence of the illumination of the two points since it is concerned with the resultant intensity only. Hence, the Sparrow criterion has been used several times for the evaluation of partially coherent systems.

The object under consideration consists of two equally bright points separated by a distance 2**b** (see Fig. 13.1). The mutual intensity in the object $\Gamma_{ob}(\xi_1, \xi_2)$ is given by Eq. (11.35), evaluated under the condition of stationary phase, which is

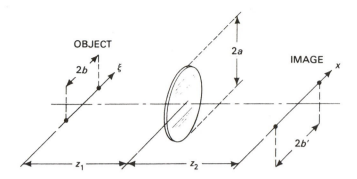

Fig. 13.1 Schematic of basic imaging system.

applicable when the object is in the near field of the lens. Setting the constant D equal to unity then yields

$$\Gamma_{ob}(\xi_1,\xi_2) = I_o\,\gamma(\xi_1,\xi_2)\,\{[\delta(\xi_1 - \mathbf{b}) + \delta(\xi_1 + \mathbf{b})]$$

$$\times [\delta(\xi_2 - \mathbf{b}) + \delta(\xi_2 + \mathbf{b})]\}\,, \tag{13.1}$$

where ξ is the coordinate in object space, $\gamma(\xi_1,\xi_2)$ is the complex degree of coherence, and I_o is a normalized intensity. The mutual intensity in the image $\Gamma_{im}(\mathbf{x}_1,\mathbf{x}_2)$ is then given by

$$\Gamma_{im}(\mathbf{x}_1,\mathbf{x}_2) = \int\!\!\int_{ob} \Gamma_{ob}(\xi_1,\xi_2)\,K\!\left(\frac{\mathbf{x}_1}{z_2} + \frac{\xi_1}{z_1}\right) K^*\!\left(\frac{\mathbf{x}_2}{z_2} + \frac{\xi_2}{z_1}\right) d\xi_1\,d\xi_2\,, \tag{13.2}$$

where $\mathbf{x}$ is the coordinate in image space, z_1 and z_2 are the object and image distances, respectively, and $K(\mathbf{x}/z_2 + \xi/z_1)$ is the amplitude impulse response of the imaging system giving a real inverted image, i.e., a single-lens imaging system. The intensity in the image is determined by setting $\mathbf{x}_1 = \mathbf{x}_2 = \mathbf{x}$ in Eq. (13.2) and, hence,

$$I_{im}(\mathbf{x}) = \int\!\!\int \Gamma_{ob}(\xi_1,\xi_2)\,K\!\left(\frac{\mathbf{x}}{z_2} + \frac{\xi_1}{z_1}\right) K^*\!\left(\frac{\mathbf{x}}{z_2} + \frac{\xi_2}{z_1}\right) d\xi_1\,d\xi_2\,. \tag{13.3}$$

Substituting Eq. (13.1) into (13.3) yields

$$I_{im}(\mathbf{x}) = I_o \int\!\!\int_{ob} \gamma(\xi_1,\xi_2)\{[\delta(\xi_1 - \mathbf{b}) + \delta(\xi_1 + \mathbf{b})]\,[\delta(\xi_2 - \mathbf{b}) + \delta(\xi_2 + \mathbf{b})]\}$$

$$\times K\!\left(\frac{\mathbf{x}}{z_2} + \frac{\xi_1}{z_1}\right) K^*\!\left(\frac{\mathbf{x}}{z_2} + \frac{\xi_2}{z_1}\right) d\xi_1\,d\xi_2\,. \tag{13.4}$$

Solving Eq. (13.4) gives [since $\gamma(\mathbf{b},\mathbf{b}) = \gamma(-\mathbf{b},-\mathbf{b}) = 1$]

$$I_{im}(\mathbf{x}) = I_o\{|\,K(\mathbf{x} + \mathbf{b}')|^2 + |\,K(\mathbf{x} - \mathbf{b}')|^2$$

$$+ 2\,\mathrm{Re}[\gamma(\mathbf{b},-\mathbf{b})\,K(\mathbf{x} + \mathbf{b}')\,K^*(\mathbf{x} - \mathbf{b}')]\}\,, \tag{13.5}$$

where Re denotes a real part and $|\,K(\mathbf{x})|^2$ is, of course, the intensity impulse response of the imaging system and $2|\mathbf{b}'| = 2z_2|\mathbf{b}|/z_1$, the separation of the two Gaussian image points. It should be noted that the ratio of $2|\mathbf{b}'|$ to $2|\mathbf{b}|$ is the magnification equal to z_2/z_1.

13.3   ONE-DIMENSIONAL SYSTEM

A one-dimensional system that is diffraction limited and of width $2a$ has an amplitude impulse response

$$K(x) = 2a \operatorname{sinc}\left(\frac{kax}{z_2}\right) \tag{13.6}$$

and, hence, Eq. (13.5) reduces to

$$I_{im}(x) = 4a^2 I_o \left[\operatorname{sinc}^2 \frac{ka}{z_2}(x - b') + \operatorname{sinc}^2 \frac{ka}{z_2}(x + b) \right.$$

$$\left. + 2\operatorname{Re} \gamma(b, -b) \operatorname{sinc} \frac{ka}{z_2}(x - b') \operatorname{sinc} \frac{ka}{z_2}(x + b') \right] . \tag{13.7}$$

13.3.1   Incoherent Limit

The limiting form discussed in Chapter 6 is recovered from Eq. (13.2) by setting $\gamma(b, -b) = 0$; hence,

$$I_{im}(x)\Big|_{\mathrm{incoh}} = 4a^2 I_o \left[\operatorname{sinc}^2 \frac{ka}{z_2}(x - b') + \operatorname{sinc}^2 \frac{ka}{z_2}(x + b') \right] . \tag{13.8}$$

13.3.2   Rayleigh Criterion

The two-point resolution condition using the Rayleigh criterion gives a separation $2b'_r$ which from Eq. (13.8) is

$$2b'_r = \frac{3.142 z_2}{ka} . \tag{13.9}$$

13.3.3   Sparrow Criterion

The Sparrow criterion uses the second derivative of the resultant image illuminance set equal to zero at the midpoint between the two Gaussian image points, i.e.,

$$\frac{\partial^2 I(x)}{\partial x^2}\bigg|_{x=0} = 0 . \tag{13.10}$$

When this criterion is imposed on Eq. (13.8), the separation of the two resolved image points is

$$2b'_s = \frac{2.606 z_2}{ka} . \tag{13.11}$$

The result is from Chapter 6 [Eq. (6.12)] and is shown here for comparison with other results to be derived in this chapter.

13.3.4 Coherent Limit

To discuss a similar result for the coherent limit, we set $\gamma(b, -b) = 1$, and Eq. (13.7) becomes

$$I_{im}(x)\big|_{coh} = 4a^2 I_0 \left[\text{sinc}^2 \frac{ka}{z_2} (x - b') + \text{sinc}^2 \frac{ka}{z_2} (x + b') \right.$$

$$\left. + 2\,\text{sinc}\, \frac{ka}{z_2} (x - b')\,\text{sinc}\, \frac{ka}{z_2} (x + b') \right] . \qquad (13.12)$$

Again applying the Sparrow criterion, we note that

$$\frac{\partial^2}{\partial x^2} \left\{ 4a^2 I_0 \left[\text{sinc}\, \frac{ka}{z_2} (x - b') + \text{sinc}\, \frac{ka}{z_2} (x + b') \right]^2 \right\} \Bigg|_{x=0} = 0$$

$$= 2 \left[\text{sinc}\, \frac{ka}{z_2} (x - b') + \text{sinc}\, \frac{ka}{z_2} (x + b') \right] \qquad (13.13)$$

$$\times \left[\frac{\partial^2}{\partial x^2} \text{sinc}\, \frac{ka}{z_2} (x - b') + \frac{\partial^2}{\partial x^2} \text{sinc}\, \frac{ka}{z_2} (x + b') \right] \Bigg|_{x=0} = 0 ,$$

because of the symmetry of the sinc function. We have to evaluate

$$\frac{\partial^2}{\partial \theta^2} (\text{sinc}\,\theta) \bigg|_{\theta=0} = 0 . \qquad (13.14)$$

We note that

$$\frac{\partial}{\partial \theta} (\text{sinc}\,\theta) = \frac{\theta \cos\theta - \sin\theta}{\theta^2} , \qquad (13.15)$$

and, hence,

$$\frac{\partial^2}{\partial \theta^2} (\text{sinc}\,\theta) = \frac{-\theta \sin\theta - \cos\theta}{\theta^2} - \left(\frac{\theta^2 \cos\theta - 2\theta \sin\theta}{\theta^4} \right)$$

$$= -\frac{\sin\theta}{\theta} - \frac{2\cos\theta}{\theta^2} + \frac{2\sin\theta}{\theta^3} .$$

Therefore,

$$\frac{\partial^2}{\partial x^2} I(x) \bigg|_{x=0} = 2\,\text{sinc}\, \frac{kab'}{z_2} - 2\cos \frac{kab'}{z_2} - \frac{kab'}{z_2} \sin \frac{kab'}{z_2} = 0 . \qquad (13.16)$$

Solving Eq. (13.16) by iteration gives

$$2b'_s = \frac{4.164 z_2}{ka} . \qquad (13.17)$$

13.4 DISCUSSION: ONE-DIMENSIONAL SYSTEM

A comparison between Eqs. (13.11) and (13.17) shows that the two-point resolution is better (i.e., the two points may be set closer together) in the incoherent case

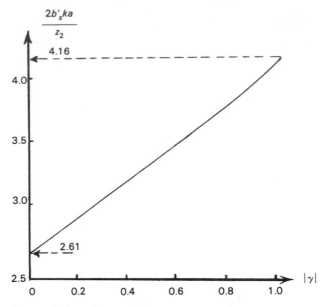

Fig. 13.2 Two-point resolution with partially coherent light for a one-dimensional system using the Sparrow criterion (from Ref. 2, Chapter 11).

by a factor of 1.59, the ratio of the two limiting cases. The intermediate states of coherence have been calculated from Eq. (13.7) by Rojak[1] and a discussion is to be found in Beran and Parrent.[2] Basically, the change is almost linear as the degree of coherence is varied. This result is indicated in Fig. 13.2 (previously shown in Fig. 11.13) where the value $2b_s'ka/z_2$ is plotted against the degree of coherence. The end points of the curve are the values calculated earlier.

It is interesting to plot the actual intensity distributions in the image for a few typical examples. These calculations have been discussed in detail by Nyyssonen and Thompson.[3] Here, we limit ourselves to plotting the intensity distributions in the two limits from Eqs. (13.8) and (13.12). Figure 13.3 shows the curves for various values of the parameter $\delta = 2b'ka/z_2$. Clearly, these results can be obtained by varying either b' or a for a given wavelength and a given image distance z_2. In Fig. 13.3(a), the separation of the image points indicated by

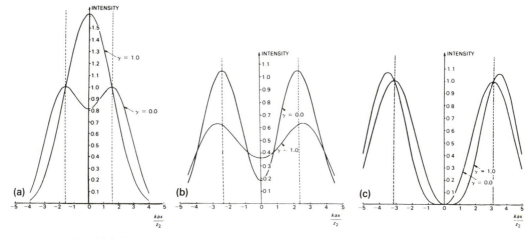

Fig. 13.3 Resultant intensity distribution in the one-dimensional image of a two-point object for both coherent and incoherent points: (a) $\delta = 3.14$, (b) $\delta = 4.71$, and (c) $\delta = 6.28$ (from Ref. 3).

vertical dotted lines is $\delta = 3.14$. The incoherent situation ($\gamma = 0$) shows quite good resolution of the two objects but in the coherent limit ($\gamma = 1$) they are not resolved. Figures 13.3(b) and (c) illustrate two other examples for $\delta = 4.71$ and $\delta = 6.28$, respectively.

13.5   TWO-DIMENSIONAL SYSTEM

The amplitude impulse response of the equivalent two-dimensional system of diameter $2a$ is

$$K(\mathbf{x}) = \pi a^2 \left[\frac{2J_1\left(\frac{ka|\mathbf{x}|}{z_2}\right)}{\left(\frac{ka|\mathbf{x}|}{z_2}\right)} \right] = \pi a^2 \Lambda_1\left(\frac{ka|\mathbf{x}|}{z_2}\right) \tag{13.18}$$

and Eq. (13.5) then becomes

$$I_{im}(\mathbf{x}) = \pi a^4 I_o \left\{ \Lambda_1\left[\frac{ka}{z_2}(|\mathbf{x}| - b')\right]^2 + \Lambda_1\left[\frac{ka}{z_2}(|\mathbf{x}| + b')\right]^2 \right.$$
$$\left. + 2\operatorname{Re}\gamma(b, -b)\,\Lambda_1\left[\frac{ka}{z_2}(|\mathbf{x}| - b')\right]\Lambda_1\left[\frac{ka}{z_2}(|\mathbf{x}| + b')\right] \right\}. \tag{13.19}$$

13.5.1   Incoherent Limit

The incoherent limit is derived from Eq. (13.19) by setting $\gamma(b, -b) = 0$, obtaining [see also Eq. (6.14)]:

$$I_{im}(\mathbf{x})\big|_{\text{incoh}} = \pi a^4 I_o \left(\left\{ \Lambda_1\left[\frac{ka}{z_2}(|\mathbf{x}| - b')\right] \right\}^2 \right.$$
$$\left. + \left\{ \Lambda_1\left[\frac{ka}{z_2}(|\mathbf{x}| + b')\right] \right\}^2 \right). \tag{13.20}$$

13.5.2   Rayleigh Criterion

The two-point resolution limit using the Rayleigh criterion gives a separation $2b'_r$ given by

$$2b'_r = \frac{3.832 z_2}{ka}. \tag{13.21}$$

13.5.3   Sparrow Criterion

Evaluating the Sparrow criterion defined by Eq. (13.10) gives a separation $2b'_s$ defined by

$$2b'_s = \frac{2.976 z_2}{ka}. \tag{13.22}$$

Again, this result was given in Chapter 6. The derivation of this result requires the differentiation of Eq. (13.20) twice and involves Bessel functions of orders 2 and 3. Because of this complexity, the analysis is not given in detail here.

13.5.4   Coherent Limit

To compare the results for the coherent situation with that derived above, we evaluate Eq. (13.19) with $\gamma(b, -b) = 1$. Hence,

$$I_{im}(\mathbf{x})\big|_{\text{coh}} = 4a^2 I_o \left\{ \Lambda_1 \left[\frac{ka}{z_2} (|\mathbf{x}| - b') \right] + \Lambda_1 \left[\frac{ka}{z_2} (|\mathbf{x}| + b') \right] \right\}^2 . \quad (13.23)$$

The result for the Sparrow criterion is

$$2b'_s = \frac{4.600 z_2}{ka} . \quad (13.24)$$

13.6   DISCUSSION: TWO-DIMENSIONAL SYSTEM

As expected, the two-dimensional incoherent two-point resolution is better than the corresponding coherent case; this time by a factor of 1.55. The basic features of the resultant incoherent and coherent intensity distributions are shown in Fig. 13.4.

13.7   CONCLUSIONS

The two-point resolution results are summarized in Table 13.I. While these criteria can be useful, they do, in fact, overlook significant effects that occur in the image. In the image of two points, the only measurable quantity is the separation of the two points as determined by the separation of the maxima in the resultant image intensity distribution. This can be misleading as indicated by Fig. 13.4(b). In that example, when $\gamma = 1$, the result shows two peaks but the separation is incorrect. This is clearly a result of the coherent addition of the two distributions.

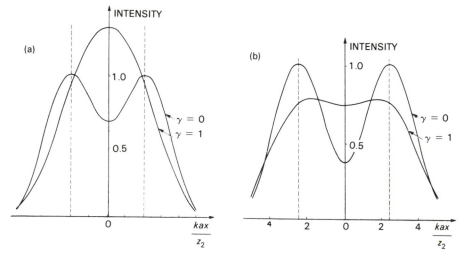

Fig. 13.4 Resultant intensity distribution in the two-dimensional image of a two-point object for both coherent and incoherent points: (a) $\delta = 4.0$ and (b) $\delta = 4.8$ (from Ref. 3).

TABLE 13.I
Two-Point Resolution Results

	Incoherent	Coherent
One-dimensional Rayleigh	$2b'_r = \dfrac{3.142z_2}{ka}$	—
One-dimensional Sparrow	$2b'_s = \dfrac{2.606z_2}{ka}$	$2b'_s = \dfrac{4.164z_2}{ka}$
Two-dimensional Rayleigh	$2b'_r = \dfrac{3.832z_2}{ka}$	—
Two-dimensional Sparrow	$2b'_s = \dfrac{2.976z_2}{ka}$	$2b'_s = \dfrac{4.600z_2}{ka}$

REFERENCES

1. F. Rojak, "Two-point resolution with partially coherent light," MS Thesis, Lowell Technological Institute, Lowell, Mass. (1961).
2. M. J. Beran and G. B. Parrent, Jr., *The Theory of Partial Coherence,* p. 124, Prentice Hall, Englewood Cliffs, N.J. (1964).
3. D. Nyyssonen and B. J. Thompson, "Two-point resolution with partially coherent light," *J. Opt. Soc. Am.* 57(4), 580A (1967); see also D. N. Grimes and B. J. Thompson, "Two-point resolution with partially coherent light," *J. Opt. Soc. Am.* 57(11), 1330–1334 (1967).

14 Coherent Imaging: Examples

14.1 INTRODUCTION

The literature contains very few examples of discussions of image formation with partially coherent light. The reason for this is probably that the calculations are somewhat difficult because of the nonlinear nature of the problem. Furthermore, the need for understanding partially coherent image formation was not essential since most imaging systems and image evaluation techniques were considered incoherent processes—this is no longer the case.

Thus far in our discussions of coherent imaging, we have placed the main emphasis on a comparison between the incoherent image forming process and the coherent image forming process (see Chapter 12) since this could provide some insight into the more general problems associated with the partially coherent situation. However, it must be stressed that coherent imaging is important in its own right because of the great interest in coherent optical data processing and holography. In both of these examples, the systems used provide images that are essentially coherent.

In this chapter, we wish to discuss some further examples of incoherent image formation for transilluminated objects and describe the effects produced in image formation in reflected light. Both types of image forming processes are basic to a variety of areas in optical data processing and holography.

14.2 IMAGE OF AN EDGE OBJECT

The edge object has become useful in image evaluation and assessment. The transfer function of the imaging system may be determined from the image of an edge by first differentiating and then performing a Fourier transformation of the intensity distribution of the edge image. The value of this procedure is that it can be used in evaluating systems employed in both acquiring and assessing image quality. High-quality man-made edges (chrome-on-glass) are extremely useful in instrument evaluation, particularly for microdensitometers and other scanning systems. Naturally occurring edges in photographic records can often be used successfully to evaluate active photographic systems.

In this section, we consider the image of a perfect edge described by

$$I_{ob}(\xi) = \begin{cases} I_o \ , \ \xi \geq 0 \\ 0 \ , \ \xi < 0 \end{cases} . \tag{14.1}$$

The object intensity distribution $I_{ob}(\xi)$ has a discontinuity or edge at $\xi = 0$ of magnitude I_o. The incoherent real image formed by a single one-dimensional diffraction limited lens was given in Eqs. (12.21) and (13.2). Hence,

$$I_{im}(x) = \int_{-\infty}^{\infty} I_{ob}(\xi) K\left(\frac{x}{z_2} + \frac{\xi}{z_1}\right) K^*\left(\frac{x}{z_2} + \frac{\xi}{z_1}\right) d\xi \ , \tag{14.2}$$

where $K(x/z_2 + \xi/z_1)$ is the amplitude impulse response of the imaging lens; z_1 and z_2 are the object and image distance, respectively; and $k = 2\pi/\lambda$. The normalized intensity impulse response of a lens of aperture $2a$ is, of course, sinc^2 (kax/z_2). Hence, Eq. (14.2) becomes

$$I_{im}(x) = \int_{-\infty}^{\infty} I_{ob}(\xi) \text{sinc}^2 ka\left(\frac{x}{z_2} + \frac{\xi}{z_1}\right) d\xi \ . \tag{14.3}$$

Integrating Eq. (14.3) under the conditions of Eq. (14.1) gives

$$I_{im}(x) = \frac{1}{2} + \frac{1}{\pi}\left[Si\left(\frac{kax}{z_2}\right) - \frac{1 - \cos\left(\frac{kax}{z_2}\right)}{\left(\frac{kax}{z_2}\right)} \right] . \tag{14.4}$$

Here, the resultant intensity has been normalized so that the asymptotic value of the intensity in the illuminated region is unity. The term $Si(\theta)$ is the sine integral function defined by

$$Si(\theta) = \int_0^{\theta} \frac{\sin\theta}{\theta} d\theta \ . \tag{14.5}$$

The image formed with coherent illumination with a constant phase of the same edge is given by [see Eqs. (12.23) and (13.2)]:

$$I_{im}(x) = \left| \int_{-\infty}^{\infty} \sqrt{I_{ob}(\xi)} \ K\left(\frac{x}{z_2} + \frac{\xi}{z_1}\right) d\xi \right|^2 . \tag{14.6}$$

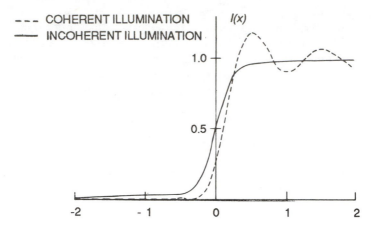

Fig. 14.1 Theoretical intensity distribution in the coherent and incoherent image of a high-contrast edge (after Ref. 1).

Substituting the one-dimensional form of the amplitude impulse response gives

$$I_{im}(x) = \left| \int_{-\infty}^{\infty} \sqrt{I_{ob}(\xi)} \, \text{sinc} \, ka \left(\frac{x}{z_2} + \frac{\xi}{z_1} \right) d\xi \right|^2 , \tag{14.7}$$

where we have assumed that there is no phase variation associated with the edge so that the amplitude distribution can be written as the square root of the intensity distribution across the edge. Solving Eq. (14.7) with the condition of Eq. (14.1), we obtain, using the same normalization as in Eq. (14.4),

$$I_{im}(x) = \left[\frac{1}{2} - \frac{1}{\pi} \, Si \left(\frac{kax}{z_2} \right) \right]^2 . \tag{14.8}$$

Clearly, Eqs. (14.4) and (14.8) will not yield identical intensity distributions. The theoretical curves generated by Eqs. (14.4) and (14.8) are shown in Fig. 14.1 for comparison. Two effects should be noted: the coherent image intensity distribution shows a pronounced ringing or fringing caused by the sharp cutoff of the amplitude transfer function and the position of the edge appears to shift. The amount of the shift depends on the method used for choosing the position of the edge. If the edge is assumed to be at the half-intensity point, then the shift is given by

$$\text{shift} = \frac{0.212 \lambda z_2}{a} , \tag{14.9}$$

i.e., this is about one fifth of the width of the impulse response of the system. For degrees of coherence lying between 0 and 1, the curves for the resultant intensity fall between the limiting curves of Fig. 14.1 and the shift of the edge varies with the degree of coherence.

The edge ringing and edge shifting can be observed experimentally. Figure 14.2 shows in (a) a photograph of an incoherent edge image to be compared with (b) a photograph of the coherent image of the same edge taken with the same

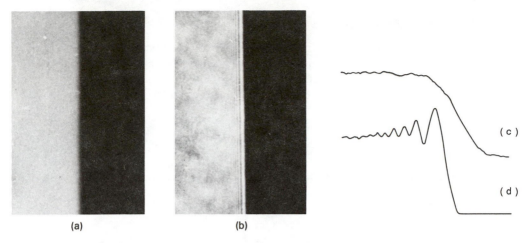

(a)   (b)

Fig. 14.2 (a) Incoherent image of an edge, (b) coherent image of an edge, (c) microphotometer trace of (a), and (d) microphotometer trace of (b) (after Ref. 1)

imaging system. In (c) and (d), microphotometer traces of the photographic results of (a) and (b) are shown. These are curves of the log intensity. A more detailed discussion of the exact experimental details can be found in a paper by Considine[1]. Excellent theoretical discussions are contained in a thesis by Skinner.[2]

The edge, of course, is not always of high contrast. When light is transmitted through both regions that specify the edge, the ringing in the image is apparent on both sides of the edge. As the contrast is decreased, the shift in the edge gets smaller. The intensity distributions in the coherent and incoherent images of a low-contrast edge are shown in Fig. 14.3.

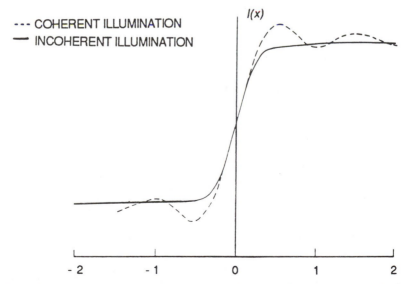

--- COHERENT ILLUMINATION
— INCOHERENT ILLUMINATION

Fig. 14.3 Theoretical intensity distribution in the coherent and incoherent images of a low-contrast edge (after Ref. 1).

14.3   IMAGE OF A SLIT OBJECT

The use of bar targets for measuring resolution of optical systems is perhaps the most common and rapid method of evaluating an optical system. In this section, we compare the image of a slit (or a bar) under the coherent and incoherent limits. We define the one-dimensional object intensity distribution as

$$I_{ob}(\xi) = \begin{Bmatrix} I_o & , |\xi| \leq b \\ 0 & , |\xi| > b \end{Bmatrix} \tag{14.10}$$

for a slit of width $2b$. The incoherent image of this slit in terms of the normalized impulse response is then given by

$$I_{im}(x) = \int_{-b}^{b} I_o \operatorname{sinc}^2 ka \left(\frac{x}{z_2} + \frac{\xi}{z_1} \right) d\xi . \tag{14.11}$$

Integrating Eq. (14.11) with $I_o = 1$ gives

$$I_{im}(x) = \frac{1}{\pi} \left[\frac{2\beta}{\beta^2 - \alpha^2} + \frac{\cos^2(\alpha + \beta)}{\alpha + \beta} \right.$$
$$\left. - \frac{\cos^2(\alpha - \beta)}{\alpha - \beta} + \int_{\alpha+\beta}^{\alpha-\beta} \frac{\sin^2 \sigma}{\sigma} d\sigma \right] , \tag{14.12}$$

where

$$\alpha = \frac{kax}{z_2} ,$$

$$\beta = \frac{kab}{z_1} ,$$

$$\sigma = ka \left(\frac{x}{z_2} + \frac{\xi}{z_1} \right) . \tag{14.13}$$

In the coherent situation, we assume again that there is no phase variation associated with the slit (or bar). Hence,

$$I_{im}(x) = \left[\int_{-b}^{b} \sqrt{I_o} \operatorname{sinc} ka \left(\frac{x}{z_2} + \frac{\xi}{z_1} \right) d\xi \right]^2 . \tag{14.14}$$

Performing the integration with the change of variables used previously,

$$I_{im}(x) = \frac{1}{\pi^2} \left(\int_{\alpha-\beta}^{\alpha+\beta} \frac{\sin \sigma}{\sigma} d\sigma \right)^2 . \tag{14.15}$$

A typical result for a well-resolved high-contrast slit is shown in Fig. 14.4(a). The ringing occurs at both edges as does the shift, resulting in a narrower

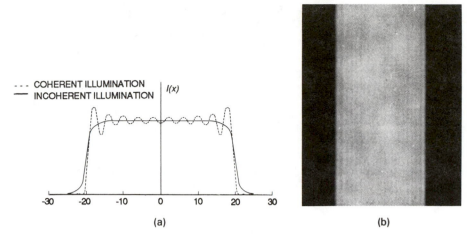

(a)   (b)

Fig. 14.4 (a) Theoretical intensity distribution in the coherent and incoherent images of a slit and (b) photograph of the coherent image of a slit.

measured slit width in the coherent image. A photograph of the image of a well-resolved slit is shown in Fig. 14.4(b). For an opaque bar in a transparent field, a similar type of result would be apparent with the bar edges shifting outward making it appear too wide.

The examples discussed here have deliberately ignored any phase effects that might be present in the object. The phase might well affect the coherent image significantly; for example, in using the standard bar target (either USAF or long line), the object is then on film and an optical thickness variation would be associated with the silver image. Figure 14.5 shows a comparison between the incoherent and coherent image of a high-contrast bar target in the region where the incoherent system is producing considerable contrast reduction. Most of the major effects that can be seen in the coherent image, Fig. 14.5(b), are almost certainly caused by the phase effects in the bar target object.

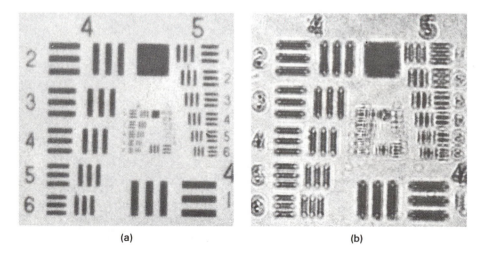

(a)   (b)

Fig. 14.5 (a) Incoherent image of a high-contrast bar target and (b) coherent image of the same bar target (after Ref. 1).

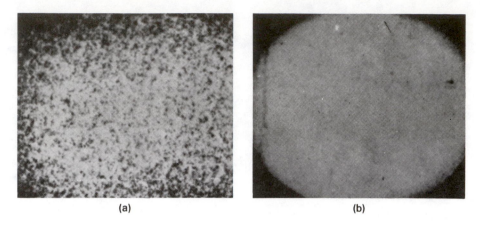

(a)   (b)

Fig. 14.6 (a) Photograph of a portion of a wall illuminated with a laser beam and (b) photograph of the same portion of wall with spatially coherent light with limited coherent length.

14.4   REFLECTED LIGHT IMAGING

The effects detailed above can be anticipated, of course, in image formation in reflected light. However, the nature of the surface of the object will now play an important role in the image formation. The surface only has to be irregular on the order of a wavelength of the incident light for significant path differences to be introduced. When the object is illuminated coherently, these path differences will give rise to interference effects in the image. This is the cause of the well-known speckling effects produced when a surface is illuminated with a laser beam. Figure 14.6(a) shows this effect. The photograph is of a portion of a blank wall illuminated with light from a He-Ne gas laser. The speckling completely obliterates any direct information about the surface. For comparison, Fig. 14.6(b) shows a similar photograph taken with the same imaging system in which the illuminating beam was spatially coherent but the coherence length of the radiation was much less than the average path difference introduced by the surface; hence, the image is essentially incoherent. These figures were also shown in Chapter 11.

A further example is shown in Fig. 14.7. A standard bar target was printed onto a matte photographic paper. This photograph then served as an object to be

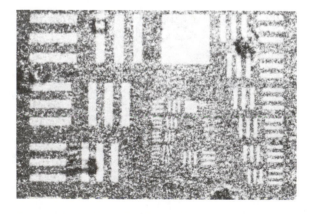

Fig. 14.7 Coherent image of a photograph of a resolution target printed on a matte paper (after Ref. 1).

rephotographed when coherently illuminated. Again, the characteristic speckling is observed.

It is worth noting that the characteristic size of the speckling is determined by the impulse response of the imaging system. The speckles are much larger with, say, an f/64 system than they are with the same system used at f/9. Speckle noise is covered in more detail in Chapters 20 and 21.

14.5   CONCLUSIONS

The comparison given here and in the previous chapter of coherent and incoherent imaging provides some insight and feeling for the problems involved. Basically, the differences observed in the coherent images are caused by the inherent nonlinearity of the problem. Since we are accustomed to seeing and evaluating incoherent images—and hence thinking linearly—the coherent image quality cannot be immediately approximated intuitively.

The effects of coherent imaging are usually considered to be of a deleterious nature and a number of workers have tried various techniques to remove the coherence effects. For instance, rotating ground glass and colloidal solutions of milk and gold have been used resulting in a time-averaged result that looks like an incoherent image. It is interesting to note that there might be occasions when the coherence effects are advantageous. In Fig. 14.1, the coherent edge image has a steeper slope and hence looks "sharper." The edge ringing, particularly in the low-contrast example (Fig. 14.3), might make possible the detection of the edge that is below the threshold of the detector in the incoherent imaging situation.

REFERENCES

1. P. S. Considine, "Effects of coherence on imaging systems," *J. Opt. Soc. Am.* 56(8), 1001–1008 (1966).
2. T. J. Skinner, "Energy considerations propagation in random medium and imaging in scalar coherence theory," PhD Thesis, Boston Univ. (1965); see also "Surface texture effects in coherent imaging," *J. Opt. Soc. Am.* 53(11), 1350 (1963).

15   Coherence Theory Solution to the Pinhole Camera*

15.1   INTRODUCTION

To illustrate some of the techniques used in solving propagation problems, the formalism of coherence theory developed in Chapters 10 and 11 is applied to a simple image formation system—a pinhole camera.[1] Theoretical predictions are compared with typical experimental results. As discussed in Chapter 11, the quantity of interest in coherence theory is a cross-correlation function between the complex amplitude of the optical field at two separate space-time points. For many problems (including those of image formation) in which polarization effects are not important, the optical field can be treated as a scalar quantity. In this case, the complex cross-correlation function is defined by Eq. (11.1) as

$$\Gamma_{12}(\tau) = <V_1(t)V_2^*(t + \tau)> \, , \tag{15.1}$$

where the angle brackets denote a long time average, $V_1(t)$ is the complex scalar field amplitude at the space point 1 and time t, and $V_2^*(t + \tau)$ is the complex conjugate of the scalar amplitude at point 2 and time $t + \tau$. Equation (15.1) defines the mutual coherence function $\Gamma_{12}(\tau)$, which has the following three important properties:

1. The value of $\Gamma_{12}(\tau)$ is a physically measurable quantity.

2. Once $\Gamma_{12}(\tau)$ is known for all pairs of points over a source surface, it can be propagated, i.e., $\Gamma_{12}(\tau)$ can be determined everywhere in the field.

3. When $\tau = 0$ and points 1 and 2 are the same, Γ_{12} reduces to the time-averaged intensity at a point in the optical field.

---

*Reprinted with permission from "Coherence Theory Solution to the Pinhole Camera," by George O. Reynolds and John H. Ward, *J. SPIE* 5(1), 3–8 (October/November 1966).

If the physical problem is described by the application of the quasimono-chromatic approximations defined by Eqs. (11.7) and (11.8) as

$$\Delta\nu \ll \bar{\nu} \tag{15.2a}$$

and the path length requirement as

$$\tau \ll \frac{1}{\Delta\nu} , \tag{15.2b}$$

the mutual coherence function is measured readily over a plane by doing a series of Young's two-pinhole experiments. As discussed in Chapter 11, the visibility of the interference fringes can be directly related to the magnitude of the mutual coherence function, and the relative phase is related to the position of the central fringe. The quasimonochromatic approximation states that the mean frequency $\bar{\nu}$ must be much greater than the bandwidth $\Delta\nu$ and that all time delays τ between points 1 and 2 must be much smaller than the reciprocal of the bandwidth.

Also, when the quasimonochromatic approximation holds and $\Gamma_{12}(0)$ is known over a surface, the solution for $\Gamma_{12}(\tau)$ at two points in the field can be written in the integral form [see Eq. (11.31)]:

$$\Gamma(\mathbf{P}_1,\mathbf{P}_2,\tau) = e^{-2\pi i\bar{\nu}\tau} \iint \Gamma(\mathbf{S}_1,\mathbf{S}_2,0) \frac{\partial G_1}{\partial n_1}(\mathbf{S}_1,\mathbf{P}_1)$$

$$\times \frac{\partial G_2}{\partial n_2}(\mathbf{S}_2,\mathbf{P}_2)\, d\mathbf{S}_1, d\mathbf{S}_2 , \tag{15.3}$$

where G_1 and G_2, which are the appropriate Green's function solutions to the Hemholtz equation, vanish over the source surface; $\partial/\partial n_1$ and $\partial/\partial n_2$ are derivatives taken along the normal to the surface at points 1 and 2. Figure 15.1 shows the geometry and coordinates used in the propagation problem between two planes.

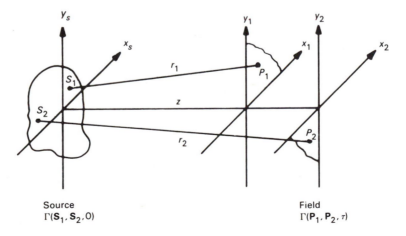

Fig. 15.1 Geometry used in setting up the propagation problem for $\Gamma(S_1,S_2;0)$.

The procedure used to solve for $\Gamma(\mathbf{P}_1,\mathbf{P}_2,\tau)$ starts by either measuring or assuming a form for $\Gamma(\mathbf{S}_1,\mathbf{S}_2,0)$ over the source plane labeled by coordinates (ξ_s,η_s). The mutual coherence function is then propagated: Eq. (15.3) is solved and $\Gamma(\mathbf{P}_1,\mathbf{P}_2,\tau)$ is computed for field points $\mathbf{P}_1,\mathbf{P}_2$. Again, Eq. (15.3) as it stands is only a solution for the quasimonochromatic case when small time delays and short path differences are used.

15.2   PINHOLE CAMERA WITH INCOHERENT ILLUMINATION

Equation (15.3) can now be used to show how an image is formed by a pinhole camera. Figure 15.2 is a diagram of the experimental arrangement. The object—a photographic transparency of a resolution target—was a distance z_1 in front of the pinhole. Either photographic film or a viewing eyepiece was placed a distance z_2 behind the pinhole. A tungsten light bulb illuminated a ground glass screen, and the scattered light then transilluminated the object. Although the ground glass was not necessary to produce an image of the negative transparency, it was used to prevent the pinhole camera from producing an image of the light bulb.

Since the principal characteristics of the pinhole camera can be demonstrated by a one-dimensional analysis, we will treat the problem theoretically as if the object, pinhole, and image were all one dimensional. In the case of the pinhole camera, Eq. (15.3) must be used twice in succession: first to propagate from object plane to pinhole plane and then, after the pinhole aperture is accounted for, to propagate from the pinhole plane to the image plane.

We assume that the field at the object is incoherent. That is,

$$\Gamma(\xi_1,\xi_2) = I(\xi_1)\delta(\xi_1 - \xi_2) . \tag{15.4}$$

Here $I(\xi_1)$ is merely the intensity of the light across the object plane, which is determined by the light bulb and the optical density of the transparency, and $\delta(\xi_1 - \xi_2)$ is the Dirac delta function having the property

$$\int_{-\infty}^{\infty} I(\xi_1)\delta(\xi_1 - \xi_2)\,d\xi_1 = I(\xi_2) . \tag{15.5}$$

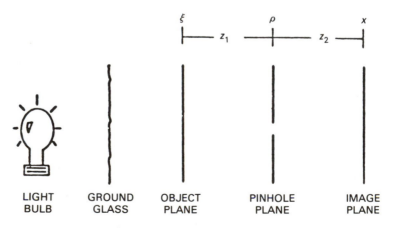

Fig. 15.2 Experimental setup for one-dimensional pinhole camera problem.

Thus, in Eq. (15.4) when $\xi_1 \neq \xi_2$, $\Gamma(\xi_1, \xi_2)$ is zero and there is no correlation of the field at two separate points.

Equation (15.3) for the propagation of $\Gamma(\xi_1, \xi_2)$ can now be written

$$\Gamma(x_1, x_2) = \int\int_{-\infty}^{\infty} \int\int_{-d/2}^{d/2} I(\xi_1)\delta(\xi_1 - \xi_2)$$

$$\times \frac{\partial G_1}{\partial n_1}(\xi_1, \rho_1) \frac{\partial G_2^*}{\partial n_2}(\xi_2, \rho_2) \frac{\partial G_1}{\partial n_1}(\rho_1, x_1)$$

$$\times \frac{\partial G_2^*}{\partial n_2}(\rho_2, x_2)\, d\xi_1\, d\xi_2\, d\rho_1\, d\rho_2 \ , \tag{15.6}$$

where we have four Green's functions: two for the propagation from points (ξ_1, ξ_2) to points (ρ_1, ρ_2) in the pinhole plane and two for the propagation from points (ρ_1, ρ_2) to (x_1, x_2) in the image plane. Integration over the object has the limits $\pm\infty$. The effect of the pinhole aperture is included by limiting the integration over the pinhole plane to $\pm d/2$, where d is the pinhole diameter. Equation (15.6) can be simplified by doing the ξ_2 integration and using the sifting property of the δ function given in Eq. (15.5). Another simplification of Eq. (15.6) is achieved by letting $x_1 = x_2$ so that the mutual intensity function reduces to the intensity in the image plane. Once these operations are carried out, Eq. (15.6) becomes

$$I(x) = \int\int\int_{-\infty}^{\infty} I(\xi)\, \frac{\partial G_1(\xi, \rho_1)}{\partial n_1}\, \frac{\partial G_2^*(\xi, \rho_2)}{\partial n_2}$$

$$\times \frac{\partial G_1(\rho_1, x)}{\partial n_1}\, \frac{\partial G_2^*(\rho_2, x)}{\partial n_2}$$

$$\times \mathrm{rect}(\rho_1 \,|\, d/2)\,\mathrm{rect}(\rho_2 \,|\, d/2)\, d\rho_1\, d\rho_2\, d\xi \ , \tag{15.7}$$

where $\xi_1 = \xi$. In Eq. (15.7), the limits of the integration over (ρ_1, ρ_2) have been changed and the functions $\mathrm{rect}(\rho_1 \,|\, d/2)$ and $\mathrm{rect}(\rho_2 \,|\, d/2)$ are used, where

$$\mathrm{rect}(\rho \,|\, d/2) = \begin{cases} 1 & |\rho| \leq d/2 \\ 0 & |\rho| > d/2 \end{cases} . \tag{15.8}$$

Now if the distances from the optical axis are small compared with the lengths (z_1, z_2), the normal derivatives of the Green's functions have the form

$$\frac{\partial G_1}{\partial n_1}(\xi, \rho_1) \cong \frac{ik}{2\pi r_1}\, e^{ikr_1} \ , \tag{15.9}$$

where $k = 2\pi/\overline{\lambda}$ ($\overline{\lambda}$ is the mean wavelength of the radiation) and r_1 is the distance between ξ and ρ_1:

$$r_1 = [z_1^2 + (\xi - \rho_1)^2]^{1/2} \ . \tag{15.10}$$

Thus, $\partial G_1/\partial n_1$ is essentially a diverging spherical wave of radius r_1. If we now restrict our attention to the vicinity of the optical axis, the paraxial approximation can be made and r_1 binomially expanded to give

$$r_1 \cong z_1 + \frac{\xi^2}{2z_1} + \frac{\rho_1^2}{2z_1} - \frac{\xi\rho_1}{z_1} \ . \tag{15.11}$$

The other Green's functions have the same form except that proper coordinates must be used and $G_1^*(\xi_1, \rho_1) = G_2(\xi_1, \rho_1)$, where the asterisk denotes a complex conjugate.

When Eqs. (15.9) and (15.11) are substituted into (15.7), the intensity in the image plane is

$$I(x) = K \iiint\limits_{-\infty}^{\infty} I(\xi) \left\{ \mathrm{rect}(\rho_1 \,|\, d/2) \right.$$

$$\times \exp\left[ik\rho_1^2 \left(\frac{1}{2z_1} + \frac{1}{2z_2} \right) \right] \Big\}$$

$$\times \left\{ \mathrm{rect}(\rho_2 \,|\, d/2) \exp\left[-ik\rho_2^2 \left(\frac{1}{2z_1} + \frac{1}{2z_2} \right) \right] \right\}$$

$$\times \exp\left[-ik \left(\frac{\xi}{z_1} + \frac{x}{z_2} \right)(\rho_1 - \rho_2) \right] d\rho_1 \, d\rho_2 \, d\xi \ . \tag{15.12}$$

At this point, it is most convenient to define two functions:

$$A_1(\rho_1) = \mathrm{rect}(\rho_1 \,|\, d/2) \exp\left[ik\rho_1^2 \left(\frac{1}{2z_1} + \frac{1}{2z_2} \right) \right] ,$$

$$A_2(\rho_2) = \mathrm{rect}(\rho_2 \,|\, d/2) \exp\left[-ik\rho_2^2 \left(\frac{1}{2z_1} + \frac{1}{2z_2} \right) \right] . \tag{15.13}$$

These functions have Fourier transforms that we denote by $\tilde{A}_1(\mu)$ and $\tilde{A}_2(\mu)$. Using these, we write Eq. (15.12) as

$$I(x) = K \int\limits_{-\infty}^{\infty} I(\xi) \left\{ \int\limits_{-\infty}^{\infty} A_1(\rho_1) \right.$$

$$\times \exp\left[-ik\rho_1 \left(\frac{\xi}{z_1} + \frac{x}{z_2} \right) \right] d\rho_1 \int\limits_{-\infty}^{\infty} A_2(\rho_2)$$

$$\times \exp\left[ik\rho_2 \left(\frac{\xi}{z_1} + \frac{x}{z_2} \right) \right] d\rho_2 \Big\} d\xi \ . \tag{15.14}$$

Thus, in Eq. (15.14), we have Fourier transforms of $A_1(\rho_1)$ and $A_2(\rho_2)$ when the integrations of $d\rho_1$ and $d\rho_2$ are carried out. Also $A_1(\rho_1)$ and $A_2(\rho_2)$ are complex symmetric functions such that $[\tilde{A}_1(\mu)]^* = \tilde{A}_2(-\mu)$. Equation (15.14) may then be written

$$I(x) = K \int_{-\infty}^{\infty} I(\xi) \mid \tilde{A}(x - m\xi)\mid^2 d\xi \ , \qquad (15.15)$$

where $m = z_2/z_1$ is the linear magnification of the pinhole camera. We now have the solution in a recognizable form. The solution states that the image $I(x)$ is a linear superposition of the object with some function $\mid\tilde{A}\mid^2$. Each point in the object gives rise to an intensity impulse response $\mid\tilde{A}\mid^2$. All such impulse responses are weighted with the object intensity distribution and incoherently added to obtain the intensity distribution in the image plane. Suppose that the object was only a point source of light located on the optical axis. This can be expressed by

$$I_{obj}(\xi) = \delta(\xi) \ . \qquad (15.16)$$

Using the property of the Dirac delta function, we can write the point image as

$$I_{im}(x) = \mid\tilde{A}(x)\mid^2 \ . \qquad (15.17)$$

The function $\mid\tilde{A}(x)\mid^2$, usually called the *impulse response* or *point spread function* of the instrument, is the squared diffraction pattern of the system from a point source of light.

If we employ the far-field conditions of Eq. (2.1),

$$z_1 > \frac{d^2}{\lambda} \quad \text{and} \quad z_2 > \frac{d^2}{\lambda} \ , \qquad (15.18)$$

it is easy to solve for $\mid\tilde{A}(x)\mid^2$. In this case, the quadratic phase factors in $A_1(\rho_1)$ and $A_2(\rho_2)$ disappear. The impulse response function becomes just the far field (or Fraunhofer diffraction pattern) of the pinhole. For a slit source on-axis, this is

$$\mid A(x)\mid^2 = d^2 \left| \frac{\sin\left(\frac{\pi x d}{\lambda z_2}\right)}{\left(\frac{\pi x d}{\lambda z_2}\right)} \right|^2 . \qquad (15.19)$$

For a point source and a two-dimensional circular aperture camera, the impulse response is

$$\mid\tilde{A}(x)\mid^2 = \left(\frac{\pi d}{4}\right)^2 \left[\frac{2J_1\left(\frac{\pi x d}{\lambda z_2}\right)}{\left(\frac{\pi x d}{\lambda z_2}\right)} \right]^2 . \qquad (15.20)$$

It is now possible to talk about resolution of the pinhole camera when both the object and image planes are in the far field of the aperture. In the case of a camera with a circular aperture, the first zero of the impulse response function occurs at the radius $x_0 = 1.22\,\bar{\lambda}z_2/d$. According to the Rayleigh resolution criterion, two point images will be resolved if they are separated by a distance of x_0 or greater. To relate this distance to spatial frequency, the value $1/x_0$ can be called the *spatial frequency resolution limit* in the image plane:

$$\text{resolution limit (image space)} = \frac{d}{1.22\,z_2\bar{\lambda}}\cdot \tag{15.21}$$

To find the resolution limit in the object plane, Eq. (15.21) is multiplied by m:

$$\text{Rayleigh resolution limit (object space)} = \frac{d}{1.22\,z_1\bar{\lambda}}\cdot \tag{15.22a}$$

This can be written more simply by using the relation

$$z_1 = \frac{nd^2}{\bar{\lambda}}, \tag{15.22b}$$

where n is the number of far-field distances between the object and pinhole planes. By substituting Eq. (15.22b) into (15.22a), the object resolution becomes

$$\text{resolution (object space)} = \frac{1}{1.22\,nd}\cdot \tag{15.23}$$

Thus, resolution in the object plane is inversely related to the product of pinhole diameter and number of far-field distances to the object. This assumes, of course, that both the object and image are at distances of at least one far field.

In a pinhole camera, increasing the aperture d reduces the resolution if the far-field number n is kept fixed. To verify this experimentally, a three-bar resolution target was recorded through pinholes of several diameters. As described earlier, this object was illuminated by the light bulb through the

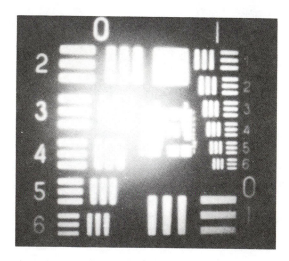

Fig. 15.3 Pinhole camera image of a resolution target. The pinhole was 200 μm in diameter, the object and image were at one far-field distance, and the resolution is 4 lines/mm.

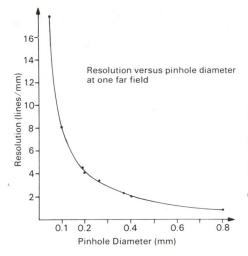

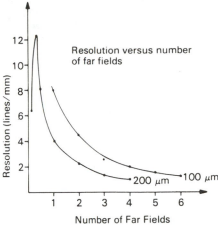

Fig. 15.4 Pinhole camera resolution as a function of pinhole diameter. Experimental points fall very close to the curve predicted by assuming the Rayleigh resolution criterion.

Fig. 15.5 Resolution as a function of the number of object far-field distances is shown for cameras using 200- and 100-μm-diam pinholes.

ground glass screen. Figure 15.3 is a typical image of the bar target obtained using a 200-μm pinhole at one far-field distance. Resolution here is about 4 lines/mm.

Figure 15.4 plots resolution at one far-field distance for a series of pinholes. Experimental results (denoted by the dots) follow very closely the theoretical curve given by Eq. (15.23). With the object at one far-field distance, the image plane was moved back to successively greater distances. Resolution of the object did not change as long as the far-field conditions were obeyed, as would be expected from Eq. (15.23).

Figure 15.5 plots the resolution for pinholes of both 200- and 100-μm diameters as a function of n. The curves follow the predicted values closely. For distances between one and one third of a far-field distance, however, resolution continues to increase as if the far-field conditions were still obeyed. Between one third and one fourth of a far-field distance there is considerable loss of resolution. The explanation for this increase in resolution at far-field distances less than one is that, even though the impulse response function is no longer the far-field diffraction pattern, it is still structurally similar to this pattern. For less than one third of a far-field distance, the diffraction pattern becomes Fresnel-like in appearance, indicating that the quadratic phase factors in Eq. (15.13) become larger.

To substantiate this conjecture, the impulse response of a 200-μm pinhole was recorded at several positions (Fig. 15.6) and microdensitometer traces were made of the patterns. Figure 15.7 is the characteristic trace for Fraunhofer diffraction of a circular aperture. The center maximum is much higher than the first ring. As the pattern is recorded at successive near-field distances, the relative height of the first diffraction ring increases, but the shape is not changed appreciably. Finally, characteristic Fresnel diffraction begins to predominate at about one third of a far-field distance. These resolution effects are described in terms of the transfer function in the Appendix, Sec. 15.5, at the end of this chapter.

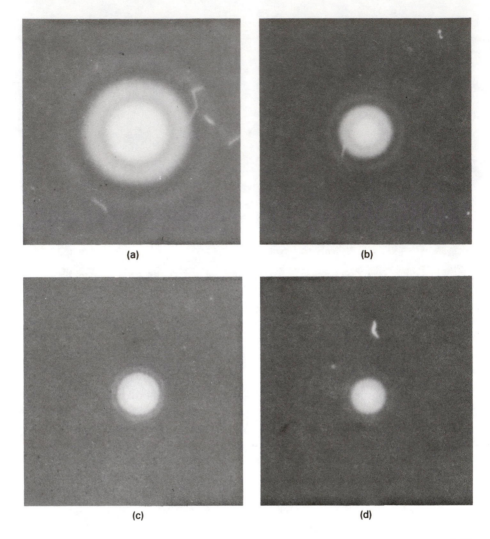

(a)   (b)

(c)   (d)

Fig. 15.6 Impulse response of a 200-μm-diam pinhole camera at object distances of (a) $d^2/\bar{\lambda}$, (b) $d^2/2\bar{\lambda}$, (c) $d^2/3\bar{\lambda}$, and (d) $d^2/4\bar{\lambda}$.

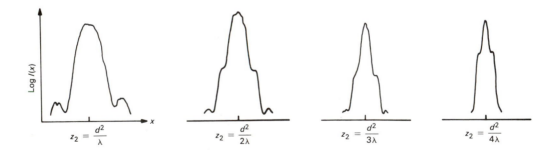

Fig. 15.7 Microdensitometer traces of the impulse response of a 200-μm-diam pinhole camera at different object distances.

Fig. 15.8 Image of a fan-shaped resolution target with a 200-μm-diam pinhole camera. Both object and image distance were at one third of a far-field distance.

Spurious resolution also occurs in the pinhole camera when either the object or image distance is less than one far field. Figure 15.8 shows a fan-shaped resolution target imaged through a 200-μm pinhole. Both the object and image distance were one third $d^2/\overline{\lambda}$. The bars are resolved at the large end of the fan. Then there is an area at slightly higher spatial frequencies where resolution is lost. Resolution appears to recover at an area of yet higher frequencies. However, close examination shows that there is spurious resolution in this area—white bars are black. Spurious resolution can also be observed by defocusing a lens-produced image. In both cases, spurious resolution results from the appreciable contributions of quadratic phase terms in Eq. (15.13).

15.3   PINHOLE CAMERA WITH COHERENT ILLUMINATION

If the object is coherently illuminated, then $\Gamma(\xi_1, \xi_2)$ may be written in the form [see Eq. (11.28)]:

$$\Gamma(\xi_1, \xi_2) = U_1(\xi_1) U_2^*(\xi_2) , \tag{15.24}$$

and it is only necessary to solve a single instead of a double-wave equation. The intensity in the image plane is then found by squaring and time-averaging the resulting amplitude distribution. If the object and image planes are in the far field of the pinhole, the quadratic phase factors in the function $A(\rho)$, defined in Eq. (15.13), are negligible and the intensity distribution in the image plane is

$$<|U(x)|^2> = \left\langle \left| K' \int U_1(\xi) \right. \right.$$
$$\left. \left. \times \exp \frac{ik\xi^2}{2z_1} \operatorname{sinc} \frac{\pi d}{\lambda z_1} (x - m\xi) d\xi \right|^2 \right\rangle , \tag{15.25}$$

where, again, $m = -z_2/z_1$, the angle brackets denote a long time average, and K' denotes the appropriate constants. If we now demand that

$$\frac{\xi^2}{\lambda z_1} \ll 1 , \tag{15.26}$$

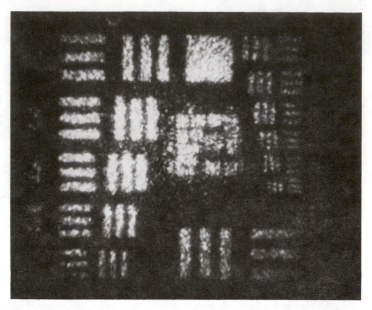

Fig. 15.9 Image of resolution target illuminated by ground glass scattered laser light. Pinhole diameter was 200 μm and the object and image distances were at one far field.

then Eq. (15.25) describes a coherent image as a coherent superposition of object points. Equation (15.26) is just a field-of-view limitation in the object planes, i.e., the coherent pinhole camera has a field of view in which only a part of the object about the size of the pinhole itself is actually imaged. If the object is illuminated with a diffuse coherent beam, however, then light from all parts of the object is collected by the pinhole and an image with an unrestricted field of view can be observed. The image appears speckled because of the diffusing surface. Figure 15.9 is an example of an image formed with a diffuse coherent beam.

15.4   CONCLUSIONS

We have solved the pinhole camera problem using the formalism of coherence theory, considering both incoherent and coherent illumination. The incoherent image was found to be a superposition of object points convolved with an intensity impulse response or a point spread function. The impulse response for incoherent objects, when object and image distances are both greater than $d^2/\overline{\lambda}$, was shown to be just the Fraunhofer diffraction pattern of the pinhole.

Resolution was predicted by using the Rayleigh criterion. Best resolution was achieved at one third $d^2/\overline{\lambda}$; however, using the camera at such short object and image distances produces spurious resolution effects. When coherent illumination was used, a recognizable image was observed only when a diffuser was placed between the object and the coherent source. An image can be formed without the diffuser, but the field of view is so restricted that the image strongly resembles the diffraction pattern of the pinhole itself.

15.5   APPENDIX: TRANSFER FUNCTION OF THE PINHOLE CAMERA

The transfer function for a pinhole camera has been investigated by various authors.[2-4] The results are in quantitative agreement with those shown in Figs. 15.4 and 15.5 in that the cutoff frequency of the transfer function varies inversely

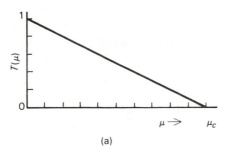
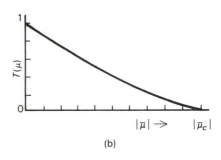

(a)   (b)

Fig. 15.10 Transfer functions for the pinhole camera: (a) one-dimensional pinhole ($\mu_c = 1/nd$) and (b) two-dimensional pinhole ($|\mu_c| = 1/1.22nd$), radial slice of two-dimensional, circularly symmetric, intensity transfer function shown in Chapter 12.

with both the size of the pinhole and the number of far-field distances from the pinhole to the image plane. This result is obtained mathematically by Fourier transforming the impulse responses of Eq. (15.19) or (15.20), respectively.

The Fourier transform of Eq. (15.19)—the impulse response for the one-dimensional pinhole—was previously discussed in Chapters 2 and 8 and the Fourier transform of Eq. (15.20) was discussed in Chapter 12. Using these results, the transfer functions for the one- (1-D) and two-dimensional (2-D) pinhole cameras are shown in Fig. 15.10. Note that they are diffraction limited since there are no aberrations in a pinhole camera.

The cutoff frequencies of the two transfer functions are given by

$$\mu_c^{\text{1-D}} = 1/\bar{\lambda}f^{\#} \ ,$$

$$|\mu_c^{\text{2-D}}| = 1/1.22\bar{\lambda}f^{\#} \tag{15.27}$$

for the two cases, respectively. The $f^{\#}$ of the pinhole camera in the image plane is

$$f^{\#} = z_2/d \ . \tag{15.28}$$

The condition on z_2 expressed by Eq. (15.18) can be written as

$$z_2 = nd^2/\bar{\lambda} \ , \tag{15.29}$$

where n is the number of far-field distances and d is the pinhole size. Substituting Eq. (15.29) into (15.28) and (15.28) into (15.27) yields

$$\mu_c^{\text{1-D}} = 1/nd \ , \tag{15.30}$$

$$|\mu_c^{\text{2-D}}| = 1/1.22nd \ . \tag{15.31}$$

Equations (15.30) and (15.31) show that the cutoff frequency of the pinhole camera transfer function is inversely proportional to both the pinhole size and the number of far-field distances. When transformed to the variables of object space with the system magnification, $m = -z_2/z_1$, the results of Eqs. (15.30) and (15.31) are in agreement with Eq. (15.23) and the discussion concerning resolution effects of the pinhole camera. Thus, even though the pinhole camera is diffraction limited, high-resolution performance is obtained only by decreasing the pinhole size. This severely limits the light gathering capability—and increases the exposure time—of the camera. For these reasons, the applications of pinhole cameras are extremely limited.

REFERENCES

1. P. E. Boucher, *Fundamentals of Photography,* 2nd ed., pp. 5–7, D. Van Nostrand Co., Inc., New York (1947).
2. K. Sayanagi, "Pinhole imagery," *J. Opt. Soc. Am.* 57(9), 1091–1099 (1967).
3. R. E. Swing and D. P. Rooney, "General transfer function for the pinhole camera," *J. Opt. Soc. Am.* 58(5), 629–635 (1968).
4. H. V. Soule, "Versatile multispectral imaging systems," *Electro-Optical Systems Design,* 3(4), 27–33, Milton S. Kiver, Chicago (April 1971).

16   Diffraction and Interference with Partially Coherent Light

16.1   INTRODUCTION

The ideas of partially coherent light were introduced by discussing a two-beam interference experiment in which the two beams were isolated from the field by using two small apertures (Chapters 10 and 11). In all of these discussions, it was assumed that the light was coherent over either of the two apertures when considered independently. Hence, the envelope function for the interference fringes was characterized by the Fraunhofer diffraction pattern of the individual aperture. But, in fact, the light is not always completely coherent across the individual apertures if they are of finite size. One of the questions we wish to investigate in this chapter is the effect on the characteristic Fraunhofer diffraction pattern when the difffracting aperture is not completely coherent. This is an important consideration when using diffraction techniques for measurement purposes, such as in certain techniques for measuring small particles and fiber filaments.

As a further example in the understanding of partially coherent light, we will extend the two-beam interference analysis to multiple-beam interference.

16.2   DIFFRACTION WITH PARTIALLY COHERENT LIGHT

For this discussion, we limit ourselves to quasimonochromatic light and spatially stationary systems. Hence, if we consider the simplified system shown in Fig. 16.1 with a point source on the optical axis, the lens L_1 would produce a coherent collimated beam of light incident on the aperture, and the lens L_2 would produce the Fraunhofer diffraction pattern characteristic of the aperture geometry in the diffraction plane. The intensity distribution seen is, of course, the square of the Fourier transform of the uniformly illuminated aperture. In general, it is the square of the Fourier transform of the aperture distribution.

In the situation illustrated in Fig. 16.1, each point in the uniform incoherent source will give rise to a plane wave illumination of the aperture with the plane wave tipped with respect to the aperture plane. The angle depends on the distance

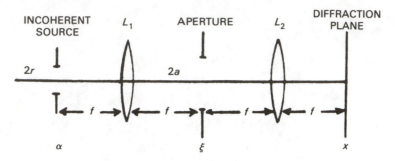

Fig. 16.1 Diffraction with a finite-sized source.

of the point from the optical axis (see Fig. 16.2). Hence, the optical disturbance incident on the aperture plane must be multiplied by the aperture; e.g.,

$$A(\boldsymbol{\xi}) = \text{rect}(\boldsymbol{\xi}|a) \; , \tag{16.1}$$

where $\text{rect}(\boldsymbol{\xi}|a)$ is the rectangular function of width $2a$ characterizing the aperture. We may write

$$\psi(\boldsymbol{\xi}) = \exp\left(\frac{-ik\boldsymbol{\alpha}\cdot\boldsymbol{\xi}}{f}\right)\text{rect}(\boldsymbol{\xi}|a) \; . \tag{16.2}$$

The Fourier transform of this distribution is then

$$\tilde{\psi}(\mathbf{x}) = \int_{-a}^{a} \exp\left(\frac{-ik\boldsymbol{\alpha}\cdot\boldsymbol{\xi}}{f}\right)\exp\left(\frac{-ik\mathbf{x}\cdot\boldsymbol{\xi}}{f}\right)d\boldsymbol{\xi} \tag{16.3}$$

$$= \int_{-a}^{a} \exp\left[\frac{-ik\boldsymbol{\xi}\cdot(\mathbf{x}+\boldsymbol{\alpha})}{f}\right]d\boldsymbol{\xi} = 2a\,\text{sinc}\,\frac{ka|\mathbf{x}+\boldsymbol{\alpha}|}{f} \; . \tag{16.4}$$

This integral is a sinc function centered at the point $x + \alpha$ and the resultant intensity is, of course, the modulus squared of Eq. (16.4), i.e.,

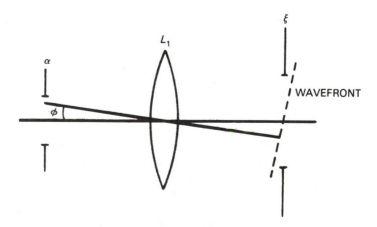

Fig. 16.2 Plane wave illumination from a typical source point.

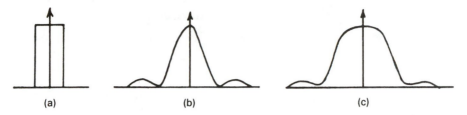

Fig. 16.3 The source intensity distribution (a) is convolved with the sinc^2 diffraction pattern (b) to produce (c).

$$I(\mathbf{x}) = 4a^2 \operatorname{sinc}^2 \left[\frac{ka|\mathbf{x} + \boldsymbol{\alpha}|}{f} \right] \tag{16.5}$$

for each point in the object. Since the wave equations are linear and may be treated as stationary for small angles, the resultant intensity for all such points in the incoherent source is

$$I_R(\mathbf{x}) = \int_{-r}^{r} 4a^2 \operatorname{sinc}^2 \left[\frac{ka|\mathbf{x} + \boldsymbol{\alpha}|}{f} \right] d\boldsymbol{\alpha} \; . \tag{16.6}$$

We may rewrite this equation using the rectangular function notation $\operatorname{rect}(\boldsymbol{\alpha}|r)$:

$$I_R(\mathbf{x}) = 4a^2 \int_{-\infty}^{\infty} \operatorname{rect}(\boldsymbol{\alpha}|r) \operatorname{sinc}^2 \left[\frac{ka|\mathbf{x} + \boldsymbol{\alpha}|}{f} \right] d\boldsymbol{\alpha} \; , \tag{16.7}$$

which is a convolution integral and represents the source function convolved with the sinc^2 function characteristic of the Fraunhofer intensity distribution of the aperture. Figure 16.3 shows this result diagramatically.

Equation (16.7) may be expressed again in terms of the Fourier transforms of the two functions in this equation using the convolution theorem. Thus,

$$I_R(\mathbf{x}) = \frac{4a^2}{\lambda f} \int \left[\int \operatorname{rect}(\boldsymbol{\alpha}|r) \exp \left(\frac{-ik\boldsymbol{\alpha} \cdot \boldsymbol{\xi}}{f} \right) d\boldsymbol{\alpha} \right]$$

$$\times \left[\int \operatorname{sinc}^2 \frac{ka\boldsymbol{\alpha}}{f} \exp \left(\frac{-ik\boldsymbol{\alpha} \cdot \boldsymbol{\xi}}{f} \right) d\boldsymbol{\alpha} \right] \exp \left(\frac{-ik\mathbf{x} \cdot \boldsymbol{\xi}}{f} \right) d\boldsymbol{\xi} \; . \tag{16.8}$$

Note that the integral in the first bracket is the Fourier transform of the uniform source, which by the van Cittert-Zernike theorem is the mutual intensity function of the illumination in the aperture plane. The other bracketed term is the autocorrelation function of the aperture amplitude, which in this example is a triangular function. With suitable normalization, we may write the first bracketed term as

$$\frac{1}{2r} \int \operatorname{rect}(\boldsymbol{\alpha}|r) \exp \left(\frac{-ik\boldsymbol{\alpha} \cdot \boldsymbol{\xi}}{f} \right) d\boldsymbol{\alpha} = \gamma_{12}(0) = \operatorname{sinc} \frac{kr\boldsymbol{\xi}}{f} \; . \tag{16.9}$$

Finally, we may write the normalized result:

$$I_R(\mathbf{x}) = \int \gamma_{12}(0)\, C(\boldsymbol{\xi}) \exp\left(\frac{-ik\mathbf{x}\cdot\boldsymbol{\xi}}{f}\right) d\boldsymbol{\xi} \;, \tag{16.10}$$

where $C(\boldsymbol{\xi})$ is the autocorrelation function of the aperture amplitude.

16.3   ONE-DIMENSIONAL APERTURES

Substituting the appropriate forms into Eq. (16.8) for the one-dimensional case, i.e., a slit incoherent source illuminating a slit aperture, we obtain the normalized result:

$$I_R(x) = \int \operatorname{sinc}\left(\frac{kr\xi}{f}\right)$$

$$\times \left\{ \int \left[\operatorname{sinc}^2\left(\frac{ka\alpha}{f}\right) \exp\left(\frac{-ik\alpha\xi}{f}\right) \right] d\alpha \right\} \exp\left(\frac{-ikx\xi}{f}\right) d\xi \;. \tag{16.11}$$

If we define $u = 2ka\alpha/f$ and $v = kar/f$, then

$$I_R(x) = \frac{1}{v}\left[Si(v+u) + Si(v-u) \right.$$

$$\left. - \frac{1-\cos(v-u)}{v-u} - \frac{1-\cos(v+u)}{v+u} \right] , \tag{16.12}$$

where Si is the sine integral defined by Eq. (14.5).

Figure 16.4 shows theoretical curves for the Fraunhofer intensity distribution for this example for various values of the parameter $c = kar/f$. The results show a gradual loss in contrast of the characteristic side lobe structure and a shift of the maxima and minima. A more detailed discussion of these results can be found in the literature.[1]

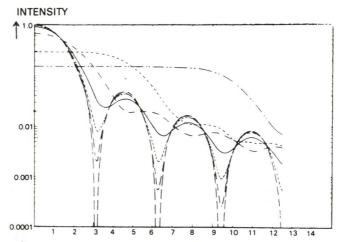

INTENSITY

Fig. 16.4 Theoretical Fraunhofer intensity distribution for a slit aperture illuminated by a slit source for various values of c: $c = 0 (- \cdot -)$, $c = 0.5 (- \cdot \cdot -)$, $c = 1.0 (----)$, $c = 2.0 (—)$, $c = 4.0$ $(- \cdot \cdot \cdot -)$, $c = 10.0 (- \cdot - \cdot -)$, and $c = 10.0 (- \cdot -- \cdot -)$ (after Shore et al.[1]).

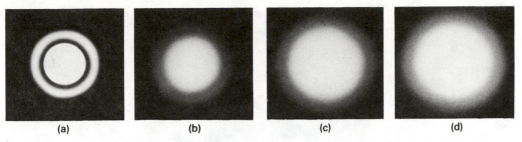

Fig. 16.5 Fraunhofer diffraction by a circular aperture illuminated by a circular incoherent source: (a) $c = 0$, (b) $c = 1.5$, (c) $c = 2.8$, and (d) $c = 4.6$ (after Shore et al.[1]).

16.4   TWO-DIMENSIONAL APERTURES

The two-dimensional case of a circular diffracting aperture illuminated by a circular incoherent source cannot be solved in closed form. However, Eqs. (16.7) and (16.8) still hold. This time the image of the circular source is convolved with the square of the Besinc function characteristic of the Fraunhofer diffraction pattern of the circular diffracting aperture. Figure 16.5 illustrates this result in a series of photographs.

Clearly, the exact form of the complex degree of coherence existing in the diffracting aperture is important for both the qualitative and quantitative discussion of the resulting diffraction pattern.

16.5   MULTIPLE-BEAM INTERFERENCE WITH PARTIALLY COHERENT LIGHT

We recall the two-beam inteference law from Chapter 11, Eq. (11.15), as

$$I_R = I_1 + I_2 + 2\sqrt{I_1 I_2}\,|\gamma_{12}(0)|\cos[\beta_{12}(0) + \delta_{12}]\ , \qquad (16.13)$$

where I_1 and I_2 are the intensities produced by each beam individually and $|\gamma_{12}(0)|$ is the modulus of the complex degree of coherence,

$$\left.\begin{array}{l} \beta_{12}(0) = \arg\gamma_{12}(0) \\[2ex] \Phi_{12} = \delta_{12} = \dfrac{2\pi}{\lambda}\,(r_2 - r_1)\ , \end{array}\right\} \qquad (16.14)$$

where r_1 and r_2 are the actual paths from the two source points to the observation point and λ is the mean wavelength.

For the more general case of a one-dimensional regular array of N apertures, we have the resultant intensity given by the double summation:

$$I_R = \sum_{n=1}^{N}\sum_{m=1}^{N}(I_n I_m)^{1/2}\,|\gamma_{nm}(0)|\cos[\beta_{nm}(0) - \delta_{nm}]\ , \qquad (16.15)$$

$$\left.\begin{array}{l} |\gamma_{nm}(0)| = 1 \text{ when } n = m \\[2ex] |\gamma_{nm}(0)| = |\gamma_{mn}(0)| \end{array}\right\}\ . \qquad (16.16)$$

In the arrangement of Fig. 16.1, we replace the aperture in the ξ plane by an

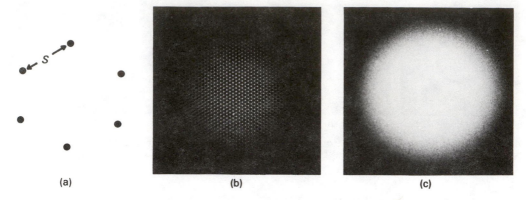

(a)   (b)   (c)

Fig. 16.6 Multiple-beam interference with partially coherent light. The array (a) is illuminated (b) coherently and (c) coherently over individual apertures but the coherence interval is much less than S (after Thompson[2]).

array of smaller apertures of width $2a$ and separation $2b$. It is possible to illuminate such an array from the one-dimensional source so that the illumination is coherent over a single aperture but incoherent with respect to other apertures in the array. This situation arises since the coherence function (in this case, a sinc function) has zero values that occur at equally spaced intervals. Hence, the intensities of the individual diffraction patterns are added, i.e.,

$$|\gamma_{nm}| = \begin{cases} 0 \text{ for } n \neq m \\ 1 \text{ for } n = m \end{cases}.$$

Therefore,

$$I_R = \sum_{n=1}^{N} I_n . \tag{16.17}$$

The analysis can be extended to a two-dimensional array illuminated by a circular incoherent source. Here, however, the Besinc coherence function does not have equally spaced zeros. Figure 16.6 shows an example. The array (a) is illuminated coherently to produce the photograph shown in (b). Figure 16.6(c) is produced by illuminating the individual apertures coherently but with a separation that is greater than the coherence interval—only the central maximum of the diffraction pattern is shown here.

16.6   ANALYSIS OF A PARTIALLY COHERENTLY ILLUMINATED ARRAY

Considering again the one-dimensional array problem, we recall the array theorem of Chapter 3. Let the array function be $A(\xi)$; hence, the optical disturbance in the aperture plane becomes, by Eq. (16.2),

$$\psi(\xi) = \exp\left(\frac{-ik\alpha\xi}{f}\right) \text{rect}(\xi|a)\, A(\xi) ,$$

where
$$A(\xi) = \sum_{n=0}^{N-1} \Delta(\xi - 2nb) ,$$

(16.18)

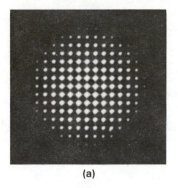
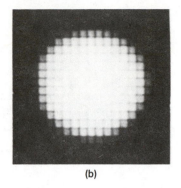

(a)   (b)

Fig. 16.7 Interference pattern of a two-dimensional array with (a) illumination provided by a circular incoherent source so that the coherence interval is much greater than the array spacing and (b) illumination provided by a square incoherent source so that the coherence interval is slightly less than the array spacing (after Thompson[2]).

for an array of N apertures of width $2a$ and separation $2b$. Hence, the resultant intensity in the diffraction plane for a single point in the source plane is given by [see Eq. (3.21)]

$$I(x) = 4a^2 \left[\frac{\sin^2 \dfrac{2\pi Nb(x + \alpha)}{\lambda f}}{\sin^2 \dfrac{2\pi b(x + \alpha)}{\lambda f}} \right] \mathrm{sinc}^2 \frac{2\pi a(x + \alpha)}{\lambda f} . \tag{16.19}$$

Therefore, the resultant intensity produced by all such points in the incoherent source is

$$I_R(x) = 4a^2 \int_{-\infty}^{\infty} \mathrm{rect}(\alpha|r) \left[\frac{\sin^2 \dfrac{2\pi Nb(x + \alpha)}{\lambda f}}{\sin^2 \dfrac{2\pi b(x + \alpha)}{\lambda f}} \right]$$

$$\times \mathrm{sinc}^2 \frac{2\pi(x + \alpha)}{\lambda f} \, d\alpha . \tag{16.20}$$

Again, this result shows that the resultant intensity is the characteristic intensity pattern associated with the diffraction pattern of the array convolved with the source function. For a very large array, the diffraction pattern consists of essentially a set of delta functions—or more precisely a set of diffraction patterns corresponding to the envelope of the diffracting array (see Fig. 3.5). As the illuminating source gets larger, the source function starts to predominate as a unit of the array in the diffraction pattern. This result is illustrated in the two examples of Fig. 16.7, the details of which are given in the caption. Notice how the square source in Fig. 16.7(b) is put down at each point on the diffraction pattern array. Further examples can be found in Ref. 2.

REFERENCES

1. R. A. Shore, B. J. Thompson, Jr., and R. E. Whitney, "Diffraction by apertures illuminated with partially coherent light," *J. Opt. Soc. Am.* 56(6), 733–738 (1966).
2. B. J. Thompson, Jr., "Multiple-beam interference with partially coherent light," *J. Opt. Soc. Am.* 56(9), 1157–1160 (1966).

17 Elementary Theory of Optical Coherence: Part II*

17.1 EXAMPLES OF SPATIAL COHERENCE EFFECTS IN OPTICAL INSTRUMENTS

Nonlinearities arising from spatial coherence effects in optical instruments occur when the coherence interval is comparable in size to the optical spread function of the system as measured in the object plane. This is the situation depicted schematically in Fig. 11.16(c). Since the spread function of the system determines the size of a resolution element in object space, we expect nonlinear effects from spatial coherence whenever the object is sufficiently small to diffract light outside of the collection angle of the instrument or when the coherence interval is large enough to enable two or more object resolution elements to interfere in image space, causing effects similar to those seen in Figs. 11.14 and 11.15. Therefore, to ensure that an instrument is operating in a linear mode, it is necessary to design or operate the instrument such that all of the diffracted light containing information about the object spatial frequencies of interest is collected. In general, these effects are of importance in the use of high-resolution analyzing instruments such as microdensitometers, microscopes, contact printers, enlargers, projection printers, and viewers. These instruments can be operated in a linear mode for low-resolution applications. Instruments such as cameras used with normal sunlight illumination, aerial cameras, the human eye, and conventional magnifiers generally operate in an incoherent linear mode.

17.1.1 Apparent Transfer Functions

High-resolution imaging systems that exhibit nonlinearities (object-dependent transfer functions) due to spatial coherence effects have been analyzed and compared in terms of the ratio of the spatial coherence interval to the diameter of the point spread function of the optical system, both measured in the object plane.[1-3] Equation (11.42) with $x_1 = x_2$ yields the image intensity. This image intensity is a function of the coherence of the radiation transilluminating the object, $\Gamma_{ob}(\xi_1 - \xi_2, 0)$, which is assumed to be spatially stationary; the transmittance function of the object, $\tau(\xi)$; and the amplitude impulse response of the lens as given by Eq. (11.36). Additional assumptions made in deriving Eq. (11.42) are

---

*Reprinted with permission from "Review of optical coherence effects in instrument design: part II," by G. O. Reynolds and J. B. DeVelis, *Opt. Eng.* 20(4), SR-124–SR-130 (July/August 1981). Part I is reprinted as Chapter 11 of this book and part III appears as Chapter 18.

that the radiation is quasimonochromatic and that the object is in the near field of the lens [Eq. (11.43)]. Equation (11.42), modified in this manner, has been evaluated to give the image intensity for the case of edge objects[2] and sine wave objects.[2,3] These solutions assumed one-dimensional, diffraction-limited optical systems and uniform one-dimensional incoherent sources.

The solution was obtained by taking the Fourier transform of the image intensity distribution to relate the image spectrum to the object spectrum through the system transfer function. When the system is linear, these functions are related by the simple product relationship

$$\tilde{I}_{im}(\mu) = \tilde{I}_{ob}(\mu)\tau(\mu) \, , \tag{17.1}$$

where

$\sim$ denotes a spatial Fourier transform

$\mu =$ spatial frequency variable

$\tilde{I}_{im}(\mu) =$ image spatial frequency spectrum

$\tilde{I}_{ob}(\mu) =$ object spatial frequency spectrum

$\tau(\mu) =$ transfer function of the optical system.

However, because of the nonlinearities of partially coherent radiation, the simple product relationship given by Eq. (17.1) is not obtained when such light is used. For instance, the procedure of differentiating an edge image to obtain the impulse response results in an asymmetric function. For sine wave objects, the nonlinearity introduces higher order harmonics. If the procedures used to measure transfer functions in linear systems are applied to systems using partially coherent light, the resulting function, which has been termed the apparent transfer function T, must be used with extreme caution.

For the case of sine wave objects, an *apparent transfer function* has been defined as the ratio of the output modulation of the fundamental to the input modulation and ignores the presence of the higher order harmonics in the output. An example of an apparent transfer function for the fundamental of an intensity sine wave object for various values of R (the ratio of the coherence interval to the diameter of the diffraction pattern of the imaging system in object space) is shown[3] in Fig. 17.1. This figure illustrates that the system is linear when $R \ll 1$ and becomes increasingly nonlinear as R becomes larger than unity. For very large values of R, a transfer function whose cutoff is determined directly by the lens aperture rather than by its autoconvolution is obtained.

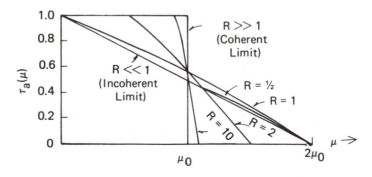

Fig. 17.1 Apparent transfer function at an input modulation of 0.95 for the fundamental of an intensity sine wave object (from Ref. 3).

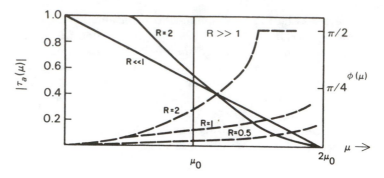

Fig. 17.2 Apparent transfer function $\tau(\mu)$ for an edge object; the modulus $|\tau(\mu)|$ is shown in solid lines and the phase $\phi(\mu)$ in dashed lines (from Ref. 2).

For the case of edge objects, the apparent transfer function was defined as the Fourier transform of the impulse response of the edge. This yields the results shown[2] in Fig. 17.2.

Again, for the values of $R \ll 1$, the system behaves incoherently. For values of R greater than unity, the system becomes increasingly more nonlinear, approaching the coherent limit for $R \gg 1$. The difference in the modulation values between Figs. 17.1 and 17.2, as well as the presence of phase shifts in Fig. 17.2, illustrates the strong object-dependence of these nonlinear systems. In fact, objects of the same kind having various contrasts also yield different apparent transfer functions.[2,3] However, in the limit of low-contrast objects with no phase, nonlinearities due to partial coherence of the radiation become minimal and the systems become linear.[4]

The results of Figs. 17.1 and 17.2 can be interpreted in terms of the schematic model of Fig. 11.16; that is to say, when the size of the coherence interval is much smaller than the size of the diffraction spot of the imaging lens in object space, the system may be considered to be an incoherent, linear imaging system. When the spatial coherence interval is comparable to, or greater than, the size of the diffraction spot of the imaging system in object space, then the effects of system nonlinearities are present for high spatial frequencies.

To relate these results to instrument design, the following procedure is suggested:

1. Select the maximum object frequency of interest for observation with a linear system.

2. Choose imaging optics capable of resolving this maximum frequency of interest, which necessarily collects all of the diffracted light associated with this maximum spatial frequency.

3. Design an illumination system having a coherence interval in the object plane (smallest distance d in Fig. 11.5 for which $|\gamma_{12}(0)| = 0.88$)[5] less than the object resolution element (size of the imaging system impulse response in object space). This prevents the occurrence of interference effects between adjacent resolution elements in object space.

The apparent transfer functions defined in this section should not be confused with the transfer function as normally defined for either incoherent or coherent systems. In those cases, the techniques of linear system analysis may be employed and, therefore, knowledge of the transfer function is useful for calculating images. The apparent transfer function on the other hand is of little use in calculating an image from a knowledge of the object that created it. It is simply the result that can be expected when making a specific measurement for a specific object and image. This distinction is of considerable concern since calculation of the system transfer function from edge trace analysis is a widely used procedure. It is always possible to make a microdensitometer trace of an edge image and process that trace to produce an apparent transfer function. However, as we have just seen, that apparent transfer function cannot be used to calculate the image of other objects, for instance sine waves. Examples of these nonlinear effects due to spatial coherence in various high-resolution analyzing instruments are discussed in the following sections.

17.1.2   The Microdensitometer

One of the first instruments analyzed for the effects of system nonlinearities due to partial coherence was the classical microdensitometer.[6-8] This instrument was designed to scan photographic film and plot the film microdensity versus distance along the scan direction. The term "microdensity" usually refers to the density measured for an area larger than the grain size and smaller than 1 mm^2. A schematic of the optical system for a typical microdensitometer showing the influx and efflux optics relative to the source, sample, and detector planes is shown in Fig. 17.3.

When a microdensitometer is used to scan high-resolution imagery, and the coherence of the illumination striking the sample (photographic film) becomes comparable in size to an object pixel [Fig. 11.16(c)], system nonlinearities will occur in the output. These nonlinearities manifest themselves as slight changes in edge position or density values that are difficult to quantify. Moreover, photographic film has a phase image (due to emulsion swelling) proportional to the density image, which can introduce additional nonlinearities into the system.

System nonlinearities arising from the use of partially coherent light are illustrated by the appearance of a density output when scanning phase objects with a classical microdensitometer, because systems linear in intensity should produce a constant response to phase objects. This example was one of the first demonstrations used to illustrate that a coherence problem existed in micro-densitometers.[7] An illustration of this nonlinear effect due to scanning a phase

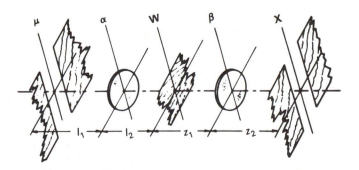

Fig. 17.3 Typical (simplified) microdensitometer optical system showing the location of the coordinate systems and the object and image distances for the optical elements: μ = source aperture, α = influx optics, **w** = sample plane, β = efflux optics, and **x** = image plane and sensor aperture (from Ref. 9).

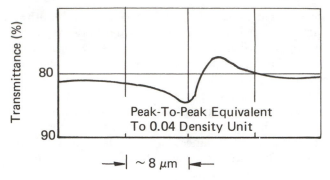

Fig. 17.4 Typical response of an NBS phase edge measured on a Mann microdensitometer with a source aperture size of 41 ×41 μm, a 0.4 numerical aperture (NA), 40× influx objective, and a 0.25 NA, 20× efflux objective (from Ref. 10).

edge with a classical microdensitometer is shown in Fig. 17.4. Rather than obtaining the expected linear system response (i.e., a constant), density differences of $\Delta D = 0.04$ were observed for coherence intervals on the order of 3 μm in the object plane. Further measurements revealed that classical microdensitometers exhibited coherence intervals of a few micrometers in diameter in the object plane.[7]

Typical coherence measurements made by using a shearing interferometer on the Joyce-Loebl and the Mann-Data instruments are shown in Fig. 17.5. Additional measurements of the coherence interval for the Kodak Model III and the PDS instruments showed comparable results.[12] Therefore, from Fig. 11.16(c), one would expect that system nonlinearities would result when the optical system was analyzing samples having spatial frequencies in excess of 100 to 150 lp/mm.

A thorough analysis, extending earlier efforts,[9,13–15] directed at determining the linear operational modes for classical instruments has been performed.[8] The

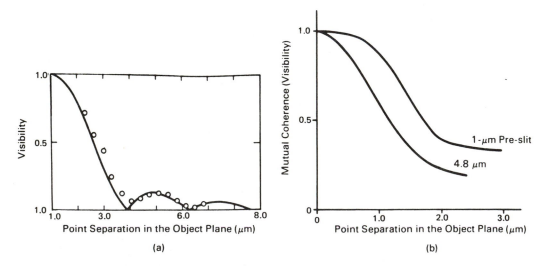

Fig. 17.5 Typical coherence measurements on commercial microdensitometers: (a) measurements of normalized visibility versus two-point separation in the Joyce-Loebl microdensitometer for a condenser objective of nominal numerical aperture: NA = 0.08; and scanning objective: NA = 0.45 (from Ref. 11); (b) coherence measurements on a Mann microdensitometer with a 0.25 NA for the efflux optics with 1-μm pre-slit and a 4.8-μm source aperture (from Ref. 7).

results of this analysis indicate that microdensitometers can be built to operate in eight possible modes. Classical instruments utilizing incoherent sources make up four of these modes—two image scanning modes and two sample scanning modes. The other four modes use coherent illumination; one was utilized to design a linear microdensitometer.[16]

The image scanning modes can be interpreted in terms of partially coherent imaging systems, such as those depicted in Fig. 11.16. In such systems, it is essential that the efflux optics collect all the light transmitted through and diffracted by the sample in order to have a linear system (i.e., $R \ll 1$). For underfilled operation [i.e., the numerical aperture (NA) of the influx optics is less than the NA of the efflux optics], the influx optics determines the system performance; for overfilled operation (i.e, the NA of the influx optics is greater than the NA of the efflux optics), the efflux optics determines system performance. In both cases, the coherence interval at the sample plane is related to the maximum spatial frequency that can be passed linearly by the system. This relationship can be seen in the following physical argument.

If we assume that the object is a sine wave intensity distribution, then in order for partially coherent diffraction to occur, a reasonable number of sine wave periods must be contained within the dimension of the spatial coherence interval, i.e.,

$$CI = NP , \tag{17.2}$$

where

$CI =$ the coherence interval at the object (sample) plane

$P =$ the period of the sine wave

$N =$ the number of cycles of the sine wave contained within the coherence interval.

Since P is reciprocally related to spatial frequency, as shown above, the coherence interval determines the maximum linear spatial frequency through the choice of N. Published data on partially coherent diffraction effects indicate that values of N between 4 and 10 are reasonable choices.[17] The analysis in the literature, which is conservative, assumes $N = 1$, accounting for the high values predicted for the limit of linearity.[9]

Calculation of the limit of linearity for the Joyce-Loebl microdensitometer operating in an underfilled image scanning mode, using the formula derived in Ref. 8, was performed. The mode used in the measurements[7] of Fig. 17.5 (NA of condenser $= 0.08$, NA of objective $= 0.45$) was used in conjunction with the above criteria ($4 < N < 10$) to yield a limit of linearity in the range of 80 to 200 lp/mm in agreement with experimental observation. Further measurements were made to compare these various modes of microdensitometer operation.[8] However, accurate linearity tests, including target calibration, which could experimentally determine the actual limit of linearity, have yet to be developed.[8]

In sample scanning modes of operation, the efflux optics acts simply as a light collector and not as an imaging system. In these situations, the physical argument schematically shown in Fig. 11.16 does not apply. Thus, the limit of linearity is determined solely by the light-collecting capability of the efflux optics, i.e., the system is linear up to that spatial frequency which is diffracted beyond the collection angle of the efflux optics.

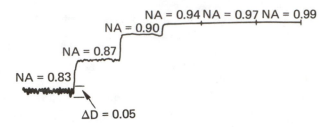

Fig. 17.6 Output responses to a 300 lp/mm replica phase grating for various values of efflux NAs in an LMD. The system demonstrates constant response at this frequency for efflux NAs greater than 0.94 (from Ref. 21).

In general, microdensitometers operate best in sample scanning modes as opposed to image scanning modes. In addition, operation with matched NAs of influx and efflux optics should be avoided in high-resolution applications, and small sampling apertures are preferable to reduce the effects of flare light.[8]

Since total collection of the light requires an efflux NA of unity, the highest limit of linearity will be achieved by sample scanning in an underfilled mode of operation. This typifies the linear microdensitometer (LMD) whose system performance is characterized by the influx optics. Such an instrument, based on these principles, useful for analyzing high-resolution data in a linear operational mode, was conceived[7,18-20] and analyzed.[16,21,22] This instrument focuses the input light from a laser source onto the sample and collects all the light with a nonimaging collector. This is a sample scanning instrument operating in the underfilled mode with coherent illumination.

The linear response of this instrument is demonstrated in Fig. 17.6 where the numerical aperture of the collector is varied from 0.83 to 0.99, while scanning a 300 lp/mm phase grating.[21] The result shows that a constant output is obtained for collector NAs greater than 0.94. The frequency response up to 500 lp/mm was measured by using a Diffraction Limited, Inc., 15-bar target[21] and is shown in Fig. 17.7. Other performance data measured with this instrument are given in Table 17.I. Its measured limit of linearity occurs at spatial frequencies greater than 500 lp/mm,[17,21] which is in agreement with theoretical predictions for a microdensitometer operating in the scanning mode with coherent illumination.[9] With liquid gates between the sample and the platen, a linear resolution limit on the order of 900 lp/mm has been predicted theoretically.[21]

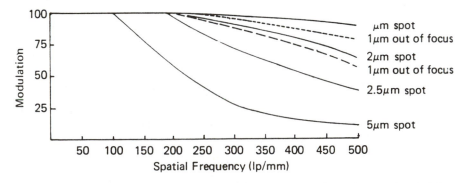

Fig. 17.7 Frequency response of an LMD measured by scanning a Diffraction Limited, Inc., 15-bar test target with various spot sizes (from Ref. 21).

TABLE 17.I

LMD Performance/Specifications*

Specification	Initial Requirement	Actual Performance
Square wave response:		
100%	200 lp/mm	456 lp/mm
Cutoff	>400 lp/mm	>912 lp/mm
Linearity limit:	200 lp/mm	>550 lp/mm
Radiometric repeatability:		
0–2 D	±0.01 D	±0.005 D
2–3 D	±0.02 D	±0.075 D
3–4 D	±0.02 D	±0.01 D
Mensuration accuracy:		
Static	±0.2 μm	±0.05 μm
0.4 mm/s	± ——	±0.2 μm
4 mm/s	±0.5 μm	±0.5 μm
Scan specifications	200 to 4000 μm/s	400 to 20,000 μm/s
Sample rate	8 kHz	8 kHz

*From Ref. 22. Here D denotes optical density.

17.1.3   The Microscope

Nonlinear effects due to spatial coherence of the illumination in a microscope can also be understood in terms of the ratio R, which can be described in terms of the parameters of Fig. 11.16. Since microscopes are generally classified by their NAs (which are inversely related to their f/number), a form of the ratio R useful in analyzing microscopes is

$$R = \left(\frac{\text{coherence interval}}{\text{spot size of image system}} \right)_{\text{object}}$$

$$= \frac{(\text{NA of objective})_{\text{efflux}}}{(\text{NA of condenser})_{\text{influx}}} . \tag{17.3}$$

It should be noted that some authors prefer to use the inverse of R when describing coherence.[23–27]

In most microscope systems in use, the illumination is derived from a primary incoherent source. Generally, the light is directed onto the object by a condenser. Some microscopes employ a light table rather than using condenser illumination, but a light table can be regarded as having an NA of unity, and so can be fitted into the general analysis.

Values of $R \ll 1$ correspond to incoherent illumination, and values of $R \gg 1$ correspond to coherent illumination. Generally, the value of R can be determined fairly easily for an imaging system. Some care must be exercised, however, as objectives and condensers are often used at NAs different from the nominal values indicated on them. The NA for a microscope objective is equal to the value stamped on it only when it is used at the proper working distance. As stated previously, the effects of coherence on imaging depend on the value of R and the spatial frequency of the object as defined by Eq. (17.2).

An experiment has been performed to demonstrate the relationship between these various parameters. The linearity is achieved by mismatching the NAs of

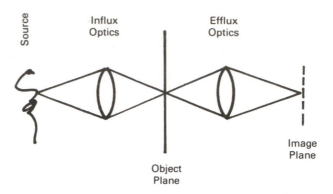

Fig. 17.8 Schematic of microscope linearity experiment.

the influx and efflux optics such that the condition $R < 1$ is realized. The optical system of Fig. 17.8 consisted of an NA/1.0 condenser system (influx optics) that imaged the source onto the target. A thermoplastic replica of a phase image-diffraction grating having a fundamental frequency of 200 lp/mm was used as a target. This target had very weak higher harmonics when observed in an optical diffractometer. The efflux optics was a $40\times$, NA/0.65 microscopic objective that imaged the target onto the detector in the image plane. A stop was included in the condenser assembly so that various values of R could be realized. The images obtained at the image plane for various values of R are shown in Figs. 17.9, 17.10, and 17.11. In Fig. 17.9, the condenser was set to NA/0.83, but the modulation of the phase image is not visible. This is the expected result for a linear system. The black smear on the left side is an ink line on the target put there to facilitate visual focusing. The magnification on the photomicrographs is approximately $500\times$. Figure 17.10 shows the image when the condenser NA is matched to the objective NA, i.e., NA/0.65. In this case, the phase image appears visible, but has very low

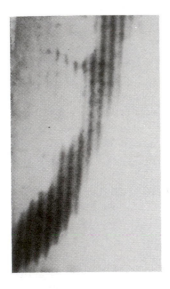

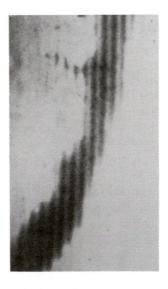

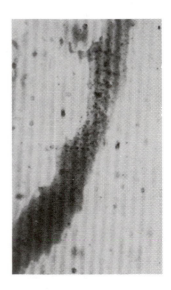

Fig. 17.9 Image of a 200 lp/mm phase grating obtained through the optical system of Fig. 17.8 for overfill condition: $(NA)_{influx}$ = 0.83; $(NA)_{efflux}$ = 0.65. (Courtesy of P. F. Mueller.)

Fig. 17.10 Image of a 200 lp/mm phase grating obtained through the optical system of Fig. 17.8 for matched condition: $(NA)_{influx}$ = 0.65; $(NA)_{efflux}$ = 0.65. (Courtesy of P. F. Mueller.)

Fig. 17.11 Image of a 200 lp/mm phase grating obtained through the optical system of Fig. 17.8 for underfill condition: $(NA)_{influx}$ = 0.085; $(NA)_{efflux}$ = 0.65. (Courtesy of P. F. Mueller.)

contrast. This result is in agreement with that predicted in the literature.[2,3] Figure 17.11 represents very coherent illumination and the phase grating is easily visible. This experiment pictorially illustrates the three regions of linearity described in Fig. 11.16 for a microscope system.

A similar optical system used to image low-frequency objects onto a scanning detector to digitize the object information in real time has been discussed in the literature.[28] To preserve photometric linearity with such a system, it is important to operate the instrument with optics having $R \ll 1$.

For imaging small objects (1 to 10 μm) in a microscope, the partial coherence of the illumination creates undesirable nonlinear edge shifts in the image. This results in mensurational errors that are serious in applications such as in integrated circuit (IC) manufacture.[5,24,29] By restricting the problem to binary edge objects, which characterize objects in the manufacture of ICs, two modes of operation were investigated[29]: $R \leqslant 0.5$ and $R \geqslant 1.5$. It was shown that $R \leqslant 0.5$ corresponds to incoherent linear operation with the corresponding edge profile, as shown in Fig. 17.12(a) whereas $R \geqslant 1.5$ corresponds to the coherent mode of operation defined by edges shifting, as illustrated in Fig. 17.12(b).

Since measurement of narrow IC lines requires NAs of 0.65 or greater, the incoherent condition cannot be met with dry condenser systems (maximum NA equal to 1.0). For this reason, operation in the coherent mode was chosen, and accurate mensurational results were obtained.[29] The value of the edge transmission at $x = 0$ in the coherent case is described by

$$T_c = 0.25(I_m + I_0 + 2\sqrt{I_m I_0} \cos \phi) , \qquad (17.4a)$$

where

$I_m =$ the maximum transmittance threshold

$I_0 =$ the minimum transmittance threshold

$\phi =$ the phase difference between light rays passing through the clear and semiopaque areas of the object.

In the incoherent case, the value of the edge transmission at $x = 0$ is given by

$$T_c = 0.5(I_m + I_0) . \qquad (17.4b)$$

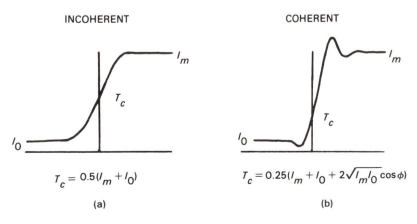

Fig. 17.12 Edge profile and threshold formulas for optical edge detection in (a) incoherent and (b) coherent illumination (from Ref. 29).

Mensurational data are obtained by using the coherent edge location and measurements of I_m, I_0, and ϕ to characterize the coherence effects. In essence, the region of partial coherence that results in a variable edge location[17] has been avoided and measurements are made relative to the coherent edge shift position.[30] With this approach, line widths of 0.5 μm with an estimated uncertainty of 0.05 μm have been measured.[24,29]

17.1.4   The Contact Printer

In contact printing, it is normally assumed that the original transparency and the duplicating material are placed in contact with each other and transilluminated by the source radiation. In many practical situations, a small air gap exists between the two media and, in addition, multiple sources are used to create uniform illumination and high-flux density, as shown schematically in Fig. 17.13.

System losses due to the partial coherence of the illumination system exist when one attempts to copy high-resolution information. In essence, when the object size is comparable to the spatial coherence interval in the plane of the original transparency [the situation depicted in Fig. 11.16(c)], then the partial coherence causes a resolution loss due to the diffraction spreading when finite propagation distances are involved. When multiple sources are used, the coherence effects associated with each individual source combine to create additional losses. These losses are illustrated by the examples in Figs. 17.14 through 17.17.

In these experiments, a collimator was set up as the exposure source using a 5-in. focal length lens and a 100-μm pinhole. The light source was a 100-W mercury arc filtered to the 0.546-μm spectral line and focused onto the pinhole. First a moderate contrast (8:1) three-bar target was contact-printed onto AHU Microfile 5400 microfilm. This microfilm, which has a nominal high-contrast resolution of 650 lp/mm, was reversal-processed to a gamma of -2, resulting in the image shown in Fig. 17.14. The bars are well resolved to 180 lp/mm (group 7, element 4) and barely resolved at the limit of the target at 228 lp/mm (group 7, element 6).

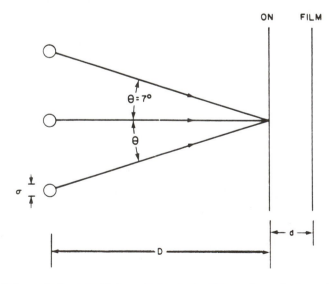

Fig. 17.13 Schematic for a multiple-source contact printer where distance d is the optical distance between the original negative (ON) and the recording film.

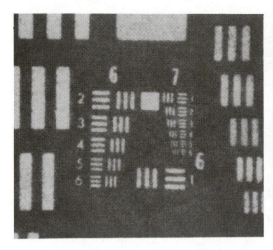

Fig. 17.14 Photomicrograph (280×) of AHU contact print with collimated light and without cover coat. (Courtesy of P. F. Mueller.)

Fig. 17.15 Photomicrograph (280×) of AHU contact print with collimated light and 3-μm cover coat illustrating no loss in resolution to 228 lp/mm compared to Fig. 17.14. (Courtesy of P. F. Mueller.)

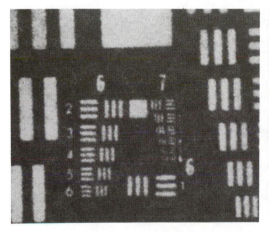

Fig. 17.16 Photomicrograph (280×) of AHU contact print with +20° and −20° divergence and without cover coat, illustrating a 33% resolution loss compared with Fig. 17.14. (Courtesy of P. F. Mueller.)

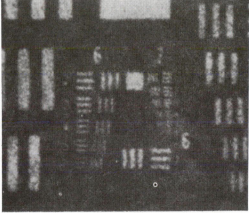

Fig. 17.17 Photomicrograph (280×) of AHU contact print with +20° and −20° divergence and 3-μm cover coat, illustrating a combined resolution loss of 50% compared to Fig. 17.14. (Courtesy of P. F. Mueller.)

The original bar target was overcoated with 3- to 4-μm-thick photoresist. The same contact print exposure was then repeated with collimated light. The results are shown in Fig. 17.15 and indicate no loss in resolution due to the overcoat at frequencies up to 228 lp/mm.

The target was contact-printed again, but this time instead of keeping the contact printing frame normal to the collimator axis, the collimating beam was rocked ±20° during exposure to simulate divergence in the illumination system. The results of this contact print are shown in Fig. 17.16. The bars are well resolved to 128 lp/mm (group 7, element 1) and resolved to 171 lp/mm (group 7, element 3). This illustrates a 33% resolution loss due to the source geometry. What is clear from these results is that divergence in the contact printer illumina-

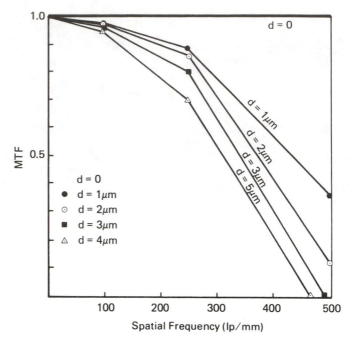

Fig. 17.18 MTF of contact printing as a function of the spatial frequency for various values of the air gap *d*. The source angle used in the printer was +7.5° (from Ref. 31).

tion system caused undercutting of the exposure and subsequent loss in resolution. Figure 17.17 shows the result when the overcoated target was contact-printed with the simulated divergent (±20°) illumination system. The last well-resolved frequency is 90 lp/mm (group 6, element 4), while the last barely resolved frequency is 128 lp/mm (group 7, element 1).

The combined loss due to divergent illumination and propagation due to the presence of the gap is approximately 50%, illustrating the serious effect of undercutting. At higher spatial frequencies, the diffraction losses due to coherence become of the same order of magnitude as the geometrical losses.

These effects in contact printers have been quantified in the literature.[31,32] The results characterize the contact printing process in terms of a transfer function as shown in Fig. 17.18. These results illustrate the dramatic loss in print quality at high spatial frequencies when an air gap exists between the original target and the printing film. The predominant losses are due to diffraction by propagation and undercutting due to source geometry. Experimental results indicate that these losses can be avoided by compensating for diffraction effects on a layer-by-layer basis.[33]

REFERENCES

1. B. J. Thompson, *Image Formation with Partially Coherent Light, Progress in Optics*, E. Wolf, ed., Vol. 7, pp. 171–230, North Holland Publishing Co., Amsterdam (1969).

2. R. J. Becherer and G. B. Parrent, Jr., "Nonlinearity in optical imaging systems," *J. Opt. Soc. Am.* 57(12), 1479–1486 (1967).

3. R. E. Swing and J. R. Clay, "Ambiguity of the transfer function with partially coherent illumination," *J. Opt. Soc. Am.* 57(10), 1180–1189 (1967).

4. R. E. Kinzly, "Role of object contrast in partially coherent image formation," *J. Opt. Soc. Am.* 56(4), 526–528 (1966).

5. D. Nyyssonen, "Linewidth measurement with an optical microscope: the effect of operating conditions on the image profile," *Appl. Opt.* 16(8), 2223-2230 (1977).

6. D. Galburt, R. A. Jones, J. W. Bossung, "Critical design factors affecting the performance of a microdensitometer," *Photog. Sci. Eng.* 13(4), 205–209 (1969).

7. G. O. Reynolds and A. E. Smith, "Experimental demonstration of coherence effects and linearity in microdensitometry," *Appl. Opt.* 12(6), 1259–1270 (1973).

8. R. E. Swing, "Microdensitometer optical performance: scalar theory and experiment," *Opt. Eng.* 15(6), 559–577 (1976).

9. R. E. Swing, "The optics of microdensitometry," *Opt. Eng.* 12(6), 185–198 (1973).

10. R. E. Kinzly, "Experimental evaluation of efflux optics for linear microdensitometry," *Opt. Eng.* 12(6), 218–225 (1973).

11. B. Justh, "Measurement of the spatial coherence of microdensitometer illumination," *J. Opt. Soc. Am.* 58(5), 714 (1968).

12. D. Nyyssonen, "Partial coherence in imaging systems," *Opt. Eng.* 13(4), 362–367 (1974).

13. R. E. Swing, "Conditions for microdensitometer linearity," *J. Opt. Soc. Am.* 62(2), 199–207 (1972).

14. R. E. Kinzly, "Partially coherent imaging in a microdensitometer," *J. Opt. Soc. Am.* 62(3), 386–394 (1972).

15. R. E. Swing, "The sampling aperture for linear microdensitometry," *Opt. Eng.* 13(5), 460–470 (1974).

16. J. P. Fallon, "Design considerations for a linear microdensitometer," *Opt. Eng.* 12(6), 206–212 (1973).

17. R. A. Shore, B. J. Thompson, R. E. Whitney, "Diffraction by apertures illuminated with partially coherent light," *J. Opt. Soc. Am.* 56(6), 733–738 (1966).

18. D. N. Grimes, "Linear microdensitometry," *J. Opt. Soc. Am.* 61(9), 1263–1264 (1971).

19. I. Weingartner, W. Miraude, E. Menzel, "Linearitat im Mikrodensitometer," *Optik* 34(1), 53–60 (1971).

20. I. Weingartner, "Laser-Mikrodensitometer," *Optik* 32(5), 508–511 (1971).

21. D. J. Cronin and G. O. Reynolds, "Optical design considerations and test results for a linear microdensitometer," *Opt. Eng.* 12(6), 201–205 (1973).

22. J. P. Fallon, R. J. Larsen, and G. O. Reynolds, "Use of the linear microdensitometer in astronomy," paper presented at the Topical Meeting on Imaging in Astronomy, Cambridge, Mass., June 1975.

23. M. Born and E. Wolf, *Principles of Optics,* Pergamon Press, New York (1964).

24. D. Nyyssonen, "Optical linewidth measurements on silicon and iron-oxide photomasks," in *Developments in Semiconductor Microlithography II,* J. W. Giffin and B. Ruff, eds., Proc. SPIE 100, 127–134 (1977).

25. D. A. Markle, "New projection printer," *Solid-State Tech.* 17(6), 50–53 (1974).

26. J. W. Bossung and E. S. Muraski, "Optical advances in projection photolithography," in *Developments in Semiconductor Microlithography III,* R. L. Ruddell et al., eds., Proc. SPIE 135, 16–23 (1978).

27. A. Offner and J. Meiron, "The performance of optical systems with quasi-monochromatic partially coherent illumination," *Appl. Opt.* 8(1), 183–188 (1969); see also "The influence of partial coherence on the microprojection of images," paper presented at OSA Topical Meeting on Uses of Optics in Microelectronics, Las Vegas, Nevada, January 25–26, 1971.

28. P. S. Considine and T. A. Lianza, "Image digitizer system design considerations," in *Contemporary Optical Systems and Components Specifications,* R. E. Fischer, ed., Proc. SPIE 181, 33–41 (1979).

29. D. Nyyssonen, "Optical linewidth measurements on wafers," in *Developments in Semiconductor Microlithography III,* R. L. Ruddell et al., eds., Proc. SPIE 135, 115–119 (1978).

30. T. J. Skinner, "Energy considerations propagation in random medium and imaging in scalar coherence theory," PhD Thesis, Boston Univ. (1965).

31. F. G. Kaspar, "Computation of light transmitted by a thick grating, for application to contact printing," *J. Opt. Soc. Am.* 64(12), 1623–1630 (1974).

32. C. B. Burckhardt, "Diffraction of a plane wave at a sinusoidally stratified dielectric grating," *J. Opt. Soc. Am.* 56(11), 1502–1508 (1966).

33. H. I. Smith, N. Efremow, and P. L. Kelley, "Photolithographic contact printing of 4000 Å linewidth patterns," *J. Electro Chem. Soc., Solid State Sci. Tech.* 121(11), 1503–1506 (1974).

18 Elementary Theory of Optical Coherence: Part III*

18.1 AN EMPIRICAL APPROACH FOR USE IN OPTICAL INSTRUMENT DESIGN

18.1.1 Coherence Nomograph

We have discussed partially coherent imaging and the effects of partial coherence on the operation of various instruments. Because of the basic nonlinearity in partially coherent imaging systems, it is only in very restrictive cases that it is possible to state with mathematical precision that partial coherence creates no problem in a particular imaging system. Through analysis of the experimental results, one can construct a nomograph describing the conditions of partial coherence under which the imaging system is linear.

The effects of partial coherence on an imaging system depend on both the value of R, the ratio of the numerical aperture of the objective to the numerical aperture of the condenser, and on the extent to which the object contains frequencies beyond the cutoff of the imaging objective. The cutoff of the imaging objective depends on its numerical aperture (NA). Coherence effects thus depend on R, the spatial frequency content of the object, and the NA of the imaging objective. These variables have been combined in an empirically determined nomograph (shown in Fig. 18.1) to give the conditions under which coherence effects are essentially absent in an imaging system. In this figure, L denotes the limiting spatial frequency in the target. Two sets of values of L are shown in Fig. 18.1. One set corresponds to the choice of $N = 4$ in Eq. (17.2), and the other set corresponds to the more conservative choice of $N = 10$.

To use this nomograph, the values of R and the NA of the imaging objective are first determined. A straight line is drawn to connect these values in the R and NA columns, and this line is extended to intersect the L column. The value of L at this intersection point is then the maximum spatial frequency the object can

---

*Reprinted with permission from "Review of optical coherence effects in instrument design: part III," by G. O. Reynolds and J. B. DeVelis, *Opt. Eng.* 20(5), SR-166–SR-172 (September/October 1981). Part I is reprinted as Chapter 11 of this book and part II appears as Chapter 17.

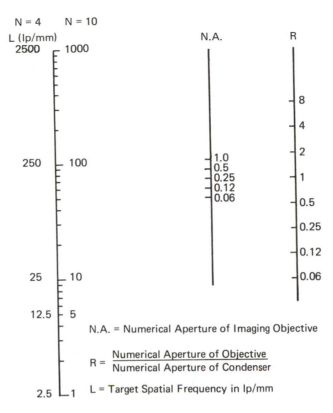

Fig. 18.1 Nomograph for finding conditions under which coherence effects are present in an imaging system when $\lambda = 1/2 \ \mu$m; $N = 4$ and $N = 10$ correspond to the resolution criterion determined from Eq. (17.1). (Courtesy of B. Justh for $N = 1$.)

contain for coherence effects to be absent, i.e., for the system to be linear in intensity.

If we wish to image an object with a known maximum frequency so that coherence effects are absent, the nomograph can be used in reverse. After selection of a value for N, a line is drawn between this maximum frequency in the L column and the value of the NA of the imaging objective in the NA column. The intersection of this line with the R column then gives the maximum permissible value of R that will assure the absence of coherence effects.

One additional point concerning this nomograph should be noted. Experiments have shown that $R = 8$ essentially corresponds to the coherent limit in an imaging system. (Similar effects are illustrated in Fig. 18.3 for a projection printer.) Thus, in applying this nomograph to imaging systems for which R is greater than 8, such as an imaging system using fully coherent laser illumination, the $R = 8$ point on the nomograph should be used. However, detailed investigations of binary targets in a microscope have shown that $R = 1.5$ is sufficient to define the coherent edge crossing for mensurational purposes.[1]

In the next sections, we discuss the consideration of coherence effects in the design of other optical instruments using the nomograph shown in Fig. 18.1.

18.1.2   Microcamera

A microcamera is a device that operates at high minifications (100 to 400$\times$) to image a target of known characteristics onto a photosensitive material. These devices are used in various photographic research and development applications.

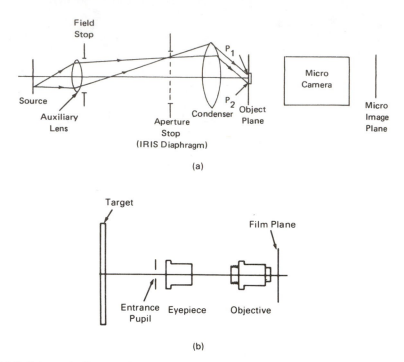

Fig. 18.2 Schematic diagram of a microcamera system: (a) Köhler's method of illumination for microcamera influx optics and (b) simplified configuration of efflux optics of a microcamera.

The key optical element in a microcamera is usually a high NA microscope objective having a high aberration correction feature and performing close to the diffraction limit. These large NAs are associated with small depths of focus on the imaging side which require that the film be held very flat during the exposure period.

In designing a microcamera, it is important that the imaging system be devoid of coherence effects so that a linear transformation exists between the object and image planes. For discussion purposes, we consider the microimage camera with a Köhler's illumination system, shown schematically in Fig. 18.2. The condenser in this system essentially images the field stop onto the object plane. A Köhler's illumination system may be operated in either the coherent or incoherent mode,[2] depending on the ratio R defined in Eq. (17.3).

To ensure linear transfer of information through the microcamera, we must have $R \leqslant 1/2$, as was shown in Fig. 17.1. Application of this R criterion and use of the nomograph in Fig. 18.1 (with the $N = 10$ criterion) shows that coherence effects can be avoided in the design of microcamera systems in the 100 to $400\times$ range for microimage frequencies of less than 3000 lp/mm. If a low-minification microcamera is designed, then coherence effects can be important if high-resolution microimages are desired.

As an example illustrating the incoherent mode of a microcamera, consider the following system. A $10\times$ microscope objective having an NA of 0.32 is used in tandem with a $10\times$, f/30, demagnifying lens to form a $100\times$ microcamera imaging system. The NA of the efflux optics is determined from the input side of the objective. For the $10\times$, 0.32 objective described above, the input NA is 0.032. To record 3000 lp/mm with this design linearly, the nomograph in Fig. 18.1 shows that an R ratio of less than 0.12 is necessary. A more realistic resolution goal of 1000 lp/mm with this system would require an R ratio of <0.25. Thus, we see how the R ratio must be considered in designing a linear microimaging system.

18.1.3   Projection Printers

Projection printers can be designed to either magnify or minify imagery. The magnification case is similar to the microscope problem discussed earlier. When used to minify by $10\times$ or less, coherence effects can be important when working with high-resolution targets. The most serious case arises when attempting to design a 1:1 projection printer, because smaller targets are needed to realize a given image resolution than in a 2 or $4\times$ projection printer. For purposes of accurate reproduction, it is necessary to operate such printers in a linear imaging mode.

A 1:1 projection imaging system has been described in the literature.[3] This optical system has an f/number of 1.5, which corresponds to an objective NA of 0.33. The region of linear operation with this system can be determined from the nomograph of Fig. 18.1, as shown for several cases in Table 18.I. The condenser NAs in the example of Table 18.I were chosen by using $R = 0.5$ as the definition of the incoherent limit.[1] Successively larger NAs up to the practical limit of $(NA)_{condenser} = 1$ make the system more incoherent and increase the resolution limit of linearity. As the numbers in Table 18.I indicate, system nonlinearities can be anticipated due to spatial coherence with this system when line widths on the order of 1.5 μm or less are desired. These results were obtained using the stringent criterion ($N = 10$) of Fig. 18.1.

Use of the less stringent criterion ($N = 4$) would set the limit of linearity at 0.6 μm. However, this limit is highly suspect since it approaches the wavelength of the light. If a 1:1 reflection projection printer is designed[4] with ultraviolet illumination having $\lambda = 0.248$ μm, then the limit of linearity for the criterion $N = 10$ becomes 0.74 μm.

In the fabrication of integrated circuits where high-resolution binary targets are projection-printed at a 1:1 scale, the printer is optimized to work in an underfilled mode, $R = 1.43$. This is well into the nonlinear region of operation.[3,5] In this case, the increased contrast and line width narrowing caused by the introduction of partial coherence, as illustrated in Fig. 18.3, are found to be useful in maintaining a high-contrast image. Thus, knowledge of the target characteristics (binary) and the nonlinear mode of operation of the optical system are combined to create an advantage in this particular situation. This mode of operation would be objectionable in wide dynamic range photography.[3] Thus, the introduction of partial coherence to sharpen edge gradients,[4,5,7] which is achieved at the expense of system linearity, should be used with caution.

TABLE 18.I
Region of Linearity for a 1:1 Projection Printer

$(NA)_{objective}$	Resolution Limit of Target ($N = 10$) (lp/mm)	Target Line Width (μm)	R	$(NA)_{condenser}$
0.33	200	2.5	0.52	0.63
0.33	300	1.65	0.4	0.83
0.33	400	1.25	0.33	1.0

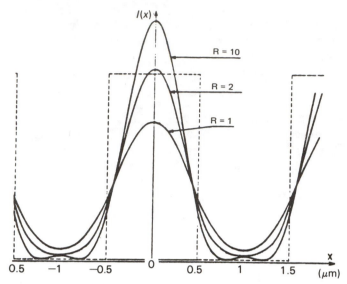

Fig. 18.3 Aerial image intensity $I(x)$ of a grating having 1-μm lines and spaces for three different values of R illustrating the linewidth narrowing and contrast increase in a projection printer as the coherence of the illumination is increased (from Ref. 6).

In this example, we focused our attention on the coherence considerations in the optical design and ignored all other optical parameters in the system. These parameters have been discussed elsewhere in the literature.[3,5,7-9]

18.1.4   Viewers

These devices, which are usually used to enlarge an image for viewing by a human observer, are normally operated in a linear mode, unless extremely high magnifications are needed. In this case, the problem is analyzed in the same manner as a microscope.

18.2   COHERENT IMAGING SYSTEMS [a]

Some instruments are specifically designed to utilize the complex amplitude superposition property of coherent radiation to preserve optical phase information, e.g., phase contrast microscopes and holographic instruments. Such instruments also exhibit image degradation when the degree of coherence is reduced, i.e., the radiation is partially coherent. In addition, as was shown in Fig. 11.8(f), utilization of radiation, which is both spatially and temporally coherent, gives rise to the phenomenon of speckle in the image (a visually disturbing random noise superimposed on the image).

Speckle in the image is caused by interference arising from a random phase distribution in the object plane, the autocorrelation length of which is less than the characteristic size of the system optical spread function measured in the object plane.

Physically, each random phase change within the spread function appears to have originated from a different point source. The coherent optical system images each plane wave as a separate amplitude spread function, and the result-

[a]This subject was covered in Chapters 12, 13, and 14 and is covered further in Chapters 29 through 35.

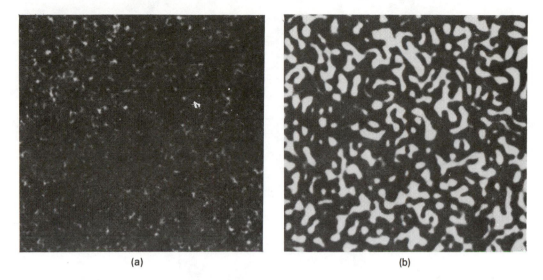

(a)   (b)

Fig. 18.4 Speckling density in a coherent image of ground glass with (a) an f/9 and (b) an f/64 imaging system; the illumination system was a collimated mercury arc source filtered at 5460 Å with a bandwidth of 100 Å (from Ref. 10).

ing intensity is the mutual interference of all the amplitude point images. An illustration of the change in speckle size with f/number is shown in Fig. 18.4. In general, the larger the temporal coherence of the source, the more predominant the resulting speckle pattern.

Speckle patterns, such as those shown in Fig. 18.4, produced with thermal sources are easier to reduce in a partially coherent imaging system because the bandwidth can be increased easily, thereby reducing the coherence length of the radiation. This effect was previously shown in Figs. 11.8(e) and (f). Speckle can be reduced or eliminated in an optical system by controlling the degree of spatial and/or temporal coherence of the system. A comprehensive review of these techniques was given previously[11] and appears in Chapter 21.

The effects of partially coherent radiation in systems specifically designed to utilize coherent radiation are demonstrated with a series of examples in the next sections.

18.2.1   Phase Contrast Microscope

The phase contrast microscope is a coherent instrument for visualizing optical phase variations on the order of a wavelength or less that would not be observable with incoherent illumination. The coherent radiation is obtained by using either a laser source or a point source. Attempts to reduce speckle noise in the system by reducing the degree of spatial coherence (i.e., by replacing the point source with an extended source) result in a loss of resolution directly dependent on the source size, as illustrated in Fig. 18.5. This figure is obtained by modifying the two-point analysis described in the literature[12] to incorporate a single point source. When the source size becomes comparable in size to the entrance pupil of the system [approximating the situation depicted in Fig. 11.16(b)], phase objects are no longer visible. Thus, a slight coherence reduction can be utilized to optimize the system performances, whereas a large amount of reduction in the degree of coherence is detrimental and severely degrades the system.

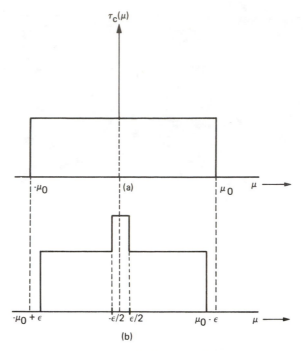

Fig. 18.5 One-dimensional coherent transfer function $\tau_c(\mu)$ for a phase contrast microscope (a) with a point source illustrating the cutoff frequency μ_0 as determined by the lens diameter and (b) with an extended source of width ϵ illustrating that the loss in resolution is directly dependent on the width of the source ϵ.

18.2.2   Holographic Systems

In holography, the use of coherent radiation is essential for storing optical phase information. The technique allows three-dimensional objects to be recorded and reconstructed in a two-step imaging process. Many applications of holography have been described[13,14] and are discussed in Chapters 26 and 27. It has been shown that the reduction of spatial and/or temporal coherence in the hologram formation process degrades the resolution of the image.[15] In essence, the introduction of partial coherence reduces the region over which interference takes place and, hence, reduces the size of the hologram. The resulting loss of interference fringes due to a reduction in the spatial coherence in a Fraunhofer hologram is shown in Fig. 18.6. The loss of resolution in a Fraunhofer hologram due to a

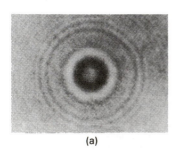
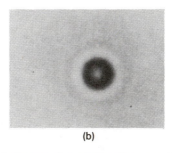
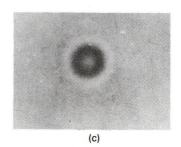

(a)   (b)   (c)

Fig. 18.6 Fraunhofer hologram degradation due to a decrease in spatial coherence. The holograms were formed at a few far-field distances of the particle $z_1 = 20$ cm. The particle size was 214 μm, and the light source was a mercury arc lamp focused onto a pinhole aperture located in the front focal plane of a 90-cm collimator lens. Aperture size and coherence interval, respectively, were as follows: (a) 0.1 mm and 1440 μm, (b) 2 mm and 72 μm, and (c) 6 mm and 24 μm (from Ref. 15).

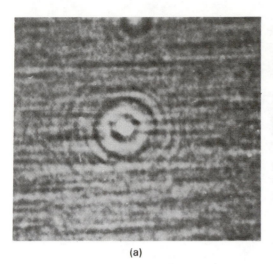

(a)   (b)

Fig. 18.7 Effects of temporal coherence in a Fraunhofer hologram. Reconstruction at 1 far-field distance is shown in (a) and at 3 far-field distances in (b). The holograms of a small opaque object were made with collimated light from a mercury arc ($\lambda' = 5461$ Å, $\Delta\lambda = 80$ Å) at 1 and 3 far-field distances from the object, respectively. Since the coherence length was fixed by an 80-Å filter, reconstruction of the hologram made at 3 far-field distances demonstrates deterioration due to temporal coherence effects (from Ref. 15).

reduction in temporal coherence of the radiation in the hologram formation process is shown in Fig. 18.7. From these results, we note that the loss of coherence in the formation of the hologram is detrimental to the holographic imaging process. In the hologram reconstruction process, some loss of coherence can be tolerated to reduce image speckle. These techniques are reviewed in Chapter 21 and elsewhere.

18.2.3   Speckle Photography and Interferometry

Speckle photography and interferometry are techniques utilizing the presence of speckle noise in a coherent system to perform precise measurements.[16] A slight loss of coherence in these systems will reduce the speckle contrast and thereby degrade the measuring capability of the system. These techniques are excellent examples of using a generally deleterious effect to gain an advantage in a particular application. These techniques also offer competition for holographic interferometric systems that are often more complicated experimentally.

18.2.4   Aberration Balancing

Coherent radiation has been used to compensate for various aberrations in optical systems.[16-22] In these systems, the aberration is multiplied by its complex conjugate to create an aberration-free wavefront. A slight reduction of the coherence in these systems rapidly degrades the system performance. An example illustrating the compensation of degradation through a specific random medium by phase aberration balancing[23] is shown in Fig. 18.8.

18.2.5   Image Processing

Coherent optical image processing uses an imaging system in which the Fourier transform (source image plane) is accessible. Coherent light is used so that a

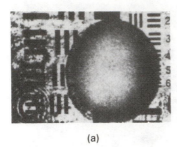 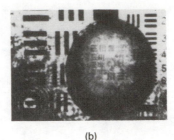 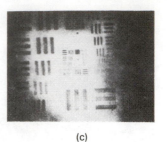

(a)   (b)   (c)

Fig. 18.8 Illustration of holographic phase compensation of an aberrating medium (excised human cataract): (a) resolution target placed directly behind cataract, (b) resolution target observed through a holographic correction filter, and (c) magnified portion of Fig. 17.10 showing high-resolution imagery observed through cataract and filter (from Ref. 23).

Fourier transform relationship exists between the front and back focal planes in a two-lens coherent processing system. Such systems, which have been used extensively for spatial filtering and pattern recognition, are discussed in the literature[24-26] and in Chapters 29 through 33. When a laser is used as the coherent source in such systems, speckle noise is normally observed. Special design procedures can be utilized to alter the spatial and/or temporal coherence and thus reduce speckle noise. This is usually accompanied by a resolution loss.[11] Spatial filtering systems can be designed with thermal sources to minimize speckle noise by reducing the coherence length, while simultaneously producing sufficient spatial coherence to preserve the capability of the system to Fourier transform and filter the object. Procedures to be considered in performing such experimentation with minimal noise effects are reviewed in Chapter 21.

An example of a coherent imaging system using Fourier processing is carrier-modulated imagery designed to achieve color image retrieval from monochrome transparencies[27] (see Chapter 34). The image of interest is incoherently recorded with a camera having a spatial multiplexing system that utilizes theta modulation (achieved by angular displacement of diffraction gratings in the image plane) to encode the color on black-and-white film. The image is retrieved by performing color mixing in the object plane achieved by using appropriate color filters in the Fourier. The control of coherence is very important in the reconstruction system in effecting high-quality color imagery that has good fidelity with a maximum amount of energy and also in minimizing speckle. This can be achieved by using high grating frequencies and a large white light source with appropriate color filters. The same speckle reduction effect has been observed in a multiple imaging application where different images are stored in each separate angularly oriented channel. An example of an image retrieved from such a system is shown in Fig. 18.9.

18.3   TEMPORAL COHERENCE CONSIDERATIONS IN OPTICAL SYSTEM DESIGN [b]

Our earlier discussions demonstrated that both spatial and temporal coherence effects exist in radiation fields and combine to define a coherence volume. Characteristic dimensions for some coherence volumes can be determined from

---

[b]This subject is discussed in more detail in Chapter 21.

Fig. 18.9 Black-and-white copy of a color image retrieved from a carrier-modulated image on a monochrome transparency demonstrating the minimization of speckle noise in a coherent imaging system. (Courtesy of P. F. Mueller.)

the data that were presented in Tables 11.I and 11.II. Radiation emanating from any pair of points within the coherence volume can give rise to interference effects that can degrade instrument performance. The loss of interference fringes caused by the lateral separation of the points outside the coherence volume (spatial coherence) was illustrated in Fig. 11.5(b), while the loss of interference fringes from the two points lying outside the coherence volume in the longitudinal direction (temporal coherence) was illustrated in Fig. 11.8. The primary concern for those instruments already discussed was the effect of spatial coherence on instrument performance. The phenomenon of speckle, which was illustrated in Figs. 11.8(e) and (f), was seen to be a generic noise source in the coherent instruments. Speckle is predominantly a temporal coherence effect that can be controlled and/or reduced by the techniques described in the literature[11,16] and in Chapter 21.

In the next sections, we discuss instruments in which the control of temporal coherence is of primary importance in the performance of the instrument.

18.3.1   Scanning Systems

Laser sources, which can concentrate large amounts of energy in a specific direction, have been utilized successfully to increase the speed of optical scanning systems.[28,29] Since these systems process the information sequentially one point at a time, spatial coherence effects are usually not present. An example of a scanning instrument used in a mensuration mode is the microdensitometer, where diffraction resulting from high-resolution elements within a given scanning spot can lead to nonlinearities in the system output when all the diffracted light from the target is not collected. In the design of an optical scanning system to record information for subsequent retrieval (e.g., film writers,[30,31] laser printing,[32] video disk recorders,[33] and holographic memories[34]), the control of temporal coherence in the optical system design is important to minimize laser system noise and prevent banding effects from occurring in the image. In some high-resolution applications,[33,34] temporal coherence considerations are necessary to minimize the spot size to control packing density and the bit error rate of the recorded data.

18.3.2   Fourier Spectrometers

Fourier spectrometers utilize the temporal coherence of the radiation to create fringes in an amplitude division interferometer, e.g., a Michelson interferometer as discussed in Chapter 23.[35] The resolution of the instrument is determined by the length of the recorded interferogram, which is formed by measuring the intensity of the central fringe as one of the mirrors is moved varying the path difference between the interfering beams. The spectrum of the sample usually consists of a broad envelope containing many narrowband peaks. When the instrument is in its zero path difference position, all of the radiation interferes and contributes to the intensity of the central fringe. As the mirror moves and the path difference increases to values greater than the coherence length of the broadband envelope, the characteristic strength of the interferogram decreases. The only radiation contributing to the intensity of the central fringe in the wings of the interferogram is the individual spectral peaks. Still further motion of the mirror eliminates the contribution of the broader peaks and ultimately causes the intensity of the central fringe from the few very narrow band lines to compete with the instrument noise. This noise limit fixes the useful length of the interferogram and, hence, instrument resolution. An example of an interferogram and its spectrum (which is obtained by Fourier transforming the interferogram) exhibiting these features is shown in Fig. 18.10. The fine structure (or high-frequency information) of the quartz powder is contained in the wings of the interferogram. A Fourier transform of a truncated interferogram containing 50 to 100 sample points would show none of the fine structure in the spectrum, but rather the broad envelope. This illustrates the influence of temporal coherence on the resolution capability of the instrument.

Improved instrument design using currently available high-quality infrared detectors and a phase-locked servo loop to control the angle between the mirrors[36] has led to improved resolution capability of the instruments from the range[35] of 0.3 to 10 cm^{-1} down to 0.003 cm^{-1}. Thus, in effect, temporal coherence considerations in the design of this instrument increased the spectral resolution by two orders of magnitude. Other instruments achieving comparable resolution have also been reported.[37]

18.3.3   Holography and Interferometry

The effects of temporal coherence are very important in the design of holographic and interferometric systems. As previously shown in Fig. 18.7, inadequate temporal coherence in the formation of Fraunhofer or Fresnel-type holograms limits the number of fringes recorded on the hologram and, hence, causes a resolution loss in the reconstructed image. Obviously, temporal coherence is necessary if off-axis reference beams are considered. This same lack of temporal coherence in the formation of image holograms and Fourier transform holograms will cause a field-of-view limitation in the reconstructed image.

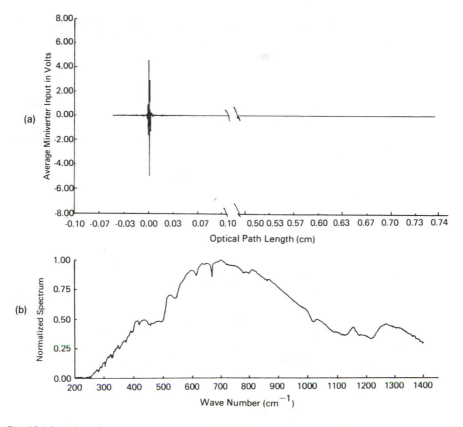

Fig. 18.10 (a) Interferogram and (b) corresponding spectra of quartz powder as measured on a low-resolution (2 cm⁻¹) Idealab Fourier spectrometer. The sharp band at 667 cm⁻¹ is atmospheric CO_2. The sharp features between 300 and 450 cm⁻¹ are water vapor. The broader bands are characteristic of the quartz powder. Significant data were present at an optical path length of 0.5 cm in expanded versions of this original interferogram. The sampling interval used was 2.5 μm. (Courtesy of J. R. Aronson.)

Again, excess temporal coherence in the holographic reconstruction process produces speckle noise that becomes superimposed on the image. Various techniques for reducing the speckle effect in holography by controlling the temporal coherence are reviewed in Chapter 21.

In interferometry, the temporal coherence of the source limits the amount of path difference that can be utilized in the experimental arrangement so that meaningful measurements can be made. An example illustrating the use of these principles is the fabrication of a holographic optical Schlieren system.[38] The instrument design equalizes the optical paths of the object beam and the reference beam to within the short coherence length of the 500-mW silver laser source by utilizing an optical delay line in the path of the reference beam. This system is useful for qualitatively and quantitatively measuring a wide variety of optical phase distributions, some of which require short exposure times.

18.3.4 Laser Contact Printing

Collimated laser beams have been proposed as a source in a contact printer,[39] both to deliver the large energy flux necessary for rapid exposure and to prevent undercutting in the contact printing process, as discussed earlier. The large coherence length of the laser introduces interference effects due to multiple reflections within the optical train. These noise effects were designed out of the instrument by using a prism and a laser line source in conjunction with the film transport system to produce noise averaging. A Kodak Concord roll-to-roll contact printer was retrofitted with an argon laser source using this design concept. Higher resolution images were obtained on Kodak 8430 and 3404 films using this modified system than in the conventional mode of operation of the instrument using a mercury arc source.[40]

18.3.5   Speckle Effects in Fiber Optic Communication Systems

Interference between various modes in an optical fiber communication system occurs when the path length differences of the various modes are less than the coherence length of the source.[41] This effect gives rise to a modal noise that seriously affects digital systems and completely disables analog systems, as illustrated in Fig. 18.11. Modal noise results in energy loss at a fiber joint and also increases the alignment tolerances of the joint. The future directions in fiber optic communication systems of developing single longitudinal mode lasers and low-dispersion fibers to increase communication bandwidth will be affected adversely by the presence of this speckle effect. Proposed solutions to this modal noise problem include the use of broad-bandwidth sources (small coherence length), single-mode fibers, improved joint alignment, and avoidance of bending in the fiber cable. Practical limitations to fiber optic communication systems caused by this effect are currently being investigated.[41]

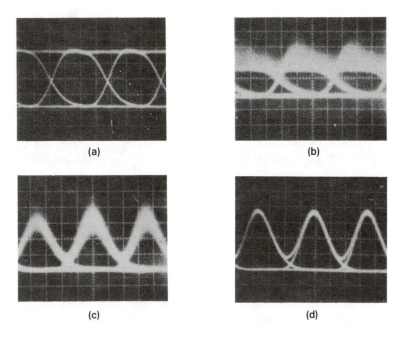

(a)   (b)

(c)   (d)

Fig. 18.11 The reduction in modal noise that is achieved in a practical 140 Mbit/s system by the use of spectrum broadening: (a) laser output, direct NRZ, (b) modal noise induced by misaligned joint (NRZ), (c) modal noise still present with RZ modulation above threshold, and (d) negligible modal noise with RZ modulation from below threshold (from Ref. 41).

18.4   SUMMARY

In Chapters 11, 17, and 18, we have discussed the effects of partial coherence, which are important in the design of optical instruments. The effects were categorized into four basic classes according to their predominant degree and type of coherence: spatial incoherence, partial spatial coherence, complete spatial coherence, and temporal coherence.

In the incoherent limit, the system is linear in intensity and instrument performance is completely characterized by the optical transfer function of the system. In the case of partial spatial coherence, instrument performance depends on the ratio of the spatial coherence interval to the size of the system resolution element, both measured in the object space (see Fig. 11.16). The ratio was used to define three regions of operational interest for various instruments. These three regions are: (1) the incoherent linear region, (2) the partially coherent, object-dependent nonlinear region, and (3) the coherent, nonlinear region where the nonlinearity is fixed for each given object. To determine the linear region of operation for the various instruments, a nomograph was empirically constructed. The limits of linearity for various instruments predicted from the nomograph were in agreement with available published data. The instruments considered were microdensitometers, microscopes, contact printers, micro-cameras, projection printers, and viewers.

In Chapters 17 and 18, it was shown that some instruments require complete spatial coherence to achieve their optimum performance. Reducing the degree of spatial coherence in these instruments results in a performance that is less than optimum. These losses were illustrated for phase contrast microscopes, holographic instruments, speckle photography and interferometry, phase aberration balancing, and image processing systems.

The primary effect of temporal coherence in optical instruments is the introduction of speckle noise. In addition, some systems, such as scanning systems, Fourier spectrometers, holographic and interferometric systems, laser contact printers, and fiber optic communication systems, must utilize the coherence length of the source radiation in their design to achieve optimum performance. The temporal coherence in these systems can be controlled to reduce the speckle noise at the expense of system performance.

REFERENCES

1. D. Nyyssonen, "Optical linewidth measurements on wafers," in *Developments in Semiconductor Microlithography III*, R. L. Ruddell et al., eds., Proc. SPIE 135, 115–119 (1978).
2. M. Born and E. Wolf, *Principles of Optics*, Pergamon Press, New York (1964).
3. D. A. Markle, "New projection printer," *Solid-State Tech.* 17(6), 50–53 (1974).
4. J. W. Bossung and E. S. Muraski, "Optical advances in projection photolithography," in *Developments in Semiconductor Microlithography III*, R. L. Ruddell et al., eds., Proc. SPIE 135, 16–23 (1978).
5. A. Offner and J. Meiron, "The performance of optical systems with quasi-monochromatic partially coherent illumination," *Appl. Opt.* 8(1), 183–188 (1969); see also "The influence of partial coherence on the microprojection of images," paper presented at OSA Topical Meeting on Use of Optics in Microelectronics, Las Vegas, Nevada, January 25–26, 1971.

6. M. Lacombat and G. M. Dubroeucq, "Coherent illumination improves step-and-repeat printing on wafers," in *Developments in Semiconductor Microlithography IV,* Jim Dey, ed., Proc. SPIE 174, 28–36 (1979).

7. J. D. Cuthbert, "Optical projection printing," *Solid-State Tech.* 20(8), 59–69 (1978).

8. J. W. Bossung, "Projection printing characterization," in *Developments in Semiconductor Microlithography II,* J. W. Giffin, B. Ruff, eds., Proc. SPIE 100, 80–84 (1977).

9. J. Roussel, "Step and repeat wafer imaging," in *Developments in Semiconductor Microlithography III,* R. L. Ruddell et al., eds., Proc. SPIE 135, 30–35 (1978).

10. P. S. Considine, "Effects of coherence on imaging systems," *J. Opt. Soc. Am.* 56(8), 1001–1008 (1966).

11. J. B. DeVelis, Y. M. Hong, and G. O. Reynolds, "Review of noise reduction techniques in coherent optical processing systems," in *Coherent Optical Processing,* Proc. SPIE 52, 55–81 (1974); reprinted here as Chapters 20 and 21.

12. D. J. Cronin, J. B. DeVelis, and G. O. Reynolds, "Equivalence of annular source and dynamic coherent phase contrast viewing systems," *Opt. Eng.* 15(3), 276–278 (1976).

13. R. J. Collier, C. B. Burckhardt, and L. H. Lin, *Optical Holography,* Academic Press, New York (1971).

14. *Handbook of Optical Holography,* H. J. Caulfield, ed., pp. 379–632, Academic Press, New York (1979).

15. G. O. Reynolds and J. B. DeVelis, "Hologram coherence effects," *IEEE Trans. Antennas Propag.* AP-15(1), 41–48 (1967).

16. A. E. Ennos, *Speckle Interferometry, Progress in Optics,* E. Wolf, ed., Vol. 16, pp. 233–288, North Holland Publishing Co., Amsterdam (1978).

17. E. N. Leith and J. J. Upatnieks, "Holographic imagery through diffusing media," *J. Opt. Soc. Am.* 56(4), 523 (1966).

18. J. W. Goodman, W. H. Huntley, Jr., D. W. Jackson, and M. Lehmann, "Wavefront-reconstruction imaging through random media," *Appl. Phys. Lett.* 8(12), 311–313 (1966).

19. J. Upatnieks, A. VanderLugt, and E. Leith, "Correction of lens aberrations by means of holograms," *Appl. Opt.* 5, 589–593 (1966).

20. H. Kogelnik and K. S. Pennington, "Holographic imaging through a random medium," *J. Opt. Soc. Am.* 58(2), 273–274 (1968).

21. G. O. Reynolds, J. L. Zuckerman, W. A. Dyes, and D. Miller, "Holographic phase compensation techniques applied to human cataracts," *Opt. Eng.* 12(1), 23–35 (1973).

22. J. W. Hardy, "Active optics: a new technology for the control of light," *Proc. IEEE* 66(5), 651–697 (1978); see also *Proc. IEEE* 66(10), 1287 (1978).

23. G. O. Reynolds, J. L. Zuckerman, W. A. Dyes, and D. Miller, "Phase aberration balancing of simulated cataracts in the reflection mode," *Opt. Eng.* 12(2), 80–82 (1973).

24. D. Casasent, "Coherent optical pattern recognition," *Proc. IEEE* 67(5), 813–825 (1979).

25. Special Issue on Optical Computing, H. J. Caulfield, ed., *Proc. IEEE* 65(1) (1977).

26. *Optical Data Processing: Applications,* D. Casasent, ed., Vol. 23 of *Topics in Applied Physics,* Springer-Verlag, Heidelberg (1978).

27. P. F. Mueller, "Color image retrieval from monochrome transparencies," *Appl. Opt.* 8, 2051–2057 (1969).

28. *Laser Scanning Components and Techniques: Design Considerations/Trends,* L. Beiser, G. F. Marshall, eds., Proc. SPIE 84 (1976).

29. Special Issue on Laser Recording/Laser Systems, L. Beiser, ed., *Opt. Eng.* 15(2) (1976).

30. N. Gramenopoulos and E. D. Hartfield, "Advanced laser image recorder," *Appl. Opt.* 11, 2778–2782 (1973).

31. M. Hannah and W. Harris, "Laser color recording directly on nine-inch color film," *Opt. Eng.* 15(2), 119–123 (1976).

32. *Laser Printing,* S. T. Dunn, ed., Proc. SPIE 169 (1979).

33. G. C. Kenney et al., "An optical disk replaces 25 mag tapes," *IEEE Spectrum* 16(2), 33–38 (1979).

34. H. N. Roberts, J. W. Watkins, and R. H. Johnson, "High speed holographic digital recorder," *Appl. Opt.* 13(4), 841–850 (1974).

35. G. Vanasse and H. Sakai, *Fourier Spectroscopy, Progress in Optics,* E. Wolf, ed., Vol. 6, pp. 259–330, North Holland Publishing Co., Amsterdam (1967).
36. H. L. Buigs, "Fourier transform spectroscopy as a step to laser spectroscopy," in *Impact of Lasers in Spectroscopy,* Proc. SPIE 49, 31–34 (1974).
37. P. Connes, "High resolution and high information Fourier spectroscopy," in *Aspen Int. Conf. on Fourier Spectroscopy,* G. A. Vanasse et al., eds., sponsored by Air Force Cambridge Research Labs. (1970).
38. R. L. Kurtz and L. M. Perry, "A holographic optical Schlieren system," *Opt. Eng.* 18(3), 243–248 (1979).
39. H. Heckscher and B. J. Thompson, "Contact printing with coherent light," *Photo. Sci. Eng.* 8(5), 260–265 (1964).
40. E. L. Bouche and D. Servaes, Final Report Contract AF30-602-6842, Tech/Ops, Burlington, Mass. (1965).
41. R. E. Epworth, "The phenomenon of modal noise in fiber systems," in *Technical Digest on Optical Fiber Communication,* sponsored by OSA and IEEE, Washington, D.C., March 6–8, 1979, IEEE 79CH 14316 QEA.

19 Selected Criteria for Image Analysis

19.1 INTRODUCTION

In Chapters 1 through 18, we have attempted to discuss some of the various aspects of image formation and the background that is needed for a quantitative treatment of this subject. Indeed, it is a fairly obvious statement to say that the problem of image formation and evaluation appears to be a perennial one in optics. Historically, the very early developments in the theory of image formation were in geometrical descriptions of the process, and the techniques of ray tracing were devised. The Gaussian lens formula is indeed simple and useful, and we showed how this result can be obtained from the general treatment of image formation in terms of wave optics. More accurate descriptions of image formation depend on the discussion of the problem in terms of the wave nature of light—a purely monochromatic approximation leading to the coherent imaging result, and the statistical nature of light leading to the more familiar incoherent image formation. The subject took on an entirely new character in the early 1950s with the application of communication theory to optics. This resulted in a new vocabulary being introduced into the optical literature; image formation was now described in terms of the impulse response of the imaging system or in terms of the system transfer function. Frequency domain analysis has thus become a powerful tool in the description and evaluation of optical systems; so much so that lens manufacturers provide modulation transfer function curves with their lenses, and film handbooks now contain modulation transfer functions for films.

The communication theory approach, however, also led to a more detailed study of coherent image formation and to the field of study now known as optical data processing of information, both analog and digital. While this latter is a reasonably new development, it has its roots in the early work of Abbe[1] in his discussions of image formation in a microscope. Basically, Abbe's theory was that an optical image is formed by the superposition of interference fringes. It is worthwhile to discuss here the implications of Abbe's theory in a little more detail.

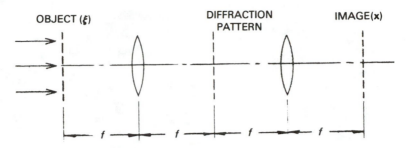

Fig. 19.1 In-line optical system.

19.2   IMAGE FORMATION

The basic ideas of Abbe's theory were very nicely illustrated in a series of experiments by Porter,[2] a reference well worth reading. We can illustrate the concepts involved here and at the same time illustrate some of the basic concepts of coherent optical data processing or spatial filtering. Consider the optical system shown schematically in Fig. 19.1. This system is discussed in detail and utilized in Chapters 29 through 37 as an analog computing and/or spatial filtering system. The object is illuminated coherently with a collimated beam of quasimonochromatic light; the first lens produces a Fraunhofer diffraction pattern of this object in the focal plane of the lens. This diffraction pattern is, of course, the Fourier transform of the object. If a second lens is used as indicated, then an image of the original object is formed in the focal plane. This is basically the optical system used by Maréchal and Croce,[3] O'Neill,[4] Thompson,[5] and specifically that used by Cutrona et al.[6] in spatial filtering experiments. Clearly, the position and magnification of the image can be obtained by purely geometrical considerations; however, the subtleties of this optical system certainly are appreciated only in terms of a physical optical description.

We consider here a simple but illustrative example taken from the work of Thompson. The original object shown in Fig. 19.2(a) consisted of a two-dimensional array of small circular holes; note that one hole in the upper left-hand corner is missing. From Chapters 2 and 3, it is fairly easy to predict the form of the Fraunhofer diffraction pattern formed by the first lens. This is illustrated in Fig. 19.2(b)i; the diffraction pattern shown here extends only to the edge of the central maximum of the Airy pattern produced by one of the holes alone. Since the system depicted in Fig. 19.1 is an in-line system, the light-producing diffraction pattern propagates on to form the image of the original object. Hence, by allowing various portions of the diffraction pattern to pass through the suitable aperture placed in the diffraction plane, the nature of the image can be changed in a controlled manner. For example, if all of the pattern shown in Fig. 19.2(b)i is allowed to pass, then the image is as shown in (b)ii. If the central nine diffraction peaks are allowed to pass, then the image is still a fairly good representation of the original object [see (c)i and ii]. In (d), the three horizontal diffraction peaks produce an image of vertical bars, with horizontal resolution preserved but vertical resolution lost. The position of the missing hole, however, is indicated quite clearly. The three vertical diffraction peaks produce an image comprising horizontal bars only as in (e). The resolution is retained in the vertical direction but not horizontally. A weak subsidiary fringe is evident in the image between the main horizontal bars. This subsidiary information in the image is completely spurious. In (f), the zero order is also removed and the image consists of

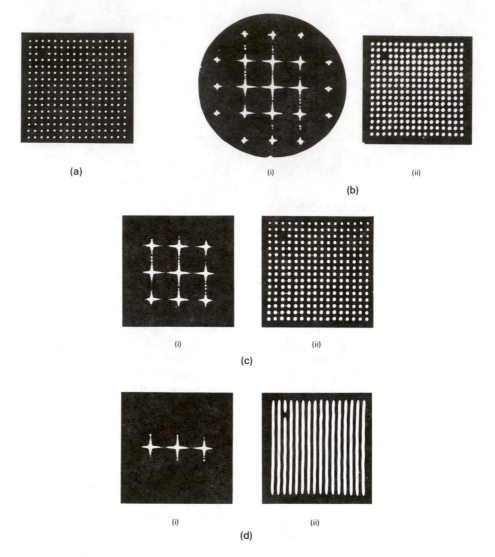

Fig. 19.2 (left) An illustration of the Abbe theory of image formation: (a) the original object, (b) diffraction pattern (i) of the original object used to form the image of (ii). In (c) and (d), the "filtered" patterns are shown in (i) and the resultant images are labelled (ii).

horizontal bars of half the original spacing of the holes. In this sense, the image is totally spurious. Finally, in (g), only the zero diffraction order (often termed the *dc component*) is allowed to contribute to the image, resulting in loss of resolution both vertically and horizontally—the only remaining information is the envelope function of the original object array, but still the missing hole position is preserved.

It is possible to use Figs. 19.2(c) through (g) in the reverse order to see how the image builds up in terms of adding more interference fringes, which are produced by allowing more diffraction peaks to contribute to the image. In Chapter 30, Fig. 30.2 illustrates buildup of the image by including more spatial frequencies from the object.

A number of other examples of this type can be found in an excellent text by Taylor and Lipson.[7] For applications of these basic ideas to spatial filtering, a good source is O'Neill[8] and Chapters 29 through 37 of this book.

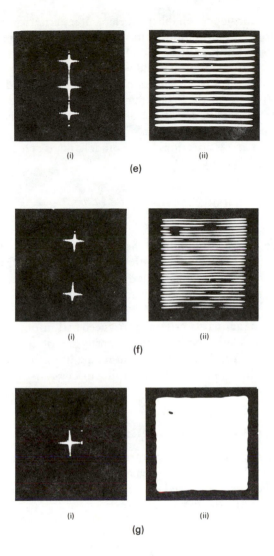

Fig. 19.2 (right) An illustration of the Abbe theory of image formation continued. In (e)–(g), the "filtered" patterns are shown in (i) and the resultant images are labelled (ii).

19.3   IMAGE QUALITY CRITERIA

The problem of the definition and determination of an image quality criterion has long been, and still is, a major problem in the field of image evaluation and assessment. This statement is true even when considering a purely incoherent situation. The values of the various quality criteria and their meanings and interpretations for partially coherent or coherent situations have not been evaluated. One would like to think that an image quality criterion could be equally useful and meaningful in the general partially coherent case as well as in the two limiting situations. At present, the usefulness and meaning of the various quality criteria in incoherent light are not fully determined. This situation arises mainly because of the difficulties associated with the experimental measurement of the criteria.

If we consider even the most elementary criterion—the two-point resolution criterion—the difficulties immediately become apparent. Quite clearly, the Rayleigh criterion cannot be usefully applied to anything but the incoherent case. However, the definition could be modified for the coherent situation so that the criterion applies to the addition of the amplitude and phase distributions of the two patterns; i.e., the first minimum of one amplitude distribution coincides with the central maximum of the other. Immediately, the difficulty arises that the amplitude is not a directly measurable quantity.

The Sparrow criterion, on the other hand, is at least definable in all situations since it is given by

$$\frac{\partial^2}{\partial x^2}[I(x)]\bigg|_{x=0} = 0 \; ; \tag{19.1}$$

i.e., the second differential of the resultant intensity distribution $I(x)$ is evaluated at $x = 0$, where x is the spatial coordinate. In fact, this particular resolution criterion has been evaluated and compared for the incoherent and coherent limits for the one- and two-dimensional optical system and discussed in detail in Chapters 6 and 13.

The image quality criteria introduced by Linfoot[9] attempt to define measures of the "similarity between the intensity distributions in the image and in the object." We briefly review the Linfoot quality criteria here and comment on their applicability in light of the preceding discussions.

19.3.1   Fidelity Defect and Fidelity

Let use define an ideal image of an object to be $I_{ob}(x)$ and the actual image of that object as $I_{im}(x)$. The first measure is determined basically by the mean square difference of the idealized image intensity and the actual image intensity suitably normalized by the factor

$$\frac{1}{\text{area}} \iint\limits_{-\infty}^{\infty} |I_{ob}(x)|^2 \, dx \; , \tag{19.2}$$

where the area here refers to the area of the image. With this normalization, the *fidelity defect* ϕ of the image is given by the defining equation

$$\phi = \frac{\displaystyle\iint\limits_{-\infty}^{\infty} [\,|I_{ob}(x) - I_{im}(x)|^2]\,dx}{\displaystyle\iint\limits_{-\infty}^{\infty} |I_{ob}(x)|^2 \, dx} \; . \tag{19.3}$$

In this form, we have assumed stationarity over the whole image field. Actually, this is not always appropriate, but areas of the image field can be defined over which the impulse response of the system is constant. However, it is not very convenient to operate in this manner, and perhaps a position-dependent ϕ can be defined. The *fidelity* $\Phi = 1 - \phi$ and can be stated (as can the fidelity defect) in terms of the transfer function of the system, which—again assuming stationarity—we find to be

$$\Phi = \frac{1 - \displaystyle\iint_{-\infty}^{\infty} |1 - \tau(\mu)|^2 \, |\tilde{I}_{ob}(\mu)|^2 \, d\mu}{\displaystyle\iint_{-\infty}^{\infty} |\tilde{I}_{ob}(\mu)|^2 \, d\mu} . \tag{19.4}$$

Here, $\tau(\mu)$ is the transfer function, $\tilde{I}_{ob}(\mu)$ is the Fourier transform of $I_{ob}(\mathbf{x})$, and μ is the spatial frequency variable.

19.3.2 Relative Structural Content

Consider a field that is large compared to the resolution limit of the system and, again, use the complete image field. A structural density factor SD_{ob} is defined by

$$SD_{ob} = \frac{\displaystyle\iint_{-\infty}^{\infty} |I_{ob}(\boldsymbol{\xi})|^2 \, d\boldsymbol{\xi}}{\left[\displaystyle\iint_{-\infty}^{\infty} I_{ob}(\boldsymbol{\xi}) \, d\boldsymbol{\xi}\right]^2} \; : \; SD_{im} = \frac{\displaystyle\iint_{-\infty}^{\infty} |I_{im}(\mathbf{x})|^2 \, d\mathbf{x}}{\left[\displaystyle\iint_{-\infty}^{\infty} I_{ob}(\boldsymbol{\xi}) \, d\boldsymbol{\xi}\right]^2} \tag{19.5}$$

and, hence, the *relative structural content* is defined as

$$SD_{rel} = \frac{SD_{im}}{SD_{ob}} = \frac{\displaystyle\iint_{-\infty}^{\infty} |I_{im}(\mathbf{x})|^2 \, d\mathbf{x}}{\displaystyle\iint_{-\infty}^{\infty} |I_{ob}(\boldsymbol{\xi})|^2 \, d\boldsymbol{\xi}} . \tag{19.6}$$

Rewriting Eq. (19.6) in terms of the transfer function gives

$$SD_{rel} = \frac{\displaystyle\iint_{-\infty}^{\infty} |\tau(\mu) \, \tilde{I}_{ob}(\mu)|^2 \, d\mu}{\displaystyle\iint_{-\infty}^{\infty} |\tilde{I}_{ob}(\mu)|^2 \, d\mu} . \tag{19.7}$$

19.3.3 Correlation Quantity

Here, the zero ordinate of the correlation between the idealized image and the actual image with the same normalization as in the previous factors is given in

$$Q = \frac{\iint\limits_{-\infty}^{\infty} I_{ob}(\mathbf{x})\, I_{im}(\mathbf{x})\, d\mathbf{x}}{\iint\limits_{-\infty}^{\infty} |\, I_{ob}(\mathbf{x})|^2\, d\mathbf{x}} \tag{19.8}$$

or

$$Q = \frac{\iint\limits_{-\infty}^{\infty} \tau(\mu)\, |\, \tilde{I}_{ob}(\mu)|^2\, d\mu}{\iint\limits_{-\infty}^{\infty} |\, \tilde{I}_{ob}(\mu)|^2\, d\mu}\; . \tag{19.9}$$

The three quality factors discussed here are not completely independent since they are related by the following equation

$$\Phi + \mathrm{SD}_{rel} = 2Q\; . \tag{19.10}$$

19.4   DISCUSSION

Essentially, the three quality factors require the measurement or evaluation of the following expressions:

$$\iint\limits_{-\infty}^{\infty} |\, I_{ob}(\mathbf{x}) - I_{im}(\mathbf{x})|^2\, d\mathbf{x}\; , \tag{19.11a}$$

$$\iint\limits_{-\infty}^{\infty} |\, I_{im}(\mathbf{x})|^2\, d\mathbf{x}\; , \tag{19.11b}$$

$$\iint\limits_{-\infty}^{\infty} I_{ob}(\mathbf{x})\, I_{im}(\mathbf{x})\, d\mathbf{x}\; , \tag{19.11c}$$

$$\iint\limits_{-\infty}^{\infty} |\, I_{ob}(\boldsymbol{\xi})|^2\, d\boldsymbol{\xi}\; . \tag{19.11d}$$

Or, in terms of the transfer functions, the important parameters are, along with Eq. (19.11d),

$$\iint\limits_{-\infty}^{\infty} |1 - \tau(\mu)|^2\, \tilde{I}_{ob}(\mu)\, d\mu\; , \tag{19.12a}$$

$$\iint\limits_{-\infty}^{\infty} |\tau(\mu)\, \tilde{I}_{ob}(\mu)|^2\, d\mu\; , \tag{19.12b}$$

$$\iint\limits_{-\infty}^{\infty} \tau(\mu)|\tilde{I}_{ob}(\mu)|^2\, d\mu \ . \tag{19.12c}$$

One point that is immediately obvious is that the relative structural content SD_{rel} takes no direct account of the phase of the transfer function since it depends only on $|\tau(\mu)|$. In particular, contrast reversal will not be taken into account.

These results can be summarized by writing:

$$\Phi, SD_{rel}, Q = \frac{\displaystyle\iint\limits_{-\infty}^{\infty} \Psi[\tau(\mu)]\,|\tilde{I}_{ob}(\mu)|^2\, d\mu}{\displaystyle\iint\limits_{-\infty}^{\infty} |\tilde{I}_{ob}(\mu)|^2\, d\mu}\ , \tag{19.13}$$

where $\Psi(\tau)$ has the special forms

$$1 - |1 - \tau(\mu)|^2 \ , \quad |\tau(\mu)|^2 \ , \quad \tau(\mu) \ . \tag{19.14}$$

The discussions above specify the various quality factors for the aerial image formed by an imaging system. The problem of measurability still remains. The transfer function of the system is, of course, measurable in principle, but it is much more difficult to specify the Fourier transform of the object intensity distribution for a real object. However, for the special case when the input to a system is a photographic negative, then it is possible to measure the object intensity distribution or its spectrum. These comments are only true, however, for the incoherent situations. How should the coherent or partially coherent cases be handled? A number of possibilities exist. For the coherent case, the quantities $I_{ob}(\xi)$ and $I_{im}(\mathbf{x})$ are still definable, but perhaps the meaningful parameters might be the complex amplitudes $A_{ob}(\xi)$ and $A_{im}(\mathbf{x})$ and, hence, the amplitude transfer function. In general, the important and measurable quantity would be the mutual intensity function in the object and in the image.

19.4.1 Application to Photographic Images

If the criteria discussed above are now modified to include the photographic film, then we have to modify the transfer function discussed to include the transfer function of the film. The transfer function of such a combined system can be shown to be multiplicative; hence, using $\psi[\tau_s(\mu), \tau_F(\mu)]$ to denote the dependence on the system transfer function and the film transfer function, Eq. (19.13) may be rewritten as

$$\Phi, SD_{rel}, Q = \frac{\displaystyle\iint\limits_{-\infty}^{\infty} \Psi(\tau_s, \tau_F)|\tilde{I}_{ob}(\mu)|^2\, d\mu}{\displaystyle\iint\limits_{-\infty}^{\infty} |\tilde{I}_{ob}(\mu)|^2\, d\mu}\ . \tag{19.15}$$

Unfortunately, this expression ignores granularity and photographic γ. A possible way of including γ is to replace $\psi[\tau_s(\mu), \tau_F(\mu)]$ with $\psi[\gamma\tau_s(\mu), \tau_F(\mu)]$, which changes the quality factors Q and SD_{rel} into γQ and $\gamma^2\mathrm{SD}_{rel}$ and Φ into $2\gamma Q - \gamma^2\mathrm{SD}_{rel}$. This can still be misleading, since the photographic noise is not taken into account. However, when noise does become important, the incoherent interpretation of these factors is certainly no longer valid.

19.4.2   Application to Photographic Lenses

To assist in the evaluation of lenses, the quality criteria are redefined again so as to measure the optical quality of the lens relative to a given object and a given emulsion and not of the whole photographic system (lens-plus-emulsion) relative to a given object. For these new definitions, we use the normalized spatial frequency power spectrum of the object intensity distribution, defined as

$$\chi(\mu) = \frac{|\tilde{I}_{ob}(\mu)|^2}{\displaystyle\iint_{-\infty}^{\infty} |I_{ob}(\mu)|^2 \, d\mu} \quad ; \tag{19.16}$$

hence,

$$\Phi_\chi, \mathrm{SD}_\chi, Q_\chi = \frac{\displaystyle\iint_{-\infty}^{\infty} \psi[\tau_s(\mu), \tau_F(\mu)] \, \chi(\mu) \, d\mu}{\displaystyle\iint_{-\infty}^{\infty} \psi[\tau_0(\mu)] \, \chi(\mu) \, d\mu} \quad , \tag{19.17}$$

where $\tau_0(\mu)$ is the transfer function for an unaberrated system and no photographic spread. If χ is constant, then

$$\Phi_{\chi_c}, \mathrm{SD}_{\chi_c}, Q_{\chi_c} = \frac{\displaystyle\iint_{-\infty}^{\infty} \psi[\tau_s(\mu), \tau_F(\mu)] \, d\mu}{\displaystyle\iint_{-\infty}^{\infty} \psi[\tau_0(\mu)] \, d\mu} \quad . \tag{19.18}$$

By this definition, the perfect system has a quality factor of unity. When the contrast reversal occurs, Φ can go negative.

REFERENCES

1. E. Abbe, "Beiträge zur Theorie des Mikroskops und der Mikroskopischen Wahrnehmung," *Archiv. Mikroskop. Anat.* 9, 413–480 (1873).
2. A. B. Porter, "On the diffraction theory of microscopic vision," *Phil. Mag.* II, 154–166 (1906).
3. A. Maréchal and P. Croce, "Un filtre de frequences spatiales pour l'amelioration du contraste des images optiques," *Compt. Rend.* 237, 607–609 (1953).
4. E. L. O'Neill, "Spatial filtering in optics," *IRE Trans. Info. Theory* IT-2(2), 56–65 (1956).
5. B. J. Thompson, *The Study of Partial Coherence and Diffraction Phenomena,* PhD Thesis, Univ. of Manchester, England (1959).
6. L. J. Cutrona, E. N. Leith, C. J. Palermo, and L. J. Porcello, "Optical data processing and filtering systems," *IRE Trans. Info. Theory* IT-6(3), 386–400 (1960).
7. C. A. Taylor and H. Lipson, *Optical Transforms,* Bell, London (1964).
8. E. L. O'Neill, *Introduction to Statistical Optics,* Addison-Wesley Publishing Co., Reading, Mass. (1963).
9. E. H. Linfoot, *Fourier Methods in Optical Image Evaluation,* Focal Press, Ltd., London (1964).

20   Photographic Films *

20.1   INTRODUCTION

Here and in Chapter 21, the various sources of noise in coherent imaging systems are reviewed and classified. This classification includes film granularity noise,[1-10] film processing noise (including both phase images and film processing non-linearities),[11-15] dust noise, and speckle noise.[16-26]

In Sec. 20.2, we review those film parameters that are necessary to define a maximum achievable system performance figure in terms of system resolution. The important film parameters are granularity, film MTF, and threshold modulation.[27-31]

In Chapter 21, we continue our study of photographic films with discussions of the types and sources of noise in coherent imaging systems, numerous speckle-averaging techniques, and a summary.

20.2   REVIEW OF PHOTOGRAPHIC FILMS

A photographic emulsion is a suspension of silver halide crystals (called *grains*) in a gelatin medium. When a silver halide grain absorbs a photon, a hole electron pair is formed. On the average, at least four silver atoms are needed to form a stable silver speck in a grain, which means that a grain must absorb at least four photons to create such a stable silver speck. These stable silver specks are referred to as the *latent image*. The process of development reduces those grains that contain a latent image to metallic silver; the other grains (which do not contain a latent image) are not reduced to silver. In addition, the reaction rate determined by the strength of the developer used in the processing can cause spontaneous reduction of a grain having no latent image speck to metallic silver, thus forming a random or diffuse image called *fog*. This diffuse image is one source of photographic noise. In addition to the effects of spontaneous reduction, it is possible for the metastable latent image to undergo spontaneous decay back to silver halide. This phenomenon is referred to as *latent image fade*.

---

*Reprinted with permission from "Review of Noise Reduction Techniques in Coherent Optical Processing Systems," by J. B. DeVelis, Y. M. Hong, and G. O. Reynolds, in *Coherent Optical Processing,* H. J. Caulfield, ed., SPIE Proc. 52, 55–81 (1975). This article is continued in Chapter 21.

The photon absorption process occurs at a microscopic level, whereas the resulting processed image is macroscopic. Therefore, the photographic process is equivalent to an amplifier. For example, the photons are absorbed over a cross-sectional area of approximately 4 Å^2. After development, the entire silver grain has an area of perhaps 1 μm^2 (10^8 Å)^2, the development process having effectively caused an amplification in area of the order of 10^8.

Some of the parameters characterizing the photographic process that are of interest in optical processing are photographic speed, a measure of film sensitivity, exposure dynamic range, spectral response, and image quality. To obtain high photographic speed (the ability to record a low photon flux), the grains must be of relatively large area to increase the probability of collecting four photons to allow the formation of a latent image. Smaller grain sizes produce higher resolution but require more intense exposures to produce an image. To obtain a wide exposure dynamic range, a distribution of grain sizes is needed to produce a broad range of grain sensitivities. In practice, this is realized in emulsion manufacturing by varying the physical grain shape and size or by varying the sensitivities of the individual grains. The resolution capability of a film is related to the reciprocal of the average grain diameter. The spectral sensitivity of an emulsion extends from the ultraviolet cutoff of the gelatin base (about 2200 Å) to the energy gap cutoff of silver halides (about 4900 Å). To extend this sensitivity further into the visible and the infrared regions, various dyes are adsorbed onto the surfaces of the grains, and these subsequently affect the image quality. Thus, as the average grain size is increased, the resulting resolution capability lowers, the speed generally increases. Similarly, fine-grain emulsions tend to be of lower speed, narrower exposure dynamic range, and higher resolution capability. Other important film effects such as reciprocity failure and intermittency, adjacency effects, Herschel effects, solarization effects, and Clayden effects are not considered in the discussion presented here and the interested reader is referred to the literature.[1]

When photographic film is used in optical systems, regardless of the degree of coherence, the film properties described above are used to determine the following useful parameters in system design: sensitometry, modulation transfer function (MTF), resolution limit, granularity, and spectral sensitivity. One other parameter of importance in the design of optical systems is the signal detection threshold of the film, which is not usually specified by the manufacturer since it varies according to the application. We attempt in this chapter to obtain a phenomenological measure of this parameter as it applies to coherent optical systems.

Various models have been developed in an attempt to predict these parameters from fundamental principles.[4,5,32,33] Rather than review these models, we take an alternative approach by describing the measurements associated with the known film parameters of interest in order to determine empirically an indicator of achievable system performance useful in coherent optical systems.

20.2.1 Pertinent Film Parameter Measurements

20.2.1.1 Kelly's Model of Effective Exposure

A linear model has been proposed by Kelly[34] that relates the aerial image incident on the film to the developed image. This model directly includes effects of sensitometry and film MTF on the developed image. Recently, this effective

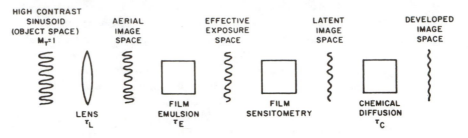

Fig. 20.1 Kelly's system for describing effective exposure.[34]

exposure model has been extended to include nonlinear film effects.[35,36] The nonlinear model has proved useful in removing the nonlinear adjacency effect in computer processing of imagery.[36] In this chapter, we use the Kelly model as a basis for determining the achievable optical processing system performance, since this is consistent with available data.

To clarify the terminology and measurements described here, we briefly review Kelly's concept of effective exposure, which is illustrated schematically in Fig. 20.1. An object (a high-contrast sine wave of unit modulation index or visibility) is imaged by the lens (having a transfer function denoted by τ_L) into aerial image space. This aerial image is presented to the film, which is considered a three-stage process. The aerial image is first acted on by the transfer function of the isotropic emulsion (τ_E) to create an effective exposure distribution that is, in turn, subjected to the sensitometric effects of the film to create a latent image distribution. The chemical diffusion process within the film (denoted by a chemical transfer function τ_C) acts on the latent image to create a processed image in developed image space. The processed image is where all transmittance measurements (viewing, microdensitometry, macrodensitometry, etc.) are made so that the data correspond to developed image space. If we remove the effects of film sensitometry and chemical diffusion from the developed image, we speak of effective exposure space. If the effective exposure distribution is divided by the emulsion MTF (τ_E), then we are working in aerial image space.

In the optical processing case, we are concerned directly with developed image space and its relationship to aerial image space. Therefore, we assume that the film emulsion is characterized by a transfer function and that the nonlinearity of the film sensitometry is experimentally controlled. This linear process is usually described by

$$I_E(\mathbf{x}) = \int I_{AI}(\mathbf{x}') S(\mathbf{x} - \mathbf{x}') \, dx' \,, \qquad (20.1)$$

where

$I_E(\mathbf{x})$ = intensity distribution in effective exposure

$I_{AI}(\mathbf{x}')$ = intensity distribution in aerial image

$S(\mathbf{x}')$ = intensity impulse response of the film under the above assumptions.

20.2.1.2   Film Sensitometry

To understand the assumption of film linearity, we must discuss film sensitometry.

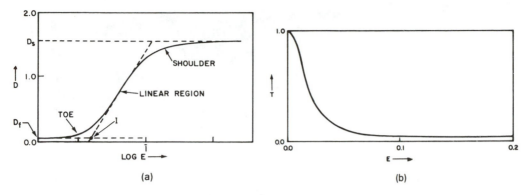

Fig. 20.2 Sensitometric curves for photographic film (from Ref. 1, pp. 163–166): (a) H&D curve for a photographic negative and (b) transmittance versus exposure data for the same photographic negative.

Film Characteristic Curves: It is well known that the characteristic Hurter-Driffield (H&D) curve relates the photographic density to the input exposure. Density is defined as the negative logarithm of transmittance, i.e.,

$$D = \log 1/T_I , \tag{20.2}$$

where T_I is the intensity transmittance.

We use density rather than transmittance because it corresponds to the known logarithmic response of the human eye. The H&D curve relates density to the logarithm of exposure (log E) and is often referred to as the D–log E curve. Figure 20.2(a) illustrates an H&D curve for a photographic negative.

Below the toe region of the curve, the exposure is weak enough that the image density is nondetectable visually. This is the base fog level D_f of the film. Increased exposure renders a visible density, and the logarithm of the exposure remains fairly linear with density until reaching the shoulder of the curve. Exposures beyond this point give no significant density increase[37]; i.e., the film becomes overexposed. This density is termed the saturation density D_S.

The usable exposure region where increased exposure renders a detectable density difference is called the *exposure dynamic range* of the film. This range can be increased by specialized processing techniques that are useful for some applications. At larger exposures, a density decrease is sometimes observed, an effect known as *solarization*.

The data of Fig. 20.2(a) are sometimes plotted as transmittance versus exposure [Fig. 20.2(b)] by using Eq. (20.2). This curve is useful in situations where extremely low exposures are encountered because it gives a higher gradient for these low-exposure values.

Film Gamma: The slope of the straight line portion of the D–log E curve is called the *photographic gamma*. It varies depending on the film processing chemistry used as well as the grain size distribution. Some film gammas are quite insensitive to processing (aerial films), whereas others are quite variable. Film image contrast is related to gamma: high or steep gammas mean high contrast (small exposure dynamic range) and low gammas imply low-contrast film imagery (large exposure dynamic range).

The film gamma is measured by using a known illumination source to contact print a step wedge of known transmittance onto the film. After processing, the

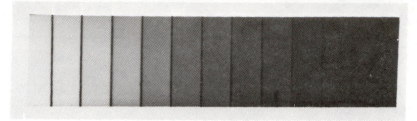

Fig. 20.3 Example of step wedge exposed on film. Only 13 of the 21 steps are visible in the figure because of the limited dynamic range of the photographic paper used.

densities of the resultant steps on the film (see Fig. 20.3) are measured in a macrodensitometer and a plot of measured density versus the logarithm of the known exposure is made. The slope of the linear portion of this curve is defined to be the gamma of the process.

Film Speed: In general, there are two techniques for measuring film speed. In the first, the fog level of the film and an extension of the straight line portion define an inertial point I [see Fig. 20.2(a)]. It locates a value of log E, which is that exposure required to cause a detectable density above the base fog of the film. The reciprocal of the exposure value determined by the inertial point is a measure of the film speed. This quantity is measured relative to a standard film and is typically rated in ASA units. The ASA values of some typical Kodak films are Pan-X (32), Tri-X (400), 3414 (1.6), and 649F plate (0.025).

In the second technique—the fractional gradient method—the criterion chosen is to specify the slope of some part of the D–log E curve. This point is used to determine an exposure from which the film speed is determined by its reciprocal or some appropriately chosen fraction of this number, e.g., in aerial films the half gamma speed criterion is commonly used.[38,39] In general, speed = K/E, where E is the exposure for the desired imagery and K is a constant.

Since film speed and film resolution capability are reciprocally related, the two parameters have to be optimized according to the recording requirements of the system.

20.2.1.3   *Film Spectral Sensitivity*

Another film parameter that must be considered before performing an optical experiment is the spectral response of the photographic film.[38,39] Because of the energy gap cutoff of silver halides (approximately 4900 Å), the grains are inherently sensitive to the ultraviolet and blue regions of the spectrum. To obtain green sensitivity, sensitizing dyes are added to the emulsion; such films are called *orthochromatic.* Sensitizing the emulsion for the red region of the spectrum requires the use of different sensitizing dyes in the emulsion; such films are called *panchromatic* (they respond to most of the radiation in the visible spectrum). Addition of still other sensitizing dyes extends the sensitivity into the near-infrared portion of the spectrum. The spectral sensitivity data of emulsions are supplied by film manufacturers and should be used when choosing a film for a given experiment once the source of radiation is known. For example, when recording holograms with a He-Ne laser that radiates at 6328 Å, one should select a film that is red sensitized rather than merely picking a film based on its resolution capability. The spectral response curves of various high-resolution emulsions are shown in Fig. 20.4.

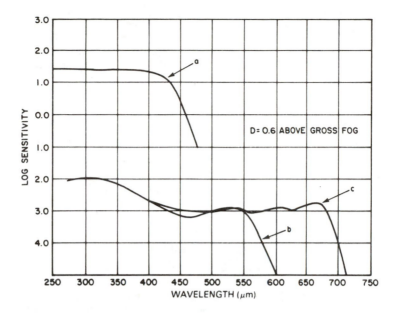

Fig. 20.4 Spectral response curves of high-resolution photographic films, D-19 development, 5 min at 68°F (from Eastman Kodak film data sheet): (a) blue sensitive Kodak spectrum analysis film No. 1 (2-s exposure), (b) orthochromatic 649-GH film (10.4-s exposure), and (c) panchromatic 649F film (10.4-s exposure).

20.2.1.4   Resolution-Exposure Characteristic of Density Images

Another inherent characteristic of a photographic emulsion is that the resolution capability builds up from a low point of the toe region of the characteristic curve to a maximum at about the midregion of the linear portion of the curve (the optimum exposure region) yielding maximum image information transfer. At higher exposure values, the resolution decreases from its peak value back to values similar to those in the toe region of the characteristic curve. Resolution exposure curves ideally have zero values of resolution at the extremes and a peak value characteristic of the particular emulsion. Resolution capability is usually expressed in units of cycles or line-pairs per millimeter. A typical resolution versus exposure curve for 3401 film is shown in Fig. 20.5.

The D versus log E and resolution versus log E curves are plotted on the same graph in Fig. 20.5, enabling the parameters of interest for a particular emulsion to be obtained easily.

20.2.1.5   Film MTF

The film MTF given in most film data sheets is a measure of the light scattering effects within the unprocessed emulsion instead of a transmittance measure of the processed photographic image. By its very nature, the film MTF is an approximate linear measure of one step in a complex nonlinear process that has proven to be useful in many different applications. The film MTF measures the change in the amount of modulation experienced by a unit modulation sinusoid of any spatial frequency exposed onto the given film. On published MTF curves, the frequency value at which the transfer function reaches 20% of its maximum value usually corresponds to the film resolution limit, as determined indepen-

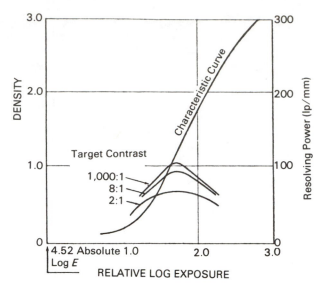

Fig. 20.5 Resolving power versus log exposure for 3401 film, D-19 development, 6 min at 68°F. The densities on the D–log E curve refer to large areas having the same luminance as the target bars (e.g., the square patch on the three-bar target) in images exposed for maximum resolution. The mean densities of just resolved target images are lower, depending on the target contrast (from Ref. 39, p. 61).

dently in a resolution measurement sequence. Since film MTF is a linear system measure, it is obtained by utilizing the concept of effective exposure as shown diagrammatically in Fig. 20.1 and described by Eq. (20.1).

To measure the MTF of a particular type of film, its D–log E curve is first determined. Variable-frequency unit modulation intensity sine waves are then carefully exposed onto the film. A reference step wedge is also exposed on the same film to ensure that the development process employed does not significantly change the D–log E curve. The film is processed and the various developed sine waves are measured with a microdensitometer. From these measurements, density values for the maximum and minimum of the sine wave are obtained. These values are then projected through the measured D–log E curve to obtain effective exposure values. The modulation of the effective exposure sine wave of frequency ν is obtained from the formula

$$m_E(\nu) = \frac{E_{max} - E_{min}}{E_{max} + E_{min}} ,$$

(20.3)

where E_{max} is the maximum effective exposure and E_{min} is the minimum effective exposure. A plot of the various normalized measured modulation versus the input spatial frequencies of the sine wave targets is the emulsion MTF for the film as specified by the manufacturer. A typical film MTF is shown in Fig. 20.6.

The most stringent restrictions on this measurement procedure are the requirements that:

1. the exposure be limited to the straight line portion of the D–log E curve

2. the same D–log E curve (measured macroscopically) is assumed valid for all spatial frequencies

3. the derived MTF is only indicative of the unprocessed emulsion.

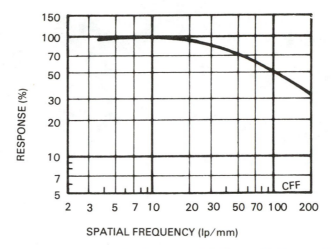

Fig. 20.6 Modulation transfer curve for Kodak high-definition aerial film type SO-243 (from Eastman Kodak film data sheet). D-19 development for 8 min at 68°F in a sensitometric machine.

In the actual use of film systems, the degree to which the first two of these restrictions is fulfilled determines the validity of the film MTF concept in an experimental analysis; e.g., if a given film is many stops overexposed, then we would not expect the film MTF analysis to be valid.

By suitably controllng the chemical development, the linear model can be extended to include an MTF relating the effective exposure to the transmittance of the developed image. This would yield the transmission MTF of the film as used in any given experiment. The modulation of the film transmission is given by

$$m_T = \frac{E_{max}^{\gamma} - E_{min}^{\gamma}}{E_{max}^{\gamma} + E_{min}^{\gamma}} , \qquad (20.4)$$

which when plotted as a function of spatial frequency and suitably normalized gives the transmission MTF. A comparison of Eqs. (20.4) and (20.3) shows that the two MTFs are identical when γ equals 1. When γ is greater than 1, then the transmission MTF is less than the emulsion MTF; when γ is less than 1, the transmission MTF is greater than the emulsion MTF.[40] Thus, in applications of spatial filtering where the input film is processed to a γ of -2, use of the emulsion MTF to predict system performance will merely determine a bound for the expected performance of the system.

Film MTFs have also been measured by edge gradient methods,[41,42] but the details are omitted here.

20.2.1.6   Granularity

Thus far, we have considered film parameters that are macroscopic and are measured with classical techniques. The ultimate limit of any detector, in this case photographic film, is the noise resulting from the physical nature of the structure of the detector. As we have already indicated, developed photographic film is comprised of opaque silver grains that are on the order of 1 μm or less in maximum dimension embedded in a transparent medium. These grains are randomly distributed at the micro level, but their macrodistribution gives rise to

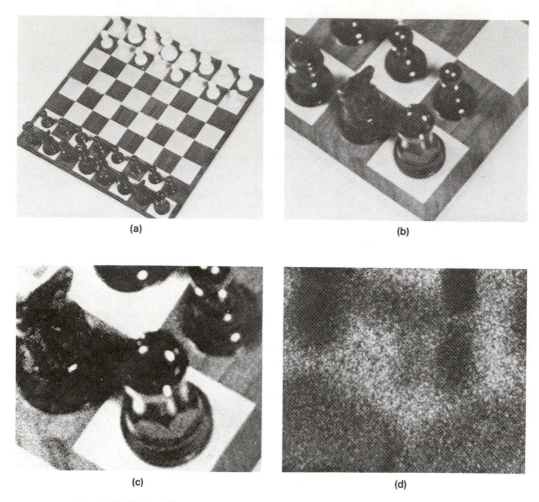

Fig. 20.7 Effect of film grain noise on image structure: (a) original photograph under normal viewing conditions, Tri-X film, (b) 3× magnification, (c) 6× magnification, and (d) 20× magnification.

the developed image. A statistical fluctuation of the image transmittance is expected when the observations of the developed image are made with microprobing devices. Because of the presence of this statistical fluctuation (film noise), in any subsequent measurement, a method is needed for the specification of the influence of this noise on the developed image. The effect of this statistical fluctuation is illustrated in Fig. 20.7, which shows the finite amount of information contained in a developed image.

Figure 20.7(a) is shown at a scale where the grain is not observable with the human eye. As we enlarge a section of the photograph in a microscope [Fig. 20.7(b)], the noise begins to influence our ability to interpret the smaller details of the image. Further increasing the magnification of the microscope [Fig. 20.7(c)] reduces the field of view but, more importantly, increases the amount of noise in the photograph compared to the signal present, further decreasing our ability to interpet the image. Finally, when the image is viewed at a very high magnification [Fig. 20.7(d)], the noise dominates and we are unable to observe an apparent signal. This demonstrates the noise structure (present to a different degree in every film) in relation to the image structure as a function of viewing condition and illustrates the necessity for quantifying the effect. To quantify the effect of this subjective graininess, we will introduce the concepts of granularity (a quantitative measure of the degree of the grain noise) and the signal detection threshold (a quantitative measure of the influence of the noise on the observed image). It should be remembered that the grain noise is signal dependent. That is, the size and statistical distribution of the grains vary with exposure level. As we shall see, this requires that the quantitative and qualitative measures of grain noise include a mean transmission term.

To measure the statistical fluctuations present in a developed film, it is scanned with an instrument (e.g., a microdensitometer) possessing a finite but variable size scanning aperture that measures the amount of light transmitted through the film. The transmission value measured in such an instrument is determined by the ratio of the area of the grains contained in the scanning aperture to the total area of the scanning aperture. The value of the transmission about the mean obtained in such a measurement is

$$t(x,y) = T(x,y) - \overline{T} .$$
(20.5)

In Eq. (20.5),

$$\overline{T} = \lim_{A \to \infty} \frac{1}{A} \int_A T(x,y)\, dx\, dy ,$$
(20.6)

where

$$T(x,y) = \int T_3(x,y,z)\, dz$$
(20.7)

and A is the area of the film being measured. The function $T_3(x,y,z)$ in Eq. (20.7) includes the effect of the three-dimensional distribution of the grains located within the emulsion. This three-dimensional grain distribution is observed by slicing an emulsion and viewing the cross section of the developed image in a microscope (a process called *microtoning*). An example of a microtoned section of an image of a diffraction grating is shown in Fig. 20.8.

When measuring a film possessing uniform density with a large area scanning aperture, a plot of transmittance as a function of distance is constant as shown in

Fig. 20.8 Microtoned section of a typical thin emulsion (EK3404) containing an image of a rectangular wave target (from Ref. 43).

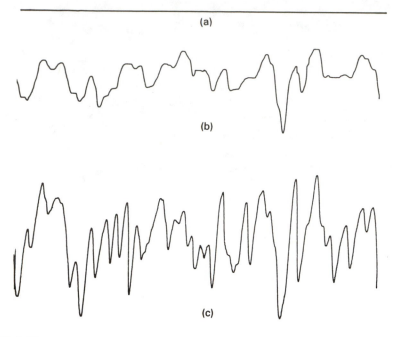

Fig. 20.9 Microdensitometer traces of a uniformly exposed field on film measured with square scanning apertures of various sizes: (a) 45 μm, (b) 10 μm, and (c) 5 μm. (The average density of the processed film is $\bar{D} = 0.8$.)

Fig. 20.9(a). This scanning aperture is so large that the average value of the transmittance is measured and no fluctuations are observed. As we reduce the size of the scanning aperture, we notice that statistical fluctuations caused by the variable number of black grains within the aperture at different positions increases dramatically as seen in Figs. 20.9(b) and (c). The existence of such fluctuations in a microdensitometer trace is referred to as the *granularity*. However, the quantity relating to granularity is σ, the standard deviation of the fluctuations usually obtained by a simple root mean square averaging process. To determine σ, let us assume that the film granularity in transmission can be represented by a function $T(x,y)$ that is a stationary random process with mean $\bar{T}$, and an autocorrelation function given by[4]

$$R(x_0, y_0) = [t(x,y)\, t(x + x_0, y + y_0)] \ , \tag{20.8}$$

where the brackets define the ensemble average, and $t(x,y)$ is defined by Eq. (20.5).

The grain noise power spectrum and the autocorrelation function are Fourier transform pairs; therefore, we have

$$N(\nu) = \int R(\mathbf{x}_0)\, e^{-2\pi i \nu \cdot \mathbf{x}_0}\, d\mathbf{x}_0 \ , \tag{20.9}$$

$$R(\mathbf{x}_0) = \int N(\nu)\, e^{2\pi i \nu \cdot \mathbf{x}_0}\, d\nu \ , \tag{20.10}$$

where

$$R(\mathbf{x}) = R(x_0, y_0) \ \text{and} \ N(\nu) = N(\nu_x, \nu_y) \ . \tag{20.11}$$

The variance of $t(x, y)$ is given by

$$\sigma^2 = \frac{1}{A} \iint_A [T(\mathbf{x}) - \overline{T}]^2 \, d\mathbf{x}$$

$$= R(0)$$

$$= \int N(\boldsymbol{\nu}) \, d\boldsymbol{\nu} \ . \tag{20.12}$$

When a developed film is scanned by a microdensitometer with an aperture of area $a = \Delta x \Delta y$, the output is given by the following convolution integral:

$$g_a(x, y) = B \iint T(x', y') \, \mathrm{rect}(x - x' | \Delta x / 2)$$

$$\times \mathrm{rect}(y - y' | \Delta y / 2) \, dx' \, dy' \ , \tag{20.13}$$

where

$$\mathrm{rect}(x - x' | \Delta x / 2) = \begin{cases} 1 & \text{if} \quad |x| \leq \Delta x / 2 \\ \\ 0 & \text{otherwise} \end{cases} , \tag{20.14}$$

$$\mathrm{rect}(y - y' | \Delta y / 2) = \begin{cases} 1 & \text{if} \quad |y| \leq \Delta y / 2 \\ \\ 0 & \text{otherwise} \end{cases} , \tag{20.15}$$

and $B = 1 / \Delta x \Delta y$. The output transmission $g_a(x)$ has an autocorrelation function given by

$$R_a(x_0, y_0) = \langle [g_a(x_0, y_0) - \overline{T}']$$

$$\times [g_a(x + x_0, y + y_0) - \overline{T}'] \rangle \ , \tag{20.16}$$

where

$$\overline{T}' = B \iint \overline{T} \, \mathrm{rect}(x - x' | \Delta x / 2) \, \mathrm{rect}(y - y' | \Delta y / 2) \, dx' \, dy'$$

$$= \overline{T} \ . \tag{20.17}$$

The power spectrum of $g_a(x)$ is given by

$$N_a(\boldsymbol{\nu}) = \int R_a(\mathbf{x}_0) \, e^{-2\pi i \boldsymbol{\nu} \cdot \mathbf{x}_0} \, d\mathbf{x}_0$$

$$= N(\nu, \mu) \, T_{\Delta x}(\nu) \, T_{\Delta y}(\mu) \ , \tag{20.18}$$

where

$$T_{\Delta x}(\nu) = \left(\frac{\sin \pi \nu \Delta x}{\pi \nu \Delta x}\right)^2 , \tag{20.19a}$$

$$T_{\Delta y}(\mu) = \left(\frac{\sin \pi \mu \Delta y}{\pi \mu \Delta y}\right)^2 . \tag{20.19b}$$

We should note that $T_{\Delta x}(\nu)$ has zero crossings at $\nu = \pm(1/\Delta x, 2/\Delta x, \ldots)$ and a main lobe of radius $1/\Delta x$.

The variance of $g_a(x)$ is given by Fourier transforming Eq. (20.18) to yield

$$R_a(0) = \sigma_a^2 = \iint N(\nu, \mu) \operatorname{sinc}^2(\nu \Delta x) \operatorname{sinc}^2(\mu \Delta y) \, d\nu \, d\mu$$

$$\leq \iint N(\nu) \, d\nu$$

$$= \sigma^2 . \tag{20.20}$$

If $N(\nu) = C$, a constant (i.e., the noise is a white noise), then we have

$$\sigma_a^2 = C \iint \operatorname{sinc}^2(\nu \Delta_x) \operatorname{sinc}^2(\nu \Delta_y) \, d\mu \, d\nu$$

$$= \frac{C}{a} \tag{20.21}$$

or

$$\sigma_a = \sqrt{\frac{C}{a}} , \tag{20.22}$$

which is the result obtained by Selwyn,[38] and it is true only for white noise. This equation is usually given in the form of density fluctuations (to correspond with the data supplied by the film manufacturers) as follows:

$$\sigma_D = \frac{G}{\sqrt{2a}} , \tag{20.23}$$

where

$$\sigma_D \cong 0.434 \frac{\sigma_a}{\overline{T}} , \tag{20.24}$$

which is valid for $\sigma_a/\overline{T} \leq 0.5$. The constant G corresponds to the granularity data supplied by film manufacturers. Some typical film granularity values are shown in Table 20.I.

TABLE 20.I
Resolution Capability and Granularity Data for Various Films*

Film	Resolution	RMS Granularity[a]
Tri-X Panchromatic Type B	~60 lp/mm (1000:1 contrast)	30×10^{-3}
Pan-X aerial 3400	200 lp/mm	16×10^{-3}
SO-243	465 lp/mm (D-19) 525 lp/mm (D-76) (1000:1 contrast)	7.4×10^{-3}
649GH	Above 2000 lp/mm (1000:1 contrast)	$<5 \times 10^{-3}$

*From Eastman Kodak film data sheet.

[a]Standard deviation is density measured for a net density of 1.0 (uniformly exposed and developed) with a 48-μm circular scanning aperture.

20.2.1.7   *Threshold Modulation*

Threshold modulation (TM) or aerial image modulation (AIM) has been introduced to describe the minimum signal modulation necessary at a given spatial frequency to create a detectable sinusoidal signal on the film. In the analysis of aerial photographic systems, the TM curve is crossed with the lens MTF curve whose crossing point projected to the spatial frequency axis is used as an indicator of the resolution limit of the system (see Fig. 20.10). This measuring

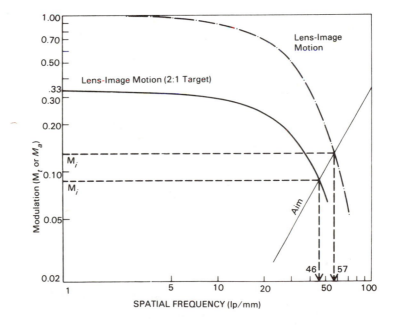

Fig. 20.10 Resolution prediction by the intersection of a system MTF and AIM curve. Modulation of the target on the ground (M_t) or in aerial image at camera lens (M_a) is shown by vertical axis; modulation required for resolution is given by M_i, also read from the vertical axis (from Ref. 44).

technique has a high degree of correlation with subjective readings of control targets by human observers.[27] The TM concept acknowledges that resolution in the sense of recognition or detection depends on contrast and frequency. Clearly, as the contrast of an image decreases, so does our ability to detect its presence because of the existence of the granular structure of the recording medium (film). Thus, any aerial image must have a minimum contrast below which it is undetectable. The minimum value is determined by the film being used. The TM curves are usually measured[28,29] by exposing targets of known modulation onto film with a lens having a known MTF. The measured minimum modulation (subjectively measured by human observers and statistically averaged) as a function of spatial frequency is the TM curve for the photographic process under consideration. In each measurement, images of optimum exposure are viewed. In some instances, the data measured in this manner are fitted to a parabolic curve by using the least-squares technique to obtain the film TM curve.[29]

Variability of TM Curves: Since the TM curve is an attempt to determine the effect of film grain noise on the detectability of a minimum modulation signal, the method of detection and the detection threshold criterion used in an experiment can lead to a variability of the resulting curve. For example, if three-bar targets are used to measure the TM curve, the visual detection criterion used (i.e., the last resolved three-bar group may not have sharp edges) has a strong influence on the resulting curve.

As an example of this effect, the variability of various TM curves for 3404 type aerial film (the SO-243 film used by Scott[28] is the same emulsion as 3404 on a different base material) as reported in the literature is shown in Fig. 20.11. All the curves are contained within the one σ limits of the Kellen et al. data.[46] The limiting curves of Kellen et al. were the one σ limits of the statistical fluctuations of the film grain noise about Scott's data[28] for the mean curve. Scott's value of the TM curve at 100 lp/mm is ≈ 0.066, and the Lauroesch value[29] is given as ≈ 0.049. One interpretation of the statistical data of Fig. 20.11 as given by Kellen et al. was that signals having modulations greater than the upper one σ curve have a high probability of being resolved; signal modulations lying between the upper and lower one σ curves are sometimes resolved; and signal modulations lower than the lower one σ curve have a low probability of being resolved. Thus, the statistical nature of TM data must be recognized whenever one attempts to obtain a resolution criterion for photographic systems.

Procedure for Estimating the Threshold Modulation Curve: The variability of the TM data of photographic film as illustrated in Fig. 20.11 is a strong indication of the differences in the signal-to-noise (S/N) criterion used in each different experiment. Because of this variability, the TM curve is not included as standard data from film manufacturers. As we have seen, such data can be very useful in the choice of film to be used in an optical experiment or system. In this section, we describe a method for estimating the TM curve from data that are supplied (i.e., film MTF, granularity, and film γ) by film manufacturers.

For films where Selwyn's law is valid, the standard deviation of the density fluctuations is given by Eq. (20.23). In the remainder of this chapter, we use ω to denote a scalar spatial frequency. If we define

$$1/\sqrt{a} = 2\omega ,$$

(20.25)

then substitution of Eq. (20.25) into Eq. (20.23) yields

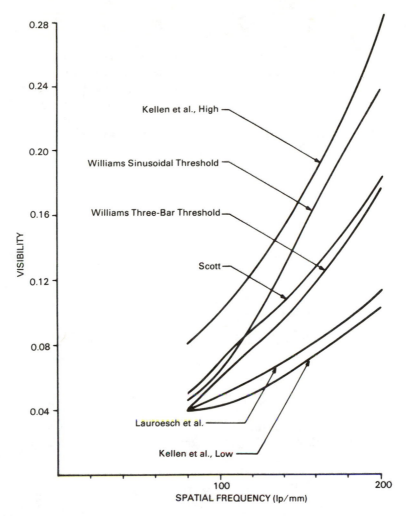

Fig. 20.11 Comparison of various TM curves for 3404 film measured by different techniques (see Table 20.II).

TABLE 20.II
Parameters Used in Measurement Techniques Employed to Produce the Curves of Fig. 20.11

Technique	Reference Source	Parameter
Lauroesch et al.	Ref. 29, Fig. 22	3404 film; D-19, D-76 process; average of five readings; three-bar targets
Scott	Ref. 28, Fig. 5	SO-243 film; D-19; 8 min at 68°F; three-bar targets
Williams (theoretical extrapolation)	Ref. 45, Table 3	3404 film; D-19; 6 min at 68°F; sinusoidal threshold; three bars using peak bar value
Kellen et al.	Ref. 46	3404 film; D-19; 8 min at 68°F; $\overline{D} = 0.7$, sinusoidal

$$\sigma_D = \sqrt{2} \, G\omega \, , \tag{20.26}$$

where ω is the spatial frequency variable measured in line pairs per millimeter and the factor of 2 is used to be consistent with the sampling theorem. Equation (20.26) assumes that no instrument noise is present when measuring granularity, which is a reasonable assumption for the range of validity of Selwyn's law.

To determine the minimum detectable threshold for a sinusoidal signal on a film whose grain noise fluctuations have an average amplitude given by σ_D as expressed by Eq. (20.24), we express the average noise amplitude as a visibility in the form

$$V_{\text{noise}} = \frac{T_{max} - T_{min}}{T_{max} + T_{min}} = \frac{1 - 10^{-2\sigma_D}}{1 + 10^{-2\sigma_D}} \, , \tag{20.27}$$

where

$$T_{max} = 10^{-\overline{D} + \sigma_D} \, , \tag{20.28}$$

$$T_{min} = 10^{-\overline{D} - \sigma_D} \, . \tag{20.29}$$

Substituting Eq. (20.26) into Eq. (20.27) gives

$$V_{\text{noise}}(\omega) = \frac{1 - 10^{-2\sqrt{2}\,G\omega}}{1 + 10^{-2\sqrt{2}\,G\omega}} \, , \tag{20.30}$$

which becomes the TM curve of the film when it is divided by the film MTF and multiplied by $(S/N)_E$, the signal-to-noise ratio in effective exposure, i.e.,

$$\text{TM}(\omega) = (S/N)_E \frac{V_{\text{noise}}(\omega)}{\text{MTF}_{\text{film}}(\omega)} \, . \tag{20.31}$$

Thus, we see that Eq. (20.31) can be used to generate a film TM curve from the granularity number, the film MTF, and the $(S/N)_E$ ratio. The determination of the $(S/N)_E$ ratio can be expressed in terms of the $(S/N)_D$, the signal-to-noise ratio in developed image space (see Fig. 20.1).

As shown in the Appendix, Sec. 20.3 at the end of this chapter, the relationship between the $(S/N)_E$ ratio in effective exposure space and the $(S/N)_D$ ratio in developed image space is given by

$$(S/N)_E = |\gamma| \, (S/N)_D \, , \tag{20.32}$$

where γ is the slope of the D–log E curve of the photographic film. If our criterion for detectability is $(S/N)_E = 1$ (a choice in reasonable agreement with existing experimental data),[45] then the TM curve of interest is

$$\text{TM}(\omega) = \frac{1}{|\gamma|} \left[\frac{1 - 10^{-2\sqrt{2}\,G\omega}}{1 + 10^{-2\sqrt{2}\,G\omega}} \right] \left[\frac{1}{\text{MTF}_{\text{film}}} \right] \, , \tag{20.33}$$

which is the average noise threshold of the film for $(S/N)_E = 1$ expressed in terms of commonly available film data. We now use this TM data for predicting system performance in photographic and holographic situations.

20.2.2   Achievable System Performance

20.2.2.1   *Photographic Applications*

To utilize these film data in a photographic application, the crossing point of the optical system MTF with the estimated TM curve of the film given by Eq. (20.33) projected to the spatial frequency axis is a reasonable indicator of achievable system performance.[27,28]

20.2.2.2   *Holographic Applications*

In holographic applications (covered in Chapters 25, 26, and 27) where the radiation is coherent and the system is not linear in intensity, the important parameter is the visibility of the carrier frequency compared with the TM curve on the film as given[47] by Eq. (20.33). Carrier frequency holograms are usually recorded to obtain a linear transmission versus exposure relationship.

The system performance attainable in these situations can be predicted by crossing a unit fringe visibility curve with the film TM curve and projecting the crossing point to the spatial frequency axis. This will merely define an upper bound on performance as it determines the performance for a point object. In any practical holographic experiment where the object has a finite size, the fringe visibility is difficult to determine since it is related to the Fourier or Fresnel transform of the object being made into a hologram. Generally, the point source result is used as a guide in experimental design and a film with as much resolution as possible is chosen. For Fresnel, Fraunhofer, and sideband Fresnel holograms, the performance indicator will directly affect the resolution attainable in the reconstructed image; in Fourier transform holograms, the performance indicator will affect the field of view of the reconstructed image within the constraint of the space bandwidth product for holograms.[48]

It is apparent from this discussion that the noise criterion discussed here is as useful in holography as it is in photography.[47] The fringe modulation in holography is dependent on our choice of object (i.e., a hologram is inherently a nonlinear system), whereas in incoherent photography the system performance is determined by the imaging system.

20.3   APPENDIX: DERIVATION OF THE RELATIONSHIP BETWEEN $(S/N)_D$ AND $(S/N)_E$

To derive Eq. (20.32) we must relate the S/N ratio in density image space to that in effective image space. If we assume that we always located on the straight line portion of the D–log E curve, then

$$D = \gamma \log E ,$$ (20.34)

where γ is slope of the straight line portion of the D–log E curve.

The noise fluctuations on the film are contained between the limits

$$D_{max} = \overline{D} + \sigma_D ,$$ (20.35)

$$D_{min} = \overline{D} - \sigma_D .$$ (20.36)

Therefore, the fluctuations along the log exposure axis are related to those along the density axis by

$$\sigma_D = \gamma(\Delta \log E) = \frac{D_{max} - D_{min}}{2} \ .$$ (20.37)

The noise visibility in density image space is

$$V_{noise} = \frac{1 - 10^{-2\sigma_D}}{1 + 10^{-2\sigma_D}} \ .$$ (20.38)

For small values of $2\sigma_D$ (a reasonable assumption for films of interest), Eq. (20.38) may be written as

$$V_{noise} \cong 2.3\sigma_D \ .$$ (20.39)

The visibility of the average noise fluctuations in effective exposure space is given by the formula

$$V_E = \frac{1 - 10^{-2(\gamma\Delta \log E)}}{1 + 10^{-2(\gamma\Delta \log E)}} \ ,$$ (20.40)

$$\cong 2.3 \, (\Delta \log E) \ .$$ (20.41)

Combining Eqs. (20.37), (20.39), and (20.41) to remove the effect of the characteristic curve of the photographic process and using magnitude signs to prevent the occurrence of negative visibilities, we obtain

$$|\gamma| \, V_E \cong V_{noise} \ .$$ (20.42)

Consistent with the assumptions made for deriving Eq. (20.31) in the text, the just detectable noise visibility in density image space is obtained by multiplying V_{noise} by $(S/N)_D$. Thus, Eq. (20.42) becomes

$$|\gamma| \, V_E(S/N)_D = V_{noise}(S/N)_D \ .$$ (20.43)

Since noise fluctuations of most films are usually numerically small ($\sigma_D < 0.1$), it is reasonable to anticipate a linear behavior between the $(S/N)_D$ ratio in developed image space and the $(S/N)_E$ ratio in effective exposure space to be of the form

$$V_E(S/N)_E = V_{noise}(S/N)_D \ .$$ (20.44)

A comparison of Eqs. (20.43) and (20.44) yields

$$|\gamma| \, (S/N)_D = (S/N)_E \ .$$ (20.45)

Thus, if we select a S/N criterion in effective exposure space, Eq. (20.45) shows the effect of film gamma on the $(S/N)_D$ ratio in developed image space.

REFERENCES

1. C. E. K. Mees and T. H. James, *The Theory of the Photographic Process*, Macmillan Publishing Co., New York (1966).
2. J. F. Hamilton, "The photographic grain," *Appl. Opt.* 11, 13–21 (1972).
3. J. W. Goodman, "Film-grain noise in wavefront-reconstruction imaging," *J. Opt. Soc. Am.* 57(4), 493–502 (1967).
4. E. L. O'Neill, *Introduction to Statistical Optics*, Chapter 7, Addison-Wesley Publishing Co., Reading, Mass. (1963).
5. S. A. Benton and R. E. Kronauer, "Properties of granularity Weiner spectra," *J. Opt. Soc. Am.* 61(4), 524–529 (1971).
6. K. Biedermann, "The scattered flux spectrum of photographic materials for holography," *Optik* 31, 367–389 (1970).
7. H. M. Smith, "Light scattering in photographic materials for holography," *Appl. Opt.* 11, 26–32 (1972).
8. D. H. R. Vilkomerson, "Measurements of the noise spectral power density of photosensitive materials at high spatial frequencies," *Appl. Opt.* 9, 2080–2087 (1970).
9. R. J. Collier, C. B. Burckhardt, and L. H. Lin, *Optical Holography*, Chapter 9, Academic Press, New York (1971).
10. C. E. Thomas, "Film characteristics pertinent to coherent optical data processing systems," *Appl. Opt.* 11(8), 1756–1765 (1972).
11. G. R. Knight, "Effect of film nonlinearities in wavefront reconstruction imaging," PhD Thesis, Stanford University (1967).
12. J. W. Goodman and G. R. Knight, "Effects of film nonlinearities on wavefront reconstruction images," *J. Opt. Soc. Am.* 58(9), 1276–1283 (1968).
13. O. Bryngdahl and A. Lohmann, "Nonlinear effects in holography," *J. Opt. Soc. Am.* 58(10), 1325–1334 (1968).
14. A. Kozma, G. W. Jull, and K. O. Hill, "An analytical and experimental study of nonlinearities in hologram recording," *Appl. Opt.* 9, 721–731 (1970); corrections, *Appl. Opt.* 9, 1947 (1970).
15 A. Kozma, "Photographic recording of spatially modulated coherent light," *J. Opt. Soc. Am.* 56(4), 428–432 (1966).
16. J. D. Rigden and E. I. Gordon, "The granularity of scattered maser light," *Proc. IRE* 50(11), 2267–2368 (1962).
17. B. M. Oliver, "Sparkling spots and random diffraction," *Proc. IEEE* 51(1), 220–221 (1963).
18. J. W. Goodman, "Some effects of target-induced scintillation on optical radar performance," *Proc. IEEE* 53(11), 1688–1700 (1965).
19. L. H. Enloe, "Noise-like structure in the image of diffusely reflecting objects in coherent illumination," *Bell System Tech. J.* 46(9), 1479–1490 (1967).
20. R. J. Collier, C. B. Burckhardt, and L. H. Lin, *Optical Holography*, Chapter 12, Adademic Press, New York (1971).
21. S. Lowenthal and H. Arsenault, "Image formation for coherent diffuse objects: statistical properties," *J. Opt. Soc. Am.* 60(11), 1478–1483 (1970).
22. J. C. Dainty, "Some statistical properties of random patterns in coherent and partial coherent illumination," *Optica Acta* 17(10), 761–772 (1970).
23. L. I. Goldfischer, "Autocorrelation function and power spectral density of laser-produced speckle patterns," *J. Opt. Soc. Am.* 55(3), 247–253 (1965).
24. T. J. Skinner, "Surface texture effects in coherent imaging," *J. Opt. Soc. Am.* 53(11), 1350A (1963).
25. P. Kirkpatrick and H. M. A. El-Sum, "Image formation by reconstructed wave fronts," *J. Opt. Soc. Am.* 46(10), 825–831 (1956).
26. E. N. Leith and J. Upatnieks, "Wavefront reconstruction with diffused illumination and three-dimensional objects," *J. Opt. Soc. Am.* 54(11), 1295–1301 (1964).
27. G. C. Brock, *Image Evaluation for Aerial Photography*, Chapter 8, Focal Press, Ltd., London (1970).
28. F. Scott, "Three-bar target modulation detectability," *Phot. Sci. Eng.* 10(1), 49–52 (1966).

29. T. J. Lauroesch, G. G. Fulmer, J. R. Edinger, G. T. Keene, and T. F. Kerwick, "Threshold modulation curves for photographic films," *Appl. Opt.* 9(4), 875–887 (1970).

30. G. C. Higgins, "Methods of analyzing the photographic system including the effects of nonlinearity and spatial frequency response," *Phot. Sci. Eng.* 15(2), 106–118 (1971).

31. O. H. Schade, Sr., "An evaluation of photographic image quality and resolving power," *Soc. Mot. Pict. Tel. Engrs.* 73(2), 81–119 (1964).

32. H. Freiser and E. Klein, "Contribution of the theory of the characteristic curve," *Phot. Sci. Eng.* 4(4), 264–270 (1960).

33. J. J. DePalma and J. Gasper, "Determining the optical properties of photographic emulsion by the Monte Carlo method," *Phot. Sci. Eng.* 16(3), 181–191 (1972).

34. D. H. Kelly, "Systems analysis of the photographic process. I. A three-state model," *J. Opt. Soc. Am.* 50(3), 269–276 (1960); see also "Systems analysis of the photograhic process. II. Transfer function measurements," *J. Opt. Soc. Am.* 51(3), 319–330 (1961).

35. C. N. Nelson, "Prediction of densities in fine detail in photographic images," *Phot. Sci. Eng.* 15(1), 82–97 (1971).

36. D. C. Ehn and M. B. Silevitch, "Diffusion model for the adjacency effect in viscous development," *J. Opt. Soc. Am.* 64(5), 667–676 (1974).

37. A. Shepp and W. Kammerer, "Increased detectivity by low gamma processing," *Phot. Sci. Eng.* 14(5), 363–368 (1970).

38. *SPSE Handbook of Phot. Sci. and Eng.*, W. Thomas, Jr., ed., Wiley Interscience Publishing Co., New York (1973).

39. *Photographic Considerations for Aerospace,* 2nd ed., H. J. Halland, H. J. Howell, eds., Itek Corporation, Lexington, Mass. (1966).

40. M. A. Kriss, "A review of old and new methods of evaluating the image structure of color film," in *Color: Theory and Imaging Systems,* A. Eynard, ed., pp. 113–166, Soc. Phot. and Eng., Washington, D.C. (1973).

41. R. A. Jones, "An automated technique for deriving MTFs from edge traces," *Phot. Sci. Eng.* 11(2), 102–106 (1967).

42. R. V. Shack, "Characteristics of an image forming system," *J. Res. Natl. Bur. Stand.* 56(5), 245–260 (1956).

43. P. F. Mueller, "Color image retrieval from monochrome transparencies," *Appl. Opt.* 8(10), 2051–2057 (1969).

44. R. Welch, "The prediction of resolving power of air and space photographic systems," *Image Tech.* 14(5), 25–32 (1972).

45. R. Williams, "A re-examination of resolution prediction from lens MTFs and emulsion threshold," *Phot. Sci. Eng.* 13(5), 252–261 (1966).

46. P. F. Kellen, J. D. Boardman, and G. O. Reynolds, paper presented at SPSE Seminar on Recent Advances in the Evaluation of the Photographic Image, Newton, Massachusetts, July 15–16, 1971.

47. K. Biedermann, "A function characterizing photographic film that directly relates to brightness of holographic images," *Optik* 28(2), 160–176 (1968).

48. J. B. DeVelis and G. O. Reynolds, *Theory and Applications of Holography,* Chapter 5, Addison-Wesley Publishing Co., Reading, Mass. (1967).

21   Sources of Coherent Noise and Their Reduction*

21.1   INTRODUCTION

We conclude the discussion we began in Chapter 20 about photographic films with a review here of the sources of coherent noise and their reduction.

In Sec. 21.2, we discuss the types and sources of noise in coherent imaging systems. These include dust noise, phase image noise, and speckle. Various techniques for reducing the effects of dust noise are described. Control of the spatial and temporal coherence of the system radiation, which reduces the effect of system speckle noise, is discussed and demonstrated.

In Sec. 21.3, several of the numerous speckle-averaging techniques that have been introduced in the literature are put into the broad categories of spatial and temporal coherence control techniques and further subclassified in terms of ensemble- and temporal-averaging techniques.

Results are summarized in Sec. 21.4 by listing five experimental procedures to be considered when designing a coherent optical system in which all noise effects will be minimized. Two experiments obeying these procedures are discussed and the results are shown.

21.2   SYSTEM NOISE CONSIDERATIONS IN COHERENT OPTICAL SYSTEMS

In addition to the ultimate noise limitation imposed by the film as discussed in Chapter 20, coherent optical systems suffer from additional system noise arising from various sources. As a result, the figure of performance discussed previously becomes an upper bound that is achievable only if the system noise effects can be appropriately reduced. Here we discuss the types and origins of system noise commonly encountered in coherent optical systems. Methods for reduction of their influence to obtain improved image quality are also discussed. The types of noise in addition to grain noise arising from the film grain structure are catego-

---

*This chapter is a continuation of the article reprinted with permission in Chapter 20: "Review of Noise Reduction Techniques in Coherent Optical Processing Systems," by J. B. DeVelis, Y. M. Hong, and G. O. Reynolds, in *Coherent Optical Processing,* H. J. Caulfield, ed., SPIE Proc. 52, 55–81 (1975).

rized as noise arising from film nonlinearities caused by the chemical processing, noise arising from the phase image of the film caused by changes in the gelatin during processing, noise arising from diffraction caused by dust particles in the system or lens imperfections, and finally speckle noise arising from interference caused by random phase disturbances within the system. In general, speckle noise tends to be the predominating noise factor in the system and considerable research has been devoted to techniques for its reduction. The other types of system noise can be minimized by careful experimental procedures. If all system noise except grain noise were absent, the system performance could be adequately predicted by utilizing the figure of performance outlined in Sec. 20.2.2. Thus, this criterion predicts the design goal for a coherent optical system experiment. The other system noise factors tend to degrade system performance further.

21.2.1   Effects of Film Nonlinearities and Linear Processing

In coherent optical systems, a linear relationship must exist between the amplitude transmittance of the developed image and the input exposure. Deviations from this linear relationship are termed *film nonlinearities,* which distort the spatial frequency spectrum of the desired image. These nonlinear effects are avoided by controlling the film sensitometry, i.e., linear processing.[a]

If the input exposure is on the straight line portion of the D–log E curve [refer to Fig. 20.2(a)] then

$$D = \gamma \log E .\tag{21.1}$$

Combining Eqs. (20.2) and (21.1), we see that

$$T_I = E^{-\gamma} .\tag{21.2}$$

The amplitude transmittance of the film is given by

$$T_A = T_I^{1/2} = E^{-\gamma/2} .\tag{21.3}$$

Thus, if one desires an amplitude transmittance proportional to the input exposure, then Eq. (21.3) tells us that we must process the film such that $\gamma = -2$. The negative sign of gamma means that a reversal is required so that the resulting film that is placed into the coherent processing system is a photographic positive. In the field of coherent processing, the $\gamma = -2$ process is commonly called *linear processing.*

To implement linear processing, pre- or post-fogging techniques are usually required. These techniques reduce the contrast of the object exposure making it possible to fit a scene with a wide exposure range onto the linear portion of the D–log E curve.[1] An example illustrating the effect of nonlinear processing is shown in Fig. 21.1.

In Fig. 21.1(a) the film containing encoded multiple images was processed linearly, resulting in high-quality reconstructions. In Fig. 21.1(b), the effect of deliberate nonlinear processing of the film resulted in mixing between the various images causing a noticeable effect on image quality. In a single image, this effect is much more subtle. An example demonstrating this subtle effect for a square

---

[a]Linear processing is covered further in Chapter 34.

Fig. 21.1 Images retrieved from one of four channels of a multiplexed recording: (a) $\gamma = -2$ linear processing and (b) $\gamma = +2$ nonlinear processing (from Ref. 1).

wave object is shown in Fig. 21.2. In Fig. 21.2(a), the film was processed linearly and the Fourier transform exhibited the absence of the even harmonics, as expected. In Fig. 21.2(b), the film was nonlinearly processed resulting in the generation of even harmonics.

Linearity may also be achieved by utilizing a low-contrast object. An example illustrating the absence of film nonlinearities in the low-contrast limit is shown in Fig. 21.3. Figure 21.3(a) is the Fourier transform of a high-contrast ($\approx 1000:1$) cosine fringe field made in an interferometer and processed nonlinearly. All the orders are present indicating the nonlinearity of the process since only the dc and the first harmonics should be present. In Fig. 21.3(b), the fringe contrast was reduced to $\approx 2:1$ and processed with the same gamma used in (a). The resulting transform shows the presence of only first-order harmonics, illustrating the linearization by lowering object contrast.

Low-contrast objects are often met in holographic applications. In such cases, the linear relationship between transmission and exposure is best described by the T versus E curve, where the background intensity behaves as a bias.[2-7] Requiring that the bias point occur in the middle of the linear portion of the T versus E curve maximizes the dynamic range of the holographic experiment.

(a)

(b)

Fig. 21.2 Fourier power spectrum: (a) square wave and (b) rectangular wave showing the absence of the even harmonics for the square wave. (Courtesy of P. Mueller.)

Fig. 21.3 Fourier power spectra of (a) high-contrast and (b) low-contrast ($\cos^2$ fringes at a fundamental frequency of 15 lp/mm). Recorded on AHU microfilm and processed nonlinearly.

An example of a D–log E curve described by Kozma[7] for a 649F Kodak plate exposed with a He-Ne laser (6328 Å) and developed in D-19 for 12 min at 68°F is shown in Fig. 21.4. The O denotes the operating bias point (scaled to maximize the near dynamic range of the signal) and is seen to occur in the far region of the D–log E curve. The measured T versus E curve for this experiment is shown in Fig. 21.5. The operating bias point O is seen to lie in the middle of the linear portion of the T versus E curve.

In holography, the film nonlinearities are sometimes avoided by spatially band limiting the signal and using a spatial filter to select a single harmonic order of the reconstructed image.[5] This same effect could be achieved by using a spatial filter to reconstruct the spectrum shown in Fig. 21.3(a).

21.2.2   Phase Image Noise

The noise arising from the phase image consists of two components.[8,9] One component is a phase change arising from a relief image on the film caused by a differential hardening of the gelatin when fixing due to the presence of the density image on the film. The other component of the phase image is an internal phase change caused by index variations of the gelatin (due to the presence of a density image).[8,9] The surface phase image can be removed by inserting the film in an index matching liquid (xylene, balsam, freon) commonly referred to as a *liquid*

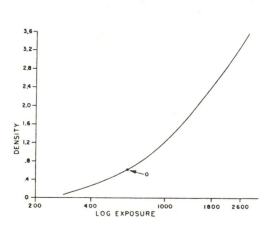

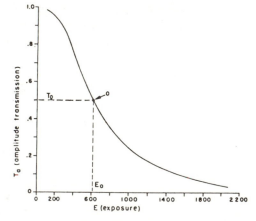

Fig. 21.4 Log exposure versus density for a 649F Kodak plate exposed with He-Ne light (6328 Å) and developed in D-19 for 12 min at 68°F (from Ref. 7).

Fig. 21.5 Exposure versus amplitude transmittance curve for the 649F plate used in Fig. 21.4, where O is the operating bias point (from Ref. 7).

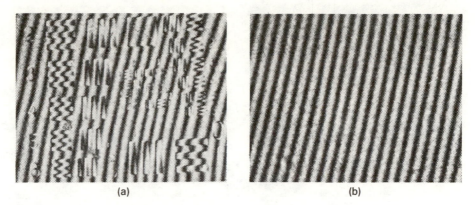

(a)   (b)

Fig. 21.6 Phase relief image suppression by use of liquid gating: phase image of bar targets viewed in reflection mode (a) in an interference microscope and (b) in an interference microscope using xylene as a liquid gate. (Courtesy of W. Dyes.)

gate. The effect of the internal phase image is completely removed by performing the optical processing experiment in the reflection mode. In the transmission mode, the effect of the internal phase image is minimal since its magnitude is small and its frequency response is nearly constant.[8] An example illustrating the suppression of the relief phase image by liquid gating is shown in Fig. 21.6. The phase images shown here were made in an interference microscope operating in a reflection mode. The reference mirror in the instrument was tilted to generate fringes across the field of view which shift at the position of the phase image as seen in Fig. 21.6(a). The suppression of the surface phase image by using a liquid gate in the interference microscope is observed in Fig. 21.6(b).

21.2.3   Cleanliness of Optical Components

The noise arising from dirt and dust deposited on various elements of the optical systems or from lens inclusions appears as diffraction patterns superimposed on the image. This source of noise is best removed by ensuring that the experiments are performed in a clean area with good quality optics that are cleaned before performing the experiment. This requirement can sometimes be relaxed by utilizing a rotating piece of tilted flat glass[10] or a rotating prism[11] in the entrance pupil of the optics. This causes the source image to rotate in a small circle in the Fourier transform plane. This modification in the optical system creates a coherent noise suppression by a time-averaging mechanism, which in turn reduces image resolution. Similar results can be obtained by rotating the entire imaging system about the optical axis in a time short compared to the exposure.[12] All of these techniques make the noise time dependent in the image plane when the image is stationary. Therefore, a time average causes the noise components to spread out over the entire image plane, reducing their objectional effects when viewing the image. Results using the rotating flat plate technique are illustrated in Fig. 21.7 where the incoherent, coherent, and time-averaged coherent images of a seismic section are shown.

The predominant noise effect in coherent optical systems is speckle noise. Many techniques have been proposed to reduce the effects of speckle noise; these techniques are systematically discussed in the next section.

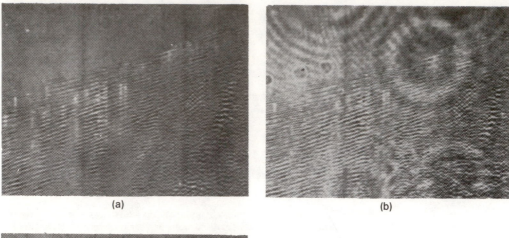

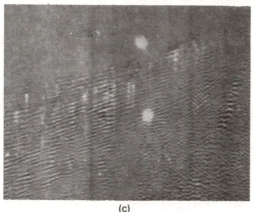

Fig. 21.7 Example of reduction of coherent optical noise by the time-averaging method: (a) incoherent photograph of seismic section, (b) seismic section photographed with coherent laser light, and (c) seismic section photographed with laser light and rotating glass plate to time average coherent noise (from Ref. 10).

21.3   SPECKLE NOISE REDUCTION TECHNIQUES

21.3.1   Speckle and Holographic Systems

Diffusers were originally introduced in holographic systems to reduce noise effects caused by dust particles and/or lens imperfections. In addition, the diffuser also increased the field of view of the hologram, sharpened the depth of focus of the three-dimensional image, and reduced the dynamic range requirement on the recording film. The object diffracts the illumination, spreading the information over the entire holographic film plane (redundancy), so that a small piece of the hologram can reconstruct the object with a corresponding loss of resolution (as determined by the space bandwidth product)[13] and increase its depth of field. It was also observed that the diffuser introduced a granular noise pattern superimposed on the image.[14] This noise pattern is referred to as *speckle noise*. The size of the speckle in the reconstructed image is determined by the size of the hologram used to reconstruct the image. These effects are nicely demonstrated in Fig. 21.8. In this experiment, quasimonochromatic coherent radiation from a laser at 6328 Å was reflected from an object consisting of chessmen to form the hologram. The reconstructions shown here use the same source and show a decrease in speckle and increased resolution—but with a loss of depth of focus—as greater portions of the hologram are illuminated. The smallest beam size used in this experiment was almost 0.5 mm in diameter.

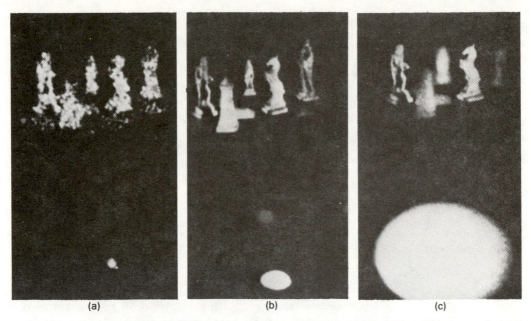

(a)   (b)   (c)

Fig. 21.8 Reconstructions from various size pieces of a sideband Fresnel hologram formed with diffuse illumination (from Ref. 15).

21.3.2   Depth-of-Focus Noise in Holographic Systems

Because of the three-dimensional nature of reconstructed holographic images, depth-of-focus noise can be a serious problem. The depth-of-focus advantage of diffuse illumination has been observed in a holographic particle sizing system. A reconstructed image from a typical particle sizing experiment is shown in Fig. 21.9. In this figure, the noise created by the out-of-focus particles makes it difficult to measure the particle size distribution.

In Figs. 21.10(a) and (b), the reconstructed images of two objects approximately 1 cm apart in a diffusely illuminated sample volume are shown. Figure 21.10(a) shows the reconstruction of a three-bar target and (b) shows the recon-

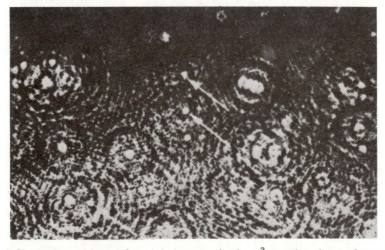

Fig. 21.9 Reconstructed image from a hologram of a 1-cm^3 sample volume of suspended 30-μm lycopodium powder. The arrows indicate focused images and the noise is from out-of-focus particles (from Ref. 16).

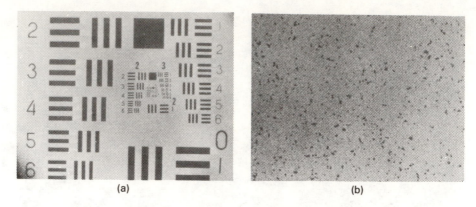

Fig. 21.10 Reconstructed images from a diffuse hologram: (a) reconstruction of resolution target plane from diffuse hologram and (b) reconstruction of particle plane from diffuse hologram showing that the system has a sharp depth of focus. (The out-of-focus bar target causes a slight nonuniformity of the background.) (Courtesy of W. Dyes.)

struction of a plane of small particles. In Fig. 21.10(b), the only indication of the out-of-focus bar target is that a slight nonuniformity in the background is present. The resolution in the reconstructed image is speckle limited, and additional speckle reduction methods would be necessary to achieve higher resolution.

21.3.3   Speckle in Coherent Imaging Systems [b]

In any coherent optical system, a random phase aberration in the object plane, with an autocorrelation length that is less than the characteristic size of the system impulse response measured in the object plane, gives rise to a speckle pattern in the image plane. Physically, each random phase inclusion within the impulse response appears to have originated from a different point source. The coherent imaging system images each phase inclusion as a separate amplitude impulse response, and the resulting intensity distribution reflects the mutual interference of all the amplitude point images giving rise to speckle. An illustration of the change in speckle size with f number is shown in Fig. 21.11. In general, the larger the temporal coherence of the source, the more predominant the resulting speckle pattern.

Speckle patterns such as those in Fig. 21.11 produced with the thermal sources are easier to reduce in a partially coherent imaging system because the bandwidth can be easily increased, thereby reducing the coherence length of the radiation. This effect is demonstrated in Fig. 21.12 where coherent images of a cement block were made. In Fig. 21.12(a), a collimated beam from a mercury arc source with a coherence length of less than the depth of the surface roughness was used. In Fig. 21.12(b), a He-Ne gas laser having a coherence length much greater than the depth of the surface roughness was used as the source. This result was also shown as Fig. 11.8 in Chapter 11. Both illuminating fields have approximately the same degree of spatial coherence. The absence of speckling in Fig. 21.12(a) shows that the coherence length is a useful parameter for controlling speckle in coherent imaging experiments (also see Table 11.I in Chapter 11).

---

[b]These effects were introduced and illustrated previously in Chapters 11, 14, and 18.

(a)   (b)

Fig. 21.11 Speckling density in a coherent image of ground glass with an (a) f/9 and (b) f/64 imaging system. The illumination system was a collimated mercury arc source filtered at 5460 Å with a bandwidth of 100 Å (from Ref. 17).

The coherence length of a source is defined as

$$\Delta l_c = \frac{c}{\Delta \nu} = \frac{\bar{\lambda}^2}{\Delta \lambda} \, , \qquad\qquad (21.4)$$

where $\bar{\lambda}$ is the mean wavelength, $\Delta\lambda$ is the bandwidth of the radiation, and c is the speed of light. Some typical values of coherence lengths for various sources are listed in Table 21.I.

The speckle may also be reduced by controlling the ensemble-averaged spatial coherence of the illuminating beam. This is illustrated qualitatively in Fig. 21.13. In Fig. 21.13(a), the interference pattern from a two-pinhole experiment with a random diffuser over the pinholes and a laser illumination source is shown. High-contrast fringes exist over the field of view of the instrument indicating a high degree of spatial and temporal coherence. In Fig. 21.13(b), the diffuser was continuously moved during the exposure, reducing the fringe visibility considerably. Since the temporal coherence was not varied in this experiment, the resulting loss of fringe visibility is directly related to a decrease in the ensemble-averaged spatial coherence.[19] Thus, we see that a reduction in the spatial coherence of the radiation also averages speckle noise.

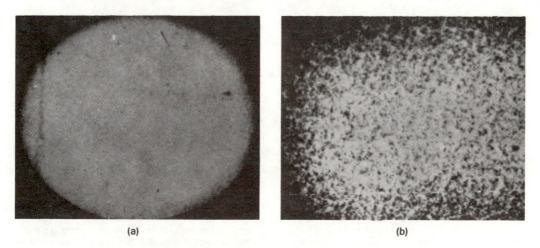

(a)   (b)

Fig. 21.12 Photographs of a cement block illuminated with (a) a collimated merury arc source and (b) a He-Ne gas laser (from Ref. 18).

TABLE 21.I
Coherence Length for Various Types of Sources

Source Type	Mean Wavelength (Å)	Bandwidth (Å)	Coherence Length
Mercury arc	5461	100	<0.03 mm
Mercury arc	5461	10	<0.3 mm
Silver laser	5145	$\sim 9 \times 10^{-2}$	<3 cm
He-Cd laser	4416	$\sim 7 \times 10^{-3}$	<27.5 cm
He-Ne laser	6328	10^{-4} to 10^{-5}	$\approx$100 m

The effect of decreasing the ensemble-averaged spatial coherence in a coherent imaging situation is demonstrated in Fig. 21.14. In this experiment, a bar target object was illuminated with diffuse coherent radiation from a He-Ne laser and ensemble-averaged images were observed. Figure 21.14(a) resulted from one exposure, (b) from an ensemble of six exposures where the diffuser was moved between exposures, (c) from an ensemble of ten exposures, and (d) was obtained from an ensemble of thirty exposures (a large member ensemble). These results demonstrate the improvement in image quality by increasing the number of ensemble members used to average the speckle noise to a uniform background. This averaging process reduces the image contrast and increases the apparent resolution.

Another interpretation of this result that is useful in describing speckle reduction techniques is the principle that the image remains stationary and the noise does not. Thus, on the average, the images of the individual ensemble members add while the noise is averaged over the ensemble yielding a constant value in the large member ensemble limit.

Having determined that the source of speckle is the coherence of the illuminating radiation used in an experiment and having discussed the principle of the reduction mechanism, we now categorize several of these various speckle reduction methods that have appeared in the literature as spatial and/or temporal coherence reduction methods.

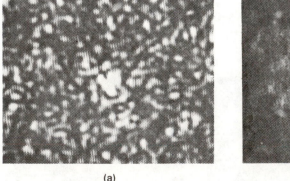
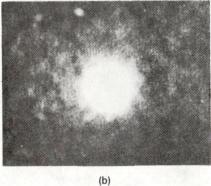

(a)   (b)

Fig. 21.13 Two-pinhole interference experiment performed with laser light and 300-μm pinholes at a separation of 2 mm: (a) fringe pattern resulting from illumination with a diffuse coherent field and (b) time-averaged fringe pattern resulting from illumination with a time-varying diffuse coherent field (from Ref. 17).

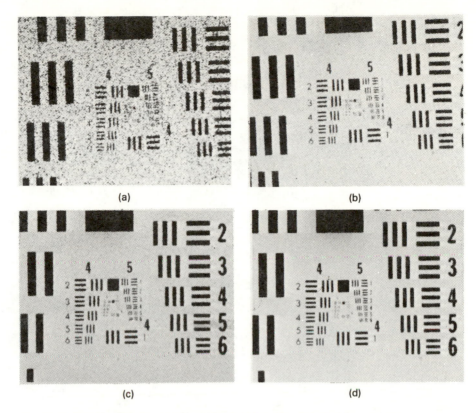

(a)   (b)

(c)   (d)

Fig. 21.14 Demonstration of the principle of speckle reduction by ensemble-averaging: (a) one exposure, (b) six sequential exposures, (c) ten sequential exposures, and (d) thirty sequential exposures.(Courtesy of W. Dyes.)

21.3.4   Speckle Reduction by Control of Temporal Coherence

Two techniques have appeared in the literature for reducing speckle by control of the temporal coherence. One is to use a broadband source as demonstrated in Fig. 21.12. The other technique[20] is to perform a sequential or simultaneous average over an ensemble of images, each of which was formed with different narrow bandwidth radiation. The result of this experiment is shown in Fig. 21.15.

This technique can be applied to coherent imaging systems or to holographic microscopy. In both cases, speckle reduction can result in a significant improvement in the quality of the image.

21.3.5   Speckle Reduction by Control of Spatial Coherence

21.3.5.1   Control of Spatial Coherence

The reduction of speckle by controlling the spatial coherence of the illumination system has been observed in both spatial filtering systems[10] and holographic systems.[21] The decrease in spatial coherence required to reduce the effect of speckle implies a larger source size that in turn reduces the system resolution performance. An example showing this speckle reduction technique at the cost of resolution performance is shown in Fig. 21.16.

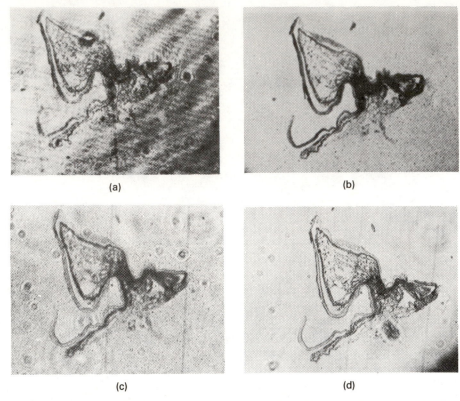

Fig. 21.15 Experiment performed to show the averaging effect of multicolor speckle. The object is an 8-μm-thick section from the optic nerve of a crayfish. The length of the nerve is approximately 1 mm. The images were obtained with (a) collimated laser light illumination, (b) collimated white light, (c) collimated source at 5500 Å with a 5-Å bandwidth, and (d) the average image obtained from six separate band-limited wavelengths spanning the spectrum from 4300 to 5800 Å (from Ref. 20).

21.3.5.2   Speckle Reduction by Spatial Ensemble-Averaging Techniques

Ensemble-averaging techniques have been proposed as methods for reducing speckle. These may be subcategorized as (a) redundancy by spatial multiplexing, (b) redundancy by spatial frequency multiplexing, and (c) redundancy by spatial and spatial frequency multiplexing.

Redundant Spatial Multiplexing: Redundant spatial multiplexing has been achieved by sequential[22,23] and simultaneous recording.[24-30] Both techniques create superpositions of redundant low spatial frequency images in a wide bandwidth recording system to average the speckle noise. An interesting example showing speckle noise averaging by the simultaneous recording technique is shown in Fig. 21.17. These techniques have been applied to both holographic[24] and coherent imaging systems.[29]

Redundant Spatial Frequency Multiplexing: Speckle has also been averaged by multiplexing (incoherently combining) low-frequency images in the spatial frequency plane in both holographic and coherent optical imaging systems. This is accomplished by scanning a hologram with an aperture[31,32] or by utilizing a diffraction grating to create a multichannel illumination system.[21,29,30] An example of speckle noise reduction with such a grating illumination system

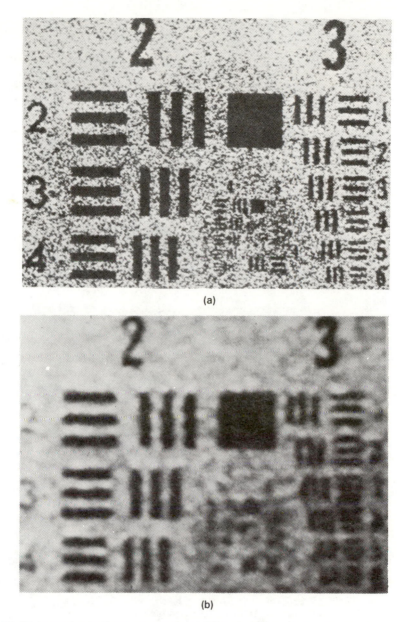

Fig. 21.16 Image formed by reconstructing a diffuse hologram of a test target: (a) reconstruction accomplished with an unresolved source of diameter d and (b) source size of $9d$ used to reconstruct the hologram (from Ref. 21).

where the channels are incoherent with respect to each other is shown in Fig. 21.18(b) compared to an image made with a single plane wave in (a). A system combining diffuse illumination and multiple incoherent channels realized by a moving diffraction grating has also been achieved in a holographic reconstruction.[33]

Redundant Spatial and Spatial Frequency Multiplexing: The two techniques described above have been combined to reduce speckle in a holographic microscope.[34] This technique shows that an average of six channels, created in this case

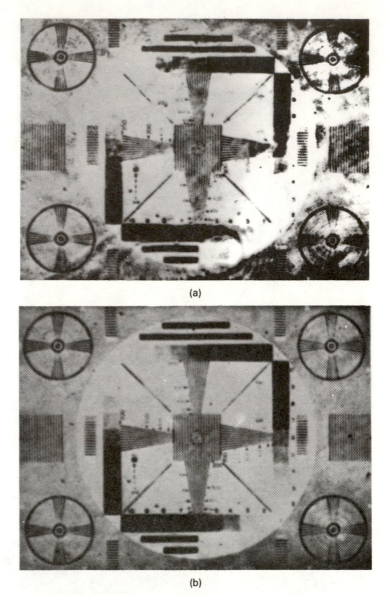

(a)

(b)

Fig. 21.17 Speckle noise averaging by simultaneous recording techniques. Reconstruction produced by (a) one of nine subholograms and (b) nine redundant holograms showing the speckle noise reduction (from Ref. 27).

by a prism arrangement, achieves a desirable statistical average. A high numerical aperture (NA) microscope is used to average low NA images to achieve the speckle reduction. The clarity of the image field (i.e., uniformity in intensity) achieved by this technique is shown in Fig. 21.19.

21.3.6   Speckle Reduction by Time-Averaging

21.3.6.1   *Coherent Imaging System*

Speckle reduction has been achieved in coherent imaging by utilizing a moving diffuser between the source and the object and a sufficiently long time average to remove speckle. In such systems, the image is stationary in space and the speckle

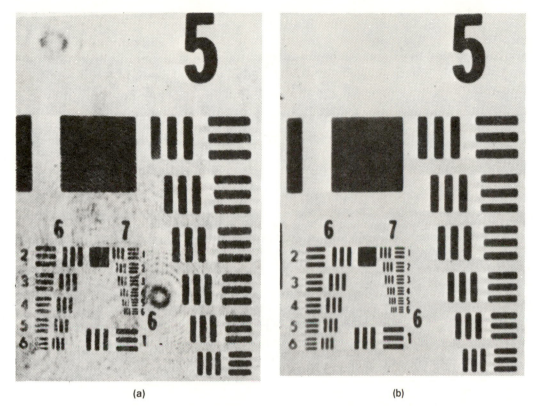

(a)   (b)

Fig. 21.18 Images from a two-lens unit magnification telescope: (a) plane wave illumination and (b) multiple incoherent wave illumination showing speckle averaging (from Ref. 21).

noise is averaged by the motion of the diffuser. A long exposure time allows the stationary image signal to increase while the noise is averaged to a constant value. Thus, the speckle noise is averaged causing the background exposure to increase giving the appearance of a lower contrast image. The time-averaged image obtained in such an experiment is equal to the ensemble-averaged image (which

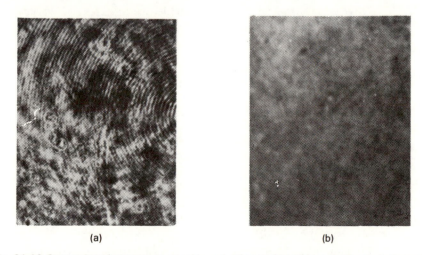

(a)   (b)

Fig. 21.19 Comparison between no speckle reduction and speckle reduction in holographic microscopy with a NA of 0.31 for each subimage: (a) microscopic objective NA = 1.25, λ = 6328 Å and (b) reconstruction with speckle removal (from Ref. 34).

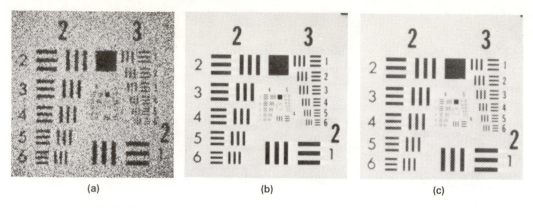

Fig. 21.20 Comparison between image of a three-bar target made (a) in reflection with a laser source (coherent image), (b) in reflection with a laser source and a moving diffuser (ensemble-averaged coherent image), and (c) with reflected incoherent illumination (incoherent image). The equivalence of ensemble-averaged coherent imaging and incoherent imaging in terms of resolution cutoff and subjective image quality is apparent in (b) and (c). (Courtesy of W. Dyes.)

reduces spatial coherence as illustrated in Figs. 21.13 and 21.14). Both the time- and/or ensemble-averaged images are the same as those obtained in an equivalent incoherent imaging system.[35] An example illustrating the equivalence of time-averaged coherent and incoherent imagery is shown in Fig. 21.20.

21.3.6.2   Image Holography

Speckle reduction by time-averaging techniques may also be achieved in image holography. An example of speckle reduction in the reconstruction of an image hologram is shown in Fig. 21.21. In this experiment, a laser beam was focused onto a moving diffuser at the focal plane of a collimator to achieve the time average. In holography, this technique is unique to image holography due to the stationarity of the image plane. An image hologram was used as a filter to negate the effect of a random phase-distorting medium (simulated cataract). Figure 21.21(a) shows the corrected image obtained without using a rotating diffuser to average speckle and (b) shows the speckle-averaged result. This system is another example of speckle-averaging by effective reduction of spatial coherence of the illumination.

21.3.6.3   Dynamic Coherent Imaging Systems

Time-averaging techniques in which the illumination is systematically varied have been used to average speckle noise.[10,11] These systems require longer exposure times to obtain speckle noise reduction. They may use a rotating prism effect to average many different speckle patterns onto a stationary image. These systems also utilize the time average to change the ensemble-averaged spatial coherence (effective source size) of the system at a slight cost of resolution, i.e., the source size appears to be larger. The effective prism angle determines the resolution performance of the system. The maximum performance of the system is achieved when the prism angle matches the half-angle of the extreme ray from the object that is used to specify the NA of the lens.

    An example comparing incoherent, coherent, and dynamic coherent photographs made through the same system is shown in Fig. 21.22. These photographs

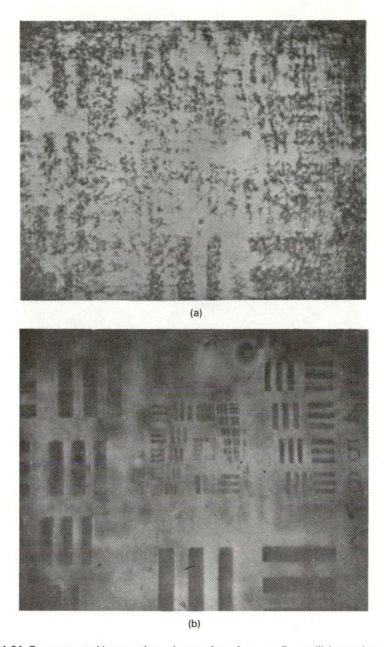

(a)

(b)

Fig. 21.21 Reconstructed images through a random phase medium utilizing an image holo-gram correction filter: (a) image of resolution target after holographic correction without a rotating diffuser and (b) same as (a) but with a rotating diffuser (from Ref. 36).

demonstrate the speckle noise reduction, the comparable resolution perfor-mance between incoherent and dynamic coherent systems, and the differences in the image contrast at higher resolution. The technique has also been applied to achieve dark field and Zernike phase contrast imagery.[11]

21.4   DESIGN CONSIDERATIONS FOR COHERENT OPTICAL SYSTEMS

We have seen that noise in coherent imaging systems arises from a variety of sources. The theoretical limit to system performance imposed by all these noise

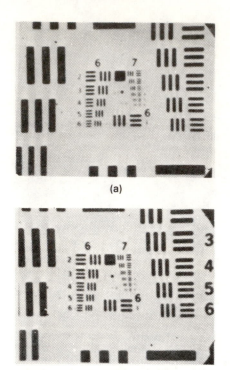

(a)

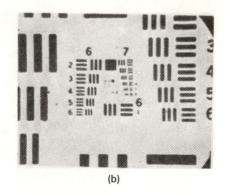

(b)

(c)

Fig. 21.22 Comparison of images produced in the same optical imaging system
with (a) incoherent illumination (incoherent image), (b) collimated mercury arc
illumination (laser coherent image), (c)
dynamic coherent illumination with an
off-axis mercury source and matched
prism angle (dynamic coherent) (from Ref.
11).

sources is determined by the photographic film if the film used in a given
experiment was chosen by using the achievable system performance criterion
outlined in Chapter 20. In practice, this limit is rarely achieved.

We have also seen that the most serious and objectionable kind of coherent
system noise is speckle, which may be eliminated by a variety of techniques. All
these techniques effectively control the temporal and/or spatial coherence of the
illuminating radiation, thereby creating many independent speckle patterns
superimposed on a stationary image that can be effectively averaged to reduce
the speckle noise effect.

From this discussion, we conclude that the following procedures should be
considered when designing a coherent optical system in order to minimize all
noise effects:

1. Choose a film whose achievable performance criterion is the system resolu-
    tion limitation.

2. Minimize film nonlinearities by careful control of film processing.

3. Minimize film relief image noise by appropriate liquid gating[37] or by bleach-
    ing the film in carrier-modulated systems.[38]

4. Use good quality, clean optics.

5. Minimize the coherence of the source radiation by any appropriate technique
    compatible with the given experiment to reduce speckle noise.

An example of a coherent imaging system designed by following these procedures is carrier-modulated imagery[c] designed to achieve color image retrieval

---

[c] Also discussed in Chapters 18 and 34.

Fig. 21.23 Color image, shown here in black and white, retrieved from a carrier-modulated image on a monochrome transparency. (Courtesy of P. Mueller.)

from monochrome transparencies.[39] In this system, the recording film is chosen such that the spatial carrier frequency is well below the spatial frequency cutoff predicted by the achievable performance criterion. The film nonlinearities that create intermodulation noise are minimized by processing the film to $\gamma = -2$, and clean optics capable of resolving the carrier frequency well are used in the retrieval system. The phase noise is reduced by either liquid gating or by bleaching the film to create a modulated phase image. The control of coherence is very important in the reconstruction system in order to obtain high-quality color imagery having good fidelity with a maximum amount of energy. This was achieved by using a large white-light source with appropriate color filters, thereby avoiding speckle. An example of an image achieved from this system (printed here in black and white) is shown in Fig. 21.23.

Another example of a coherent imaging system that produces good quality imagery is a white light or thick hologram.[38] These holograms are recorded with three different color lasers on films that will resolve the carrier frequency. Film nonlinearities are avoided by controlling the photographic processing, and dirt effects are removed by the redundancy of the hologram. The surface relief image noise effects are minimized by this method. The white light required for reconstruction at the Bragg angle minimizes the source coherence, thereby yielding speckle-free imaging. An example of an image reconstructed from such a hologram is shown in Fig. 21.24. This figure demonstrates a speckle-free image when a white source is used to illuminate the hologram.

(a)   (b)

Fig. 21.24 Image retrieved from a thick hologram using (a) white light and (b) argon laser sources. (Courtesy of W. Dyes.)

REFERENCES

1. P. F. Mueller, "Linear multiple image storage," *Appl. Opt.* 8, 267–273 (1969).
2. C. E. Thomas, "Film characteristics pertinent to coherent optical data processing systems," *Appl. Opt.* 11, 1756–1765 (1972).
3. G. R. Knight, "Effect of film nonlinearities in wavefront reconstruction imaging," PhD Thesis, Stanford Univ. (1967).
4. J. W. Goodman and G. R. Knight, "Effects of film nonlinearities on wavefront reconstruction images," *J. Opt. Soc. Am.* 58(9), 1276–1283 (1968).
5. O. Bryngdahl and A. Lohmann, "Nonlinear effects in holography," *J. Opt. Soc. Am.* 58(10), 1325–1334 (1968).
6. A. Kozma, G. W. Jull, and K. O. Hill, "An analytical and experimental study of nonlinearities in hologram recording," *Appl. Opt.* 9, 721–731 (1970); corrections, *Appl. Opt.* 9, 1947 (1970).
7. A. Kozma, "Photographic recording of spatially modulated coherent light," *J. Opt. Soc. Am.* 56(4), 428–432 (1966).
8. R. L. Lamberts, "Characterization of a bleached photographic material," *Appl. Opt.* 11, 33–41 (1972).
9. H. Smith, "Photographic relief images," *J. Opt. Soc. Am.* 58(4), 533–539 (1968); see also "Production of photographic relief images with arbitrary profile," *J. Opt. Soc. Am.* 59(11), 1492–1495 (1969).
10. C. E. Thomas, "Coherent optical noise suppression," *Appl. Opt.* 7, 517–522 (1968).
11. D. J. Cronin and A. E. Smith, "Dynamic coherent optical system," *Opt. Eng.* 12(2), 50–55 (1973).
12. R. Kirkpatrick and H. M. A. El-Sum, "Image formation by reconstructed wave fronts. I, physical principles and methods of refinement," *J. Opt. Soc. Am.* 46, 825–831 (1956).
13. J. B. DeVelis and G. O. Reynolds, *Theory and Applications of Holography,* Chapter 5, Addison-Wesley Publishing Co., Reading, Mass. (1967).
14. E. N. Leith and J. Upatnieks, "Wavefront reconstruction with diffused illumination and three-dimenstional objects," *J. Opt. Soc. Am.* 54(11), 1295–1301 (1964).
15. E. N. Leith and J. Upatnieks, "Photography by laser," *J. Sci Am.* 212(6), 24–35 (1965).

16. B. J. Thompson, G. B. Parrent, Jr., J. H. Ward, and B. Justh, "A readout technique for the laser fog disdrometer," *J. Appl. Meteorol.* 5(3), 343–348 (1966).
17. P. S. Considine, "Effects of coherence on imaging systems," *J. Opt. Soc. Am.* 56(8), 1001–1009 (1966).
18. B. J. Thompson, "Advantages and problems of coherence as applied to photographic situations," *SPIE J.* 4(1), 7–11 (1965).
19. M. J. Beran and G. B. Parrent, Jr., *The Theory of Partial Coherence,* Prentice Hall, Englewood Cliffs, N.J. (1964).
20. N. George and A. Jain, "Speckle reduction using multiple tones of illumination," *Appl. Opt.* 12(6), 1202–1212 (1973).
21. J. Upatnieks and R. W. Lewis, "Noise suppression in coherent imaging," *Appl. Opt.* 12(9), 2161–2166 (1973).
22. H. J. Caulfield, "Speckle averaging by spatially multiplexed holograms," *Opt. Comm.* 3(5), 322–323 (1971).
23. W. Martienssen and S. Spiller, "Holographic reconstruction without granulation," *Phys. Lett.* 24A(2), 126–127 (1967).
24. H. J. Gerritsen, W. J. Hannan, and E. G. Ramberg, "Elimination of speckle noise in holograms with redundancy," *Appl. Opt.* 7(11), 2301–2311 (1968).
25. E. N. Leith and J. Upatnieks, "Imagery with pseudorandomly diffused coherent illumination," *Appl. Opt.* 7(10), 2085–2090 (1968).
26. J. Upatnieks, "Improvement of 2-D image quality in coherent optical systems," *Appl. Opt.* 6(11), 1905–1910 (1967).
27. R. Bartolini, W. Hannan, D. Karlsons, and M. Lurie, "Embossed hologram motion pictures for television playback," *Appl. Opt.* 9(10), 2283–2290 (1970).
28. P. K. Katti and M. Singh, "Speckle free redundant holography," *Opt. Comm.* 8(4), 345–347 (1973).
29. G. B. Brandt, "Spatial frequency diversity in coherent optical processing," *Appl. Opt.* 12(2), 368–372 (1973).
30. D. Gabor, "Laser speckle and its elimination," *IBM J. Res. and Dev.* 14(5), 509-514 (1970).
31. J. C. Dainty and W. T. Welford, "Reduction of speckle in image plane hologram reconstruction," *Opt. Comm.* 3(5), 289–294 (1971).
32. F. T. S. Yu and E. Y. Wang, "Speckle reduction in holography by means of random spatial sampling," *Appl. Opt.* 12(7), 1656–1659 (1973).
33. J. Upatnieks and B. J. Chang, "Noise reduction in holographic images," *Opt. Comm.* 9(4), 348–349 (1973).
34. R. F. Van Ligten, "Speckle reduction by simulation of partially coherent object illumination in holography," *Appl. Opt.* 12(2), 255–265 (1973).
35. T. J. Skinner, "Energy considerations propagation in random medium and imaging in scalar coherence theory," PhD Thesis, Boston Univ. (1965).
36. G. O. Reynolds, J. L. Zuckerman, W. A. Dyes, and D. Miller, "Phase aberration balancing of simulated cataracts in the reflection mode," *Opt. Eng.* 12(2), 80–82 (1973).
37. J. W. Goodman, *Introduction to Fourier Optics,* McGraw-Hill Book Co., New York (1968).
38. R. J. Collier, C. B. Burckhardt, and L. H. Lin, *Optical Holography,* Chapter 9, Adademic Press, New York (1971).
39. P. F. Mueller, "Color image retrieval from monochrome transparencies," *Appl. Opt.* 8(10), 2051–2057 (1969).

22   Division of Wavefront
        Interferometry

22.1   INTRODUCTION

The principles of diffraction developed in preceding chapters were utilized in Chapter 3 to determine the diffracted field resulting from an array of similar diffracting apertures. This approach will lead naturally into the principles of interference, which are fundamentally important phenomena observed and utilized in physical optics. In essence, classical interferometry has been a very useful tool because of its ability to measure the relative phase of a wavefront by comparison to a calibrated reference wavefront. Table 22.I illustrates a small part of the myriad of applications for classical interferometry.

There are two general methods of creating beams of light that will interfere, and these provide a basis for classifying interference into two types. In one method, the beam is divided by passage through apertures placed side by side. This method is called *division of wavefront*. In the second method, the beam is divided at one or more partially reflecting surfaces into two or more wavefronts. This method is called *division of amplitude*. In this case, the two original wavefronts have the same width but reduced amplitudes.

In Chapters 22 through 24 we will discuss both types of interferometers and illustrate their utility in various applications.

TABLE 22.I
Typical Applications of Classical Interferometry

Field or Discipline	Measurement
Astronomy	Stellar diameters
Spectroscopy	Closely spaced spectral lines
Optics	Test lenses
Biology	Microscopic specimens
Metrology	Precise length measurement
Holography	Three-dimensional display
Physics and chemistry	Index of refraction measurements
Fluid dynamics	Index profiles of turbulence
Electronics	Phase contrast microscopy

22.2   ARRAY THEOREM

This theorem was derived and discussed in Chapter 3. Because of its importance to our present development of division of wavefront interferometry, we shall review the theorem and then use it to describe classical interferometry. The array theorem offers a method of treating interferometry problems with relative mathematical simplicity. The reason for this is that many diffraction problems are associated with diffraction by arrays of identical diffracting apertures or objects (e.g., grating spectroscopy, and multiple-beam interferometry, and x-ray diffraction by a crystal).

Consider the array of N diffracting apertures shown in Fig. 22.1. The individual diffracting apertures are identical and the positions of their respective centers of gravity are known. The distance of the n'th aperture from the center of the coordinate system is designated by ξ_n.

If the entire aperture distribution of this diffracting array is represented by $D(\xi)$, and the amplitude and phase distribution of an individual diffracting aperture by $\psi(\xi)$, then the amplitude and phase distribution of the entire aperture can be represented by

$$D(\xi) = \sum_{n=1}^{N} \psi(\xi - \xi_n) = \sum_{n=1}^{N} \int \psi(\xi - \alpha)\, \delta(\alpha - \xi_n)\, d\alpha \ . \tag{22.1}$$

If we interchange the sum and integral in Eq. (22.1), we obtain

$$D(\xi) = \int \psi(\xi - \alpha)\, A(\alpha)\, d\alpha \ , \tag{22.2}$$

where

$$A(\alpha) = \sum_{n=1}^{N} \delta(\alpha - \xi_n) \ . \tag{22.3}$$

The $A(\alpha)$ of Eq. (22.3) characterizes the locations of the centers of the various diffracting apertures; i.e., it is the array function.

To determine the Fraunhofer diffraction pattern of the array of apertures, we must Fourier transform Eq. (22.2). Thus, apart from the obliquity factors, the Fraunhofer diffraction pattern of the array is given by

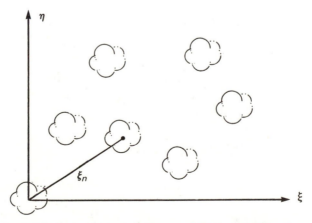

Fig. 22.1 Geometrical arrangement of an array of N identical diffracting apertures.

$$\tilde{D}(\mathbf{x}) = \int_{-\infty}^{\infty} D(\boldsymbol{\xi}) \exp\left[\frac{-2\pi i(\boldsymbol{\xi}\cdot\mathbf{x})}{\lambda f}\right] d\boldsymbol{\xi} \ . \qquad (22.4)$$

Substitution of Eq. (22.2) into Eq. (22.4) yields

$$\tilde{D}(\mathbf{x}) = \left[\int_{-\infty}^{\infty}\int_{-\infty}^{\infty} \psi(\boldsymbol{\xi}-\boldsymbol{\alpha})\, A(\boldsymbol{\alpha})\, d\boldsymbol{\alpha}\right] \exp\left[\frac{-2\pi i(\boldsymbol{\xi}\cdot\mathbf{x})}{\lambda f}\right] d\boldsymbol{\xi} \ . \qquad (22.5)$$

Equation (22.5) is the result of interest. It is the Fourier transform of the convolution integral, which was shown in a previous chapter to be the product of the respective transforms, i.e.,

$$\tilde{D}(\mathbf{x}) = \tilde{\psi}\left(\frac{\mathbf{x}}{\lambda f}\right) \tilde{A}\left(\frac{\mathbf{x}}{\lambda f}\right) , \qquad (22.6)$$

which is the array theorem, where $\tilde{\psi}$ represents the Fraunhofer diffraction pattern of an individual element of the array and $\tilde{A}$ is the Fourier transform of the array. (Note that $\tilde{A}$ is the distribution that would be obtained from an array of point sources that are all coherent with respect to each other and are in phase.) Thus, to determine the diffraction pattern of such an array, Eq. (22.6) states that we merely multiply the diffraction pattern of one individual diffracting aperture of the array with the Fourier transform of the array itself. This theorem is now demonstrated with a series of examples of division of wavefront interferometry by including two slits, three slits, M slits, and finally an infinite array of slits (i.e., a diffraction grating).

22.3   EXAMPLES OF DIVISION OF WAVEFRONT INTERFEROMETRY

22.3.1   Example 1: Two Slits

Let the diffracting aperture in this example consist of two identical slits separated by a known distance. For ease of computation, we do a one-dimensional calculation and center the slits symmetrically with respect to the origin as shown in Fig. 22.2. The individual diffracting apertures can be represented by

$$\psi(\xi) = \text{rect}(\xi|a) , \qquad (22.7)$$

and the array function is

$$A(\xi) = \delta(\xi - b) + \delta(\xi + b) . \qquad (22.8)$$

The diffraction pattern is given by the products of the respective transforms. Since transforms of these particular functions were discussed in Chapter 3, we can write the answer as

$$\psi(x) = 4a \, \text{sinc}\left(\frac{2\pi ax}{\lambda f}\right) \cos\left(\frac{2\pi bx}{\lambda f}\right) . \qquad (22.9)$$

The resulting intensity pattern in the Fraunhofer plane is given by

$$I(x) = |\psi(x)|^2 = 16a^2 \, \text{sinc}^2\left(\frac{2\pi ax}{\lambda f}\right) \cos^2\left(\frac{2\pi bx}{\lambda f}\right) . \qquad (22.10)$$

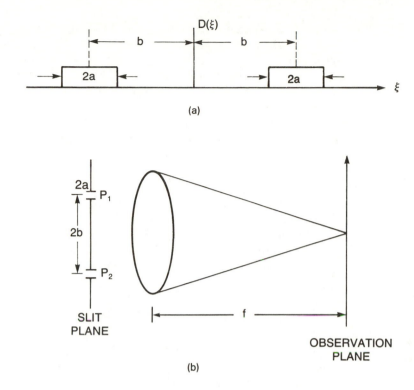

Fig. 22.2 Double-slit interferometer: (a) geometry for two-slit diffraction and (b) schematic of a two-slit interferometer.

This is the classical result obtained by Young near the beginning of the nineteenth century. The result indicates that bright and dark bands (called *fringes*) are contained within the diffraction pattern. These are in addition to the relative interference maxima and minima contained in the diffraction pattern itself. As an example of this difference, see Fig. 22.3. In Fig. 22.3(a), we show the diffraction pattern of a single slit demonstrating the secondary maxima, and in Fig. 22.3(b) we show the diffraction pattern of a double slit demonstrating the diffraction maxima and the interference fringes.[1]

The fringes and the diffraction maxima and minima may both be attributed to the same physical phenomenon, interference, which allows the radiation to superimpose constructively and destructively. When a local maximum appears,

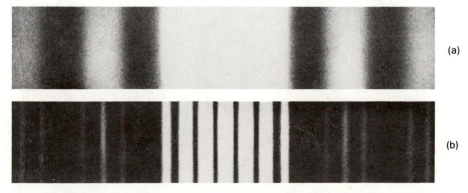

Fig. 22.3 Comparison of single-slit and double-slit diffraction: (a) single-slit diffraction pattern and (b) double-slit diffraction and interference fringes (from Ref. 1).

the two diffraction patterns are in phase and they add; this is termed *constructive interference*. When a local minimum appears, the two diffraction patterns are completely out of phase and they subtract; this is *destructive interference*. Intermediate regions between the local maxima and minima exhibit intensity values determined by the phase difference between the two diffraction patterns on a point-by-point basis.

For the problem of diffraction by a single aperture, the radiation from the different subelements of the aperture behave as secondary sources; this is an example of Huygens' principle. When the wavefronts are combined in a diffraction plane, the beams combine (interfere) in such a way that the light intensity distribution is not uniform but rather exhibits regions of maxima and minima. These maxima and minima are directly related to the size and shape of the diffracting aperture; i.e., the diffraction pattern is obtained by integrating the light distribution over the infinitesimal elements of the aperture by using the principle of superposition.

In our example of diffraction by a slit, the intensity pattern due to one slit alone is

$$I(x) = 4a^2 \operatorname{sinc}^2 \left(\frac{2\pi ax}{\lambda f} \right) . \tag{22.11}$$

The zeros (minima) of this function occur when

$$\frac{2\pi ax_n}{\lambda f} = n\pi \quad \text{or} \quad x_n = \frac{n\lambda f}{2a} . \tag{22.12}$$

Thus, relative maxima and minima occur periodically along the x axis as shown in Fig. 22.4.

As previously discussed in Chapter 2 and shown again by Eq. (22.12), the width of the central diffraction maximum is inversely related to the size of the diffracting aperture.

For the problem of two-beam interference, two entire diffraction patterns or two wavefronts are superimposed. This is easily observed if one places two apertures in front of a lens and illuminates them with a collimated beam. At the focal plane of the lens, the two diffraction patterns of the individual apertures overlap and fringes will be observed. For slit apertures, the fringes will resemble those seen in Fig. 22.3(b). As one moves the observation plane out of the focal plane, the diffraction patterns of the two apertures begin to separate and the fringe pattern changes as illustrated in Fig. 22.5 for circular apertures.

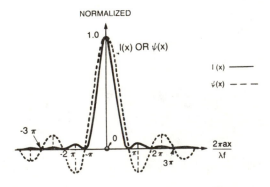

Fig. 22.4 Normalized intensity and amplitude of diffraction pattern of a slit (———— intensity; - - - - - - amplitude).

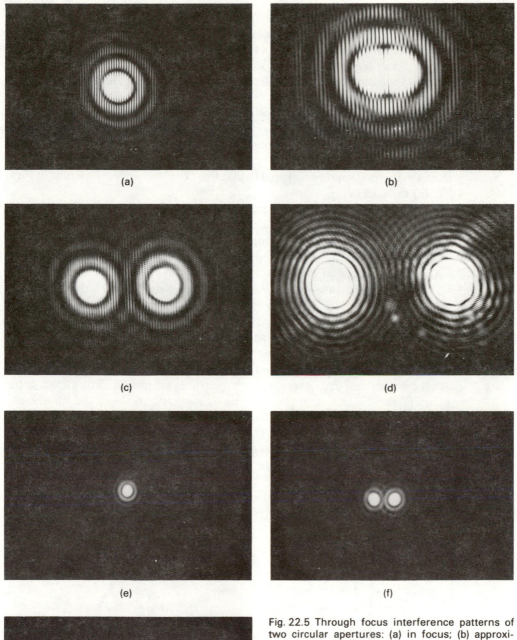

Fig. 22.5 Through focus interference patterns of two circular apertures: (a) in focus; (b) approximately one depth-of-field out of focus; (c) approximately four depths-of-field out of focus; (d) many depths-of-field out of focus; (e) same as (a) with collimated thermal radiation from a mercury arc source; (f) same as (c) with a green filter $\Delta\lambda \cong 100\,\text{Å}$; (g) same as (c) with white light spectrum. The noncircular symmetry of the central diffraction patterns is due to color dispersion of the point image of the arc source. The pinholes were ~440 μm in diameter and separated by 2140 μm and the lens had a focal length of ~200 mm. Experiments (a) through (d) utilized a He-Ne laser as the source and (e) through (g) utilized a collimated mercury arc source and a 50-μm pinhole to illustrate the effects of temporal coherence on the experiment. (Courtesy of A. Ho and D. Servaes.)

With laser illumination and circular apertures as in Fig. 22.2(b), the intensity distribution in the focal plane resulting from interference between the two Fraunhofer diffraction patterns of the apertures is shown in Fig. 22.5(a). As one moves out of focus, the individual diffraction patterns separate and make the gradual transition from Fraunhofer to Fresnel as shown in Figs. 22.5(b), (c), and (d). The additional fringe structure superimposed on the cosine fringes in these figures is caused by the out-of-focus phases of the two diffraction patterns. In Figs. 22.5(e), (f), and (g), the same effect is observed with radiation from a thermal source. These experiments show the loss of both the number of fringes and the fringe contrast due to the decrease of temporal coherence of the radiation from the thermal source compared to that from the laser source observed in Figs. 22.5(a) through (d).

When in focus, the diffraction patterns exactly overlap; see Fig. 22.5(a) or (e). Due to the shift theorem of Fourier analysis, they each have a different phase factor because they are coming from different directions. These waves interfere to give the cosine fringes exhibited in Fig. 22.3(b). These fringes are contained within the envelope of the diffraction pattern and the number of fringes contained within the central diffraction maximum is determined by the ratio of slit separation to slit width. From Eq. (22.10), the diameter of the central diffraction maximum is given by $\lambda f / a$, and the fringe period is given by $\lambda f / b$. Therefore, the number of fringes in the central maximum is given by $N = b / a$. In Fig. 22.6, we show a plot of the intensity distribution of Eq. (22.10) for $b/a = 5$.

The basic relationship between fringe frequency and aperture separation can also be determined and observed as the apertures are moved farther apart in the plane of the diffracting apertures. The light distributions in the Fraunhofer diffracting plane appear to be converging at a larger angle causing a higher frequency of the cosine fringes. Thus, a two-slit experiment may be used as a device to sequentially sample an angular spectrum of plane waves, which directly relates the spatial frequency of the cosine fringes to the angle at which the two beams come together. In Fig. 22.7, we observe the relationship between fringe frequency and slit separation for various diffracting aperture configurations. As was discussed in Chapter 11 on coherence, this concept is fundamental for the

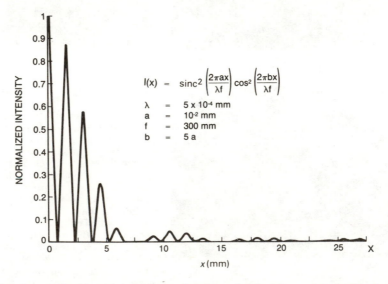

$$I(x) = \mathrm{sinc}^2\left(\frac{2\pi a x}{\lambda f}\right) \cos^2\left(\frac{2\pi b x}{\lambda f}\right)$$

λ = 5×10^{-4} mm
a = 10^{-2} mm
f = 300 mm
b = 5 a

x (mm)

Fig. 22.7 Two-slit fringe pattern exhibiting the change in fringe frequency as a function of slit separation: (a) $b = 10a$, (b) $b = 20a$, and (c) $b = 30a$; see Eq. (22.10). (Courtesy of A. Ho.)

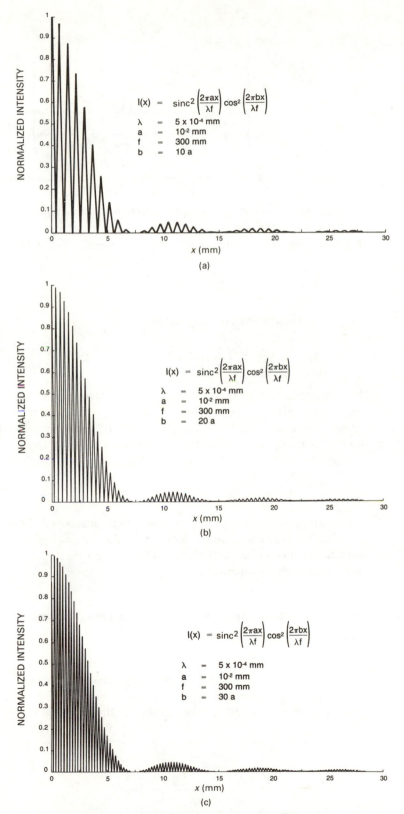

Fig. 22.7 Two-slit fringe pattern exhibiting the change in fringe frequency as a function of slit separation: (a) $b = 10a$, (b) $b = 20a$, and (c) $b = 30a$; see Eq. (22.10).

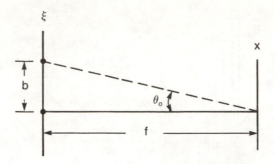

Fig. 22.8 Geometry for introduction of angular coordinates.

measurement of the degree of spatial coherence of a radiation field.

To further examine the angular spectrum perspective, we can calculate from Eq. (22.9) the diffraction pattern of the two-point array in the form

$$\bar{A}(x) = 2\cos\left(\frac{2\pi bx}{\lambda f}\right) \qquad (22.13)$$

or in terms of angular coordinates:

$$\bar{A}(x) = 2\cos 2\pi \left(\frac{b}{f}\right)\left(\frac{x}{\lambda}\right) . \qquad (22.14)$$

With reference to Fig. 22.8, we notice that $\tan^{-1}(b/f)$ is the angle θ_0 subtended at the Fourier plane by the two points. If we wish to convert the angle θ_0 into a spatial frequency (number of line pairs/ unit distance), we utilize the small-angle approximation, i.e., $\tan\theta_0 \cong \theta_0$, and introduce the notation

$$\nu = \frac{\theta_0}{\lambda} = \frac{b}{\lambda f} , \qquad (22.15)$$

where ν is the spatial frequency conjugate to the distance b. This shows the direct relationship between spatial frequency of the fringes and the angular subtent between the two points of the array. Thus, large angular subtents are associated with large spatial frequencies.

22.3.2   Phase Measurement with a Two-Slit Interferometer

22.3.2.1   Thickness Measurement

Consider a thin coating having a known index of refraction n_1, applied to a portion of a uniform substrate of the same material of index n_1, as shown in Fig. 22.9. If this coated substrate is placed in the two-slit interferometer of Fig. 22.2(b) such that the coated part is over one slit and the uncoated part covers the other,

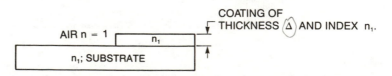

Fig. 22.9 Schematic of partially coated substrate.

then the interferometer can be used to measure the thickness of the coating Δ. In this case, the array function of Eq. (22.8) is modified because the slit with the coating in its path introduces a constant phase shift to one of the array elements:

$$A(\xi) = \delta(\xi - b) \exp\left[i\left(\frac{2\pi}{\lambda}\right) n_1 \Delta\right] + \delta(\xi + b) \exp\left(\frac{i2\pi\Delta}{\lambda}\right) . \qquad (22.16)$$

The resulting intensity pattern in the focal plane becomes

$$I(x) = 16a^2 \operatorname{sinc}^2\left(\frac{2\pi ax}{\lambda f}\right) \cos^2\left\{\frac{2\pi}{\lambda}\left[\frac{bx}{f} + (n_1 - 1)\Delta\right]\right\} , \qquad (22.17)$$

which shows that the constant phase delay creates a phase shift in the fringe pattern proportional to $(n_1 - 1)\Delta/\lambda$. This fringe shift can be used to measure Δ since n_1 and λ were assumed to be known. This effect is illustrated in Figs. 22.10(a) and (b) for an interferometer having its b/a ratio equal to 5 and coating

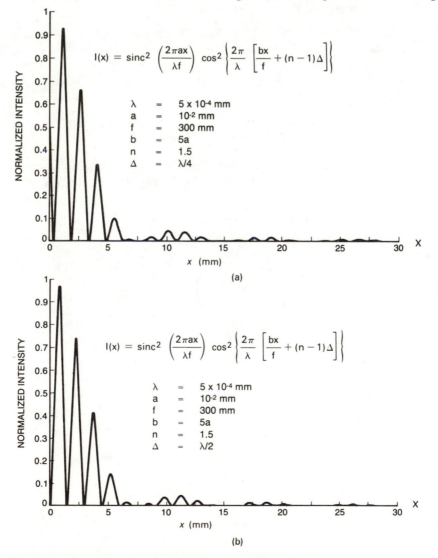

Fig. 22.10 Fringe pattern of a two-slit interferometer having delays over one slit: (a) $\Delta = \lambda/4$ and (b) $\Delta = \lambda/2$. (Courtesy of A. Ho.)

thickness of $\Delta = \lambda/4$ and $\lambda/2$, respectively. Comparison of Fig. 22.10 with Fig. 22.6 shows the fringe shifting due to the coating. Experimentally, this technique is usually limited to fringe shifts on the order of $1/10$ to $1/20$ of the fringe period.

22.3.2.2   Index Measurement

A Rayleigh refractometer is a modified double-slit interferometer that is used to measure index of refraction of an unknown sample.[1] In this instrument, also described by Eq. (22.17), Δ and λ are assumed to be known as is the index of refraction (usually air, $n_2 = 1$) over one of the slits. The fringe shift is then directly related to ($\Delta n = n_1 - n_2 = n_1 - 1$) the refractive index of the unknown medium. With a 1-cm cell, Smith and Leiderman[2] were able to measure the index of refraction fluctuations of a random medium consisting of gelatin particles in a gelatin matrix to a precision of 2.2×10^{-5} with green light.

22.3.3   Example 2: Three-Slit Array

The three-slit array[3] in one dimension is characterized by the geometry of Fig. 22.11. In this example, the elemental aperture distribution is given by

$$\psi(\xi) = \text{rect}(\xi|a) ,\tag{22.18}$$

and the array distribution becomes

$$A(\xi) = \delta(\xi) + \delta(\xi - b) + \delta(\xi + b) .\tag{22.19}$$

Notice that we merely added an on-axis aperture to the distribution shown in Fig. 22.2(a). The Fourier transform of the array given by Eq. (22.19) is

$$\tilde{A}(x) = 1 + \exp\left(\frac{ikbx}{f}\right) + \exp\left(\frac{-ikbx}{f}\right)\tag{22.20}$$

$$= 1 + 2\cos\left(\frac{kbx}{f}\right),\tag{22.21}$$

and the intensity is

$$I(x) = |\tilde{A}(x)|^2 = 1 + 4\cos\left(\frac{kbx}{f}\right) + 4\cos^2\left(\frac{kbx}{f}\right) .\tag{22.22}$$

This distribution exhibits a weak interference maximum located at the position of a strong maximum of the two-slit experiment as shown in Fig. 22.12.

The overall result of going from two to three slits is the appearance of secondary maxima. The prominent maxima become stronger and the secondary

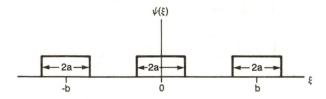

Fig. 22.11 Geometry for three-slit array.

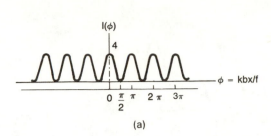
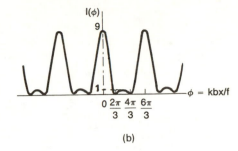

Fig. 22.12 Comparison of array intensity diffraction pattern for (a) two slits and (b) three slits (from Ref. 4).

maxima become weaker. As the number of slits is further increased, sharpening occurs until, in the limit, the well-known distribution of a diffraction grating is produced. This is discussed in a later section.

22.3.4   Phase Measurement with a Three-Slit Interferometer

We now investigate the effect of placing a constant phase step over the central slit of the three-slit array. For a zero phase step, the answer is given in Eq. (22.22) and plotted in Fig. 22.12(b). When a $\pi/2$ phase step is placed over the central slit, the intensity becomes

$$I(x) = 1 + 4\cos^2\left(\frac{kbx}{f}\right) . \tag{22.23}$$

This distribution is shown in Fig. 22.13. When the phase step θ over the central slit is given by $\theta = \pi/2 + \epsilon$ where $\epsilon < 2\pi$, Eq. (22.20) becomes

$$I(x) = \tilde{A}(x) = \exp(i\theta) + \exp\left(\frac{ikbx}{f}\right) + \exp\left(\frac{-ikbx}{f}\right) , \tag{22.24}$$

and the intensity distribution is

$$I(x) = |\tilde{A}(x)|^2 = 1 + 4\cos^2\left(\frac{kbx}{f}\right) + 4\cos\left(\frac{kbx}{f}\right)\cos\theta . \tag{22.25}$$

For small phase steps (i.e., $\epsilon \ll \pi$), Eq. (22.25) becomes

$$I(x) = |\tilde{A}(x)|^2 \cong 1 + 4\cos^2\left(\frac{kbx}{f}\right) - 4\epsilon\cos\left(\frac{kbx}{f}\right) . \tag{22.26}$$

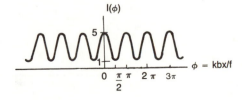

Fig. 22.13 Intensity distribution of three-slit array when the phase step θ over the central slit is $\pi/2$ (from Ref. 4).

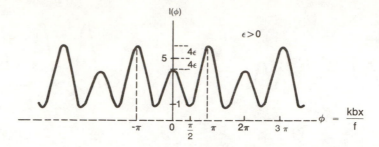

Fig. 22.14 Intensity distribution of three-slit array having a phase step $\theta = \pi/2 + \epsilon$ over the central slit (from Ref. 4).

A plot of Eq. (22.26) is given in Fig. 22.14, which shows that the small phase change θ introduced over the central slit appears as an amplitude modulation on the composite fringe pattern. Thus, the three-element interferometer differs from the two-element interferometer in that a small phase difference is measured as an amplitude modulation rather than as a phase shift. A plot of the exact solution given by Eq. (22.25) is shown in Fig. 22.15 for various values of ϵ. From this plot, we can determine the value of ϵ in two ways. First, a reference intensity measurement for one of the maximum peaks is measured when a phase step having a phase of exactly $\pi/2$ radians [i.e., an optical path difference (OPD) of $\lambda/4$] is placed over the central slit. Thus, for an unknown phase step over the central slit, the change in intensity of the central peak decreases by $4 \sin \epsilon$, thereby enabling ϵ to be measured to the accuracy of the intensity measurement. The second method would require measuring the intensity difference between two successive peaks and equating the difference to $8 \sin \epsilon$. This removes the need for the reference measurement. The $\epsilon = \pi/50$ plot in Fig. 22.15 satisfies the small-angle approximation of Eq. (22.26). Thus, we estimate that an intensity measurement having

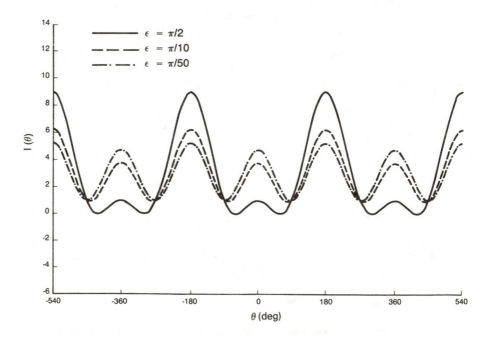

Fig. 22.15 Plot of Eq. (22.25) for $\epsilon = \pi/2$, $\pi/10$, and $\pi/50$. (Courtesy of A. Ho.)

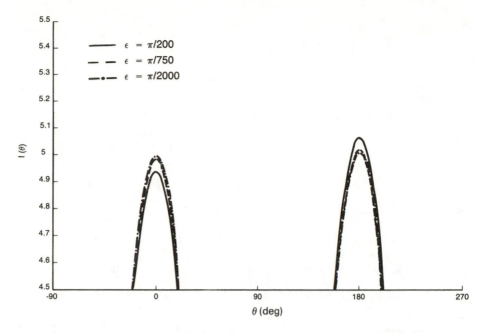

Fig. 22.16 Plot of two successive peaks from the intensity distribution of Eq. (22.25) for very small values of ϵ: $\epsilon = \pi/200$, $\pi/750$, and $\pi/2000$. (Courtesy of A. Ho.)

an accuracy of 10% is needed to measure a phase of $\pi/50$. Figure 22.16 is an expanded plot of Eq. (22.25) in the region of two successive peaks for very small values of ϵ and shows that the peaks are approaching those in Fig. 22.13. These data indicate that an intensity measurement accuracy of ~0.5% is needed to measure a phase of $\pi/2000$.

We note that for green light ($\lambda = 5000$ Å) this measurement accuracy enables a measurement of an OPD of ~10 Å since $\epsilon = (2\pi/\lambda)$OPD and the intensity difference measurement corresponds to 8ϵ as shown in Fig. 22.14.

The plots shown in Figs. 22.12, 22.13, and 22.14 for the three-slit array were fringes due to the array alone. In any experiment, the effect of the finite-sized aperture is obtained by multiplying the fringe intensities of Eq. (22.22) or (22.25) with the intensity of the slit diffraction pattern given by Eq. (22.11). In essence, this will modulate the fringe patterns of Figs. 22.12, 22.13, and 22.14 as seen below.

Experimental plots illustrating the effect of the finite aperture dimension are shown in Fig. 22.17(a) for two 100-μm slits separated by 215 μm and in Fig. 22.17(b) for three 90-μm slits separated by 212 μm. These results quantitatively agree with the theoretical results shown in Fig. 22.12. The measurement error is 0.5% for the two-slit experiment and 2.8% for the three-slit experiment.

Experimental results for finite-sized apertures showing the effect of placing a phase step over the finite-sized central aperture are shown in Fig. 22.18. Figure 22.18(a) shows the result when $\theta = 0$, i.e., Eq. (22.22) multiplied by Eq. (22.11); Fig. 22.18(b) is for $\theta = \pi/2$, i.e., Eq. (22.23) multiplied by Eq. (22.11); and Fig. 21.18(c) is for $\theta = \pi/2 + \epsilon$ where $\epsilon \ll \pi$, i.e., Eq. (22.26) multiplied by Eq. (22.11). Figure 22.18(b) shows a single frequency whereas (a) and (c) exhibit a combination of two frequencies as predicted by the analysis.

Now that we have investigated interference from two- and three-slit aperture distributions, we can make the extension to M slits and then to an infinite grating.

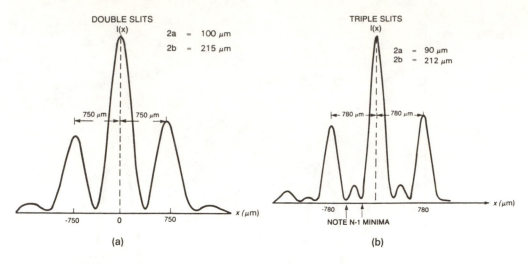

Fig. 22.17 Intensity plot of (a) two-slit experiment—two 100-μm slits separated by 215 μm, $f = 254$ mm, and $\lambda = 0.6328$ μm; (b) three-slit experiment—three 90-μm slits separated by 212 μm, $f = 254$ mm, and $\lambda = 0.6328$ μm. (Courtesy of M. Cirella.)

22.3.5   Example 3: Multiple-Slit Array

If M identical slits, each as described by Eq. (22.18), are included in an equally spaced array as shown in Fig. 22.19, the one-dimensional array function can be written as

$$A(\xi) = \sum_{n=0}^{M-1} \delta(\xi - nb) \ . \tag{22.27}$$

The Fourier transform of the array function described by Eq. (22.27) is given by

$$\tilde{A}(x) = \sum_{n=0}^{M-1} \exp\left(\frac{-2\pi inbx}{\lambda f}\right) \tag{22.28}$$

apart from a linear phase factor determined by the distance between the optical axis and the center of the array. Recalling that the geometric series $S = \sum_{n=0}^{M-1} C^n$ may be summed to yield

$$S = \frac{1 - C^M}{1 - C} \ , \tag{22.29}$$

we may reduce Eq. (22.28) to

$$\tilde{A}(x) = \frac{1 - \exp\left(\dfrac{-2\pi iMbx}{\lambda f}\right)}{1 - \exp\left(\dfrac{-2\pi ibx}{\lambda f}\right)} \ . \tag{22.30}$$

The corresponding intensity distribution of the Fourier transform is given by

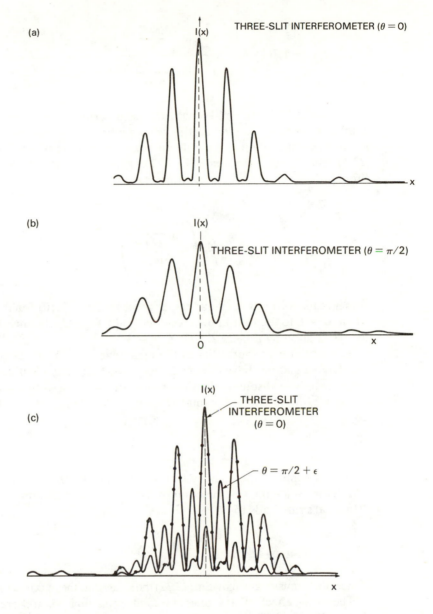

Fig. 22.18 Experimental result of three-slit interferometer of Fig. 22.17(b) with a phase step θ over the central slit: (a) $\theta = 0$, (b) $\theta = \pi/2$, and (c) $\theta = \pi/2 + \epsilon$ where $\epsilon \ll \pi$. (Courtesy of M. Cirella.)

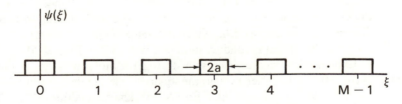

Fig. 22.19 Array of M identical, equally spaced diffracting slits.

$$I(x) = |\tilde{A}(x)|^2 = \frac{1 - \cos\left(\dfrac{2\pi Mbx}{\lambda f}\right)}{1 - \cos\left(\dfrac{2\pi bx}{\lambda f}\right)} = \frac{\sin^2\left(\dfrac{\pi Mbx}{\lambda f}\right)}{\sin^2\left(\dfrac{\pi bx}{\lambda f}\right)} . \qquad (22.31)$$

Equation (22.31) is the well-known function describing the intensity distribution of an M element diffraction grating having infinitely narrow slits. If we want to account for the finite size of the slit apertures, then we must use Eqs. (22.31) and (22.11) and the array theorem of Eq. (22.6) to give the intensity distribution in the Fraunhofer zone of the multiple diffraction as

$$I(x) = 4a^2 \,\text{sinc}^2 \frac{2\pi ax}{\lambda f} \left(\frac{\sin\dfrac{\pi Mbx}{\lambda f}}{\sin\dfrac{\pi bx}{\lambda f}}\right)^2 , \qquad (22.32)$$

which reduces to our previous result given by Eq. (22.10) for the special case when $M = 2$, to the well-known result given by Eq. (22.11) when $M = 1$, and to the result given by Eq. (22.22) when $M = 3$.

△ The addition of more slits to the array has caused two effects. First, the interference maxima have become sharper; however, they are still centered in the same place. The absolute intensity of each principal maximum is greater by the factor M^2, the square of the total number of slits. This is seen as follows. Let

$$\theta = \frac{bx}{\lambda f} , \qquad (22.33)$$

then the maxima of the function $\sin(M\pi\theta)/\sin\pi\theta$ occur when $\theta = n$, where n is an integer. Since the sinc function of Eq. (22.32) is maximum on axis, use of l'Hospital's rule leads to

$$I(x)|_{max} \cong \left(\lim_{\theta \to n} \frac{\sin M\pi\theta}{\sin\pi\theta}\right)^2 = M^2 . \qquad (22.34)$$

Hence, the intensity given by Eq. (22.32) indicates that the maxima increase by M^2.

The second effect is that a series of secondary maxima has appeared between the principal maxima. The number of secondary maxima for an array having M slits is given by $M - 2$. There are also $M - 1$ minima located between the principal maxima as shown in Fig. 22.17(b). Photographs of experimental results illustrating the $M - 2$ secondary maxima and $M - 1$ secondary minima and actual diffraction patterns were shown in Fig. 3.4 of Chapter 3 for two to six slits. Further results showing one-dimensional plots of diffraction patterns arising from one to six slits illustrating both energy buildup by M^2 and the appearance of $M - 2$ secondary maxima and $M - 1$ secondary minima are shown in Fig. 22.20.

Experimental measurements illustrating the buildup of the on-axis peak height of the diffraction pattern versus the square of the number of slits (M^2) is shown in Fig. 22.21 verifying the result of Eq. (22.34). The data points in Fig. 22.21 are taken from Fig. 22.20.

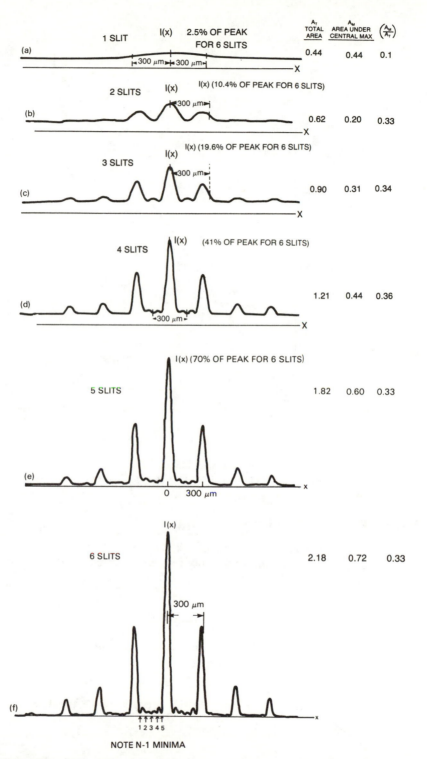

Fig. 22.20 Plots of diffraction patterns from one-dimensional slit arrays illustrating multiple-slit interference. Additional data are shown on each figure: (a) one slit 88 μm wide, (b) two slits 88 μm wide, (c) three slits 88 μm wide, (d) four slits 88 μm wide, (e) five slits 88 μm wide, and (f) six slits 88 μm wide. (Courtesy of M. Cirella.)

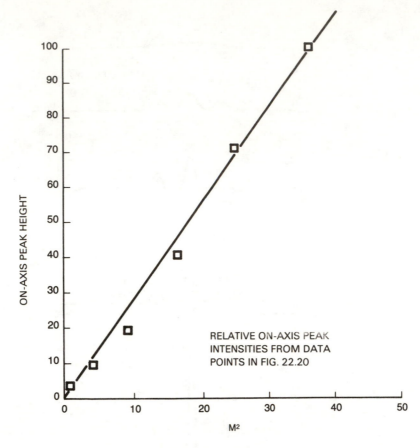

Fig. 22.21 Plot of on-axis peak of multiple-slit arrays versus the square of the number of slits *M* in the array (□ = data points from Fig. 22.20). (Courtesy of M. Cirella.)

22.3.6   Example 4: Infinite Grating

The infinite grating (see Fig. 22.22) may be represented by the aperture function $\psi(\xi)$ of Eq. (22.7) and the array function (Dirac comb) given by

$$A(\xi) = \sum_{n=-\infty}^{\infty} \delta(\xi - nb) \ . \tag{22.35}$$

From Eq. (22.2), the entire aperture distribution is

$$D(\xi) = \psi(\xi) \circledast A(\xi) \ , \tag{22.36}$$

where $\circledast$ denotes the convolution process. Since the Dirac comb is its own

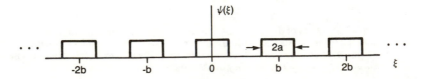

Fig. 22.22 Infinite array of identical, equally spaced diffracting slits.

Fourier transform,[5] the Fraunhofer diffraction pattern is given by

$$\tilde{D}(x) = 4ab \, \text{sinc}\left(\frac{2\pi ax}{\lambda f}\right) \sum_{n=-\infty}^{\infty} \delta\left(x - \frac{n\lambda f}{b}\right) , \qquad (22.37)$$

and the corresponding intensity pattern is given by

$$I(x) = 16a^2 b^2 \, \text{sinc}^2\left(\frac{2\pi ax}{\lambda f}\right) \sum_{n=-\infty}^{\infty} \delta\left(x - \frac{n\lambda f}{b}\right) . \qquad (22.38)$$

Thus, for the infinite grating, the diffraction orders appear as an intensity modulated Dirac comb. If we consider the special case of a square wave, i.e.,

$$b = 4a , \qquad (22.39)$$

the zeros of the envelope function occur when

$$\frac{2\pi ax}{\lambda f} = m\pi \qquad (22.40)$$

or

$$x_m = 2m\lambda f/b . \qquad (22.41)$$

We notice that when $2m$ in Eq. (22.41) equals n in Eq. (22.38), the zero of the envelope function is contiguous with the location of the diffraction orders. Thus, the even diffraction orders of the square wave grating are missing. This result was observed experimentally in Fig. 21.2 of Chapter 21 and further utilized in the discussion of phase gratings in phase contrast imaging in Chapter 35.

22.3.7   Spectral Resolvability of a Diffraction Grating

Diffraction gratings used in practical applications have a finite width. Therefore, the aperture function is described by modifying Eq. (22.36) to

$$D(\xi) = \left[\text{rect}(\xi \,|\, a) \circledast \sum_{n=-\infty}^{\infty} \delta(\xi - nb) \right] \text{rect}(\xi \,|\, A) , \qquad (22.42)$$

where $2A$ denotes the finite width of the grating. The corresponding Fraunhofer diffraction pattern given by the array theorem is

$$\tilde{D}(x) = a_n \sum_{n=-\infty}^{\infty} 2A \, \text{sinc}\left[\frac{2\pi(x - n\lambda f/b)A}{\lambda f} \right] , \qquad (22.43)$$

where the a_n obtained from Eq. (22.37) are the amplitude strengths of the individual orders of the Dirac comb. The intensity corresponding to Eq. (22.43) is plotted in Fig. 22.23 for the cases of one and three incident wavelengths where the grating frequency is high enough to prevent interference between the diffraction orders.

The ability of this diffraction grating to distinguish between two separate

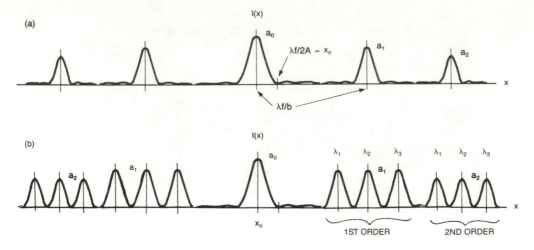

Fig. 22.23 (a) Intensity distribution of a finite width diffraction grating for a single wavelength and (b) Fraunhofer pattern for three wavelengths (after Ref. 1).

colors (wavelengths) in any order is found by utilizing the Rayleigh criterion[a] (see Fig. 22.24) to give

$$\frac{n\lambda_2 f}{b} - \frac{n\lambda_1 f}{b} = \frac{\bar{\lambda} f}{2A} , \tag{22.44}$$

where $\bar{\lambda} = (\lambda_1 + \lambda_2)/z_2$ is the mean wavelength. Since the spectral resolving power of a diffraction grating is defined by $\bar{\lambda}/\Delta\lambda$, from Eq. (22.44) we see that

$$\frac{\bar{\lambda}}{\Delta\lambda} = \frac{(2A)n}{b} = n(2A)\nu_0 , \tag{22.45}$$

where $2A$ is the width of the grating and $1/b$ is the spatial frequency ν_0. Thus, the resolving power of the grating is proportional to the total number of lines in the grating and the diffraction order used in the experiment.

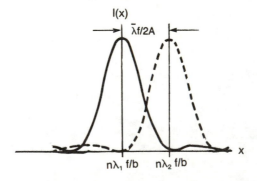

Fig. 22.24 Schematic for spectral resolution in the n'th order by Rayleigh criterion.

---

[a] In Chapter 6, the Rayleigh criterion defined two intensity diffraction patterns as just resolved when the central maximum of one coincides with the first minimum of the other.

As a numerical example, if we wanted to resolve the two lines in the sodium doublet ($\bar{\lambda} = 5893$ Å and $\Delta\lambda = 6$ Å) in the first diffraction order, we see that a grating resolving power of 982 is required. Thus, using Eq. (22.45), a grating of frequency 100 lp/mm must be at least 9.8 mm in width in order to resolve the sodium doublet in the first diffraction order.

REFERENCES

1. R. S. Longhurst, *Geometrical and Physical Optics,* 2nd ed., John Wiley and Sons, Inc., New York (1967).
2. A. E. Smith and L. Leiderman, "Laboratory simulation of a random medium," *J. Opt. Soc. Am.* 60(5), 728A (1970).
3. F. Zernike, "A precision method for measuring small phase differences," *J. Opt. Soc. Am.* 40 (5), 326–328 (1950).
4. A. Maréchal, P. Lostis, and J. Simon, "A precision interferometer with high light-gathering power," in *Advanced Optical Techniques,* A. C. S. Van Heel, ed., Chapter 12, pp. 437–446, North Holland Publishing Co., Amsterdam (1967).
5. J. W. Goodman, *Introduction to Fourier Optics,* McGraw-Hill Book Co., New York (1968).

23   Division of Amplitude
        Interferometry

23.1   INTRODUCTION

There are numerous types of interferometers that are based on a division of the wave amplitude. One or more of these wavefronts are changed before recombination, which causes them to interfere. In the most general case, there can be N wavefronts such as in a Fabry-Perot interferometer; however, in the more common interferometers, we have two ($N=2$) wavefronts. It is beyond the scope of this chapter to cover all such interferometers. Instead, we will cover the more common types and refer the reader to the literature[1-8] for the others.

23.2   GENERAL ANALYSIS

In the most general case of N wavefronts, the intensity of the interferogram is found by adding amplitudes to give

$$I(x) = \sum_{n=1}^{N} |f_n(x,y,c\Delta t)|^2 \, , \tag{23.1}$$

where c is the speed of light, Δt is the time delay, and $c\Delta t$ is the path difference. In general, the wavefront amplitudes $f_n(x,y,c\Delta t)$ are complex functions. For the common and important case of $N=2$, Eq. (23.1) reduces to

$$I(x) = |f_1(x,y,ct_1) + f_2(x,y,ct_2)|^2 \, . \tag{23.2}$$

In all of these two-beam interferometers, it is important for the path difference $[c(t_2 - t_1)]$ to be less than the coherence length of the source radiation to ensure that the two waves are coherent. In general, this means that interferometers with laser sources have relaxed path length difference requirements.

We are now prepared to examine a number of different types of interferometers whose intensity distribution in the interferogram is given by Eq. (23.2).

23.3   CASE I: WAVEFRONT PRESERVING INTERFEROMETRY FOR HOLOGRAMS

A hologram is essentially an interferogram, usually made in a division of amplitude interferometer. The hologram has an intensity distribution that is given by Eq. (23.2) where $f_1(x,y,ct_1)$ is a reference wave and $f_2(x,y,ct_2)$ is a Fresnel or Fraunhofer diffraction pattern (transform) of an object. The interferogram resulting from Eq. (23.2) stores the amplitude and phase of the complex diffraction pattern (transform) obtained from the object. An example of such a hologram is shown in Fig. 23.1. Here $f_2(x,y,ct_2)$ is the Fraunhofer diffraction pattern of an unresolved (point) source and $f_1(x,y,ct_1)$ is a plane wave. The resulting hologram of a point has been termed a *zone lens* whereas Fresnel and Fraunhofer holograms can be considered *modulated zone lenses* as discussed further in Chapter 25.

The zone lens concept is useful in that it emphasizes the built-in lens property of the hologram which causes the reconstructed image to be in focus at the appropriate image distance.[9] A hologram of a more general object may be represented as a linear superposition of such zone lenses. Many of the interesting properties of holograms may be easily deduced from such a representation.

In the case of holography, what is generally important is not the interferogram itself but the fact that one can reconstruct the original three-dimensional wavefront of the object from the hologram since the phase information is stored. This subject is covered extensively with illustrations in Chapters 25, 26, and 27 and is not discussed in any more detail here.

23.4   CASE II: WAVEFRONT MEASURING INTERFEROMETERS

Division of amplitude interferometers have proven very useful for measuring and testing wavefronts, such as optical surfaces, index profiles of turbulence, and gas flow in wind tunnels. In these applications, the interferometer is configured so that $f_1(x,y,ct_1)$ of Eq. (23.2) is a calibrated reference wave (usually plane) and $f_2(x,y,ct_2)$ is the wavefront of interest, $\exp[i\phi(x,y)]$. The wavefront of interest is superimposed onto a plane wave created by a high-quality collimator lens such that ct_2 is a fixed distance and $c(t_2 - t_1)$ is less than the coherence length of the radiation. The resulting fringe patterns on the interferograms indicate the deviation of the measured wavefront from a plane wave in units of optical wavelength. If the measured wavefront arises from an ideal surface and the collimator output is a perfect plane wave, then the interferometer yields a variable brightness distribution depending on the path length difference. If the measured beam exhibits tilt, then straight-line fringes are observed where the fringe frequency is proportional to the amount of tilt.

More complex wavefronts such as those arising from the Seidel aberrations of a lens or propagation of the test wave through a flow field give rise to more complicated fringe patterns. Analysis of the resulting fringe patterns enables one to determine the characteristic phase deviation of the wavefront of interest from a plane wave.

23.4.1   Example 1: Twyman-Green Interferometer

The Twyman-Green interferometer, shown schematically in Fig. 23.2, is essentially a fixed-path Michelson interferometer with a point source, which is

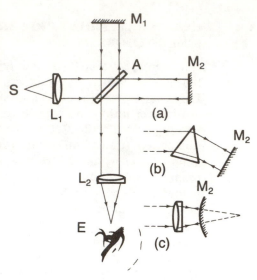

Fig. 23.1 Fraunhofer hologram (interference pattern) of a small unresolved object showing zone lens fringes (from Ref. 9).

Fig. 23.2 Schematic of a Twyman-Green interferometer for measuring (a) plane mirrors, (b) prisms, and (c) lenses and/or curved mirrors (from Ref. 2).

extremely useful for mirror, lens, or prism testing in optical fabrication shops. The reference wave $f_1(x, y, ct_1)$ of Eq. (23.2) is a plane wave and $f_2(x, y, ct_2)$, the wavefront to be measured, is a modified plane wave. With thermal sources, we must set $t_1 \cong t_2$ so that the path length difference is less than the coherence length of the radiation. This requirement is greatly relaxed by using a laser source to create the LUPI, the laser unequal path interferometer.[10,11] The Twyman-Green interferometer may be used with or without tilt between the two waves. A series of simple interferograms obtained from a Twyman-Green interferometer is shown in Fig. 23.3. Figure 23.3(a) shows the bright field output from two in-focus plane waves which are also in phase. If one of the waves were π out of phase with respect to the other, then a dark field would have been observed. When one of the waves is tilted by five wavelengths, the linear fringes of Fig. 23.3(b) occur. A defocus aberration of five waves in the measurement wave of Fig. 23.2(c) introduces a sphericity (quadratic phase factor), which yields circularly symmetric fringes as shown in Fig. 23.3(c). Finally, a combination of five waves of tilt and five waves of defocus gives the interferogram shown in Fig. 23.3(d).

The wavefront error (aberration) in the measurements arises from the difference between the aberrated wavefront and a reference spherical wavefront whose

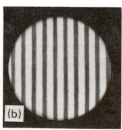

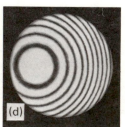

Fig. 23.3 Interferograms of a perfect lens from a Twyman-Green interferometer of the type illustrated in Fig. 23.2(c): (a) no tilt or defocusing, (b) five waves of tilt error, (c) five waves of defocusing aberration, and (d) five waves of tilt and fives waves of defocusing (from Ref. 4).

center of curvature is the Gaussian image point. This aberrated wavefront can be represented mathematically by a power series expansion[12] in the variables of the exit pupil. Various terms in this series expansion are grouped as: a constant term, which shifts the wavefront to a different reference sphere at the same center of curvature; linear phase, a tilt (transverse defocusing) due to an off-axis component of the wavefront; longitudinal defocusing (quadratic phase), which is spherically symmetric about the Gaussian image point; the five-Seidel (third-order) aberrations; and fifth and higher order aberrations. The power series expansion of the optical path difference (OPD) in polar coordinates in the exit pupil, i.e., $x = r\cos\phi$ and $y = r\sin\phi$, due to the primary aberrations can be represented by[4,12]

$$\mathrm{OPD} = Ar^4 + Br^3\cos\phi + Cr^2$$
$$\times (1 + 2\cos^2\phi) + Dr^2 + Er\cos\phi + Fr\sin\phi \; , \qquad (23.3)$$

where

A = measure of longitudinal third-order spherical aberration

B = measure of sagittal coma

C = measure of sagittal astigmatism

D = measure of defocusing

E = measure of tilt about the x axis

F = measure of tilt about the y axis.

Interferograms for some of these aberrations are shown in Fig. 23.4. In Fig. 23.3, we have already shown the interferometric effects of the constant term, transverse tilt, and longitudinal defocusing. In Fig. 23.4(a), we show the case of spherical aberration, and in (b), spherical aberration with tilt both at the paraxial focus [$D=0$ in Eq. (22.3)]. For Fig. 23.4(a), $B=C=E=F=0$ and $A=6$ waves in Eq. (23.3), and for (b), $B=C=F=0$, $E=5$ waves, and $A=6$ waves in Eq. (23.3).

In Fig. 23.5, we show the case of coma at the location of paraxial focus [$D=0$

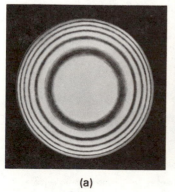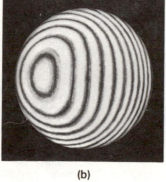

(a)   (b)

Fig. 23.4 Interferograms for a lens with six waves of spherical aberration at the paraxial focus [$D=0$ in Eq. (23.3)]: (a) without tilt and (b) five waves of tilt (from Ref. 4).

Fig. 23.5 Interferogram for a lens with fives waves of coma at the paraxial focus (from Ref. 4).

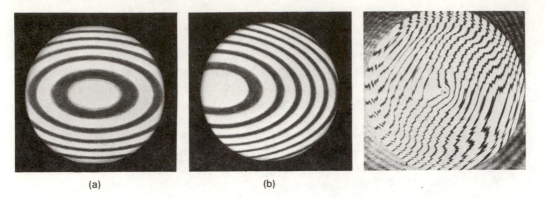

(a)   (b)

Fig. 23.6 Interferogram for a lens with two waves of astigmatism at the Petzval focus: (a) without tilt and (b) three waves of tilt (from Ref. 4).

Fig. 23.7 Laser Twyman-Green interferogram of a Moore-turned spherical mirror having a diameter of 40 cm and a radius of curvature of 1320 cm (from Ref. 13).

in Eq. (23.3)] for $A = E = F = 0$ and $B = 5$ waves of coma in Eq. (23.3).

In Fig. 23.6, we show the case of astigmatism at the Petzval focus[a] [$D = 0$ in Eq. (23.3)]. For Fig. 23.6(a), $C = 2$ waves and $A = B = E = F = 0$ in Eq. (23.3), and for (b), $A = B = F = 0$, $E = 3$ waves of tilt, and $C = 2$ waves of astigmatism.

Interferograms for various combinations of these Seidel aberrations appear in the literature and the interested reader is referred there.[4]

The Twyman-Green interferometer has been reconfigured with a laser source to make a laser unequal path interferometer (LUPI).[10,11] In Fig. 23.7, we show the LUPI interferogram of a Moore-turned spherical mirror.

23.4.2   Example 2: Mach-Zehnder Interferometer

This interferometer, shown schematically in Fig. 23.8, has been very useful in measuring wavefronts from lenses and mirrors, measuring gas flow and turbu-

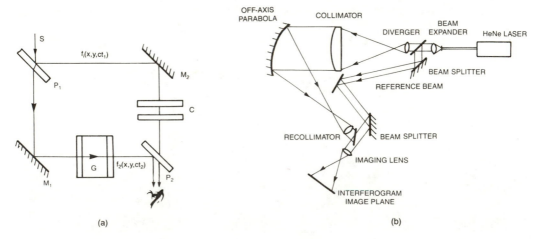

(a)   (b)

Fig. 23.8 Schematic diagram of a Mach-Zehnder interferometer: (a) for measuring refractive samples (from Ref. 2) and (b) modified for measuring reflected wavefronts (from Ref. 13).

---

[a]Astigmatic lenses focus the two perpendicular axes in different planes called the *sagittal* and *tangential* focal planes. The Petzval focus is defined to be at their midpoint.

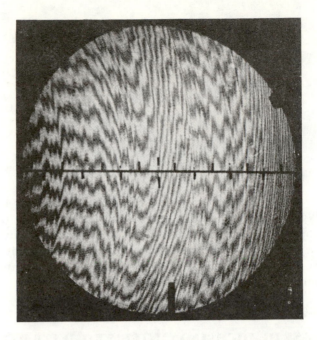

Fig. 23.9 Mach-Zehnder interferogram of preliminary cut of an off-axis parabola. This interferogram has proved to be a valuable diagnostic tool in correction of machine deficiencies (from Ref. 13).

lence, and in making holographic correlation filters. In this interferometer, the function $f_1(x,y,ct_1)$ in Eq. (23.2) is a plane or spherical test wave and $f_2(x,y,ct_2)$ is the wavefront to be tested. As seen in Fig. 23.8(a), one path of the interferometer contains a compensator C to adjust optical path lengths when thermal sources are used. The other path contains the refractive sample to be measured. In measuring gas samples, the integrated index variations over the volume of the gas modify the wavefront passing through it. High-contrast localized fringes can be realized with fast exposures by using a lens to image various planes within the volume of the sample.

In the case of making a Fourier transform holographic filter with a laser source, one would place a Fourier transforming lens in one path to obtain the transform of the desired object, and the other path creates the reference wave. This hologram can be used to make a matched filter to detect objects in a scene whose Fourier transform matches that stored in the hologram,[14] as discussed in Chapter 33. In Fig. 23.9, we show the results of using the modified Mach-Zehnder interferometer of Fig. 23.8(b) as a diagnostic tool to measure the wavefront from an off-axis parabola. Such interferograms have been useful in diagnosing and correcting machining errors in diamond-cutting machines.

23.4.3   Example 3: Watson Interference Microscope

The interference microscope, shown schematically in Fig. 23.10, is an invaluable tool for visualizing OPDs, i.e., used to visualize phase objects. In this case, the function $f_1(x,y,ct_1)$ in Eq. (23.2) is a plane wave and the function $f_2(x,y,ct_2)$ is the phase object.

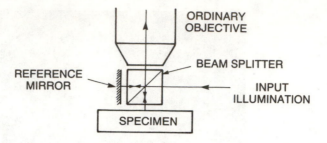

Fig. 23.10 Schematic diagram of a Watson interference microscope (from Ref. 2).

In Fig. 23.11, we show a phase image of a phase three-bar target visualized in a Watson interference microscope. Obviously, since the target was a pure phase object, a direct noninterferometric imaging system would show nothing but a clear field. Thus, the interference microscope has produced a visual image of a pure phase object as a fringe modulation. For a further discussion of phase contrast techniques, see Chapter 35.

23.5   CASE III: MICHELSON INTERFEROMETER WITH VARIABLE DELAY

23.5.1   General Discussion

In these interferometers, one wavefront is delayed relative to the other to create interference. The change in the interferogram as a function of the delay (i.e., OPD) is measured and used to determine the spectral characteristic of the wavefront as discussed in Chapter 10. This interferometer is a fundamental tool for measuring the temporal coherence of a radiation field. A schematic diagram of a Michelson interferometer is shown in Fig. 23.12.

The compensator ensures that the two paths are equal when $\Delta = 0$, i.e., it removes the path difference effects due to the finite thickness of the beam splitter (BS). The mirror M_2 is fixed and the mirror M_1 is adjusted to introduce variable path differences in the experiment. With a collimated input beam, as in Fig.

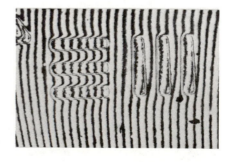

Fig. 23.11 Interferograms of phase images of a three-bar target on bleached 3404 film, taken in a Watson interference microscope. (Courtesy of D. Cronin.)

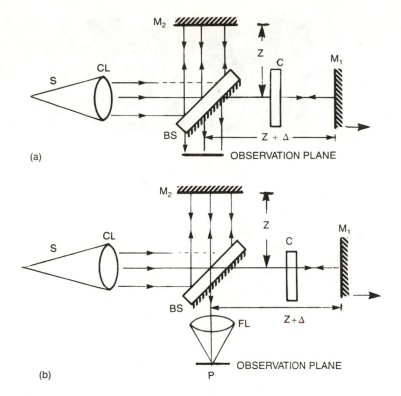

Fig. 23.12 Schematic diagram of a Michelson interferometer for (a) collimated input and output and (b) collimated input with focused output, where S = source, CL = collimating lens, BS = beamsplitter, M_1 = movable mirror, M_2 = fixed mirror, C = path compensating plate, Δ = path difference, FL = focusing lens, and P = point of observation.

23.12(a), two parallel beams emerge from the beam splitter and are observed at the observation plane. If the two beams are exactly in phase, a maximum bright field is observed in the observation plane. If the two beams are exactly out of phase, a maximum dark field is observed. Intermediate values of phases between the beams give rise to constant intensity fields between these two limits. Thus, the value of the intensity at any point P in the observation plane varies as the mirror moves. If one of the mirrors is tilted through an angle θ_0, then spatial cosine fringes having frequency $(\sin\theta_0)/\lambda$ are observed in the observation plane. As the mirror M_1 moves, the intensity of the entire fringe field varies between the maximum brightness and zero each time the mirror moves through a distance of one-half wavelength.

If a lens is used to focus the light on the observation plane as in Fig. 23.12(b), and the mirrors M_1 and M_2 are parallel, then a focused image of the source is seen. A magnifier is needed to observe the details of this source image. When $\Delta = 0$ (i.e., the path lengths are exactly equal), a dark field is observed as shown in Fig. 23.13(a). In this case, the dark field exists because the beam splitter was silvered on its back surface so that a π-phase shift is experienced in the light propagating toward mirror M_2. As the mirror moves away from the observer, circular fringes are introduced as shown in Fig. 23.13(b). The frequency of these circular Fresnel fringes increases as the path difference increases; an additional fringe appears in the field of view with each half wavelength movement of the mirror. A typical high-frequency fringe pattern resulting from a mirror motion of many wavelengths is shown in Fig. 23.13(c).

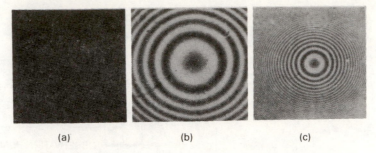

(a)   (b)   (c)

Fig. 23.13 Typical fringes in a focused image Michelson interferometer: (a) $\Delta = 0$, dark field situation, (b) mirror M_2 moved a few wavelengths, and (c) mirror M_2 moved many wavelengths (from Ref. 7).

23.5.2   Mathematical Analysis

Mathematically, in the Michelson interferometer, the function $f_1(x,y,ct_1)$ in Eq. (23.2) is the reference wave given by $f(x,y,z)$, and the function $f_2(x,y,ct_2)$ is given by $f(x,y,z+\Delta)$. The resulting intensity (long time average) at a given point $P(x,y,z)$ from Eq. (23.2) can be written as

$$I(P,\Delta) = <|f(x,y,z)+f(x,y,z+\Delta)|^2>$$
$$= <|f(x,y,z)|^2>+<|f(x,y,z+\Delta)|^2>$$
$$+<f(x,y,z)f^*(x,y,z+\Delta)>$$
$$+<f^*(x,y,z)\,f(x,y,z+\Delta)>\;, \qquad (23.4)$$

where the angle brackets denote a long time average.

Since $f(x,y,z+\Delta)$ only differs in path difference (hence, phase) from $f(x,y,z)$ in a Michelson interferometer, Eq. (23.4) may be written

$$I(P,\Delta) = I_1 + I_1 + 2I_1\,\mathrm{Re}\,\gamma_{11}(\Delta/c)$$
$$= 2I_1[1 + \mathrm{Re}\,\gamma_{11}(\Delta/c)]\;, \qquad (23.5)$$

where

$$I_1 = <|f(x,y,z)|^2> = <|f(x,y,z+\Delta)|^2>$$

$\gamma_{11}(\tau) = $ normalized self-coherence function

$\Delta/c = \tau$

$\Delta = $ path difference introduced between the two beams in a Michelson interferometer.

Equation (23.5) may also be written

$$I(P,\tau) = 2I_1\left\{1 + |\gamma_{11}(\tau)|\cos[\arg\gamma_{11}(\tau)]\right\}\;, \qquad (23.6)$$

where $\gamma_{11}(\tau) = |\gamma_{11}(\tau)|\exp[i\arg\gamma_{11}(\tau)]$. The visibility of the fringe pattern of Eq. (23.6) is

$$V = \frac{I_{max} - I_{min}}{I_{max} + I_{min}} = |\gamma_{11}(\tau)|\;, \qquad (23.7)$$

which is a measure of the magnitude of the normalized self-coherence function.

The phase term $\arg[\gamma_{11}(\tau)]$ in Eq. (23.6) is the phase of the self-coherence function which arises from asymmetries in the source power spectrum. Thus, in general, the phase of the self-coherence function does not appear as a phase modulation on a periodic carrier as it does in the measurements of spatial coherence with a two-pinhole interferometer.

For the special case of quasimonochromatic radiation,

$$\gamma_{11}(\tau) \cong \gamma_{11}(0)\exp(-2\pi i \bar{\nu}\tau) \tag{23.8}$$

so that Eq. (23.5) reduces to

$$I(P,\tau) = 2I_1\left(1 + |\gamma_{11}(0)|\cos\{2\pi\bar{\nu}\tau - \arg[\gamma_{11}(0)]\}\right) . \tag{23.9}$$

Equation (23.9) illustrates that the phase of the self-coherence function, due to geometrical source asymmetries, modulates the geometrical fringe pattern created by path differences in the interferometer; i.e., the quasimonochromatic nature of the source removes the need for source spectral measurements since all the source power is assumed to exist in the vicinity of $\bar{\nu}$ in the quasimonochromatic approximation.

Since the self-coherence function and the self-power spectrum are Fourier transform pairs,[15] the visibility of the fringe pattern and the associated phase shifts due to the phase of the self-coherence function $[\arg\gamma_{11}(\tau)]$ in Eq. (23.6) can be measured with a Michelson interferometer to determine the power spectrum of the source by Fourier transformation. This principle forms the basis for the field of Fourier spectroscopy.

23.5.3   Applications of Michelson Interferometers

23.5.3.1   Example 1: Measuring Coherence Length

As an example illustrating the use of a Michelson interferometer, suppose that the power spectrum $\hat{\gamma}_{11}(\nu)$ of a quasimonochromatic source to be measured with a Michelson interferometer is given by

$$\hat{\gamma}_{11}(\nu) = \text{rect}(\nu|\nu_0) = \begin{cases} 1 & |\nu| \leqslant \nu_0 \\ 0 & |\nu| > \nu_0 . \end{cases} \tag{23.10}$$

Since this source is symmetric, its Fourier transform, the self-coherence function, has no phase. Thus, the only phase in the term $[\arg\gamma_{11}(\tau)]$ of Eq. (23.6) is due to the geometrical path differences $(2\pi\bar{\nu}\tau)$ of Eq. (23.9).

The fringe visibility resulting from this measurement is

$$V = |\gamma_{11}(\tau)| = 2\nu_0\,\text{sinc}(2\pi\nu_0\tau)$$

$$= 2\nu_0\,\text{sinc}(2\pi\sigma\Delta) , \tag{23.11}$$

where $\sigma = \nu_0/c$ and $\Delta = c\tau$. The sinc function here has its first zero at a value $\Delta = 1/2\sigma = c/2\nu_0$. Thus, the path differences for which high-visibility fringes disappear in a Michelson interferometer are inversely proportional to the bandwidth of the radiation used in the experiment, as discussed previously in Chapter 11. Typical values of coherence lengths for various sources are illustrated in Fig. 23.14.

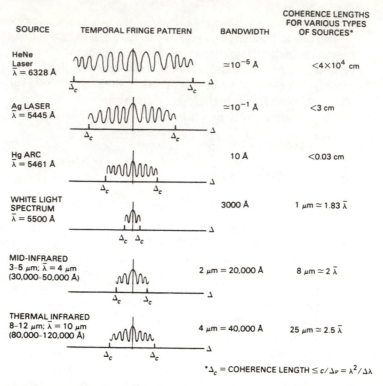

SOURCE	TEMPORAL FRINGE PATTERN	BANDWIDTH	COHERENCE LENGTHS FOR VARIOUS TYPES OF SOURCES*
HeNe Laser $\lambda = 6328$ Å		$\simeq 10^{-5}$ Å	$< 4 \times 10^4$ cm
Ag LASER $\lambda = 5445$ Å		$\simeq 10^{-1}$ Å	< 3 cm
Hg ARC $\lambda = 5461$ Å		10 Å	< 0.03 cm
WHITE LIGHT SPECTRUM $\bar\lambda = 5500$ Å		3000 Å	1 μm $\simeq 1.83\ \bar\lambda$
MID-INFRARED 3–5 μm; $\bar\lambda = 4\ \mu$m (30,000–50,000 Å)		2 μm = 20,000 Å	8 μm $\simeq 2\ \bar\lambda$
THERMAL INFRARED 8–12 μm; $\bar\lambda = 10\ \mu$m (80,000–120,000 Å)		4 μm = 40,000 Å	25 μm $\simeq 2.5\ \bar\lambda$

*Δ_c = COHERENCE LENGTH $\leq c/\Delta\nu = \lambda^2/\Delta\lambda$

Fig. 23.14 Coherence lengths and sketches of fringe patterns for various types of quasi-monochromatic sources.

23.5.3.2   *Example 2: Use of a Michelson Interferometer as a Fourier Spectrometer*

If we assume that the spectral radiation from a source is collimated and input to a Michelson interferometer, then, when a detector is placed at point P in Fig. 23.12(b), an interferogram (time profile) is recorded as the mirror M_1 is moved. Equation (23.6) describes this intensity distribution where $\gamma_{11}(\tau)$ is the Fourier transform of the source power spectrum $\hat\gamma_{11}(\nu)$, which is necessarily real. Thus, a Fourier spectrometer integrates all frequencies at each mirror position.[16] Only those frequencies that are coherent at a given mirror setting contribute to the interference term of the interferogram. Those spectral components with a coherence length that has been exceeded at a given mirror position add to the bias (or dc) level of the interferogram. The radiation into the Fourier spectrometer is spatially coherent because the instrument (which consists of a collimator mirror arrangement, imaging lens, and detector) limits the incoming radiation to that radiating from a small spatial region of the source plane. At zero path difference, all of the radiation from this effective point source is temporally coherent. This accounts for large fringe visibility near the origin of the interferogram. As the mirror moves, the path difference increases so that the coherence length of the average source spectra is exceeded, and only those narrow spectral spikes in the source contribute to the interference terms at large distances from the origin of the interferogram.

The mirror distance will eventually become so large that the modulation (due to interference from the very narrow lines in the source spectrum) will be less than the detector noise fluctuations. This limitation will determine the spectral resolution

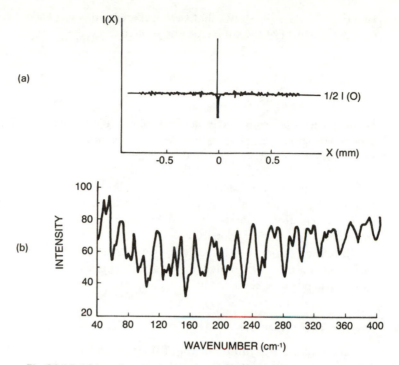

Fig. 23.15 (a) Interferogram and (b) calculated spectrum (from Ref. 17).

(length of the interferogram in time) of a given instrument and it can only be increased by increasing the aperture of the optics or the sensitivity of the detector. A typical interferogram and its corresponding spectrum are shown in Fig. 23.15. Coherence lengths for various source spectral widths were given in Fig. 23.14.

The measured output of a Fourier spectrometer is described by Eq. (23.6). If we subtract the bias term, Eq. (23.6) becomes the modulation or interferogram, i.e.,

$$\text{interferogram} = \frac{I(P, \tau)}{2I} - 1 = |\gamma_{11}(\tau)| \cos[\arg \gamma_{11}(\tau)] , \qquad (23.12)$$

showing that the interferogram is a measure of the real part of the self-coherence function.

The definition of the normalized power spectrum $\hat{\gamma}_{11}(\nu)$ as the Fourier transform of the self-coherence function is

$$\hat{\gamma}_{11}(\nu) = \int_{-\infty}^{\infty} \gamma_{11}(\tau) \exp(2\pi i \nu \tau) \, d\tau . \qquad (23.13)$$

Inverse Fourier transformation of Eq. (23.13) yields

$$\gamma_{11}(\tau) = \int_{-\infty}^{\infty} \hat{\gamma}_{11}(\nu) \exp(-2\pi i \nu \tau) \, d\nu$$

$$\gamma_{11}(\tau) = \gamma_{11}^{\text{real}}(\tau) + i\gamma_{11}^{\text{imaginary}}(\tau) . \qquad (23.14)$$

Since the interferogram produced in a Fourier spectrometer, Eq. (23.12), is a real function, Eq. (23.14) shows that the measurement is equivalent to the Fourier cosine transform of the source power spectrum, i.e.,

$$\gamma_{11}(\tau) = \int_{-\infty}^{\infty} \hat{\gamma}_{11}(\nu)\cos(2\pi\nu\tau)\,d\nu \ . \tag{23.15}$$

If the source spectrum is symmetric about a center frequency $\bar{\nu}$ (i.e., the self-coherence has no phase), then this measurement is complete and the normalized source power spectrum is obtained directly by Fourier transforming the interferogram.

If the source power spectrum is asymmetric about a center frequency $\bar{\nu}$, then the self-coherence function $\gamma_{11}(\tau)$ of Eq. (23.14) has phase, meaning that additional information is necessary to determine the asymmetric source power spectrum from the interferogram.

Since the Fourier spectrometer gives only the cosine transform of the source power spectrum, it is limited in its direct application to sources that have symmetric power spectra.

23.5.4   Asymmetric Sources

When sources are asymmetric about their mean frequency $\bar{\nu}$, the spectrum is recovered by manipulating the interferogram directly. The interferogram, for a real source containing only positive frequencies, is given by

$$I(\Delta) = \int_{0}^{\infty} \hat{\gamma}_{11}(\nu)\cos 2\pi\nu(\Delta/c)\,d\nu \ , \tag{23.16}$$

and is illustrated in Fig. 23.16(a). Since the spectrum contains no negative frequencies, we create an analytically continued symmetric source spectrum, $\hat{\gamma}'_{11}(\nu) = \hat{\gamma}_{11}(\nu) + \hat{\gamma}_{11}(-\nu)$, about the origin so that Eq. (23.16) becomes

$$I(\Delta) = \int_{-\infty}^{\infty} \hat{\gamma}'_{11}(\nu)\cos 2\pi\nu(\Delta/c)\,d\nu \ . \tag{23.17}$$

In essence, we have added zero value to the interferogram of Eq. (23.16) because $\hat{\gamma}_{11}(-\nu) = 0$. Equation (23.17) can now be cosine transformed because the analytically continued function $\hat{\gamma}'_{11}(\nu)$ is symmetric. The data $I(\Delta)$ also have to be made symmetric by experimentally calibrating the instrument [see Fig. 23.16(b)]. A cosine transformation of the symmetrized data described by Eq. (23.17) is given by

$$\hat{\gamma}'_{11}(\nu) = \int_{-\infty}^{\infty} I(\Delta)\cos 2\pi\nu(\Delta/c)\,d(\Delta/c) \ , \tag{23.18}$$

which displays the symmetrized spectrum as shown in Fig. 23.16(c). Ignoring the negative frequencies in $\hat{\gamma}'_{11}(\nu)$ of Eq. (23.18) gives rise to the original asymmetric source spectrum $\hat{\gamma}_{11}(\nu)$, which was the purpose for making the measurement.

An example of the spectrum of the atmosphere measured by a high-resolution

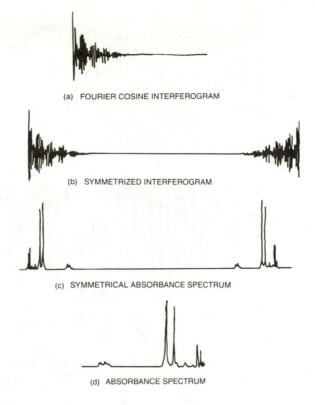

Fig. 23.16 Example of retrieving an asymmetric absorbance spectrum from a Fourier transform spectrometer: (a) Fourier cosine transform, (b) symmetrized interferogram, (c) analytically continued symmetric source spectrum obtained by taking the Fourier cosine transform of the data shown in (b), and (d) measured spectrum retrieved by ignoring the negative frequencies in (c) (from Ref. 18).

instrument having an OPD of 50 cm is shown in Fig. 23.17. Between Figs. 23.17(a) and (b), the displayed spectral resolution changes from 0.1 cm^{-1} to 0.01 cm^{-1}, whereas the displays in (c) and (d) utilize axis expansion only.

23.6   CASE IV: SHEARING INTERFEROMETRY

23.6.1   General Discussion

Shearing interferometers form an important class of measuring instruments because they are self-referencing. These interferometers have the flexibility to be used with either collimated or convergent wavefronts as well as with laser or polychromatic (white) light sources. When laser sources are used in shearing interferometers, contrast losses due to path differences, dispersion effects in the optical elements, and source size are reduced. However, care must be exercised in such instruments to avoid stray reflections and diffraction artifacts (dust, rough surfaces) that cause coherent noise effects such as extraneous high-contrast fringes and speckle. For these reasons, shearing interferometers are often used to measure the wavefront aberrations of large optical elements and systems,[4,20] turbulent or laminar flow phenomena in gases, liquids, and plasmas,[21] atmospheric turbulence for adaptive optics applications,[22] spatial coherence in high-resolution analyzing instruments,[23] and to measure the optical transfer function of optical systems.[24]

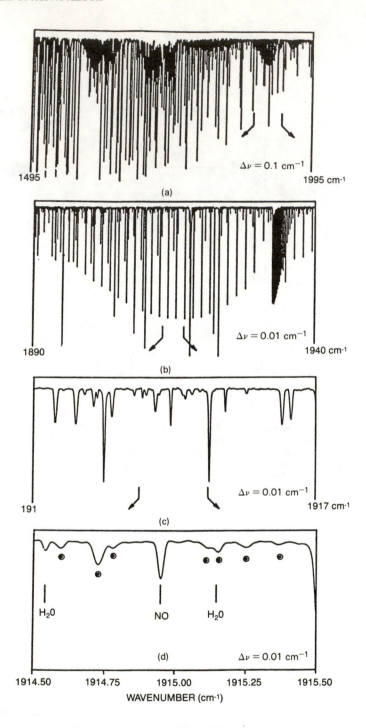

Fig. 23.17 Portion of the band 2 spectrum (1100 to 2000 cm^{-1}) of the atmosphere obtained using a 50-cm OPD in the ATMOS experiment on the space shuttle April 30, 1985, to May 6, 1985: (a) 500-cm^{-1} region of spectrum between 1495 and 1995 cm^{-1} recorded with filter 2, (b) expanded view of the 1900- to 1950-cm^{-1} region from (a), (c) 4 cm^{-1} similarly expanded from (b) in the 1913- to 1917-cm^{-1} region, and (d) 1 cm^{-1} expanded from (c) centered on 1915 cm^{-1}. This one wave number region represents one of the selected microwindows for NO (from Ref. 19).

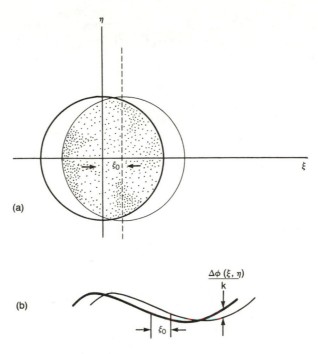

Fig. 23.18 Schematic of a sheared wavefront in a lateral shearing interferometer: (a) shear across pupil and (b) one-dimensional cross section of the wavefront.

23.6.2   Theory for Linear Shearing Interferometry

The shear in these instruments is oftentimes realized by utilizing polarization effects in appropriate crystalline devices such as Wollaston prisms, Savart polariscopes, and crystal lenses, which are all birefringent.[1,4] Alternatively, the shear can be realized by rotating parallel glass plates in a Jamin interferometer,[2,4] by using right-angle prisms in place of the plane mirrors in a Michelson interferometer, by rotating a parallel plate of glass in a triangular interferometer, or by rotating a mirror in a Mach-Zehnder interferometer when used in a converging wavefront.[4]

Mathematically, the two wavefronts in a lateral shearing interferometer (see Fig. 23.18) may be represented by $\phi(\xi, \eta)$ for the unsheared wavefront and $\phi(\xi - \xi_0, \eta)$ for the linearly sheared wavefront, where ξ_0 represents the distance that the wavefront is moved (sheared) in the ξ direction. The phase difference created by the linear shear is

$$\Delta\phi = \phi(\xi - \xi_0, \eta) - \phi(\xi, \eta) \ , \tag{23.19}$$

and the resulting interferogram is

$$I(\xi, \eta) = 2(1 + \cos \Delta\phi) \ . \tag{23.20}$$

For small values of the shear ξ_0, Eq. (23.19) becomes

$$\Delta\phi(\xi, \eta) \cong \frac{d\phi}{d\xi} \ \xi_0 \ . \tag{23.21}$$

Equation (23.21) shows that the information in the interferogram for small values of shear is a derivative of the original wavefront. To determine the wavefront quality from the measurement, an integration of the observed data is necessary.

23.6.2.1   Example 1: Defocus Measurement

As an example illustrating the use of the lateral shearing interferometer for measuring lens aberrations, consider the phase due to a defocusing aberration [see Eq. (23.3)] given by

$$\phi(\xi,\eta) = \text{phase} = k(\text{OPD}) = kD(\xi^2 + \eta^2) \ , \tag{23.22}$$

where D represents the amount of defocusing. From Eq. (23.21), we see that the shearing interferometer creates the differential wavefront given by

$$\Delta\phi(\xi,\eta) \cong 2kD\xi\xi_0 \ , \tag{23.23}$$

which corresponds to a tilted plane wave (linear phase factor) of the form

$$\exp(2ikD\xi_0\xi) \ . \tag{23.24}$$

The differential wavefront, in the region of wavefront overlap (dotted area in Fig. 23.18), as described by Eq. (23.20), gives interference fringes of the form

$$I(\xi,\eta) = 2[1 + \cos(2kD\xi_0\xi)] \ . \tag{23.25}$$

The linear fringes in the vertical direction have the frequency $\nu_0 = 2D\xi_0/\lambda$, which is linearly proportional to the amount of shear and the amount of defocus. When $D = 0$, there are no fringes present. An example of measuring a lens at different focal settings with a lateral shearing interferometer is shown in Fig. 23.19.

23.6.2.2   Example 2: Real-Time Atmospheric Compensation (RTAC)

In the field of astronomy, atmospheric turbulence limits the angular resolution performance of optical telescopes to $\sim$2 arcsec, which corresponds to the diffraction limit of a 6-cm aperture in the visible band. Therefore, the remainder of the aperture in practical telescopes is used for light collection (hence, more signal

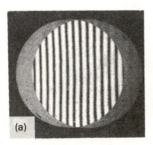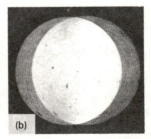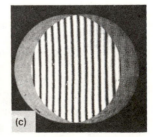

Fig. 23.19 Sequence of lateral shearing interferograms for a wavefront having no aberrations as it passes through focus. The central pattern (having no fringes) is obtained when there is no defocusing. The patterns on either side are due to slight defocusing in either direction by the same amount: (a) inside focus, (b) at focus, and (c) outside focus (from Ref. 4).

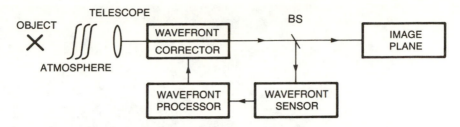

Fig. 23.20 Schematic of real-time atmospheric compensation system for a telescope.

energy) and not resolution improvement. Active optics is a method by which this seeing limitation may be overcome so that angular resolutions (λ/D) commensurate with the actual telescope diameter can be achieved.[22,25] This wavefront compensation is achieved as illustrated in Fig. 23.20. In this system, the wavefront aberration due to the atmosphere is measured by a wavefront sensing system (e.g., a shearing interferometer). The data are processed to create the conjugate wavefront, which is transposed to the wavefront correction device with a series of actuators. The real-time response of the system ($\sim$1 kHz) removes the aberration due to the atmosphere and enables the observer to realize effectively the diffraction-limited optical performance of the telescope.

One of the common methods for sensing the wavefront deformation in these systems is to utilize a double-frequency lateral shearing interferometer in which the wavefront is simultaneously sheared in two orthogonal directions.[20] The shear is introduced with a two-frequency grating in each of two orthogonal directions as shown in Fig. 23.21. From the grating equation, the amount of shear in the first order, given by $S = \lambda(\nu_1 - \nu_2)f$, is controlled by selecting the frequencies of the two gratings ν_1 and ν_2 and the focal length f of the reimaging lens.

For small values of the shear distance S, the interferogram resulting from the interference of the two sheared wavefronts in the ξ direction is given by Eq. (23.21) with $S = \xi_0$. A similar relationship exists in the η direction with ξ being replaced by η. A typical sheared wavefront in the two orthogonal directions created with this type of interferometer when used to test a lens having nonrotationally symmetric aberrations is shown in Fig. 23.22. This figure illustrates the necessity for measuring the aberrations in two directions when correcting asymmetric wavefronts such as those created by atmospheric turbulence.

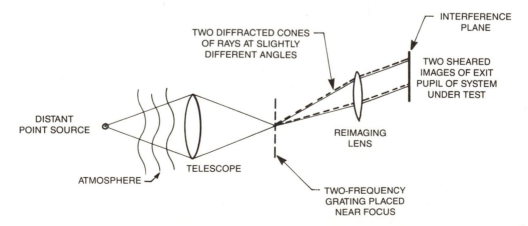

Fig. 23.21 Double-frequency grating lateral shear interferometer (from Ref. 20).

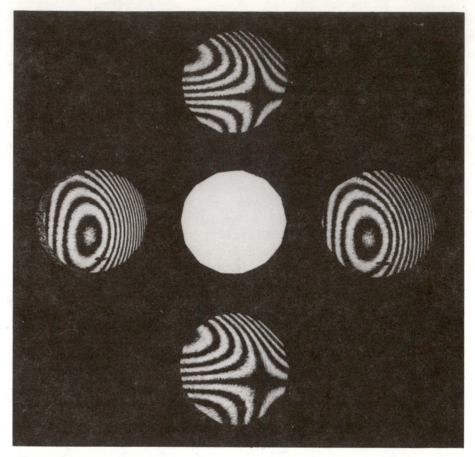

Fig. 23.22 Interferogram obtained using double-frequency crossed grating interferometer (from Ref. 20).

To reconstruct the conjugate wavefront necessary to perform RTAC from these measurements, an algorithm integrating the sheared wavefronts in each of the two orthogonal directions is used[22] to recover the atmospheric phase aberration from Eqs. (23.20) and (23.21). Both analog and digital realizations of this process have been achieved.[25,26]

The results of applying RTAC to a laboratory experiment where the wavefront disturbance was created with a glass plate having 1.5 waves of aberration peak-to-peak is shown in Fig. 23.23. Group 6-6 of the Air Force three-bar resolution target corresponds to a resolution of 114 lp/mm. In this experiment, a laser point source was used to measure the aberration of the glass plate in the shearing interferometer.

23.6.2.3   Example 3: Spatial Coherence Measurements

In measuring the spatial coherence of the field in optical instruments, it is usually impossible to use the double-slit method because of the geometries involved. This limitation is overcome by using Wollaston prisms to create a one-dimensional shearing interferometer[27] to measure spatial coherence. This technique allows small two-point separations to be obtained, uses most of the light collected by the

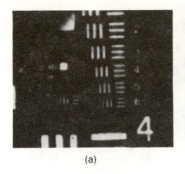
(a)

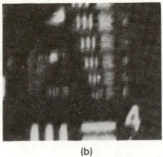
(b)

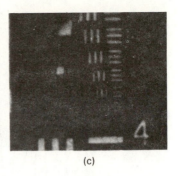
(c)

Fig. 23.23 RTAC performance with an extended image using a three-bar resolution target. The reference source is a He-Ne laser located just outside the image area. The diffraction limit of the system is 130 Hz/mm: (a) no added wavefront distortion (target group 6-67 is 114 Hz/mm); (b) image degradation due to 1.5 waves peak-to-peak wavefront distortion, RTAC off; and (c) same wavefront distortion as (b), with TRAC on. Target group 6-6 is just resolvable on the original negative (from Ref. 22, ©1978 IEEE).

system to form the fringe pattern, and also allows the effective pinhole separation to be changed easily by linearly moving the prism. This technique was used to measure the spatial coherence in the object plane of operational micro-densitometers.[23,28] These measurements were instrumental in the development of a higher performance linear microdensitometer devoid of nonlinearities due to spatial coherence.[29,30]

A schematic diagram showing the location of the prism in the optical train of the instrument is shown in Fig. 23.24. A point in the object plane images as two points in the image plane having an effective separation given by[23]

$$d = 2\theta(n_e - n_o) \left\{ [2L/(n_e + n_o)] + D \right\} , \tag{23.26}$$

where

$$\theta = \text{prism angle}$$

$$L = \text{half thickness of the prism}$$

$$n_e, n_o = \text{refractive indices (extraordinary and ordinary) of the prism}$$

$$D = \text{distance from the object to the prism.}$$

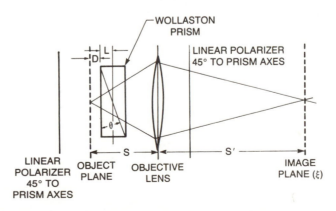

Fig. 23.24 The Wollaston prism as a shearing interferometer (from Ref. 23).

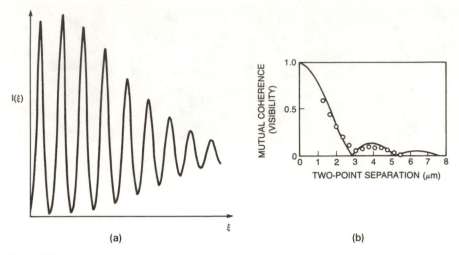

$I(\xi)$

ξ

(a)

MUTUAL COHERENCE
(VISIBILITY)

1.0

0.5

0

0 1 2 3 4 5 6 7 8

TWO-POINT SEPARATION (μm)

(b)

Fig. 23.25 Coherence measurements on a Joyce-Loebl microdensitometer: (a) typical inter-ferometer fringe pattern (logarithm of irradiance versus displacement of Wollaston prism) (from Ref. 23) and (b) measurements of normalized visibility versus two-point separation in the Joyce-Loebl microdensitometer for a condenser objective of nominal NA of 0.08 and a scanning objective NA of 0.08 (from Ref. 28).

If we define the quantity ξ as

$$\xi \equiv \frac{2L}{n_e + n_o} + D \,, \tag{23.27}$$

then the prism produces a phase tilt between the two images given by[23]

$$\phi(\xi) = \left(\frac{4\pi}{\lambda}\right) \theta(n_e - n_o) \xi \,. \tag{23.28}$$

This tilt produces cosine fringes equivalent to those produced by a pair of pinholes separated by a distance d as given in Eq. (23.26) in a double-slit experiment. By changing the distance D between the object plane and the prism face, the visibility of the resulting fringes is a direct measure of the mutual coherence function. The fringe pattern obtained in measurements on a Joyce-Loebl microdensitometer is shown in Fig. 23.25(a) and the resulting coherence function for the microdensitometer with a condenser objective numerical aper-ture (NA) of 0.08 and a scanning objective NA of 0.08 are shown in Fig. 23.25(b).

REFERENCES

1. M. Françon, *Optical Interferometry,* Academic Press, New York (1966).
2. R. S. Longhurst, *Geometrical and Physical Optics,* 2nd ed., p. 151, John Wiley and Sons, Inc., New York (1967).
3. M. Born and E. Wolf, *Principles of Optics,* Pergamon Press, New York (1964).
4. D. Malacara, *Optical Shop Testing,* John Wiley and Sons, Inc., New York (1978).
5. C. M. Vest, *Holographic Interferometry,* pp. 70–71, John Wiley and Sons, Inc., New York (1978).
6. E. Hecht and A. Zajac, *Optics,* Addison-Wesley Publishing Co., Reading, Mass. (1967).
7. F. A. Jenkins and H. E. White, *Fundamentals of Optics,* McGraw Hill Inc., New York (1976).
8. Special Issue on Novel Interferometry, G. W. Hopkins, C. L. Koliopoulos, eds., *Opt. Eng.* 19(6) (1980).
9. J. B. DeVelis and G. O. Reynolds, *Theory and Applications of Holography,* Chapters 2 and 3, Addison-Wesley Publishing Co., Reading, Mass. (1967).
10. J. B. Houston, C. J. Buccini, and P. K. O'Neill, "A laser unequal path interferometer for the optical shop," *Appl. Opt.* 6(7), 1237–1242 (1967).
11. R. J. Zielinski, "Unequal path interferometer alignment and use," *Opt. Eng.* 18(5), 479–482 (1979).
12. E. L. O'Neill, *Introduction to Statistical Optics,* Chapter 4, Addison-Wesley Publishing Co., Reading, Mass. (1963).
13. R. N. Shagam, R. E. Sladky, and J. C. Wyant, "Optical figure inspection of diamond-turned metal mirrors," *Opt. Eng.* 16(4), 375–380 (1977).
14. J. B. DeVelis and G. O. Reynolds, *Theory and Applications of Holography,* Chapter 8, Addison-Wesley Publishing Co., Reading, Mass. (1967).
15. M. J. Beran and G. B. Parrent, Jr., *The Theory of Partial Coherence,* Prentice Hall, Englewood Cliffs, N.J. (1964).
16. G. A. Vanasse and H. Sakai, *Fourier Spectroscopy, Progress in Optics,* E. Wolf, ed., Vol. 6, pp. 261–327, North Holland Publishing Co., Amsterdam (1967).
17. E. G. Steward, *An Introduction to Fourier Optics,* p. 142, Ellis Horwood, Ltd., Chichester, UK (1983).
18. J. A. deHaseth, "Mathematics of spectral treatment in the Fourier domain," in *1985 International Conference on Fourier and Computerized Infrared Spectroscopy,* J. G. Grasselli and D. G. Cameron, eds., Proc. SPIE 553, 41–46 (1985).
19. C. B. Farmer and O. F. Raper, "High-resolution infrared spectroscopy from space: a preliminary report on the results of the atmospheric trace molecule spectroscopy (ATMOS) experiments on Space Lab III," *Space Lab III Mission Review,* NASA Conf. Proc. CP-2429 (May 1986).
20. J. C. Wyant, "Double frequency grating lateral shear interferometer," *Appl. Opt.* 12(9), 2057–2060 (1973).
21. L. H. Tanner, "Design of laser interferometers for use in fluid mechanics," *J. Sci. Instrum.* 43(11), 878–886 (1966).
22. J. W. Hardy, "Active optics: a new technology for the control of light," *Proc. IEEE* 66(6), 651–697 (1978).
23. G. O. Reynolds and A. E. Smith, "Experimental demonstration of coherence effects and linearity in microdensitometry," *Appl. Opt.* 112(6), 1259–1270 (1973).
24. D. Nyyssonen and J. Jerke, "Lens testing with a simple wavefront shearing interferometer," *Appl. Opt.* 12(9), 2061–2070 (1973).
25. J. W. Hardy, J. E. Lefebure, and C. L. Koliopoulos, "Real-time atmospheric compensation," *J. Opt. Soc. Am.* 67(3), 360–369 (1977).
26. M. P. Rimmer, "Method for evaluating lateral shearing interferograms," *Appl. Opt.* 13(3), 623–629 (1974).
27. S. Mallick, "Degree of coherence in the image of a quasimonochromatic source," *Appl. Opt.* 6(7), 1403–1406 (1967).
28. B. Justh, "Measurement of the spatial coherence of microdensitometer illumination," *J. Opt. Soc. Am.* 58(5), 714A (1968).
29. J. P. Fallon, "Design considerations for a linear microdensitometer," *Opt. Eng.* 12(6), 206–212 (1973).
30. D. J. Cronin and G. O. Reynolds, "Optical design considerations and test results for a linear microdensitometer," *Opt. Eng.* 12(6), 201–205 (1973).

24   Multiple-Beam Interference

24.1   INTRODUCTION

In the division of amplitude interferometers discussed in the previous chapter, we selected $N = 2$ in Eq. (23.1) and ignored the higher order reflections from the optical components. These reflections are weak in intensity when low-reflectance surfaces are involved. This effect is illustrated in Fig. 24.1 for two parallel glass/air interfaces each having a reflectivity of 4%. The beams resulting from multiple reflections are orders of magnitude weaker than the beams resulting from the primary reflections. Thus, interference phenomena such as Newton rings are not affected by the higher order reflections.

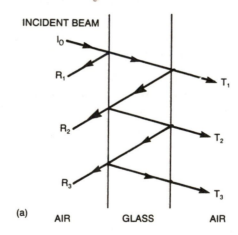

(a)   AIR   GLASS   AIR

REFLECTED BEAM INTENSITY   TRANSMITTED BEAM INTENSITY

$R_1 = 0.04\,I_o$ $T_1 = 0.92\,I_o$
$R_2 = 0.037\,I_o$ $T_2 = 1.5 \times 10^{-3}\,I_o$
$R_3 = 5.9 \times 10^{-5}\,I_o$ $T_3 = 2.4 \times 10^{-6}\,I_o$

(b)

Fig. 24.1 Reflected and transmitted multiple beams from an uncoated glass plate in air having parallel surfaces and a normal reflectance of 4%: (a) schematic of first three reflected and transmitted beams and (b) intensities of these six beams relative to the incident beam intensity I_o.

Fabry and Perot[1] recognized that these higher order beams could be exploited in interferometry by purposely increasing the reflectance R of the parallel surfaces in Fig. 24.1. The result of increasing this reflectance is to create an N-beam interference effect that sharpens the fringes.

A similar fringe-sharpening effect occurred in the consideration of N-beam division of wavefront interferometry (i.e., the diffraction grating in Chapter 22) where the N-beams were of nearly equal amplitude.

24.2 ANALYSIS

The N-beam interference effects arising from multiple reflections between highly reflecting parallel surfaces are shown schematically in Fig. 24.2. The optical path

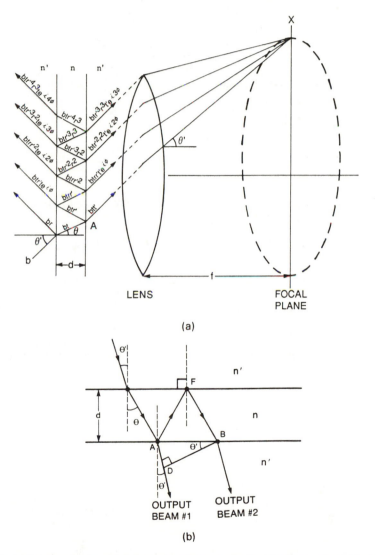

Fig. 24.2 (a) Schematic of multiple-beam interference of an incident wave of amplitude b from two highly reflecting thin surfaces having complex transmittances and reflectances (t, r) and (t', r'), respectively. The lens focuses the parallel wavefronts into the focal plane where the fringes are formed. (b) Expanded geometry for two successive transmitted waves where n' and n are the indices of refraction of the respective media.

difference between any two successively transmitted rays from Fig. 24.2(b), where AFB and AD denote the geometrical paths as indicated, is seen to be

$$\Delta = (n)AFB - (n')AD \tag{24.1}$$

and the resulting phase difference ϕ is

$$\phi = k\Delta . \tag{24.2}$$

The geometrical distance AFB is given by

$$AFB = 2d/\cos\theta \tag{24.3}$$

and the geometrical distance AD is given by

$$AD = AB\sin\theta' . \tag{24.4}$$

Use of Snell's law,

$$n'\sin\theta' = n\sin\theta , \tag{24.5}$$

in Eq. (24.4) gives

$$AD = \frac{(AB)n\sin\theta}{n'} . \tag{24.6}$$

Since $AB = 2d\tan\theta$, from Eqs. (24.1), (24.3), and (24.6), we get

$$\Delta = 2nd\ \frac{1}{\cos\theta} - \tan\theta\sin\theta = 2nd\cos\theta . \tag{24.7}$$

The phase difference of two successive rays is therefore given by

$$\phi = k\Delta = \frac{4\pi}{\lambda}nd\cos\theta . \tag{24.8}$$

The amplitude distribution in the focal plane of the lens in Fig. 24.2(a) is obtained by modifying Eq. (23.1) to give

$$A(\phi) = \sum_{m=0}^{\infty} btt'r^m r'^m e^{im\phi} = btt' + btt'rr'e^{i\phi} + btt'r^2r'^2e^{2i\phi}$$

$$+ btt'r^3r'^3e^{3i\phi} + \dots , \tag{24.9}$$

where

t,r are the complex transmittance and reflectance, respectively, of the first surface

t',r' are the complex transmittance and reflectance, respectively, of the second surface

b is the amplitude of the incident wavefront

ϕ is given by Eq. (24.8) in terms of the angle of refraction θ

θ is related to θ' through Snell's law of Eq. (24.5).

If we let the complex transmittances and reflectances of the two surfaces be identical, then

$$t = t' = |t|\,e^{i\phi'} \,,$$
$$r = r' = |r|\,e^{i\phi''} \,,$$

(24.10)

where ϕ' and ϕ'' in general are functions of wavelength depending on the dispersion characteristics of the materials. The corresponding intensity transmittance T and reflectance R are given by

$$T = tt^* \,,$$
$$R = rr^* \,.$$

(24.11)

Using Eq. (24.10), Eq. (24.9) can be written as

$$A(\phi) = bt^2 \sum_{m=0}^{\infty} r^{2m} e^{im\phi} = \frac{bt^2}{1 - r^2 e^{i\phi}} \,,$$

(24.12)

where the convergent form of the infinite series has been used.

The intensity at the focal plane of the lens in Fig. 24.2(a) is given by

$$I(\phi) = A(\phi)A^*(\phi) = \left(\frac{bt^2}{1 - r^2 e^{i\phi}}\right)\left[\frac{b^*(t^*)^2}{1 - (r^*)^2 e^{-i\phi}}\right] \cdot$$

(24.13)

Assuming $b=1$ for convenience and using Eq. (24.11), we can rewrite Eq. (24.13) as

$$I(\phi + 2\phi'') = \frac{T^2}{(1 - R)^2}\left[1 + \frac{4R}{(1 - R)^2}\sin^2\left(\frac{\phi + 2\phi''}{2}\right)\right]^{-1}$$

(24.14)

or

$$I(\phi + 2\phi'') = \frac{T^2}{(1 - R)^2}\left[1 + \frac{4N_R^2}{\pi^2}\sin^2\left(\frac{\phi + 2\phi''}{2}\right)\right]^{-1} \,,$$

(24.15)

where

$$N_R = \frac{\pi\sqrt{R}}{1 - R}$$

(24.16)

is the spatial finesse of the multiple-beam interference process and ϕ is defined in Eq. (24.8). Equation (24.14) is generally known as the *Airy formula*.

A plot of Eq. (24.14) for various values of intensity reflectance is shown in Fig. 24.3 for the case where $\phi''=0$ (i.e., no absorption in the reflecting material) and ϕ is defined by Eq. (24.8). The figure shows that the distance between the plates has to be exactly an integral number of wavelengths in order to observe a maximum

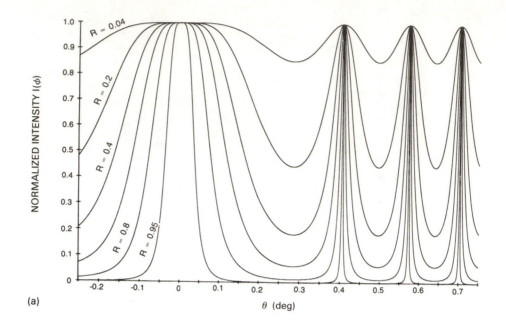

(a)

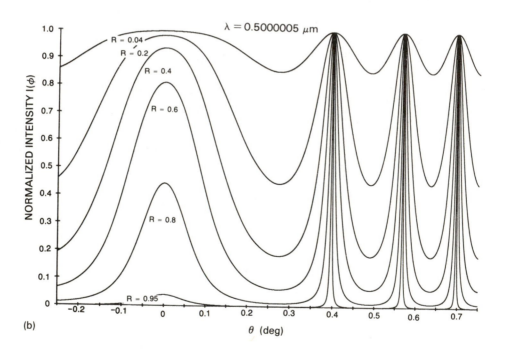

(b)

Fig. 24.3 Plot of the first few peaks of the intensity distribution in the output plane of a multiple-beam transmission interferometer [Eq. (24.14) for $\phi'' = 0$] for various values of reflectance where $d = 1$ cm and $n = 1$: (a) $\lambda = 5000$ Å $= 0.5$ μm so that the thickness d is exactly an integral number of wavelengths giving a maximum at the origin for all values of R and (b) $\lambda = 5000.005$ Å $= 0.5000005$ μm so that d is not exactly an integral number of wavelengths. This slight mismatch of 10^{-6} λ from an integral number of wavelengths causes destructive interference in the central maximum as R increases as well as a slight shifting of the higher order peaks, as seen more clearly in Fig. 24.11. (Courtesy of A. Ho.)

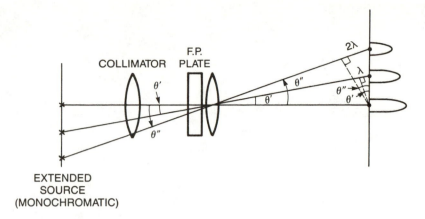

Fig. 24.4 First few maxima in the image plane of a multiple-beam interferometer where the plate separation is an integral number of wavelengths.

intensity on axis in a multiple-beam interferometer having high values of reflectance. In most multiple-beam interferometers having high reflectivity, the on-axis intensity is effectively zero because the separation of the plates is not exactly an integral number of wavelengths. This causes destructive interference such as that observed in Fig. 24.5(c). The nonperiodic characteristic of the rings seen in Figs. 24.3(a) and (b) can be described by referring to Fig. 24.4 where we see that

$$\tan\theta' = x/f \ , \tag{24.17}$$

$$\sin\theta' = j\lambda/x \ , \tag{24.18}$$

where j is an integer. In the small-angle approximation,

$$\tan\theta' \cong \sin\theta' \tag{24.19}$$

so that

$$x^2 = j\lambda f \ , \tag{24.20}$$

or

$$x = \sqrt{j\lambda f} \ . \tag{24.21}$$

Equation (24.21) shows that the positions of the maxima in the image plane of the multiple-beam interferometer obey the same equation as the diffraction rings in a Fresnel diffraction pattern. This explains the Fresnel-like nature of the maxima seen in the plots of Fig. 24.3 and the experimental result in Fig. 24.5(c).

24.3   VISIBILITY OF THE FRINGES OF AN N-BEAM INTERFEROMETER

The visibility of the transmission fringes in an N-beam interferometer can be determined from Eq. (24.15) and the definition of visibility [$V = (I_{max} - I_{min})/(I_{max} + I_{min})$] to be

$$V = 2R/(1 + R^2) \ , \tag{24.22}$$

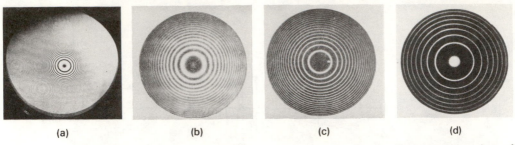

(a)   (b)   (c)   (d)

Fig. 24.5 Comparison of fringe visibility of n-beam interferometers having various values of surface reflectance: (a) Newton rings, $R \sim 4\%$ (from Ref. 2); (b) Michelson interferometer, $4\% < R < 50\%$ (from Ref. 2); (c) Fabry-Perot, $R = 80\%$, showing high-visibility fringes with a dark central fringe (from Ref. 2); and (d) Fabry-Perot fringes with central maxima. Plates are separated an integral number of wavelengths to give a bright central fringe (from Ref. 3).

where

$$I_{max} = T^2/(1 - R)^2 \; , \tag{24.23}$$

$$I_{min} = T^2/(1 + R)^2 \; . \tag{24.24}$$

Thus, as the reflectivity increases (less transmission), the visibility approaches unity and the peaks sharpen as seen in Figs. 24.3(a) and (b). Low-reflectance multiple-beam interference effects such as those observed in Newton rings or from oil films have fringes of low visibility since reflectances of these surfaces are usually less than 10%. This visibility effect is shown in Fig. 24.5 where the low-contrast fringes from Newton rings (produced between an optical flat and a convex surface) and a Michelson interferometer are compared with the high-contrast multiple-beam fringes made with parallel plates having a reflectivity of 0.8.

Multiple-beam interference fringes from highly reflecting parallel plates having a separation of exactly an integral number of wavelengths are shown in Fig. 24.5(d). Note the change in polarity of the central fringes between Figs. 24.5(c) and (d), which illustrates the sensitivity of the interferometer to plate separation. This sensitivity is shown graphically in Fig. 24.3.

If we plot Eq. (24.14) as a function of ϕ [the optical phase defined in Eq. (24.8)], rather than θ as previously done in Fig. 24.3, then the periodic structure of the maxima shown in Fig. 24.6 is obtained. In this figure, we note that a peak intensity occurs each time the optical phase ϕ changes by 2π. Thus, as shown, the distance between successive maxima is 2π.

Note that the intensity pattern plotted in Fig. 24.6 is a convenient form to use for mathematically analyzing the interferometer. However, it does not describe the spatial variations observed in a multiple-beam interferometer because it masks the Fresnel characteristic of the fringes plotted in Fig. 24.3 and shown in Figs. 24.5(c) and (d). The *finesse* of a multiple-beam interferometer is defined as the distance between peaks divided by the full width at half-power of an individual peak of the periodic output shown in Fig. 24.6. The finesse thus defined is $N_R = \pi\sqrt{R}/(1 - R)$, which agrees with the definition previously given in Eq. (24.16).

24.4   ADDITIONAL CHARACTERISTICS OF MULTIPLE-BEAM INTERFEROMETERS

There are two θ-dependent effects observed in the output of the interferometer. The first effect is the increase in spatial frequency of the intensity maxima as one

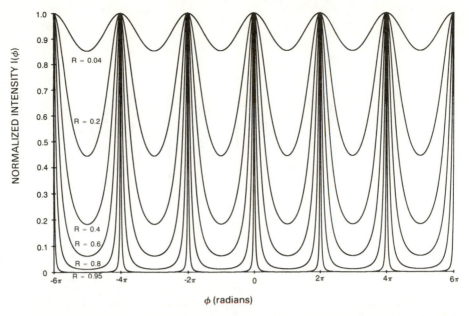

Fig. 24.6 Plot of Eq. (24.14) as a function of ϕ and various values of R indicating that the peak intensities occur each time the optical phase ϕ changes by 2π. The peak at the origin occurs because we chose $d = 1$ cm, $\lambda = 5000$ Å, and $n = 1$; i.e., d is exactly an integral number of wavelengths. (Courtesy of A. Ho.)

moves away from the origin. The second effect is that the width of each maximum becomes narrower with distance from the origin. We now describe these two characteristics of the multiple-beam interferometer in more detail.

If we let $\phi'' = 0$ in Eq. (24.14) (i.e., the interferometer has dielectric coatings with no absorption so that t and r are real), then the variable in Fig. 24.3(a) is ϕ as given by Eq. (24.8). Substitution of Eq. (24.8) into (24.14) (which was also done in plotting Fig. 24.3) yields

$$I(\theta) = \frac{T^2}{(1-R)^2} \left[1 + \frac{4R}{(1-R)^2} \sin^2(knd\cos\theta) \right]^{-1}, \tag{24.25}$$

where θ is the refraction angle of Fig. 24.2(a).

The experimental result in Fig. 24.5(c) is displayed in the variable $\theta' = x/f$ where θ' is the incidence angle as shown in Fig. 24.2(a). Use of Snell's law with $n' = 1$ [Eq. (24.5)] in Eq. (24.25) gives the output intensity as a function of θ' (the viewing angle) in the form

$$I(\theta') = \frac{T^2}{(1-R)^2} \left[1 + \frac{4R}{(1-R)^2} \sin^2 \left\{ knd\cos \left[\sin^{-1} \left(\frac{1}{n} \sin\theta' \right) \right] \right\} \right]^{-1}, \tag{24.26}$$

which is plotted in Fig. 24.7. This result demonstrates the two characteristics of the multiple-beam interference pattern (i.e., nonperiodic fringe location and higher order fringe narrowing) discussed above. The locations of the maxima in Fig. 24.7(a) were shown to be proportional to the square root of the ring number in Eq. (24.21) using the small-angle approximation for θ'.

The locations of the maxima of Eq. (24.25) [see Fig. 24.7(a)] occur when the phase of the sine squared term satisfies the condition that

$$k(nd\cos\theta) = S'\pi$$

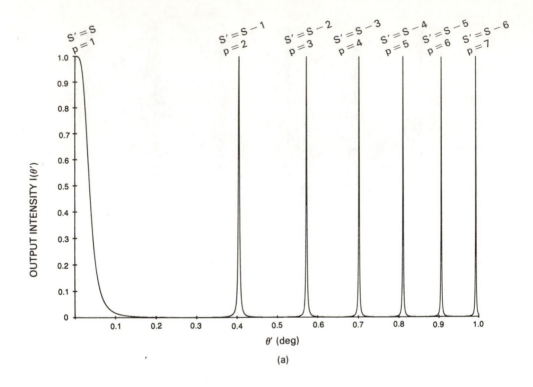

(a)

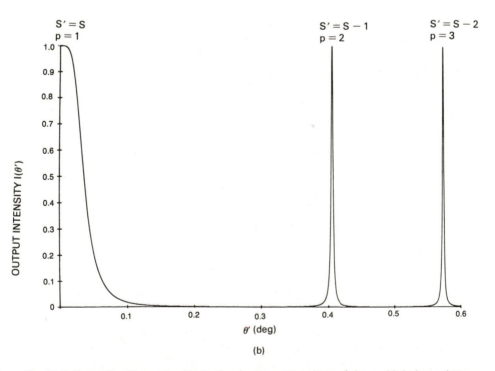

(b)

Fig. 24.7 Normalized intensity distribution in the output plane of the multiple-beam interferometer shown in Fig. 24.2(a) as a function of angle of incidence θ' for $R = 0.95$, $T = 1$, $n = 1$, $\lambda = 0.5\ \mu m$, and $d = 1$ cm: (a) The plot shows that the separation between successive maxima is decreasing and that each maximum is narrower than its predecessor. (b) The plot is on an expanded scale that illustrates these points more clearly. (Courtesy of A. Ho.)

or

$$2nd\cos\theta = S'\lambda ,$$ (24.27)

where S' is a running integer counting from the outermost maximum toward the origin starting with a value of S' determined by the separation of the plates. If we consider the maximum closest to the axis in the observation plane $\theta' \cong \theta \cong 0$, then Eq. (24.27) shows that

$$S' = S = 2nd/\lambda$$ (24.28)

and S is defined as the order of interference. If $2nd/\lambda$ is an integral number, an absolute maximum will occur on axis as shown in Fig. 24.3(a). If $2nd/\lambda$ is not an integral number, then a lower valued central maximum will occur on axis as shown in Fig. 24.3(b). When S is an integral number of wavelengths, the higher order maxima will be shifted slightly from their location by an amount ϵ. Thus, ϵ is the shift corresponding to a path difference of a fraction of a wavelength. Experimentally, this shift is usually not noticed because of the symmetric nature of the interference patterns. Therefore, we will not consider its effect further in this analysis.

In general, the positions of the maxima are counted by the algorithm

$$S' = S - (p - 1) \text{ for } p = 1,2,3... ,$$ (24.29)

where p denotes the maximum being considered starting with $p = 1$ at the origin. At the location of the maxima, the fall-off of intensity with observation angle θ' is very rapid so that the small-angle approximation may be used. Thus, substituting Eqs. (24.28) and (24.29) into (24.27) and invoking the small-angle approximation, we get

$$2nd(1 - \theta^2/2) = (S - p + 1)\lambda .$$ (24.30)

Thus, the angular location of the p'th maximum is given by [using Eq. (24.28) in (24.30)]:

$$\theta_p \cong \sqrt{\frac{(p - 1)\lambda}{nd}} .$$ (24.31)

From Snell's law, in the small-angle approximation, $\theta = \theta'/n$ when $n' = 1$. Then Eq. (24.31) becomes

$$\theta'_p = \sqrt{\frac{n(p - 1)\lambda}{d}} \text{ for } \theta \ll \pi ,$$ (24.32)

which shows that the angular locations of the maxima of the intensity pattern vary directly with $\sqrt{(p - 1)}$ (in a manner analogous to the maxima in Fresnel diffraction) and inversely with $\sqrt{d}$. The narrowing of the higher order maxima is observed in the plot of Eq. (24.25), which was shown in Fig. 24.7. Physically, this narrowing of the successive maxima occurs for two reasons. First, the reflectivity of the plates causes a narrowing in each peak as seen in Fig. 24.3. Second, the periods of the successive maxima decrease in proportion to the square root of the number of the maxima due to the Fresnel nature of the fringes, as just shown by Eq. (24.32). Thus, each successive maximum in Fig. 24.7 undergoes narrowing from both phenomena, which explains the narrow outer rings seen in Figs. 24.5(c) and (d).

24.5   CHROMATIC RESOLVING POWER OF A MULTIPLE-BEAM INTERFEROMETER

When a multiple-beam interferometer operates in a mode of high reflectance ($R \cong 1$), the interferogram can be used to resolve two closely spaced spectral components. The chromatic resolving power of such an instrument can be determined by using the periodic form of the output shown in Fig. 24.6. This approach will give valid results in the first maximum because $p = 2$ in Eq. (24.32), making the Fresnel factor of $\sqrt{(p-1)}$ unity. Thus, the small angular separation between just resolved spectral components is equal in both the θ and ϕ plots of Fig. 24.3 and 24.6, respectively.

The half-power criterion for resolving two closely spaced spectral lines in the first maximum of the periodic plot in Fig. 24.6 is shown schematically in Fig. 24.8. Two closely spaced spectral lines are considered to be resolved when their intensity distributions in the first order cross at their 50% intensity values, as shown.

To simplify the analysis, we assume that the material is not dispersive in the wavelength band of interest, such that $\phi'' = 0$ in Eq. (24.14). Application of the half-intensity criterion to the first peak in Eq. (24.14) demands that

$$\frac{4R}{(1-R)^2} \sin^2\left(\frac{\phi_0}{2}\right) = 1 \;.$$

(24.33)

In the vicinity of the first maximum (see Fig. 24.8), the intensity decrease is rapid, such that $(\phi_0/2) \ll 2\pi$. Therefore, the small-angle approximation for the sine may be used and Eq. (24.33) becomes

$$\phi_0 = \frac{1-R}{\sqrt{R}} \;,$$

(24.34)

which describes the half-angle separation between the two just resolvable peaks (i.e., the intensity addition of the adjacent spectral peaks in Fig. 24.8 yields a central dip of 17% as shown in Fig. 24.9) and, therefore, defines the minimum half-angle between two adjacent wavelengths which the instrument will just resolve. This implies that the minimum phase $d\phi$ cannot be less than $2\phi_0$. Therefore, differentiation of Eq. (24.8) with respect to the wavelength λ yields

$$d\phi = -2\pi(2nd\cos\theta)\frac{d\lambda}{\lambda^2} = 2\phi_0 \;.$$

(24.35)

Substituting Eq. (24.35) into (24.34) and using the interference condition for the central maximum (i.e., $S' = S$) given in Eq. (24.28), we obtain the chromatic resolving power of the instrument (defined as $|\lambda/d\lambda|$):

$$\left|\frac{\lambda}{d\lambda}\right| = \frac{\pi S\sqrt{R}}{(1-R)} = SN_R \;,$$

(24.36)

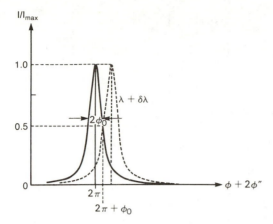

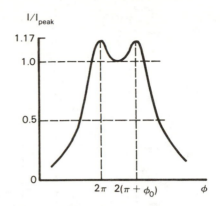

Fig. 24.8 Schematic of two spectral components, λ and $\lambda + \delta\lambda$, in the first maximum of a multiple-beam interferometer where $2\phi_0$ is the full width at half-power.

Fig. 24.9 Intensity summation of two just resolved spectral components (from Fig. 24.8) in a multiple-beam interferometer, where I_{peak} is the maximum intensity of one of the peaks alone.

where N_R was given in Eq. (24.16). Thus, the chromatic resolving power is proportional to two quantities: the order of interference S, which depends on the thickness of the refracting medium d through Eq. (24.28), and the reflectivity R.

24.6   FABRY-PEROT INTERFEROMETRY

Multiple-beam interference effects are the basis of a Fabry-Perot interferometer consisting of two highly reflecting parallel plates of variable separation and a lens to image the various source points in the image plane [see Fig. 24.2(a)]. When the plates have a fixed separation, the interferometer is called a *Fabry-Perot etalon*.

The variable separation instrument is useful for high-resolution spectroscopy, tunable filters, and surface measurements, whereas etalons are important in thin-film technology, gas index measurement, and the design of laser cavities.

24.6.1   Example 1: Spectroscopic Measurements

If a Fabry-Perot interferometer is coated such that the reflectance of each plate is 0.95 and the instrument set such that the plate separation is 1 cm, then the maximum resolving power (for a green light of mean wavelength 5000 Å, achievable with this interferometer) is given by Eq. (24.36) to be $\lambda/\Delta\lambda = 2.45 \times 10^6$, where we assume $n = 1$. This means that two wavelengths separated by 2.04×10^{-3} Å will be resolved by this instrument setting.

The chromatic resolving power given in Eq. (24.36) is illustrated in Fig. 24.10. The narrowing of the successive maxima due to reflectivity of the plates was shown previously in Figs. 24.7(a) and (b). Additional resolving power due to plate separation is shown in Fig. 24.10 for two plates differing by a factor of 10 in their separations. The output of the instrument when observing spectral lines separated by 0.05 and 0.005 Å is shown in Fig. 24.11. The lines separated by 0.005 Å are well resolved in the first order in agreement with Eq. (24.36), which was derived for the case where the Fresnel factor was unity. The periodic plot in Fig. 24.6 would lead one to believe that the chromatic resolving power was equal for all maxima. However, the plot in Fig. 24.11 shows that even though the two lines separated by 0.005 Å are well resolved in the first maximum, the Fresnel

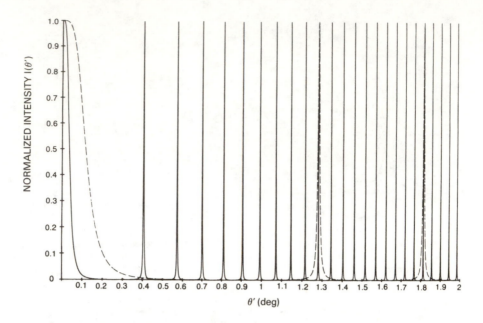

Fig. 24.10 Plot of the output of a Fabry-Perot interferometer, Eq. (24.26), for two differing plate separations, where $R = 0.95$, $\lambda = 5000$ Å, and $n = 1$ (———, $d = 1$ cm; - - - - -, $d = 1$ mm). This result demonstrates the additional chromatic resolving power obtained in a Fabry-Perot interferometer due to increasing the plate separation; i.e., the lines get narrower. The loss in FSR is also seen; i.e., the successive peaks are closer together. (Courtesy of A. Ho.)

factor brings them closer together in the second and third maxima and they are unresolved in the region of the fourteenth maximum and beyond.

The use of a Fabry-Perot interferometer for resolving two closely spaced spectral lines can be described in terms of its free spectral range (FSR)—the distance between the first two maxima as a function of the angle θ—and its chromatic resolving power in the first maximum as described by Eq. (24.36).[a] The trade-off between these two parameters and the upper limit on reflectivity caused by absorptions in the coating materials determines the practical limit of the instrument's chromatic resolving power. These trade-offs are illustrated in Fig. 24.12 for three different integral values of mirror separation. At a mirror separation of 100 μm we see a large FSR and relatively poor chromatic resolution (i.e., the first maximum is quite broad). The chromatic resolving power is improved by increasing the mirror separation to 1000 μm (1 mm) as seen by the narrowing of the central maximum. However, this increased mirror separation reduces the FSR by $\sqrt{d}$ so that the increased chromatic resolution can only be observed over the smaller bandwidth determined by the FSR. This effect is further increased, as seen in Fig. 24.12, for a plate separation of 10^4 μm = 1 cm.

In practical applications, when a Fabry-Perot interferometer is used to measure the separation of spectral lines, an additional dispersive element such as a prism or a slit is used to prevent overlapping of the lines.[2,3]

---

[a]This is equivalent to the FSR on the periodic plot of Fig. 24.6 when using the wavelength component of ϕ as the variable. With reference to the periodic plot of Fig. 24.6, the FSR is normally defined as the separation of two successive peaks measured in wave numbers (cm^{-1}).

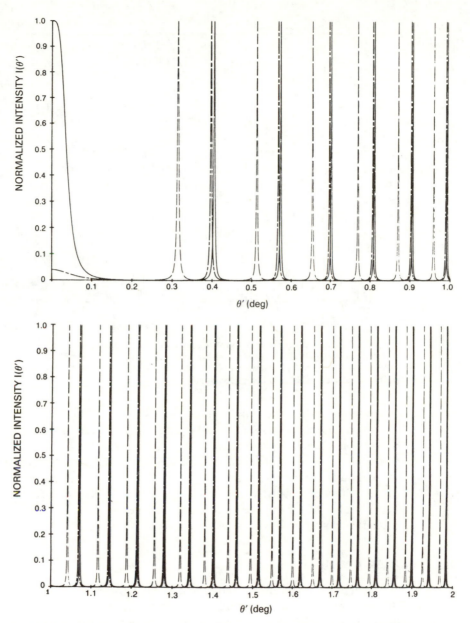

Fig. 24.11 Plot of the output maxima of a Fabry-Perot interferometer, Eq. (24.26), for three spectral lines separated by $\Delta\lambda_1 = 0.05$ Å and $\Delta\lambda_2 = 0.005$ Å, $R = 0.95$, $n = 1$, and $d = 1$ cm (———, $\lambda_1 = 5000$ Å $= 0.5$ μm; —·—·, $\lambda_2 = 5000.05$ Å $= 0.500005$ μm; ------,$\lambda_3 = 5000.005$ Å $= 0.5000005$ μm), illustrating the capability of the instrument to resolve these spectral lines easily. Since the resolving power of this instrument is 2.04×10^{-3} Å, the two lines separated by 0.005 Å are a factor of 2.5 above the resolving power of the instrument and are well separated in the plot. (Courtesy of A. Ho.)

24.6.2   Example 2: Laser Cavities

The Fabry-Perot cavity is an essential element in the design and fabrication of most laser systems. Even though we consider only the parallel plate cavity in this section, practical lasers often utilize two concave mirrors (confocal resonator)

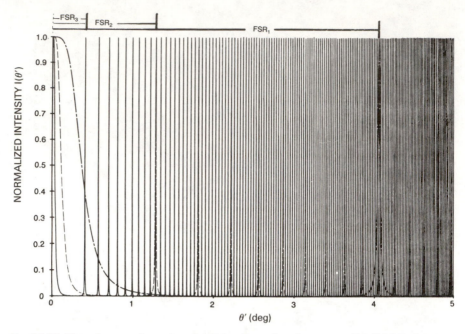

Fig. 24.12 Plot of output intensity of a Fabry-Perot interferometer, Eq. (24.26), for three different integral values of mirror separations (_____, $d = 10^4 \ \mu m = 1$ cm; ------, $d = 10^3 \ \mu m$ = 1 mm; —·—·-, $d = 10^2 \ \mu m$) where $R = 0.95$, $n = 1$, and $\lambda = 5000$ Å. This plot illustrates the trade-off between FSR and CRP. (Courtesy of A. Ho.)

separated by a distance equal to their radius of curvature, or a concave-plane mirror configuration to simplify mirror adjustment. These configurations reduce diffraction losses relative to those of the plane parallel plate cavity. The popular confocal resonator reduces the mirror alignment tolerance from the 1-arcsec value required in the plane parallel resonator to approximately one quarter of a degree. Simultaneously, the confocal resonator reduces diffraction losses by many orders of magnitude. Therefore, diffraction losses can be considered negligible in such resonator configurations.[4,5]

24.6.2.1   Plane Parallel Resonator

When Fabry-Perot parallel mirrors of high reflectivity are used as a laser cavity, the lasing bandwidth is narrowed because of the beam sharpening characteristics of the Fabry-Perot interferometer as already shown in Fig. 24.3. The separation between the mirrors d is chosen to be an integral number of half wavelengths, i.e., S = order of interference = $2nd/\lambda$ as given in Eq. (24.28). The frequencies of lasing, which occur when $nd = m\lambda/2$, where m is an integer, are given by

$$\nu_m = mv/2d \ , \tag{24.37}$$

where v is the velocity of light in the lasing medium inside the cavity. This cavity sustains lasing modes corresponding to the frequencies given by Eq. (24.37). The high-reflectivity mirrors select the stimulated emitted radiation so that it travels along the optical axis and interferes constructively to produce the on-axis Fabry-Perot peak ($\cos \theta = 0$) in Fig. 24.3. The stimulated emission at off-axis angles is rejected because it does not interfere according to Eq. (24.28). The higher order frequencies ν_{m+1}, ν_{m+2}, etc., of Eq. (24.37) are not present unless the lasing medium emits at precisely these frequencies. For example, the three strongest stimulated emission lines in the He-Ne spectrum are not integral multiples of each other. Therefore, when the cavity is tuned for the 0.6328-μm line (i.e., d is an integral multiple of 3164 μm), the 3.39- and 1.15-μm lines are rejected by the cavity. He-Ne lasers for these other two lines can be made by adjusting the cavity length d to satisfy Eq. (24.37).

24.6.2.2   Coherence Length of Laser Radiation

The mirror separation distance d is chosen to satisfy Eq. (24.37) for a selected stimulated spectral emission line of the lasing medium within the cavity. The cavity distance d also determines the value of S given by Eq. (24.28). Large values of S determine the width of the zero-order maximum as shown previously in Fig. 24.12. Thus, the longer the cavity, the higher the order S and the narrower the width of the central maximum; hence, the sharper the fall-off for mismatches of wavelength.

For example, the 1-cm cavity described by Fig. 24.11 at a mean wavelength of 5000 Å supports a bandwidth of ~0.05 Å, which corresponds to a coherence length of $l_c = \lambda^2 / \Delta\lambda = 5$ cm. This assumes that the lasing medium emits over a bandwidth of $\Delta\lambda = 0.05$ Å. Either increasing the cavity length or choosing lasing media that have extremely narrow emission linewidths has the effect of increasing the coherence length of the laser. Commercially available gas lasers have coherence lengths ranging in value from a few centimeters to kilometers.

24.6.2.3   Multimode Lasers

Some laser materials have many spectral lines in their emission spectra leading to a multiplicity of lasing frequencies. These various spectral lines have different mean wavelengths so that the Fabry-Perot cavity can support constructive interference at various angles. Thus, an angular spectrum of frequencies (wavelengths) is emitted by the cavity, each angle corresponding to a different cavity mode. A photograph of the laser output displays the interference between the various central maxima, resulting in the familiar mode structure of the laser. An example of the output from a multimode laser beam is shown in Fig. 24.13. Since each mode supports a slightly different wavelength, the bandwidth of the output is greater than that due to a single mode alone. Thus, the coherence length of such lasers is reduced. For example, the coherence length of a pulsed ruby laser, which has several modes, is on the order of a few millimeters because of the combination of the relatively short cavity length and the multimode structure supported by the cavity.

24.6.2.4   Spatial Coherence of Laser Radiation

The spatial coherence of laser radiation is normally close to unity, i.e., $|\gamma_{12}(0)| = 1$ over the spatial extent of the beam. As with any collimated beam, the spatial coherence remains unchanged if the laser beam is expanded. Lasers supporting only one mode have a higher degree of spatial coherence than multimode lasers. A two-pinhole experiment performed on the output of a single-mode laser will have a high fringe visibility for all pairs of points across the beam. A similar experiment on a multimode laser beam will yield lower values of fringe visibility because of the complex spatial phase structure and broader bandwidth of the radiation at the output of the laser cavity.

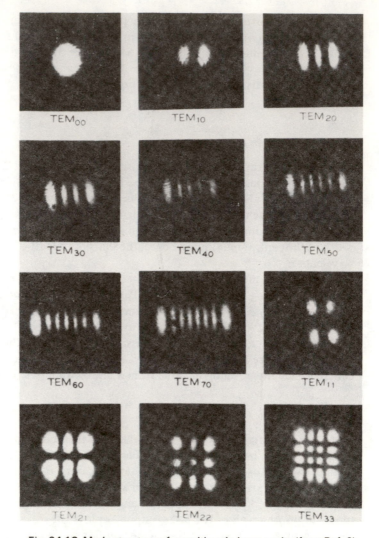

Fig. 24.13 Mode structure of a multimode laser cavity (from Ref. 6).

24.6.3   Example 3: Tunable Filters

Fabry-Perot interferometers have been used to make tunable filters in various wavelength bands.[7-9] In these applications, the value of θ in Eq. (24.14) is fixed and the multiple-beam interference effects occur as a function of wavelength. Thus, the periodic plot of Fig. 24.6 is valid where the phase difference ϕ changes by varying the wavelength.

Fabry-Perot interferometers can have large order S, as shown in Eq. (24.28), and small FSR, the distance between successive peaks, or small order and large FSR. This was seen in Fig. 24.12 as a function of the output angle θ'.

When plotted as a function of wave number (inverse wavelength), Eq. (24.14) is a period function having a frequency given by

$$\nu_0 = 2nd\cos\theta , \tag{24.38}$$

where the half-angle formula has been used. The distance between peaks, i.e., the period, called the *FSR of the interferometer,* measured in inverse length, is

$$\text{wave number FSR} = \text{period} = (\text{frequency})^{-1} = (2nd\cos\theta)^{-1} \ . \qquad (24.39)$$

Using Eq. (24.28) and letting $\theta = 0$, Eq. (24.39) becomes

$$\text{wave number FSR} = 1/\lambda S \ , \qquad (24.40)$$

which can be rewritten in terms of wavelength as

$$\text{wave number FSR} = \frac{1}{\lambda} - \frac{1}{\lambda + \Delta\lambda} = \frac{\Delta\lambda}{\lambda(\lambda + \Delta\lambda)} \ , \qquad (24.41)$$

where $\Delta\lambda$ is the wavelength between successive peaks. If $\Delta\lambda \ll \lambda$, Eq. (24.41) becomes

$$\text{wave number FSR} \cong \Delta\lambda/\lambda^2 \ . \qquad (24.42)$$

Using Eq. (24.40), Eq. (24.42) becomes

$$\Delta\lambda/\lambda^2 = 1/\lambda S \ . \qquad (24.43)$$

Thus, the wavelength FSR is given by

$$\text{wavelength FSR} = \Delta\lambda = \lambda/S \ . \qquad (24.44)$$

Equations (24.40) and (24.44) show that the FSR and the order of interference are inversely related. Therefore, these two parameters are the trade-off parameters used in designing a Fabry-Perot tunable filter.

The half-width of a peak when plotted as a function of wave number is determined by equating Eqs. (24.34) and (24.8) to give

$$\phi_0 = \frac{1 - R}{\sqrt{R}} = \frac{4\pi nd\cos\theta}{\lambda_0} \ , \qquad (24.45)$$

where $1/\lambda_0$ is the wave number at the half-intensity point of the peak of interest. Using Eq. (24.28) in (24.45), for small angles we get

$$\frac{1 - R}{\pi\sqrt{R}} = 2\lambda S\mu_0 \ , \qquad (24.46)$$

where $\mu_0 = 1/\lambda_0$. The half-width of a peak on the wave number plot is

$$\delta\mu_0 = \left(\begin{array}{c}\text{half width of peak} \\ \text{on wave number plot}\end{array}\right) = \frac{1}{\lambda} - \frac{1}{\lambda_0} = \frac{\lambda_0 - \lambda}{\lambda\lambda_0} \ . \qquad (24.47)$$

When the order of interference is large, we assume that $\lambda_0 \cong \lambda$ and define the difference in wavelength $\lambda_0 - \lambda$ as $\delta\lambda_R$. Then Eq. (24.47) becomes

$$\delta\mu_0 = \delta\lambda_R/\lambda^2 \ . \qquad (24.48)$$

The finesse of the instrument on the wave number plot is defined from Eqs. (24.40) and (24.48) as

$$\text{finesse} = \frac{\text{wave number FSR}}{\text{full width of the peak at half-power}} = N_R = \frac{\lambda}{S(2\delta\lambda_R)} , \quad (24.49)$$

where N_R is the finesse of the interferometer as defined in Eq. (24.16). From Eq. (24.49) we see that the full width of the peak at half-power is

$$2\delta\lambda_R = \lambda / SN_R . \tag{24.50}$$

This same result can be obtained in wavelength space by defining the finesse as the wavelength FSR—Eq. (24.44) divided by the full width at half-power of the peak as measured in wavelength $2\delta\lambda_R$.

The chromatic resolving power (CRP) is given by

$$\text{CRP} = \left| \frac{\lambda}{2\Delta\lambda} \right| = SN_R , \tag{24.51}$$

which agrees with Eq. (24.36) derived previously. When absorptive coatings are used on the end mirrors in a Fabry-Perot cavity, the characteristics of the interferometer are modified due to the additional phase changes that occur at the interfaces of the various coated layers. In Eq. (24.14), ϕ'' was the phase change on reflection at a single surface as defined in Eq. (24.10). This quantity was assumed to be negligible in all the previous analyses. For a mirror with multiple coatings, we define the sum of all these phase shifts to be ϵ_λ. The revised characteristics of the interferometer with these changes are given in Table 24.I.

As seen in Table 24.I, the effect of the coatings is to change the characteristics of the cavity depending on the degree of dispersion in the coatings, i.e., the value of ϵ_λ and its wavelength derivative. The important cavity characteristics that

TABLE 24.I
Summary of Characteristics of Fabry-Perot Interferometers[*]

Function	Nondispersive Cavity	Dispersive Cavity
Wavelength FSR ($\Delta\lambda$)	λ/S	$\dfrac{\lambda}{\left(S + \dfrac{\lambda}{\pi}\dfrac{d\epsilon_\lambda}{d\lambda} + \dfrac{\epsilon_\lambda}{\pi}\right)}$
Width of peak for wavelength plot ($2\delta\lambda_R$)	λ/SN_R	$\dfrac{\lambda}{N_R}\dfrac{1}{\left(S + \dfrac{\lambda}{\pi}\dfrac{d\epsilon_\lambda}{d\lambda} + \dfrac{\epsilon_\lambda}{\pi}\right)}$
Finesse ($\Delta\lambda/2\delta\lambda_R$)	N_R	N_R
CRP ($\lambda/2\delta\lambda_R$)	SN_R	$N_R\left(S + \dfrac{\lambda}{\pi}\dfrac{d\epsilon_\lambda}{d\lambda} + \dfrac{\epsilon_\lambda}{\pi}\right)$

*From Ref. 7.

change are its FSR and the half-width of a peak in wavelength, which determine the CRP of the cavity. In the periodic wave number plot of Eq. (24.14), the separation between peaks is the fundamental wavelength λ_f. All higher order peaks are harmonics of the fundamental so that a change in the FSR of the cavity alters the mean wavelength to which the cavity is tuned. Furthermore, changing the separation of the mirrors continuously changes the wavelength to which the cavity is tuned over a spectral band determined by the order of interference of the cavity.

In tunable filter applications, two Fabry-Perot cavities are used sequentially in the optical path to provide a filter capable of transmitting either wide or narrow spectral bands, centered at various mean wavelengths, over a broad spectral range. Simultaneously, the filter can provide high transmission, good spatial resolution, and wide field of view. The filter can consist of one low-order cavity followed by a high-order cavity, or two low-order cavities or two high-order cavities in series.

In the design of a tunable filter, the value of reflectivity is used to control the half-width of a peak, the dispersion is used to control the FSR, and the order of interference S is used primarily to control the bandwidth of the filter.

In addition, graded index coatings on the cavity mirrors can be designed to change the conventional Fabry-Perot output to permit transmission of both narrow and wide spectral bands over wide fields of view. Thus, the combination of coatings on each of two Fabry-Perot cavities having variable orders gives the flexibility necessary to design tunable filters. The tuning is achieved by varying the optical phase of one or both cavities during operation.

An example of a broadband, long-wave infrared (LWIR) filter fabricated by using two coated Fabry-Perot cavities in series is shown in Fig. 24.14. The broadband characteristic of the filter is achieved by using two cavities having a low order of interference. In Fig. 24.14(a) we see the filter construction for the two coated cavities, and in (b) we see the filter output when the spacing for cavity 1 was 0.150 μm and the spacing for cavity 2 was 0.175 μm.

An example of a narrowband midwave infrared (MWIR) tunable filter showing both variations in the bandwidth and shifting of the mean wavelength is shown in Fig. 24.15. Figure 24.15(a) shows the configuration of the two cavities while (b), (c), and (d) show the filter characteristics corresponding to the various order combinations of the two cavities at mean wavelengths of 2.75, 3.50, and 4.75 μm, respectively. The transmission characteristics for the various filter curves in Figs. 24.15(b), (c), and (d) are given in Table 24.II. The cavity order referred to in this table represents the harmonic used in the filter design, i.e., order zero represents the fundamental for which the cavity was designed.

24.6.4   Example 4: Interference Filters

Multiple-beam interference effects in thin films are widely used to control the reflectance and transmittance properties of optical surfaces. Some common applications are antireflection coatings on camera lenses and optical filters. Additional flexibility in filter design is obtained by combining multiple-beam interference effects in thin films with Fabry-Perot etalons to achieve filters having narrow bandwidths and low absorptions centered about the mean wavelength of interest.

(a)

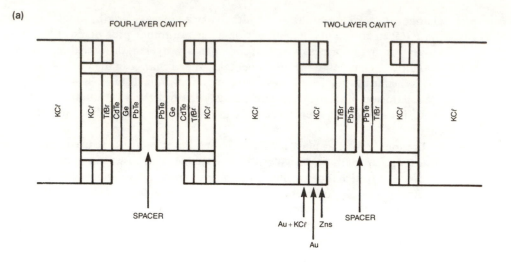

LAYER THICKNESSES ARE QUARTERWAVE AT 5 μm		CAPACITOR PADS	
MATERIAL	REFRACTIVE INDEX	MATERIAL	THICKNESS
KCℓ	1.4567	Au KCℓ	0.4 μm
TℓB$_r$	2.338	Au	0.05 μm
CdTe	2.67		
Ge	4.003	ZnS	0.1 μm
PbTe	5.64		

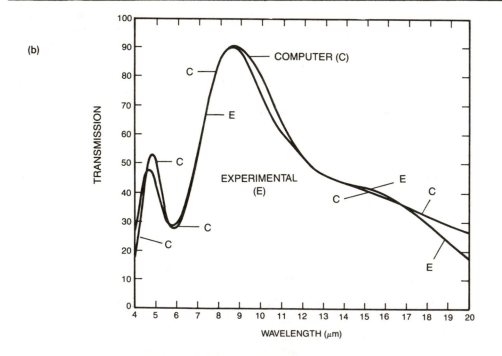

Fig. 24.14 Broadband LWIR tunable filter: (a) thin-film configuration of a LWIR dual tunable Fabry-Perot (DTFP) filter and (b) long-wave IR DTFP theoretical and experimental transmission in zero-order wideband mode. Cavity spacing: two layer, 0.175 μm; four layer, 0.150 μm (from Ref. 8).

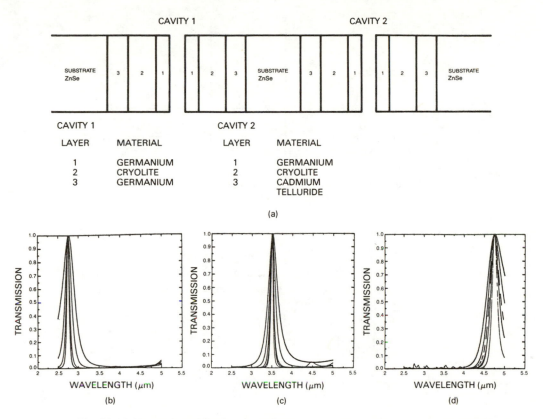

Fig. 24.15 Narrowband MWIR tunable filter: (a) thin-film configuration of MWIR DTFP, and MWIR DTFP theoretical transmission for various order combinations at (b) 2.75 μm, (c) 3.50 μm, and (d) 4.75 μm. Note the suppression of harmonics (from Ref. 8).

TABLE 24.II
Transmission Characteristics of MWIR DTFP[*]

Wavelength	Bandwidth	Cavity 1 Order	Cavity 2 Order	Peak Harmonic Leakage at λ_n	
2.75	0.36	0	0	3.1%	5.00 μm
	0.18	0	1	2.5	5.00
	0.09	1	1	5.7	5.00
	0.06	1	2	4.4	5.00
3.5	0.27	0	0	5.5%	5.00 μm
	0.18	0	1	2.2	5.00
	0.11	1	1	1.6	5.00
	0.09	1	2	3.6	5.00
4.75	0.65	0	0	<1%	
	0.41	0	1	<1	
	0.28	0	2	<1	
	0.22	0	3	1.2	2.56 μm
	0.23	1	3	1.2, 2.4	3.43, 3.06
				1.1	2.65
	0.14	2	4	1.6, 2.6	3.91, 3.54
				1.2, 2.6	3.31, 2.86
				3.7, 1.2	2.75, 2.52

[*]Design parameters are given in Fig. 24.15. From Ref. 8.

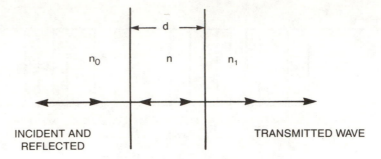

Fig. 24.16 Schematic of a normally incident wave on a single-layer dielectric thin film of index n between two infinite media of indices n_0 and n_1, respectively.

24.6.4.1   Basic Equations

Following Fowles,[4] the basic equations governing single-layer thin dielectric film interference effects for normal incidence, derived by properly matching the electric and magnetic fields at the two interfaces, are[2,4,9,10]

$$
\begin{bmatrix} 1 \\ n_0 \end{bmatrix} + \begin{bmatrix} 1 \\ -n_0 \end{bmatrix} r = \mathbf{M} \begin{bmatrix} 1 \\ n_1 \end{bmatrix} t , \tag{24.52}
$$

where

$$
\mathbf{M} = \begin{bmatrix} \cos kd & -\dfrac{i}{n}\sin kd \\ -in\sin kd & \cos kd \end{bmatrix} , \tag{24.53}
$$

$$r = \text{amplitude reflection coefficient}$$

$$t = \text{amplitude transmission coefficient}$$

$$n_0, n, n_1 = \text{indices of refraction of the various layers illustrated in Fig. 24.16}$$

$$d = \text{thickness of the thin film of index } n .$$

The information regarding the thin dielectric film of thickness d is all contained in the transfer matrix $\mathbf{M}$, whose determinant value is unity. Thus, if many thin dielectric film layers are stacked together, each has its own transfer matrix and the stack has a transfer matrix that is the product of the transfer matrices of the individual layers, i.e.,

$$
\mathbf{M}_T = \mathbf{M}_1 \mathbf{M}_2 \mathbf{M}_3 \dots \mathbf{M}_N . \tag{24.54}
$$

Substituting Eq. (24.54) into (24.52) and solving for the amplitude reflection and transmission coefficients yields

$$
r = \frac{An_0 + Bn_1n_0 - C - Dn_1}{An_0 + Bn_1n_0 + C + Dn_1} , \tag{24.55}
$$

$$
t = \frac{2n_0}{An_0 + Bn_1n_0 + C + Dn_1} , \tag{24.56}
$$

where

$$\mathbf{M}_T = \begin{bmatrix} A & B \\ C & D \end{bmatrix} .$$

The transfer matrix $\mathbf{M}_T$ is dependent on the properties of all the individual layers. The intensity reflectance R and intensity transmittance T are given by

$$R = |r|^2 , \tag{24.57}$$

$$T = |t|^2 . \tag{24.58}$$

24.6.4.2   *Highly Transmitting Thin Film Having a Single Layer*

When a single-layer thin film is designed to have low reflectance, it is known as an *antireflection coating*. This can be achieved by choosing $n_0 = 1$, n being a low-index material coated onto a substrate having a higher index n_1 and a thickness $d = \lambda/4$ ($n < n_1$ in Fig. 24.16). Under these circumstances, Eqs. (24.55) and (24.57) combine to yield

$$R = \left| \frac{n_1 - n^2}{n_1 + n^2} \right|^2 . \tag{24.59}$$

High transmittance is achieved by demanding that $R = 0$ in Eq. (24.59) to give

$$n = \sqrt{n_1} . \tag{24.60}$$

In practical coatings,[11] it is difficult to find materials that satisfy Eq. (24.60) exactly for indices of <1.9. However, in the infrared region of the spectrum, where higher index materials are necessary to achieve reasonable transmissions, single-layer coatings reduce reflection losses dramatically as illustrated in Fig. 24.17 when comparing curve 1 to curve 2. Figure 24.17 also shows that multilayer antireflection coatings achieve the effect of broadening the bandwidth over which the coating is effective, i.e., comparison of curve 2 to curve 3.

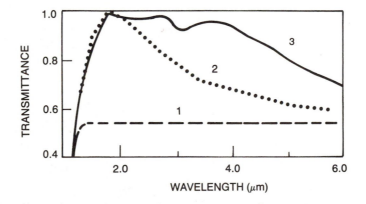

Fig. 24.17 Transmittance of a 1.5-mm-thick substrate of silicon (curve 1); coated on both sides with single quarter-wave layer of SiO_2 where $\lambda_0 = 1.8$ μm (curve 2) and multilayer quarter-wave coatings of MgF_2 and CeO_2 where $\lambda_0 = 2.2$ μm (curve 3) (from Ref. 12).

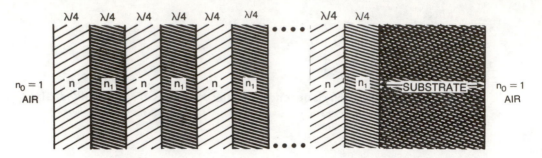

Fig. 24.18 Schematic of N pair layer of quarter-wave stacks $(n > n_1)$ for achieving high reflectance.

24.6.4.3   *Highly Reflecting Thin Films*

To achieve high values of reflectance, we let $n_0 = 1$ (air) and use a stack of quarter-wave thin films having alternate values of high and low indices, i.e., $n > n_1$ as shown in Fig. 24.18. For this case, Eqs. (24.54), (24.55), and (24.57) are combined to give

$$R = \frac{\left(\dfrac{n}{n_1}\right)^{2N} - 1}{\left(\dfrac{n}{n_1}\right)^{2N} + 1} , \tag{24.61}$$

which approaches unity for a large number of layers, $2N$.

Figure 24.19 shows the approximate reflectance variation in the visible region

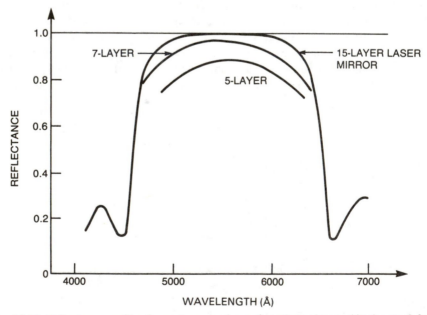

Fig. 24.19 Reflectance profiles for various numbers of coating pairs used in the stack for a typical set of materials in the visible band (from Ref. 4).

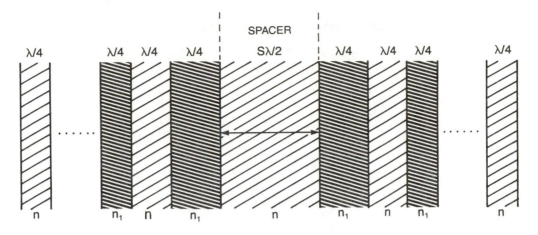

Fig. 24.20 Schematic of Fabry-Perot interference filter.

for some coating as a function of the number of pairs of quarter-wave coatings used in the stack.

24.6.4.4   *Fabry-Perot Interference Filters*

Fabry-Perot interference filters are designed by using a high-index spacer of thickness $d = S(\lambda/2)$, which is required to get a maximum from Eq. (24.14), where S is the order of interference defined by Eq. (24.28). The spacer is coated on both sides with quarter-wave plates of alternating low and high values for the refractive index as shown in Fig. 24.20. The function of the spacer is to give Fabry-Perot maxima for the wavelength of interest, and its order of interference S is chosen to place the harmonics outside the spectral region of interest. The harmonics are usually rejected by other means, e.g., additional filters, detector response, etc. The quarter-wave stacks are used to increase the reflectance R and, hence, the finesse N_R of the cavity. Dielectric quarter-wave stacks have the further advantage of keeping the filter transmittance relatively high because of their low absorption.

The bandwidth of such filters is determined from Eq. (24.36) to be

$$\Delta\lambda = \lambda/SN_R . \qquad (24.62)$$

Thus, the filter can have a narrow bandwidth depending on the values of S and N_R chosen in its design. An example illustrating various Fabry-Perot interference filters is shown in Fig. 24.21. Typical bandwidths of Fabry-Perot interference filters range from 10^{-3} to 35% of their mean wavelength. Narrow bandwidths are usually associated with lower values of mean transmittance.

The disadvantage of the Fabry-Perot interference filter is that a change in the angle of incidence lowers the mean wavelength response of the filter while broadening the bandwidth and lowering the overall transmission as shown in Fig. 24.22. Therefore, these filters are normally used in collimated radiation to avoid this shift in wavelength and resulting loss of efficiency.

Finally, we note that one could combine gradient-index coatings of the type utilized by Jain et al.[8] in the tunable filter work with the Fabry-Perot interference

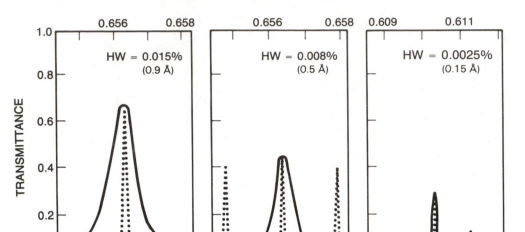

Fig. 24.21 Measured transmittance of very narrow bandpass interference filters with half-widths (HW) of <0.1%. Evaporated spacers: (a) single and (b) – (c) double quartz-spacer interference filters. The dotted curves in (a), (b), and (c) represent the transmittance of the filter plotted over an extended spectral region (upper wavelength scales) (from Ref. 11).

filter technology to construct filters having a wider field of regard, i.e., less angular sensitivity and higher transmissions.

24.6.5   Example 5: Bistable Devices for Optical Switching

The use of the electronic transistor as a binary switch forms the basis of present-day electronic computers. The practical limit of switching speeds with such devices now in use is on the order of 10^{-9} s. Optical transistors have been shown to switch at speeds nearly 1000 times faster than this, i.e., on the order of 10^{-12} s.

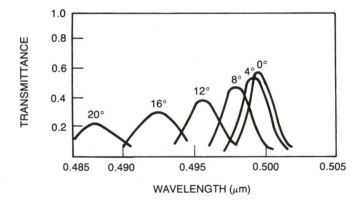

Fig. 24.22 Effects of change in angle of incidence on characteristics of a typical commercial Fabry-Perot interference filter (after Ref. 13).

This increase in switching time, coupled with the parallelism possible with optical systems, suggests that optical computers have the potential of being 10^3 to 10^4 times faster than the best electronic computers being suggested at the current time.[14]

Optical bistability using nonlinear optical media was first suggested[15] by Szoke in 1969 and first observed by Gibbs et al.[16] in 1976. These bistable devices are created by utilizing a nonlinear medium (whose refractive index changes with intensity) between highly reflecting Fabry-Perot mirrors. The mirror separation is chosen such that the optical path length is an integral number of wavelengths of the incident laser beam when the index is at its high value. In this situation, the Fabry-Perot cavity is in phase and maximum transmission of the Airy function is achieved. When the incident intensity is reduced slightly, the nonlinear index change causes a change in the optical path so that the transmission is determined by a point between the peaks of the Airy curves, i.e., near zero transmission. The resulting transmission as a function of incident intensity can be made to switch between two levels I_L and I_H as shown in Fig. 24.23 for a very small change in incident intensity, $I_P - I_0$.

Abraham et al.[14] have realized optical switches using the semiconductor material indium antimonide at 77 K with the carbon monoxide laser line adjusted in the vicinity of 5 μm (the mid-infrared region of the spectrum). Crystal lengths of a few hundred microns to a few millimeters were used and switching speeds on the order of 10^{-12} s have been observed. At 77 K indium antimonide has a band gap energy of ~0.2 eV (a wavelength equivalent of ~5 μm) giving it a nonlinearity of 100 to 1000 times greater than that of gallium arsenide.

More recent results at room temperatures with carbon dioxide lasers in the 10.6-μm region have also been obtained. Optical bistability with gallium arsenide has also been observed.

One configuration in which optical bistability has been observed is to have two laser beams incident on the crystal. One constant or bias beam has an intensity of I_0, the threshold, and the other beam is modulated between 0 and I_P as shown in Fig. 24.23. The Fabry-Perot interferometer goes in and out of phase giving rise to measured switching speeds on the order of picoseconds.

Current work in this field involves the investigation of materials having a high degree of nonlinearity, practical configurations of switches, connecting circuitry utilizing integrated and fiber optic technologies, appropriate architectures, and utilization of optical parallelism in the quest for a high-speed optical digital computer.[17,18]

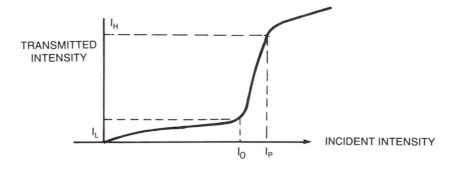

Fig. 24.23 Schematic of optical bistability utilizing a Fabry-Perot cavity and a nonlinear medium.

REFERENCES

1. C. Fabry and A. Perot, "Sur les Franges des Lames Mences Argentees et Leur Application a la Mesure de Petites Epaisseurs d'Air," *Ann. Chim. Phys.* 12, 459 (1897); see also *Ann. Chim. Phys.* 16(7), 115–144 (1899).
2. F. A. Jenkins and H. E. White, *Fundamentals of Optics,* McGraw-Hill Inc., New York (1976).
3. R. S. Longhurst, *Geometrical and Physical Optics,* 2nd ed., John Wiley and Sons, Inc., New York (1967).
4. G. R. Fowles, *Introduction to Modern Optics,* Holt, Rinehart & Winston, New York (1975).
5. O. Svelto, *Principles of Lasers,* 2nd ed., Plenum Publishing Corp., New York (1982).
6. E. Hecht and A. Zajac, *Optics,* Addison-Wesley Publishing Co., Reading, Mass. (1973).
7. P. D. Atherton, N. K. Reay, J. Ring, and T. R. Hicks, "Tunable Fabry-Perot filters," *Opt. Eng.* 20(6), 806–814 (1981).
8. A. K. Jain, D. E. Stoltzmann, G. R. Knowles, J. G. Droessler, and D. Johnson, "Dual tunable Fabry-Perot spectrally agile filter," *Opt. Eng.* 23(2), 159–166 (1984).
9. A. K. Jain, W. W. Durand, G. R. Knowles, J. G. Droessler, and M. J. Lavan, "Dual tunable Fabry-Perot: a new concept for spectrally agile filtering," in *Imaging Spectroscopy,* D. D. Norris, ed., Proc. SPIE 268, 183–189 (1981).
10. M. Françon, *Optical Interferometry,* Academic Press, New York (1966).
11. J. A. Dobrowolski, "Coatings and filters," in *Handbook of Optics,* W. G. Driscoll and W. Vaughn, eds., Sec. 8, pp. 8–81, McGraw-Hill Inc., New York (1978).
12. J. T. Cox and G. Hass in *Physics of Thin Films,* G. Hass and R. E. Thun, eds., Vol. 2, pp. 239–304, Academic Press, New York (1964).
13. I. H. Blifford, Jr., "Factors affecting the performance of commercial interference filters," *Appl. Opt.* 5(1), 105–111 (1966).
14. E. Abraham, C. T. Seaton, and S. D. Smith, "The optical computer," *Sci. Am.* 248(2), 85–93 (1983).
15. A. Szoke, U. Daneu, J. Goldhar, and N. A. Kurnit, "Bistable optical element and its applications," *Appl. Phys. Lett.* 15(11), 376–379 (1969).
16. H. M. Gibbs, S. L. McCall, and T. N. C. Venkatesan, "Differential gain and bistability using a sodium-filled Fabry-Perot interferometer," *Phys. Rev. Lett.* 36(1a), 1135–1141 (1976).
17. Special Issue on Optical Computing, J. A. Neff, ed., *Opt. Eng.* 24(1) (1985).
18. Special Issue on Optical Computing, *Proc. IEEE* 72(7) (1984).

25 Introduction to Holography

25.1 INTRODUCTION

Information about holography first appeared in published form in the late 1940s in three classic papers written by Dennis Gabor. Subsequent work involving the use of an off-axis reference wave, published in the early 1960s by Leith and Upatnieks at the University of Michigan, coupled with the invention of the laser, revitalized interest in the subject and led to a large amount of research during the 1960s and early 1970s.[1-15] The purpose of this chapter is to familiarize the reader with the principles of holography by extending the concept of the two-beam interference experiment discussed in Chapters 3, 10, and 22. In Chapters 26 and 27, we will discuss holographic interferometry and holographic applications.

25.2 RECONSTRUCTION OF A TWO-BEAM INTERFEROGRAM

In Chapter 3, it was shown that the intensity distribution along the x axis of the two-slit interferometer shown in Fig. 25.1 is given by

$$I(x) = 16a^2 A^2 \operatorname{sinc}^2 \frac{2\pi a x}{\lambda f} \cos^2 \frac{2\pi b x}{\lambda f} , \qquad (25.1)$$

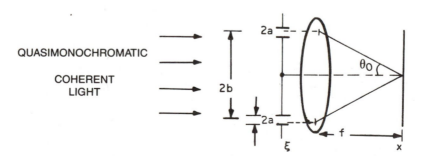

Fig. 25.1 Schematic of two-slit interferometer.

293

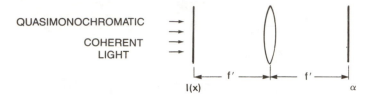

QUASIMONOCHROMATIC

COHERENT
LIGHT

I(x)

f'   f'

α

Fig. 25.2 Schematic of system for reconstructing interferograms.

where

A = amplitude of the plane wave illuminating the slit

$2a$ = slit width

$2b$ = slit separation

f = focal length of lens .

An interferogram may be prepared by photographically recording the intensity distribution, Eq. (25.1), linearly. That is, the recording is such that the amplitude transmission is proportional to the intensity given by Eq. (25.1); see Chapters 20 and 21. Throughout this chapter we will simply use the term linearly recorded to describe that photographic process. If such an interferogram is placed a distance f' in front of a lens of focal length f', and coherently illuminated (see Fig. 25.2) as discussed in Chapters 2 and 29, then apart from a quadratic phase factor, the amplitude distribution in the back focal plane of the lens is given by

$$g(\alpha) = \int I(x) \exp\left(\frac{-ik\alpha x}{f'}\right) dx .$$ (25.2)

Since $I(x)$ of Eq. (25.1) consists, apart from a constant factor, of a product of two functions of x, Eq. (25.2) can be evaluated by the convolution theorem, which gives, after normalization,

$$g(\alpha) = \text{Tr}(\alpha|2a) \circledast [\tfrac{1}{2}\delta(\alpha/m) + \tfrac{1}{4}\delta(\alpha/m \pm 2b)] ,$$ (25.3)

where $m = f'/f$. Here $\text{Tr}(\alpha|2a) = 1 - |\alpha|/2a$ with $|\alpha| < 2a$ and 0 otherwise. Figure 25.3 is a plot of Eq. (25.3) for unit magnification.

The amplitude output of the system is the replicated autoconvolution functions of an individual slit or, equivalently, the autoconvolution function of the two slits. The separation of the side order distributions from the origin is

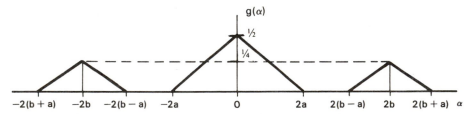

$g(\alpha)$

$\tfrac{1}{2}$

$\tfrac{1}{4}$

$-2(b + a)$ $-2b$ $-2(b - a)$ $-2a$ 0 $2a$ $2(b - a)$ $2b$ $2(b + a)$ α

Fig. 25.3 Schematic representation of Eq. (25.3), the Fourier transform of a two-slit interferogram.

proportional to the separation of the slits $2b$. There is no overlap of the light distributions from the various triangular functions as long as $b > 2a$ (a necessary criterion in order to have two separate slits in the first place). This relationship between b and a sets a kind of sampling or *resolution* criterion for the interferogram so as to guarantee separation of the various triangular functions; i.e., it determines the number of cycles of the cosine function of Eq. (25.1) contained within the first zero of the sinc function as described in Chapter 22.

25.3   RECONSTRUCTION OF IDEAL TWO-BEAM INTERFEROGRAMS

The preceding discussion is simplified by considering an ideal two-beam interferogram that consists of the interference resulting from two delta function pinholes placed symmetrically about the optical axis and in front of the lens in Fig. 25.1. The resulting normalized intensity distribution in the focal plane of the lens is given by

$$I(x) = \cos^2 \frac{2\pi bx}{\lambda f} \cdot \qquad (25.4)$$

If the intensity distribution of Eq. (25.4) is recorded on film so that the amplitude transmittance of the film is proportional to the recorded intensity and reconstructed by the experimental arrangement in Fig. 25.2, the resulting amplitude in the back focal plane of the lens in Fig. 25.2 is given by

$$g(\alpha) = [\tfrac{1}{2}\delta(\alpha) + \tfrac{1}{4}\delta(\alpha \pm 2b)] \, , \qquad (25.5)$$

as shown in Fig. 25.4. This example illustrates that the individual delta function apertures are reconstructed at the positions $\alpha = \pm 2b$ from the axis.

Interesting results will occur if we vary the size of the slits with respect to each other in the recording arrangement of Fig. 25.1.

25.4   BASIC DESCRIPTION OF A TWO-BEAM HOLOGRAM

If the two slits in Fig. 25.1 are shifted by an amount $2b$ relative to the optical axis and one slit is reduced to a delta function while the other remains unchanged, then the amplitude transmission of the slits is described by

$$T(\xi) = K\delta(\xi - 2b) + \text{rect}(\xi|a) \, . \qquad (25.6)$$

Then the intensity distribution in the back focal plane of the lens in Fig. 25.1 is

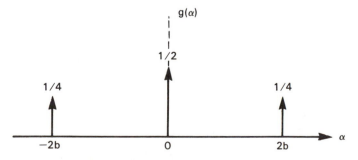

Fig. 25.4 Schematic showing the reconstruction of the ideal interferogram.

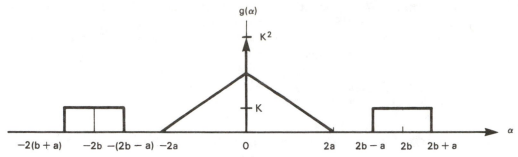

Fig. 25.5 Reconstruction of a simple hologram.

$$I(x) = K^2 + 4a^2 \operatorname{sinc}^2 \frac{2\pi a x}{\lambda f} + 4Ka \operatorname{sinc} \frac{2\pi a x}{\lambda f} \cos \frac{2\pi(2b)x}{\lambda f} . \qquad (25.7)$$

The several differences between Eq. (25.1) and Eq. (25.7) should be carefully noted. In particular the cosine term in Eq. (25.7) is not squared and it multiplies the sinc function rather than the sinc^2 function. These differences will be shown to be significant. If this intensity distribution is recorded on film and linearly processed, the resulting transparency when used in the reconstruction system of Fig. 25.2 gives the amplitude distribution in the α plane [obtained by Fourier transforming Eq. (25.7)]:

$$g(\alpha) = [K^2\delta(\alpha) + \operatorname{Tr}(\alpha|2a) + K\operatorname{rect}(\alpha \pm 2b|a)] . \qquad (25.8)$$

Thus, at the position $\alpha = \pm 2b$, an image of the original slit is reconstructed provided that $2b > 3a$. This condition between b and a prevents overlap of the rect function with the autoconvolution function of the slit $\operatorname{Tr}(\alpha|2a)$, as shown in Fig. 25.5. In this case, the light emanating from the reference point $\delta(\xi - 2b)$ in Eq. (25.6) behaves as an offset reference point for the interferogram described in Eq. (25.7). If the energy from the point source is intense enough to bias the negative portions of the sinc function [i.e., $K^2 > 4Ka$ in Eq. (25.7)], then the interferogram of Eq. (25.7) is equivalent to a *hologram,* a word coined by Gabor from the Greek word *holos,* which means the whole or entire. This means that all of the information about the object (amplitude and phase of the diffraction pattern) is preserved in the recording. Thus, in a hologram, the reference wave must bias the diffraction pattern so that the phase information of the diffraction pattern is not destroyed.

The function sinc x may be expressed in terms of a binary phase function associated with $|\operatorname{sinc} x|$. This phase function is $e^{i\pi} = -1$ when the sinc is negative; the phase is $e^{i0} = +1$ in the regions where the sinc function is positive. This binary phase behavior is shown in Fig. 25.6.

The process of recording all of the information can be demonstrated easily by reconstructing a Young's two-slit interferogram to yield an intensity distribution similar to the one illustrated in Fig. 25.3.

25.5   FORMATION AND RECONSTRUCTION OF A FOURIER TRANSFORM HOLOGRAM

If the rect function in Eq. (25.6) is replaced with an arbitrary function $f(x,y)$, e.g., a photographic transparency of a continuous-tone scene, then the same

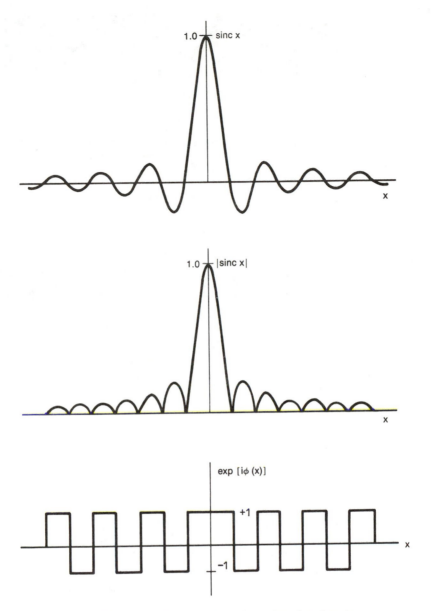

Fig. 25.6 Illustration of the complex nature of the function sinc x.

mathematical manipulations can be performed and the intensity distribution in the hologram plane (back focal plane of the lens in Fig. 25.1) is given by

$$I(\mathbf{x}) = K^2 + \left| \tilde{f} \left(\frac{x}{\lambda f} , \frac{y}{\lambda f} \right) \right|^2 + 2K \left| \tilde{f} \left(\frac{x}{\lambda f} , \frac{y}{\lambda f} \right) \right|$$

$$\times \cos \left[\frac{2\pi bx}{\lambda f} - \phi \left(\frac{x}{\lambda f} , \frac{y}{\lambda y} \right) \right] , \qquad (25.9)$$

where $|\tilde{f}|$ is the modulus of the Fourier transform of the object and ϕ is the

Fig. 25.7 Off-axis real images from a Fourier transform hologram reconstructed by the point reference method. In this experiment, laser radiation at 6328 Å was used in both steps of the process. The construction was accomplished with an object distance of 1.2 m, and the center-to-center distance between the object and the point reference was 6 cm. The reconstruction was performed with a lens. The hologram was formed in the near field of the object (from Ref. 16).

phase portion of the Fourier transform of the object; i.e., the Fourier transform can be written as

$$\tilde{f} = |f|\, e^{i\phi} . \tag{25.10}$$

By recording the hologram on film and placing the linearly developed film in the front focal plane of the lens in Fig. 25.2, the following amplitude distribution in the α plane (providing $f = f'$) is obtained:

$$g(\alpha) = K^2 + f(\alpha) \circledast f(\alpha) + f[-(\alpha + 2b),\beta] + f^*[(\alpha - 2b),\beta] . \tag{25.11}$$

Thus, the Fourier transform hologram creates two real images (one inverted with respect to the other). These images are located off axis and their centers are separated by a distance $4b$—twice the original separation between the object and the point reference in Fig. 25.1. Thus we see that Fourier transform holograms may be derived and understood as a direct extension of the reconstruction of an elementary two-slit interferogram. An example of two off-axis reconstructed images recorded in a Fourier transform hologram are shown in Fig. 25.7. Note also the on-axis distribution which is also present and expected from Eq. (25.11).

25.6   OTHER COMMENTS ON FOURIER TRANSFORM HOLOGRAMS

In any practical experiment, the delta function configuration of the above analysis cannot be realized and one must use a small pinhole with laser light focused into it from a larger lens to achieve the reference source (see Fig. 25.8). Equivalent Fourier

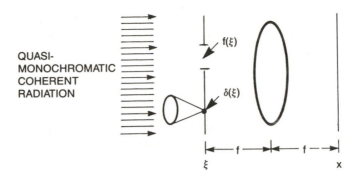

Fig. 25.8 Fourier transform recording arrangement with a finite-sized reference source.

transform holograms can also be realized by using a Mach-Zehnder interferometer (Chapter 23) with a tilted reference wave as shown in Fig. 25.9.

25.7   TYPES OF HOLOGRAMS

As we have demonstrated, a hologram is nothing more than an interferogram coded in such a way that an image may be obtained from it at a later time. In essence, this means that the hologram is made in such a manner that the phase information associated with the diffraction pattern of the object is preserved during the recording step of the process. Thus, a hologram is a recording of the intensity distribution describing the spatial interference[a] between an object field $\psi_d(x)$ (a Fresnel or Fraunhofer diffraction pattern arising from transmission through or reflection from the object) and a reference field $\psi_r(x)$ (plane or spherical wave either coaxial with or making an angle to the optical axis):

$$\text{hologram} = I(x) = |\psi_d(x) + \psi_r(x)|^2$$
$$= |\psi_d(x)|^2 + |\psi_r(x)|^2 + \psi_d(x)\psi_r^*(x) + \psi_d^*(x)\psi_r(x) \ . \quad (25.12)$$

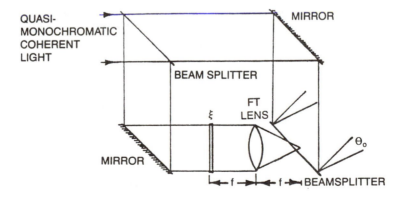

Fig. 25.9 Mach-Zehnder arrangement for recording Fourier transform holograms.

[a]The interference pattern is spatial because at optical frequencies all time variations associated with the fields are averaged out by the receiver.

Holograms have been classified in many ways and from many perspectives. In this chapter, we classify hologram types by the different ways in which they are physically formed:

1. If the hologram plane is in the Fresnel diffraction region of the object and the reference field is coaxial with the illumination field, the resulting hologram is called a *Fresnel hologram.*

2. If the hologram plane is in the Fraunhofer diffraction region of the object and the reference field is coaxial with the illumination field, the resulting hologram is called a *Fraunhofer hologram.*

3. If the hologram plane is in the Fresnel (Fraunhofer) diffraction region of the object and the reference field makes an angle with respect to the optical axis, the resulting hologram is called a *sideband Fresnel (Fraunhofer) hologram.*

4. If the hologram intensity of Eq. (25.12) is of the form $I(x) = |K + \tilde{D}(x)|^2$, where $\tilde{D}(x)$ denotes an exact Fourier transform of the object, the resulting hologram is called a *Fourier transform hologram.*

5. If an off-axis reference wave is mixed with a coherent image, the resulting hologram is called an *image hologram.*

We classify holograms in these five categories because the physical properties associated with each upon reconstruction are different. To reconstruct an image from the hologram, the hologram is placed in a coherent optical beam. The image is then formed at a position determined from the type of hologram that was produced. Upon reconstruction, the first two squared (bias) terms in Eq. (25.1) usually describe unrecognizable distributions centered on the optical axis; however, the last two cross-product terms describe the conjugate images (either real or virtual) that appear from the process.

A real and a virtual image are usually produced from the Fresnel and Fraunhofer holograms at the appropriate image planes. In the Fresnel hologram, the focused image interferes with the light associated with the second image that exists in the image plane. In the Fraunhofer hologram, however, the separation between the images along the optical axis is essentially infinite so that the fringing produced by the interference between the image and the hologram of the other image is, in effect, constant. Two images are also produced from the sideband Fresnel (Fraunhofer) holograms; since the reference field produces a carrier frequency, however, the conjugate images are separated in space without being troubled by interference from the various terms in Eq. (25.12). Finally, from the Fourier transform holograms (which inherently have higher frequency capabilities but less field of view than the other types), two real images are obtained in the same plane. Since these images are essentially in focus at infinity, their reconstruction will require Fraunhofer diffraction either by a lens or by large propagation distances.

25.8   SIMPLIFIED THREE-DIMENSIONAL HOLOGRAPHY

25.8.1   Two-Point Object

In the preceding analysis of Fourier transform holography, planar transparent objects were used and two-dimensional images were obtained. This does not illustrate the dramatic three-dimensional images containing parallax that are normally associated with laser holography.

To illustrate this three-dimensional effect, assume that two coherent point sources of strength $a\delta(\xi, z_1)$ and $a\delta(\xi, z_2)$ are located at distances z_1 and z_2 in front of the hologram plane as shown in Fig. 25.10. An off-axis reference wave that is coherent with the object waves is assumed. The intensity distribution along the x axis in the hologram plane, $I(x)$, recorded on film is given by

$$I(x) = \left| K\exp(ik\theta_0 x) + Aa \exp\left(\frac{ikx^2}{2z_1}\right) + aA \exp\left(\frac{ikx^2}{2z_2}\right) \right|^2$$

$$\cong K^2 + A^* Ka^* \exp(ik\theta_0 x) \left[\exp\left(\frac{-ikx^2}{2z_1}\right) + \exp\left(\frac{-ikx^2}{2z_2}\right)\right]$$

$$+ AK^* a\ \exp(-ik\theta_0 x) \left[\exp\left(\frac{ikx^2}{2z_1}\right) + \exp\left(\frac{ikx^2}{2z_2}\right)\right] . \qquad (25.13)$$

In Eq. (25.13), A includes the obliquity factors of the Rayleigh-Sommerfield diffraction integral, K is the amplitude of the reference wave, and $K \gg a$ so that the a^2 term is negligible. Equation (25.13) assumes that each point source gives rise to a spherical wave (parabolic in the paraxial approximation) of different radius in the hologram plane.

Developing the film and placing it in the reconstruction system (see Fig. 25.11) enables the use of Eq. (25.13) as a boundary condition in another Rayleigh-Sommerfield diffraction integral to find the amplitude distribution at a distance z_3 from the hologram in the α plane. Consider first the reconstruction of the second term in Eq. (25.13). Ignoring multiplicative constants, the reconstructed amplitude along the α axis is given by

$$g(\alpha) = \int \left\{ \exp(ik\theta_0 x) \left[\exp\left(\frac{-ikx^2}{2z_1}\right) + \exp\left(\frac{-ikx^2}{2z_2}\right)\right] \right.$$

$$\left. \times \exp\left(\frac{ik(x-\alpha)^2}{2z_3}\right) \right\} dx$$

$$= \exp\left(\frac{ik\alpha^2}{2z_3}\right) \int \left[\exp\left(\frac{-ikx^2}{2z_1}\right) + \exp\left(\frac{-ikx^2}{2z_2}\right)\right]$$

$$\times \exp\left(\frac{ikx^2}{2z_3}\right) \exp\ -ikx(\alpha/z_3 - \theta_0)\, dx . \qquad (25.14)$$

At the position $z_3 = z_1$, the first term in the integral of Eq. (25.14) evaluates to $\delta(\alpha - \theta_0 z_1)$, and at the position $z_3 = z_2$, the second term becomes $\delta(\alpha - \theta_0 z_2)$. Thus, both point sources are reconstructed at their corresponding image points at the same distance from the hologram as they were during the recording process. The images are centered off axis at distance $\theta_0 z_2$ and $\theta_0 z_3$, respectively, so that the sources are lined up one behind the other in the θ_0 direction. This example illustrates that the preservation of the phase information of the object when recording the hologram retains the three-dimensional nature of the image. Different spherical waves from the object at the hologram plane mean that the point sources were located at different distances from the hologram.

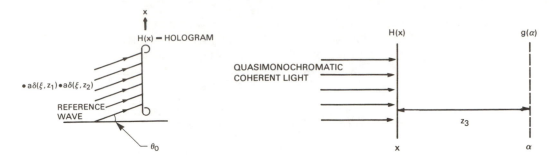

Fig. 25.10 Schematic for recording a hologram of two longitudinally separated point sources.

Fig. 25.11 Reconstruction system.

Reconstruction of the third term in Eq. (25.13) by similar techniques yields two point images, $\delta(\alpha + \theta_0 z_3)$ and $\delta(\alpha + \theta_0 z_2)$, located at the distances $z_3 = -z_1$ and $z_3 = -z_2$; i.e., virtual images located behind the hologram. (This is the image that is normally viewed in a hologram since the human eyeball refocuses the virtual image more easily.) Reconstruction of the first term in Eq. (25.13) yields an on-axis Fresnel diffraction pattern.

25.8.2   Continuum of Point Sources

The procedure outlined above can be extended to an array or continuum of point sources simulating laser light reflected from a three-dimensional object, showing that three-dimensional real and virtual images are obtained. For example, if we express the three-dimensional object as:

$$O(\xi,z) = \sum_{n=1}^{\infty} A_n \delta(\xi_n, z_n) \; , \tag{25.15}$$

where the A_n are constants. Then the intensity distribution in the hologram is given by:

$$I(x) = \left| K \exp(ik\alpha_0 x) + A \sum_{n=1}^{\infty} A_n \delta(\xi_n, z_n) \right|^2 , \tag{25.16}$$

where A and K are constants. Note that the Fourier transform of a Dirac comb is a Dirac comb. Developing the film properly and placing it in the reconstruction system (see Fig. 25.10) yields real and virtual images located off axis in the α_0 direction as before. To illustrate this, we show in Fig. 25.12 the reconstruction images of an aerosol at two different positions of focus. We readily note that the hologram stores the necessary complex wave information for each particle in the aerosol and then yields reconstructed images in each of the focal planes spanning the initial volume. Clearly, the three-dimensional aerosol object has been stored and reconstructed through the holographic process. As an example of using a continuous-tone object, we show in Fig. 25.13 a reconstruction of two continuous-tone objects separated by a distance z in object space. In this experiment, a diffuser was used in the object beam when recording the hologram to sharpen the

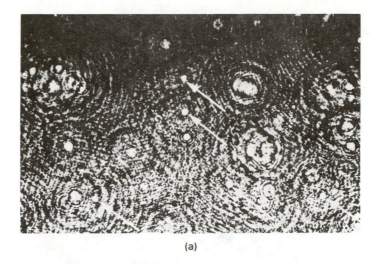

(a)

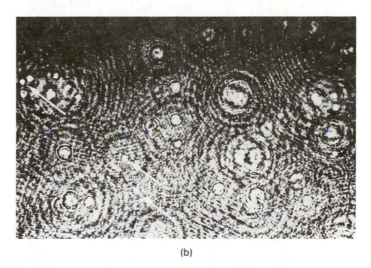

(b)

Fig. 25.12 Two reconstructions, in different focal planes, of an aerosol volume demonstrating the three-dimensional nature of the holographic reconstruction: (a) and (b) represent successive planes of the real image reconstructed from a hologram taken of a sample volume (1 cm in depth and 1 cm^2 in cross section) of suspended 30-μm lycopodium particles. The planes are separated by 5 mm along the optical axis in the volume of suspended particles. The area of the print represents a 1.7-mm^2 sampling cross section. The particle density is 2×10^3 cm^{-3}. The arrows point to focused images. (from Ref. 17).

depth of focus of the reconstructed images as described in Chapter 21. In effect, the diffuser limits the depth of focus of each individual image whereas, without the diffuser, a much greater depth of focus is available that is limited by the coherence length of the light used; the result is interference between the image and the hologram associated with the other image. This can be especially troublesome in aerosol measurements.

25.8.3   Comments on Three-Dimensional Holography

1. The use in Fig. 25.1 of the off-axis reference wave at an angle θ_0 in recording the hologram causes separation of the real and virtual images by an angle of $2\theta_0$ in the reconstructed image plane as seen in Fig. 25.5.

(a)

(b)

Fig. 25.13 (a) and (b) represent reconstructions of two semitransparent objects. Two transparencies, positioned such that neither obscured the other when viewed from the hologram recording position, were illuminated with diffuse quasimonochromatic radiation. The hologram formation distances for the stored scenes were 14 and 24 in., respectively. The diffuse illumination was obtained by placing a piece of opal glass between the 6328-Å laser source and the object. Reconstructions of the two scenes recorded in the same hologram were accomplished by placing the hologram in a coherent quasimonochromatic laser beam at 6328 Å without a diffuser. By proper positioning of the original transparencies, the reconstructed images show no interference effects (from Ref. 18).

2. Preservation of the quadratic phase factors in recording the hologram means that focusing constraints are required when viewing the reconstructed image as demonstrated in Fig. 25.12. In Fourier transform holography, the lens performs this focusing operation.

3. The conjugate terms focus on different sides of the hologram, resulting in a real and a virtual image.

4. The three-dimensional nature of the object is preserved in the image because the phase information (relative position of the various object points) is preserved when recording the hologram as shown in Figs. 25.12 and 25.13.

25.9   FRESNEL AND FRAUNHOFER HOLOGRAPHY

25.9.1   Hologram Formation

To illustrate how various system parameters affect the holographic process, we will perform an analysis of Fresnel and Fraunhofer holography. In particular, we will show the relationship between resolution and field of view of the holographic image, as well as how those items are affected by the film parameters.

Consider a one-dimensional semitransparent object $D(\xi)$ that is transilluminated with quasimonochromatic coherent radiation as shown in Fig. 25.14. The boundary condition in the object plane, ξ, is $T(\xi) = 1 - D(\xi)$. The diffracted field in the Fresnel zone, ignoring obliquity factors, is given by

$$\psi(x) = \int [1 - D(\xi)] \exp \left[\frac{ik_1(\xi - x)^2}{2z_1} \right] d\xi$$

or

$$\psi(x) = \int \exp \left[\frac{ik_1(\xi - x)^2}{2z_1} \right] d\xi - \int D(\xi) \exp \left[\frac{ik_1(\xi - x)^2}{2z_1} \right] d\xi . \quad (25.17)$$

The first term of Eq. (25.17) describes the diffracted field in the Fresnel zone of a plane wave. Physically, it is also a plane wave at the x plane since light becomes more spatially coherent by propagation.[b] Therefore, the intensity along the x axis on the film (called the Fresnel hologram) is given by

$$I(x) = \left| \left\{ 1 - \int D(\xi) \exp \left[\frac{ik_1(\xi - x)^2}{2z_1} \right] d\xi \right\} \right|^2 \quad (25.18)$$

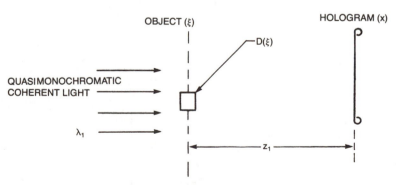

Fig. 25.14 Semitransparent object illuminated by a plane wave of wavelength λ_r.

---

[b]The derivation of this result is given in Ref. 6, pp. 34–36.

or

$$I(x) = 1 + D_F^2(x) - D_F^*(x) - D_F(x) , \tag{25.19}$$

where

$$D_F(x) = \int D(\xi) \exp\left[\frac{ik_1(\xi - x)^2}{2z_1}\right] d\xi . \tag{25.20}$$

25.9.2   Hologram Reconstruction

If we place the developed hologram in a quasimonochromatic coherent plane wave of wavelength λ_2 and step back a distance z_2 from the hologram (see Fig. 25.15) and ignore the obliquity factors, we obtain the amplitude distribution along the x axis given by

$$g(\alpha) = \int I(x) \exp\left[\frac{ik_2(x - \alpha)^2}{2z_2}\right] dx . \tag{25.21}$$

Substitution of Eq. (25.19) into (25.21) gives rise to four terms in the reconstructed image. The first term yields a plane wave as discussed previously and the second term is the *Fresnel transform* of the square of the Fresnel diffraction pattern of the object. This latter term yields an unrecognizable distribution of energy centered on axis in the α plane of Fig. 25.15. The third term yields the following amplitude distribution along the α axis:

$$g_3(\alpha) = -\int D_F^*(x) \exp\left[\frac{ik_2(x - \alpha)^2}{2z_2}\right] dx ,$$

$$= -\iint D^*(\xi) \exp\left[\frac{-ik_1(\xi - x)^2}{2z_1}\right]$$

$$\times \exp\left[\frac{ik_2(x - \alpha)^2}{2z_2}\right] dx \, d\xi ,$$

$$= -\iint D^*(\xi) \exp\left(\frac{-ik_1\xi^2}{2z_1}\right) \exp\left(\frac{ik_1\xi x}{z_1}\right) \exp\left(\frac{-ik_1 x^2}{2z_1}\right)$$

$$\times \exp\left(\frac{ik_2 x^2}{2z_2}\right) \exp\left(\frac{-ik_2 x\alpha}{z_2}\right) \exp\left(\frac{ik_2 \alpha^2}{2z_2}\right) dx \, d\xi . \tag{25.22}$$

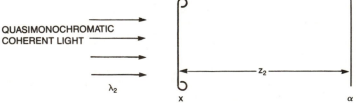

QUASIMONOCHROMATIC
COHERENT LIGHT

λ_2

x

z_2

α

Fig. 25.15 Reconstruction system for a Fresnel hologram.

Combining the exponential terms having quadratic dependence in x and demanding that the condition

$$\frac{k_2}{z_2} = \frac{k_1}{z_1} \tag{25.23}$$

exist so that the quadratic phase term in x is unity, Eq. (25.22) becomes

$$g_3(\alpha) = -\exp\left(\frac{ik_2\alpha^2}{2z_2}\right) \int D^*(\xi) \exp\left(\frac{-ik_1\xi^2}{2z_1}\right)$$

$$\times \left\{ \int \exp\left[ix\left(\frac{k_1\xi}{z_1} - \frac{k_2\alpha}{z_2}\right)\right] dx \right\} d\xi . \tag{25.24}$$

If we assume that the hologram is very large in extent, the x integral in Eq. (25.24) approximates a delta function, i.e.,

$$\int \exp\left[ix\left(\frac{k_1\xi}{z_1} - \frac{k_2\alpha}{z_2}\right)\right] dx \cong \delta\left(\xi - \frac{k_2 z_1}{k_1 z_2}\alpha\right) . \tag{25.25}$$

For convenience, we define

$$m = \frac{k_1 z_2}{k_2 z_1} . \tag{25.26}$$

Substitution of Eq. (25.26) into (25.25) and the resulting Eq. (25.25) and (25.23) into (25.24) yields

$$g_3(\alpha) = -D^*(\alpha/m) , \tag{25.27}$$

which shows that the reconstructed amplitude due to the third term of Eq. (25.19) is a scaled replica of the object (m is this scaling factor), i.e., an image located at a distance z_2 from the hologram. The reconstruction distance z_2 is given by Eq. (25.23) to be

$$z_2 = k_2 z_1 / k_1 . \tag{25.28}$$

If the same wavelength is used in both steps of the hologram process, $k_1 = k_2$, and Eq. (25.28) yields

$$z_2 = z_1 . \tag{25.29}$$

Equation (25.23) is known as the *focusing constraint* for the hologram. Physically, it is the condition among the system parameters z_2, z_1, k_1, and k_2 that removes the quadratic phase (defocusing) aberration from the holographic process. Equation (25.26) is known as the *system magnification*.

Substitution of Eq. (25.23) into (25.26) yields

$$m = k_1 z_2 / k_2 z_1 = 1 , \tag{25.30}$$

which shows that the system magnification of the in-focus image resulting from reconstructing the hologram is always unity when plane waves are used in both

TABLE 25.I
Limitations of One-Dimensional Holographic Systems[*]

Type of Hologram	Fresnel and Fraunhofer	Sideband Fresnel	Fourier Transform
Plane wave magnification	$m = 1$	$m = 1$	$m^{-1} = \dfrac{\lambda_1 z_1}{\lambda_2 z_2}$
Spherical wave magnification	$m = \left(\dfrac{R_1}{R_1 + z_1} - \dfrac{k_2 z_1}{k_1 R_2} \right)^{-1}$	$m = \left(1 - \dfrac{z_1}{R_0} - \dfrac{\lambda_1 z_1}{\lambda_2 R_2} \right)^{-1}$	$m^{-1} = \dfrac{\lambda_1 z_1}{\lambda_2 z_2}$
Plane wave resolution limit	$\text{RL}_{\text{object}} = 2\ell_1$	$\text{RL}_{\text{object}} = 2\left(\ell_1 - \dfrac{\alpha_0}{\lambda} \right)$	$\text{RL}_{\text{object}} \cong \dfrac{1}{d}$ [a]
Spherical wave resolution limit	$\text{RL}_{\text{object}} = \dfrac{2}{1 - z_1/(R_1 + z_1)}$	$\text{RL}_{\text{object}} = \dfrac{2(\ell_1 - \alpha_0/\lambda_1)}{1 - z_1/R_0}$	$\text{RL}_{\text{object}} \cong \dfrac{1}{d}$
Space bandwidth product	$\text{SBP} = 2\ell_1 L$	$\text{SBP} = 2(\ell_1 - \alpha_0/\lambda_1)L$	$\text{SBP} = 2\ell_1 L$

[*]From Ref. 6.
[a]For a two-point object separated by a distance d, the resolution limit is $\cong 1/\ell_1$.

steps of the holographic process. To achieve magnification in holography, spherical waves must be used in one or both steps of the process[6] (see Table 25.I).

It may appear as if an image has been created without the use of a lens in the hologram process. Equation (25.23) shows that the lens condition is built into the system by the process of diffraction. For this reason, the hologram of a point source has been termed a *zone lens*[6] because of its similarity to a Fresnel zone plate. The hologram described by Eq. (25.19) is a linear superposition of zone lenses. A zone lens due to a point object is shown in Fig. 25.16(b). Positive and negative Fresnel zone plates are shown in Fig. 25.16(a) for comparison purposes.

The Fourier transform hologram discussed earlier in this chapter is characterized by linear fringes rather than quadratic fringes as shown in Fig. 25.16(c). This gives rise to the higher resolution capability of Fourier transform holography at the expense of field of view as predicted by the space bandwidth product discussed in Ref. 6. Sideband Fresnel holograms have both types of fringes with the linear fringe dominating as shown in Figs. 25.16(d) and (e). This results in a loss of resolution capability in the sideband Fresnel hologram process because the film bandwidth is utilized to store the carrier frequency of the process, as shown in Ref. 6.

A similar analysis for the fourth term in Eq. (25.19) shows that a virtual image of the object, $D(\alpha)$, occurs if $z_2 = -z_1$. The image is virtual because it appears behind the hologram so that an auxiliary optical system is needed to observe it (see Fig. 25.17).

When viewing the reconstructed image from the hologram, we observe the intensity in the image plane. This means that these four amplitude terms, all centered on axis, interfere in the image plane. In Fresnel holography, this interference causes image degradation as shown in Fig. 25.18. This image deterioration, which was the limiting factor of the growth of holography in the 1950s, has been removed by means of two different techniques. The first technique

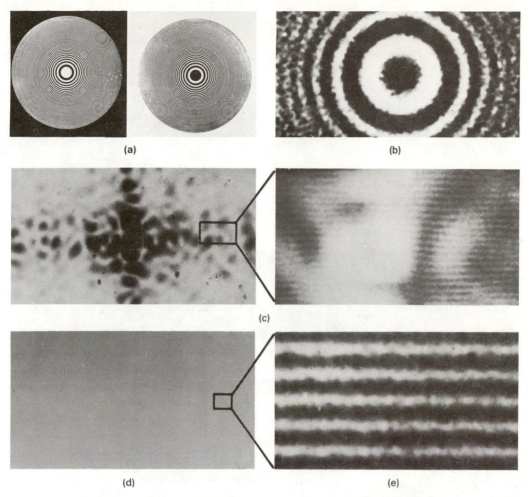

(a)

(b)

(c)

(d)

(e)

Fig. 25.16 Comparison of interference fringes of a zone-lens type hologram and a Fourier transform hologram: (a) Actual Fresnel zone plate consisting of alternate transparent and dense annular zones. (b) Fraunhofer hologram of a small object, showing zone-lens fringes. (A Fresnel hologram has a similar Fresnel fringe pattern, but with a different amplitude modulation.) (c) Fourier transform hologram and an enlarged section on the right. (d) Sideband Fresnel hologram with a 7-deg angle. (e) Enlarged section of (d), showing the 200 line/mm carrier frequency. Comparison of the fringes in (b) with those in (a) demonstrates the zone-lens characteristic of this hologram. Comparison of (c) with (b) demonstrates the difference in the average fringe frequencies of these two types. In (b) the fringe frequency increases linearly with distance in the hologram plane, and in (c) the average fringe frequency is constant with increasing distance. A combination of both these effects occurs in (d). In (e) the zone-lens fringing is not observed since a large object was used and the hologram is a coherent superposition of such fringes. [(c) courtesy of D. Raso and W. Dyes; (d) and (e) courtesy of H. Rose (from Ref. 6).]

utilizes an off-axis reference wave at an angle θ_0 to produce the linear fringe carrier shown in Fig. 25.16(e). Upon reconstruction, the various image terms of Eq. (25.19) are spatially separated so that they do not interfere as shown schematically in Fig. 25.19. Two noninterfering images from a sideband Fresnel hologram process were shown in Figs. 25.13(a) and (b).

The second technique utilizes Fraunhofer rather than Fresnel diffraction in forming the hologram. In essence, this places the out-of-focus image term at a large distance behind the hologram and removes the image degradation of Fresnel holography by replacing it with a constant background distribution. A

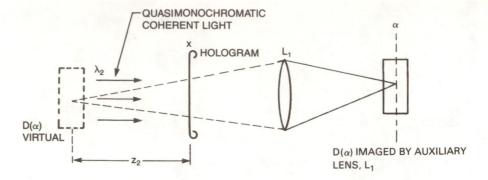

Fig. 25.17 Schematic illustrating reconstruction of the virtual image from the fourth term of Eq. (25.19).

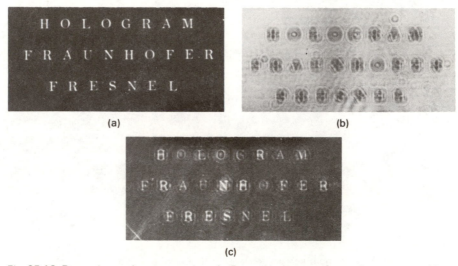

(a)

(b)

(c)

Fig. 25.18 Formation and reconstruction of a Fresnel hologram of a semitransparent object, printed page: (a) the original object, (b) the Fresnel hologram, and (c) the reconstructed image. In this experiment, the hologram is formed in a region approaching the near field of the detail of an individual letter. The formation distance $z_1 = 2$ cm, maximum letter detail dimension $d \cong 100$ μm, the reconstruction distance $z_2 = 2$ cm, and the mean wavelength of the radiation (used in both steps) $\lambda = 5461$ Å. The hologram was recorded on SO243 film and reconstructed on Pan-X. [Courtesy of J. Ward, W. Dyes, and J. Metzemaekers (from Ref. 6).]

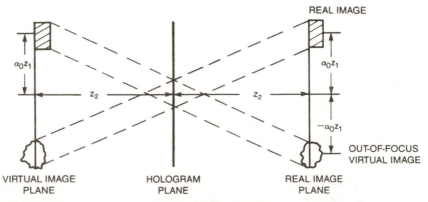

Fig. 25.19 Diagram illustrating the separation effect of real and out-of-focus virtual images in a sideband Fresnel hologram formed with transmitted light (from Ref. 6).

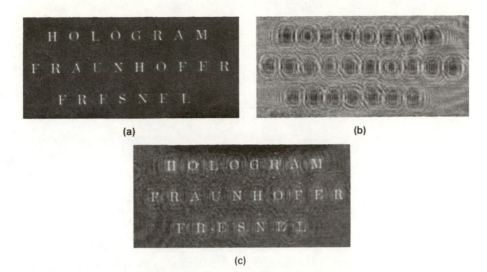

(a)

(b)

(c)

Fig. 25.20 Formation and reconstruction of a Fraunhofer hologram of a semitransparent object, printed page: (a) the original object, (b) the Fraunhofer hologram, and (c) the reconstructed image. In this experiment, the hologram is formed in the far field of an individual letter. The formation distance $z_1 = 7$ cm, the maximum letter detail dimension $d = 100$ μm, the reconstruction distance $z_2 = 7$ cm, and the mean wavelength of the radiation (used in both steps) $\lambda = 5461$ Å. The hologram was recorded on SO243 film and reconstructed on Pan-X. [Courtesy of J. Ward and W. Dyes (from Ref. 6).]

comparison between Figs. 25.20 and 25.18 shows the effect of reducing the interference from the virtual image term in Fraunhofer holography.

25.10   SPACE BANDWIDTH PRODUCT OF A FRESNEL HOLOGRAM

The analysis thus far has not considered the effects of film resolution or hologram size in deriving the reconstructed image from the hologram. If we go back to Eqs. (25.17) and (25.18) and let the object $D(\xi)$ be a point source, we see that Eq. (25.18) becomes

$$I(x) \cong \left| 1 - \exp \frac{ik_1 x^2}{2z_1} \right|^2 \quad \text{or} \quad I(x) \cong 2[1 + \sin(k_1 x^2/2z_1)] \ . \tag{25.31}$$

Equation (25.31) represents the hologram of a point object, and we notice that the quadratic phase factor in the hologram gives rise to a zone-lens function as shown in Fig. 25.16(b). The instantaneous spatial frequency of the fringes in Eq. (25.31) is given by

$$\nu_{\text{ins}}(x) = \frac{1}{2\pi} \frac{d}{dx} (\text{phase}) = \frac{x}{\lambda_1 z_1} \ . \tag{25.32}$$

Since the film on which the hologram is recorded will have a resolution limit ℓ_1, then the zone-lens fringe will no longer be recorded when

$$\nu_{\text{ins}}(x) = x_{max}/\lambda_1 z_1 = \ell_1 \ , \tag{25.33}$$

where ℓ_1 is the cutoff frequency of an ideal film having the transfer function shown in Fig. 25.21. More realistic film transfer functions and their effects on holographic systems have also been considered.[6] The maximum dimension of the

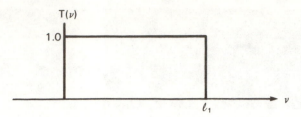

Fig. 25.21 Ideal film transfer function having cutoff frequency ℓ_1.

hologram is limited by the film cutoff frequency and is obtained from Eq. (25.33):

$$x_{max} = \lambda_1 z_1 \ell_1 \ . \tag{25.34}$$

Thus, Eq. (25.34) shows that the size of the hologram will be x_{max} because the zone-lens fringes at distances greater than x_{max} away from the origin are too high in frequency to be recorded by the film having resolution limit ℓ_1. Since the film now has a finite size, the reconstruction of the conjugate image term from the hologram of Eq. (25.31) (in one dimension) is given by

$$g_3(\alpha) = \int_{-x_{max}}^{x_{max}} \exp\left(\frac{-ik_1 x^2}{2z_1}\right) \exp\left[\frac{ik_2(x-\alpha)^2}{2z_2}\right] dx \ . \tag{25.35}$$

Invoking the focusing condition of Eq. (25.23), Eq. (25.35) evaluates to

$$g_3(\alpha) = K' \exp\left(\frac{ik_2 \alpha^2}{2z_2}\right)(2x_{max}) \, \text{sinc}\left(\frac{2\pi x_{max}\alpha}{\lambda_2 z_2}\right) \ ; \tag{25.36}$$

i.e., the reconstructed hologram of a point source becomes a diffraction spot, the size of which is determined by the size of the hologram, x_{max}. Using the Rayleigh criterion of Chapter 6, which equates the resolution limit in line pairs per millimeter to the inverse of the diffraction spot size,[c] we obtain,

$$\text{resolution limit} = \text{RL}_{\text{image}} = \frac{2x_{max}}{\lambda_2 z_2} \ . \tag{25.37}$$

Using Eq. (25.34) and the focusing condition of Eq. (25.23) in Eq. (25.37), we see that

$$\text{RL}_{\text{image}} = 2\ell_1 \ . \tag{25.38}$$

For systems with unit magnification, $\text{RL}_{\text{object}} = \text{RL}_{\text{image}}$. The resolution of the image recorded on the hologram in line pairs per millimeter is given by Eq. (25.38). In Fresnel holography, the size of the film, L, essentially determines the size of the object, L, when plane waves are used to record the hologram. Thus, the one-dimensional space bandwidth product (SBP) of a Fresnel hologram is given by

$$\text{SBP} = \text{RL}_{\text{object}}(\text{object size}) = 2L\ell_1 \ . \tag{25.39}$$

---

[c]Spot size is determined by setting the argument of the sinc function in Eq. (25.36) equal to π and solving for α as discussed previously in Chapter 6.

Notice that the product $L\ell_1$ describes the number of resolvable pixels along a one-dimensional line on the film. As we just discussed, for Fresnel and Fraunhofer holograms, the film resolution limits the resolution of the image and the film size determines the field of view. When magnification effects are considered,[6] the extent of the object contained in the image can be increased but at a lower resolution; i.e., there is a trade-off between resolution and field of view just as with a microscope. Therefore, the SBP of Eq. (25.39) determines the total information capability of the system. Fourier transform holograms are interesting in that the fringe frequency is constant [see Fig. 25.16(c)], so the resolution of the film does not limit the resolution of the image. Instead, as shown in Ref. 6, the SBP of the image is still $2L\ell_1$. However, in this type of hologram, the size of the film determines the object resolution and the film resolution determines the total field of view of the object.

The system limitations caused by the film in one-dimensional holographic processes are summarized in Table 25.I. The two-dimensional space bandwidth product has also been considered.[4]

In any given application, the space bandwidth product is very useful for examining the parametric trade-offs necessary to optimize the system performance.[4]

REFERENCES

1. N. Abramson, *The Making and Evaluation of Holograms,* Academic Press, New York (1981).
2. W. T. Cathey, *Optical Information Processing and Holography,* John Wiley and Sons, Inc., New York (1974).
3. H. J. Caulfield and S. Lu, *The Application of Holography,* John Wiley and Sons, Inc., New York (1970).
4. *Handbook of Optical Holography,* H. J. Caulfield, ed., Academic Press, New York (1979).
5. R. J. Collier, C. B. Burckhardt, and L. H. Lin, *Optical Holography,* Academic Press, New York (1971).
6. J. B. DeVelis and G. O. Reynolds, *Theory and Applications of Holography,* Addison-Wesley Publishing Co., Reading, Mass. (1967).
7. *Holographic Nondestructive Testing,* R. K. Erf, ed., Academic Press, New York (1974).
8. B. P. Hildebrand and B. B. Brenden, *An Introduction to Acoustical Holography,* Plenum Press, New York (1972).
9. W. E. Kock, *Lasers and Holography: An Introduction to Coherent Optics,* Doubleday Publishing Co., New York (1971).
10. M. Lehmann, *Holography: Technique and Practice,* Focal Press, Ltd., London (1970).
11. *Acoustical Holography,* Vol. 2, A. F. Metherell, ed., Plenum Press, New York (1971).
12. H. M. Smith, *Principles of Holography,* John Wiley and Sons, Inc., New York (1969).
13. G. W. Stroke, *An Introduction to Coherent Optics and Holography,* Academic Press, New York (1969).
14. F. T. S. Yu, *Introduction to Diffraction, Information Processing and Holography,* Massachusetts Institute of Technology Press, Cambridge, Mass. (1973).
15. F. T. S. Yu, *Optical Information Processing,* Academic Press, New York (1983).
16. G. W. Stroke, D. Brumm, and A. Funkhouser, "Three-dimensional holography with 'lensless' Fourier-transform holograms and coarse P/N Polaroid film," *J. Opt. Soc. Am.* 55(10), 1327–1329 (1965).
17. B. J. Thompson, G. B. Parrent, Jr., J. Ward, and B. Justh, "A readout technique for the laser fog disdrometer," *J. Appl. Meteorol.* 5(3), 343–348 (1966).
18. E. N. Leith and J. Upatnieks, "Wavefront reconstruction with diffused illumination and three-dimensional objects," *J. Opt. Soc. Am.* 54(11), 1295–1301 (1964).

26   Holographic Interferometry

26.1   INTRODUCTION

In Chapter 25, we described different types of holograms for creating images. In this chapter, we discuss and develop the topic of holographic interferometry, which has become a very active field of research. No attempt will be made at completeness; rather, the reader is referred to the referenced literature for complete details.

Since 1965 when the initial work on time-averaged holographic interferometry by Powell and Stetson was accomplished,[1,2] many researchers have become involved in various aspects of the subject.[3-11] Today, holographic interferometry looms as one of the major applications of holography.

The holographic technique has been applied to many areas of endeavor. In fact, many different forms of holographic interferometry have emerged. Just as in classical interferometry, the type of holographic interferometry used depends on the application. In this chapter, we treat the various types of holographic interferometers separately. The subject is introduced by extending the concept of holography that was covered in Chapter 25.

26.2   BASIC OBJECTIVE AND THE ADVANTAGES OF HOLOGRAPHIC INTERFEROMETRY

26.2.1   Basic Objective

The basic objective of holographic interferomety is to compare an object of interest to itself under different load or operating conditions. The ability of holographic interferometry to accomplish this normally difficult task has led to many advances in the ability to perform heretofore difficult measurements. In essence, the object acts as its own comparison standard and, hence, removes the need for a separate standard surface or wavefront that is so often used in optical holography.

Holographic interferometry is similar to classical wavefront-dividing interferometry[12] (see Chapter 23) if one recognizes that the hologram plays the role of the beam splitter in these interferometers. At the same time, holographic interferometry offers many advantages over classical interferometry in the making of precision measurements.

26.2.2 Advantages

The advantages of holographic interferometry—which are not available with conventional interferometry—are:

1. It is possible to perform interferometry on subjects that are hard to measure with conventional interferometry.

2. It is possible to observe optical changes in a three-dimensional volume at different viewing angles.

3. Precision optics are not required since one compares the object against its holographic image.

4. In double-exposure holographic interferometry, accurate optical alignments are not necessary and relative motions of the holographic components can be removed with pulsed lasers.

5. The desired complex wave(s) are stored so that time-lapse or differential interferometry is possible.

6. It is possible to observe both transient and steady-state events.

7. It is possible to use holographic interferometry to study small changes that occur over long periods of time, i.e., archival applications.

26.3 TYPES OF HOLOGRAPHIC INTERFEROMETRY

There are many types of holographic interferometry, including:

1. double exposure—empty volume and disturbed volume
2. double exposure—time lapse
3. single exposure—real time
4. time averaged
5. computer generated
6. sandwich
7. multiple wavelength (for contouring).

We now analyze a simple holographic interferometer to demonstrate its capability and properties. This analysis is followed by a sequence of experimental results representing the various types of holographic interferometry.

26.4 SIMPLE HOLOGRAPHIC INTERFEROMETER ANALYSIS

26.4.1 Pure Linear Phase

Consider the Fourier transform hologram made with a Fizeau interferometer (as described in Chapter 25) that results in the reconstructed holographic image shown in Fig. 26.1. The one-dimensional intensity distribution in the hologram is given by Eq. (25.7) as

$$I_1(x) = K^2 + 4a^2 \operatorname{sinc}^2 \frac{2\pi ax}{\lambda f} + 4Ka \operatorname{sinc} \frac{2\pi ax}{\lambda f} \cos \frac{2\pi(2b)x}{\lambda f}. \qquad (26.1)$$

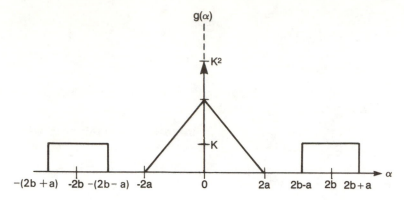

Fig. 26.1 Reconstruction of a simple hologram made in a Fizeau interferometer.

If a prism $\exp(ik\theta_0\xi)$ is placed over the wide on-axis slit in object space and a second hologram recorded on the same film, then the intensity distribution of the hologram is

$$I_{tot}(x) = I_1(x) + I_2(x)$$

$$= K^2 + 4a^2 \operatorname{sinc}^2 \frac{2\pi ax}{\lambda f} + 4Ka \operatorname{sinc} \frac{2\pi ax}{\lambda f} \cos \frac{2\pi(2b)x}{\lambda f}$$

$$+ K^2 + 4a^2 \operatorname{sinc}^2 \frac{2\pi a(x - \theta_0 f)}{\lambda f}$$

$$+ 4Ka \operatorname{sinc} \frac{2\pi a(x - \theta_0 f)}{\lambda f} \cos \frac{2\pi(2b)x}{\lambda f} . \tag{26.2}$$

Reconstruction of the double-exposure hologram described by Eq. (26.2) results in

$$g(\alpha) = 2K^2 \delta(\alpha) + 4a^2 \operatorname{Tr}(\alpha|2a) [1 + \exp(ik\theta_0\alpha)]$$

$$+ 4aK \operatorname{rect}(\alpha - 2b|a) \{1 + \exp[ik\theta_0(\alpha - 2b)]\}$$

$$+ 4aK \operatorname{rect}(\alpha + 2b|a) \{1 + \exp[-ik\theta_0(\alpha + 2b)]\} , \tag{26.3}$$

where $\operatorname{Tr}(\alpha|2a)$ is the triangle function, $k = 2\pi/\lambda$, and we have ignored the obliquity factors. The terms of interest in the intensity of the reconstructed image are of the form:

$$K^4 \delta(\alpha); \quad a^4 \operatorname{Tr}^2(\alpha|2a) \cos k\theta_0\alpha ;$$

$$a^2 K^2 \operatorname{rect}(\alpha - 2b|a) \cos k\theta_0(\alpha - 2b) ; \quad \text{and} \tag{26.4}$$

$$a^2 K^2 \operatorname{rect}(\alpha + 2b|a) \cos k\theta_0(\alpha + 2b) .$$

The cross terms, which correspond to intensity distributions outside the image, are omitted from Eq. (26.3). We note that cosine fringes of the form

$$\cos k\theta_0(\alpha \pm 2b) \tag{26.5}$$

lace both reconstructed images; see Fig. 26.2(a). Thus, the presence of the prism during the second exposure (a pure phase function) has caused interference fringes to appear in the reconstructed images. The frequency of these fringes is directly proportional to the prism angle. (It is important to note that the prism

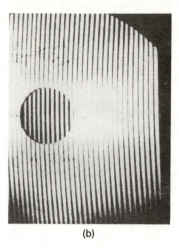

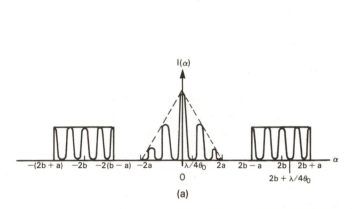

(a)   (b)

Fig. 26.2 (a) Reconstructed holographic interferogram illustrating the presence of the prism and the relationship between fringe frequency and prism angle and (b) reconstructed image of the double-exposure hologram resulting from rotating the object beam about the y axis between exposures. The object was a flat reflecting plate (from Ref. 13).

angle must be small enough to allow the transforms of the object and reference to overlap in the focal plane of the lens. This constraint can be removed if a ground glass is used over one of the beams in the hologram formation process.) In essence, the first exposure creates the reconstructed slit image (see Fig. 26.1); then, this slit image behaves as an interferometric reference for the prism that is stored in the second holographic exposure. The fringes associated with the convolution term Tr^2 of Eq. (26.4) are of no interest in holographic interferometry.

This simple example illustrates the key principle of holographic interferometry; i.e., the hologram behaves as its own reference and enables one to measure the phase difference existing in the object between the two holographic exposures.

In Fig. 26.2(b), the creation of cosine fringes from a double-exposure holographic interferogram is shown. In this experiment, the object beam was rotated about the y axis between holographic exposures to create a phase function that simulates the prism effect shown in Fig. 26.2(a). Reconstruction of the double-exposure hologram results in linear fringes parallel to the y axis.

26.4.2   Arbitrary Phase

We are now prepared to show how an arbitrary phase function $\exp[ikf(\xi)]$ at the position of the rectangular aperture will yield cosine fringes [depending on $f(\xi)$] lacing the reconstructed images; i.e., we obtain cosine fringes analogous to Eq. (26.5) given by

$$\cos[kf(\alpha \pm 2b)] \;, \tag{26.6}$$

where $f(\alpha \pm 2b)$ is the holographic image of the arbitrary phase object measured in units of wavelengths. The intensity distribution of the first hologram is given by

$$I_1(x) = \left| 2a \operatorname{sinc} \frac{2\pi ax}{\lambda f} + K \exp\left[\frac{2\pi i(2b)x}{\lambda f}\right] \right|^2 \;; \tag{26.7}$$

for the second hologram, it is

$$I_2(x) = \left| \left(2a \text{ sinc} \frac{2\pi ax}{\lambda f} \circledast F \right) + K \exp \left[\frac{2\pi i (2b) x}{\lambda f} \right] \right|^2 . \tag{26.8}$$

In Eq. (26.8),

$$F = \mathscr{F}\{\exp[ikf(x)]\} \tag{26.9}$$

is the Fourier transform of the arbitrary phase function placed over the rectangular aperture, $\circledast$ denotes the convolution process, and the exponential term arises from the off-axis delta function. If we now form the double-exposure hologram given by

$$I(x) = I_1(x) + I_2(x) , \tag{26.10}$$

we obtain the two side order images given by

$$\text{amplitude of side order } 1 = 4aK \text{ rect}(\alpha - 2b|a)$$

$$\times \{1 + \exp[ikf(\alpha - 2b)]\} , \tag{26.11}$$

$$\text{amplitude of side order } 2 = 4aK \text{ rect}(\alpha + 2b|a)$$

$$\times \{1 + \exp[ikf(\alpha + 2b)]\} . \tag{26.12}$$

When the intensity of either side order image is observed, the reconstructed image (the rectangular function) is laced by the cosine fringes predicted in Eq. (26.6). Thus, the presence of a pure arbitrary phase function has caused interference fringes to occur in the reconstructed image, the frequency of the fringes being determined by the arbitrary phase function. In general, these will be very complicated fringe patterns. As before, the first exposure creates the reconstructed slit, which then behaves as an interferometric reference for the arbitrary phase function stored in the second hologram. Hence, we have illustrated the principle of holographic interferometry for any arbitrary phase function.

An example demonstrating these results for an arbitrary phase object is shown in Fig. 26.3. In the first exposure, an off-axis reflection hologram of a flat diffuse

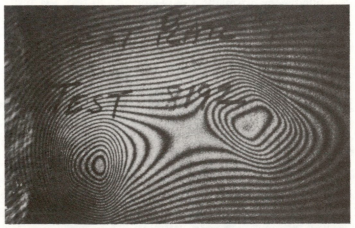

Fig. 26.3 Reconstructed image (holographic interferogram) from double-exposure hologram of a flat plate illustrating the fringes corresponding to the out-of-phase stress created on the plate between holographic exposures (from Ref. 13).

plate in a test fixture was made. Between exposures, the plate was stressed by turning two thumb screws behind the plate in the test fixture. This stress creates an arbitrary phase function $\exp[ikf(\xi, \eta)]$, which corresponds to the out-of-phase deformations of the test plate. The second exposure in the double-exposure hologram stores the Fourier transform of this arbitrary phase function. Upon processing of the film and reconstruction of the double-exposure hologram, a complicated fringe pattern (interferogram) due to the arbitrary phase function of the surface stress is observed in Fig. 26.3.

26.5   DOUBLE-EXPOSURE HOLOGRAPHIC INTERFEROMETRY

In addition to the simple example of observing the prism already discussed, the principle of double-exposure holographic interferometry can be extended to include holography of three-dimensional subjects. In this method, a hologram of the empty volume is made in one exposure and the hologram of the subject of interest placed in the volume is made in the second exposure. Then, the phase change in the volume of interest at the time of the second exposure is interferometrically compared with the reference volume. *This is a very powerful technique and it offers many experimental advantages.* This technique is already widely accepted and being used for numerous applications, some of which we now consider.

In Fig. 26.4, we show the result of a double-exposure holographic interferogram of a light bulb filament. The reference exposure was made with the filament off and then an exposure was made with the filament on. Both exposures were made with pulsed lasers that freeze any motions involved. The figure shows one reconstruction plane but the actual interferogram can be viewed in three dimensions by moving one's head relative to the hologram. A diffuse screen was used to provide back lighting and the large fringes are the result of moving the diffuser between exposures.

In Fig. 26.5, we see a reconstruction of a double-exposure interferogram of a dynamic environment, i.e., a bullet in flight. In this experiment, the laser pulse

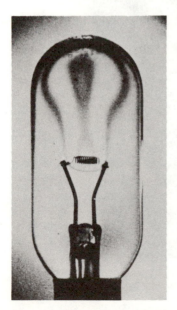

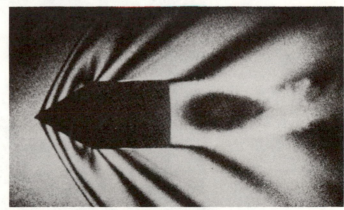

Fig. 26.4 (at left) Reconstructed image from a double-exposure hologram of a light bulb filament (after J. W. Goodman, *Introduction to Fourier Optics,* McGraw-Hill Book Co., NY, 1968).

Fig. 26.5 (above) Reconstructed image of a double-exposure holographic interferogram of a bullet in flight (after J. W. Goodman, *Introduction to Fourier Optics,* McGraw-Hill Book Co., NY, 1968).

was synchronized to the firing of the bullet so that the hologram of the bullet was recorded when the bullet was in the sample volume. The shock wave created by the bullet was also captured in the hologram since it is an arbitrary phase function.

26.6   DIFFERENTIAL OR TIME-LAPSE DOUBLE-EXPOSURE HOLOGRAPHIC INTERFEROMETRY

To perform holographic interferometry with double exposures, it is not necessary to compare the event against a known reference. *Differential* or *time-lapse* holographic interferometry is the method of comparing two three-dimensional holographic images of the same object at two different times.

In this case, the two interfering images for each side order are given in analogy to Eq. (26.11) and (26.12) by

$$\text{amplitude of side order 1} = 4aK\,\text{rect}(\alpha - 2b\,|\,a)\,\{\exp[\,ikf_1(\alpha - 2b)]$$

$$+ \exp[\,ikf_2(\alpha - 2b)]\}\ ,\qquad(26.13)$$

$$\text{amplitude of side order 2} = 4aK\,\text{rect}(\alpha + 2b\,|\,a)\,\{\exp[-ikf_1(\alpha + 2b)]$$

$$+ \exp[-ikf_2(\alpha + 2b)]\}\ .\qquad(26.14)$$

The terms $f_1(\alpha \pm 2b)$ and $f_2(\alpha \pm 2b)$ in Eqs. (26.13) and (26.14) are associated with the object at different times and will interfere to produce the fringes when the intensity associated with Eq. (26.13) or (26.14) is observed. This is demonstrated dramatically in Fig. 26.6. This figure shows the reconstruction from a double-exposure holographic interferogram of a mushroom. The exposures

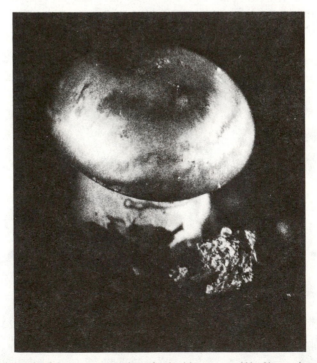

Fig. 26.6 Photograph of the reconstruction of a double-exposed He-Ne gas laser hologram of a growing mushroom. The two exposures were separated in time by 25 s. The fringes were due to the growth deformation of the surface (from Ref. 14).

were separated by 25 s, during which time the mushroom continued to grow. The interference fringes (barely visible) are due to the fact that the growth of the mushroom caused an optical phase change. The ragged fringes show that the growth is nonuniform and the low frequency of the fringes means that the mushroom did not change very much in 25 s.

As a second example, consider the result shown in Fig. 26.7. We show the reconstructed hologram of a metal elbow from a liquid rocket engine. The double-exposure hologram was recorded with a 10-mW He-Ne laser. The fringes illustrate the swelling of the elbow under pressure. The fringes trace out a locus of constant optical path length change. Neighboring fringes correspond to displacements of $\lambda/2$ or 0.3 μm.

As a final example, we show in Fig. 26.8 one of the more celebrated results of double-exposure differential holographic interferometry. This is a holographic interferogram of an 8.25×14 four-ply tire. The first hologram was made with the tire uninflated and the second with a pressure of 50 lb/cm^2. The blemish in the top right (arrow) is between the liner and first ply on the shoulder. The fringe pattern on the tread (arrow) is indicative of a separation between the first and second plies. The broad fringes represent changes in the tire between inflated and noninflated states.

26.7   SINGLE-EXPOSURE (REAL-TIME) HOLOGRAPHIC INTERFEROMETRY

In this type of holographic interferometry, a hologram of the desired object is recorded. One of the reconstructed images is then made to appear at the site of

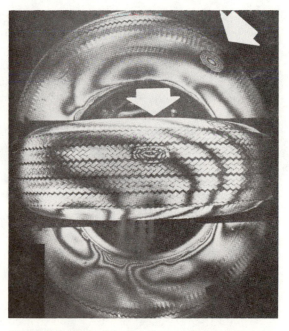

Fig. 26.7 Reconstruction from a double-exposure differential holographic interferogram of a metal elbow from a liquid rocket engine (from Ref. 14).

Fig. 26.8 Example of double-exposure holographic interferometry applied to nondestructive testing. The holographic images of a four-ply tire reveal separations between plies at the positions of the localized circular interference patterns. These flaws were brought out by natural creep after inflation to 50 psi (from Ref. 5). (Courtesy GCO Inc., Ann Arbor, Mich.)

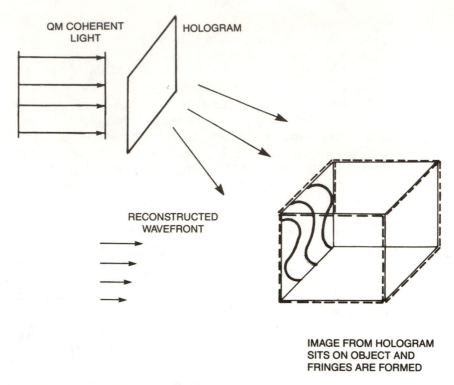

Fig. 26.9 Schematic illustrating single-exposure holographic interferometry in which the image reconstructs exactly on the object.

the original object (see Fig. 26.9), which can now be in a different physical state, i.e., deformed. The result will be an interference pattern lacing the images where the fringes are determined by the degree of deformation of the object. As the object moves, this fringe pattern can be used in real time to perform diagnosis of the object with interferometric accuracy.

Again, if we extend our simple holographic model to include three-dimensional objects, the principle of real-time single-exposure holographic interferometry evolves. This method has been used for a variety of problems in the area of nondestructive testing including objects subject to stress and strain. The major problem with such interferometry is that the fringes are the result of integrated optical path differences existing throughout the volume and are difficult to interpret. The hologram has a large depth of field and must be aligned in the original position after development.

26.7.1   Real-Time Recording Materials

The recent development of real-time or near real-time recording materials means that single-exposure holographic interferometry becomes practical because the hologram develops *in situ*, removing the need for realignment, which is necessary after chemical processing when conventional photographic plates are used to record holograms. Two types of systems are now available for these applications. One is based on the use of thermoplastic recording materials, which are deformable in direct proportion to the input exposure. Thus, deformation creates a phase hologram centered about the carrier frequency which can be reconstructed *in situ* with the reference beam such that the holographic image is interferometri-

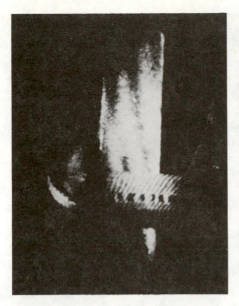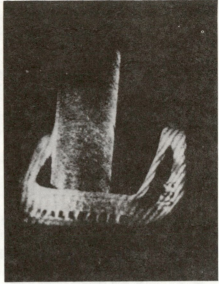

Fig. 26.10 Results of a real-time holographic interferogram of a clamp under different amounts of stress (from Ref. 15).

cally aligned with the original object. Minute motions of the object create an interferometric fringe pattern that can be observed in real time. The cycle time (record and erase) for these systems is about 1 min with times of a few seconds or less being projected.

The second system is a film-based system combined with an *in situ* monobath processing chemistry that requires a few seconds. Once processed and fixed, the hologram image can be reconstructed onto the object with interferometric accuracies so that real-time interferometry can be observed as the object changes its state. The film is not reusable and subsequent experiments are performed by advancing the film in its cassette and repeating the process.

In Figs. 26.10, we show the interferometer output with a different amount of stress applied to the clamp in each case. Note the different fringe patterns.

As a second example of this technique, consider the result shown in Fig. 26.11. In this experiment, we see the interference fringes obtained by the single-exposure technique for a turbine blade that was mechanically deformed.

26.7.2   Aberration Balancing

One last example of using single-exposure holography in an interferometric mode is to use its storage properties to cancel out wavefront aberrations.

Holographic aberrations are described in a manner similar to lens aberrations by including higher order exponential terms arising within the calculation of the diffraction integrals; e.g., third-order aberrations are defined in terms of differences in phase between the hologram wavefront and a reference sphere.[17,18] Examples of correcting astigmatism, spherical aberration, random phase from a ground glass plate, and human cataracts have appeared in the literature. These examples are shown in Figs. 26.12 through 26.14. In Fig. 26.12, we show the removal of spherical aberration using a sideband Fresnel hologram. This is accomplished by recording a hologram of the lens aberrations and then using this hologram as a compensating plate in the optical system to correct (cancel) the

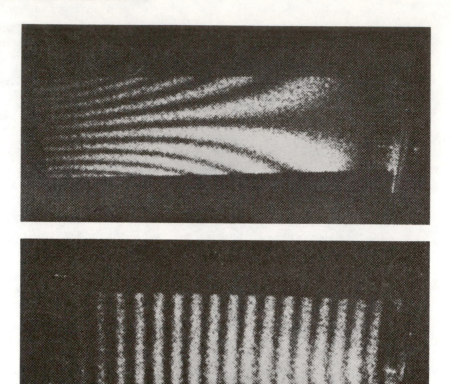

Fig. 26.11 Reconstruction from a single-exposure holographic interferogram of two states of a turbine blade that was mechanically deformed at a frequency of 2394 Hz (from Ref. 16).

undesired aberrations. In Fig. 26.13, we show an example of the removal of astigmatism in a Fraunhofer hologram. In this experiment, astigmatism was introduced into the system by tipping the hologram plane in the recording process, and then it was removed by tipping the hologram in the reconstruction process.

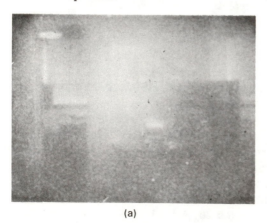
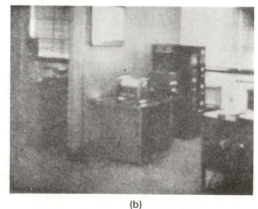

(a)   (b)

Fig. 26.12 Holographic correction over an extended field: image using lens (a) without corrector plate and (b) with corrector plate (from Ref. 19).

(a)　　　　　(b)　　　　　(c)　　　　　(d)　　　　　(e)

Fig. 26.13 Example of aberration (astigmatism) removal. In this experiment, astigmatism was introduced into the system by tipping the hologram plane in the recording process: (a) the original object; (b) the reconstructed image from the nonaberrated hologram; (c) and (d) reconstructed images from the astigmatic hologram; the different bars were focused upon so that part of the image appears in focus while the remainder of the image is out of focus; and (e) the corrected image. The astigmatism was removed by tipping the hologram ~10 deg, which corresponds to replacing the hologram in the same position in which it was recorded and reconstructing it with a plane wave. The holograms were made with 6328-Å illumination at ~1 far-field distance of the width of the crossbars. (Courtesy of J. Ward and W. Dyes.)

More difficult aberrations such as random phase can also be removed holographically. In Fig. 26.14, we show the reconstructed image from a sideband Fresnel hologram in which a diffuser was placed between the object and the hologram. If the identical diffuser is used in the reconstruction step (which requires very critical alignment techniques), the effect of the random phase error in the reconstructed image having the complex conjugate aberrated wavefront is removed. A similar result for an excised human cataract is shown in Chapter 31.

26.8 MULTIPLE-EXPOSURE OR TIME-AVERAGE HOLOGRAPHIC INTERFEROMETRY

Time-averaged holographic interferometry was the first type of holography to appear in the literature and was accomplished by Powell and Stetson.[1,2] They let the object move in time such that all holographic fringes (remember, a hologram is itself an interferogram) wash out, except those parts of the object that are stationary.

As an example of this kind of holographic interferometry, consider the results shown in Fig. 26.15. In Figs. 26.15(a) through (g), we see the mode patterns reconstructed from holograms of a vibrating object, the bottom of a 35-mm film can containing a mechanical solenoid. In these figures, the frequency is increasing.

As a second example of this technique, consider the nodal patterns formed by a vibrating guitar, shown in Fig. 26.16. In Fig. 26.16(a), the frequency is 185 Hz and in (b), the frequency is 285 Hz.

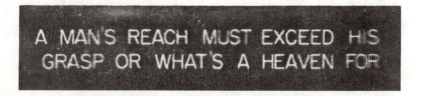

Fig. 26.14 An example of holographic imagery through a random medium. The figure shows the reconstruction of a hologram made by placing a diffuser between the object and the hologram but not in the reference beam. In the reconstruction step, the diffuser is placed in such a position that the same random phase function exists at the hologram plane (from Ref. 20).

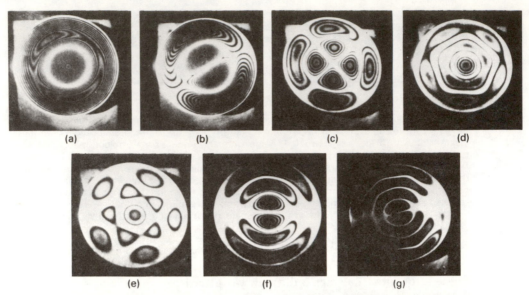

Fig. 26.15 Mode patterns reconstructed from holograms of a vibrating object. Reconstructions of seven holograms containing different resonant frequencies of the same object are shown. The experiment was performed to determine the effect of sinusoidal motion in the hologram formation process. The object was the bottom of a 35-mm film can containing a mechanical solenoid. The frequency of excitation was increased to obtain the photographs shown. The holograms were recorded on Kodak 649-F emulsion (from Ref. 2).

26.9   MULTIPLE-WAVELENGTH HOLOGRAPHY FOR CONTOURING

26.9.1   Two-Source Contouring

This method utilizes two sources separated in space, each of which forms a hologram (either sequentially or simultaneously). Thus, the two holographic

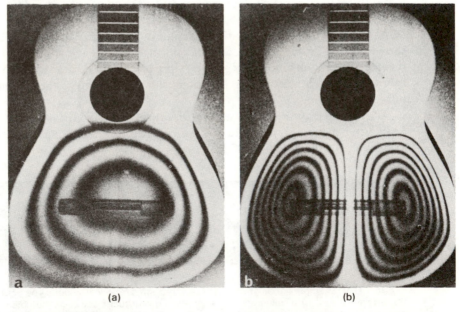

Fig. 26.16 Photographs of images of a vibrating guitar generated by time-average holograms: (a) 185 Hz and (b) 285 Hz. The fringes are characteristic of the first two vibration modes of a guitar made by George Bolin, Stockholm. The holograms were recorded by N. E. Molin and K. A. Stetson at the Institute of Optical Research, Stockholm, Sweden (from Ref. 5).

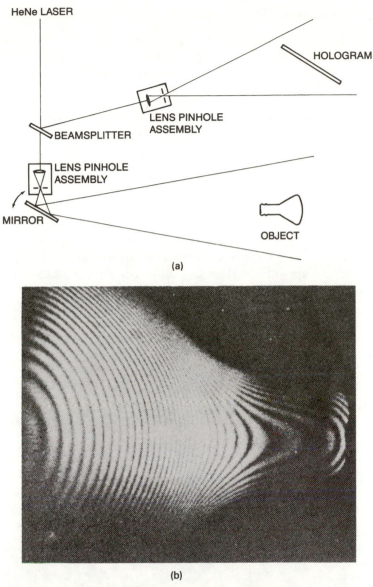

(a)

(b)

Fig. 26.17 A geometry for multiple-source holographic contour generation: (a) Schematic of a two-source system. The two source positions are realized by rotating the mirror between exposures. The geometry can usually be modified to operate in real time. (b) Multiple-source contour map of opaque, diffuse flood lamp, with contour interval of ~2.2 mm (after Hildebrand, Ref. 21).

images have a phase difference between them and the interference between these two images generates contour fringes. For straight line contours, the two sources must be moved to infinity to give two plane waves. The distance between successive contours is given by

$$\Delta h = \lambda_1 / 2 \sin(\beta/2) \ , \tag{26.15}$$

where β is the angle between the two sources and, hence, between the two plane waves.

Figure 26.17(a) shows the schematic of a two-source system and (b) shows a contour hologram of a flood lamp made by this method.

26.9.2   Two-Frequency Method

In this method, the two holograms are made with different wavelengths, λ_1 and λ_2, and the image is formed with a single wavelength, λ_1. In this case, each of the side order images has two terms that interfere and are given in analogy to Eqs. (26.11) and (26.12) by

$$\text{side order } 1 = 4aK \operatorname{rect}(\alpha - 2b|a) \{\exp[ik_1 f(\alpha - 2b, \lambda_1)]$$

$$+ \exp[ik_2 f(\alpha - 2b, \lambda_2)]\} , \tag{26.16}$$

$$\text{side order } 2 = 4aK \operatorname{rect}(\alpha + 2b|a) \{\exp[-ik_1 f(\alpha + 2b, \lambda_1)]$$

$$+ \exp[-ik_2 f(\alpha + 2b, \lambda_2)]\} , \tag{26.17}$$

where $f(\alpha \pm 2b, \lambda_1)$ and $f(\alpha \pm 2b, \lambda_2)$ are the images for the wavelengths λ_1 and λ_2, which will interfere to produce the fringes when the intensity is observed. Since the scale of the image is proportional to the wavelength, reconstructing the hologram with wavelength λ_1 causes contour fringes to appear on the surface of the image. These contours are separated by a distance

$$\Delta h = \lambda_1 \lambda_2 / 2\Delta\lambda , \tag{26.18}$$

where

$$\Delta\lambda = \lambda_1 - \lambda_2 . \tag{26.19}$$

Figure 26.18 shows the reconstruction of a contour hologram of a coin made by the two-frequency method. In this example, the contour interval is $\Delta h \cong 0.02$ mm where $\Delta\lambda \simeq 65$ Å.

Fig. 26.18 Reconstructed image from contour hologram of a coin by the two-frequency method (after Hildebrand and Haines, Ref. 22).

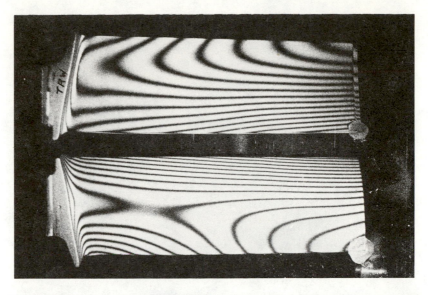

Fig. 26.19 Photograph of the reconstruction of a holographic contour map. The subjects were the concave and convex surfaces of the turbine blades from an aircraft engine. The blades were mounted on the window of a glass box. The hologram was recorded with a He-Ne laser. For the first exposure, the box was filled with air; for the second, it was filled with sulfur hexafluoride (from Ref. 14).

26.9.3   Index Mismatch Contouring

In this technique, a double-exposure hologram is made in which the object is immersed in a different index for each exposure. The distance beween contours on the reconstructed image is

$$\Delta h = \lambda / 2(n_1 - n_2) \ . \tag{26.20}$$

Figure 26.19 is a contour map of the concave and convex surfaces of the turbine blades from an aircraft engine. The first hologram was made in the presence of air and for the second hologram the blade was immersed in sulfur hexafluoride.

Even though holographic contouring at first seems intriguing, it should be pointed out that the disadvantages could outweigh the advantages. The same information is usually obtainable with incoherent sources and utilization of diffraction gratings to do moiré fringe contouring. Holographic contouring could still prove useful in the study of transient phenomena.

26.10   COMPUTER-GENERATED HOLOGRAPHIC INTERFEROMETRY

One of the interesting practical uses of holographic interferometry occurs in the testing of optical lenses after fabrication. A computer hologram of the ideal mathematical lens surface is generated and it is reconstructed and interferometrically measured against the lens being fabricated. Any inhomogeneities in the test lens appear as localized fringes and these are used in turn to define areas needing further polishing. Great improvements in measurement accuracy (factors of 2 to 4) have been made possible by the application of this technique.

In Fig. 26.20(a), we show a master computer-generated hologram (CGH)

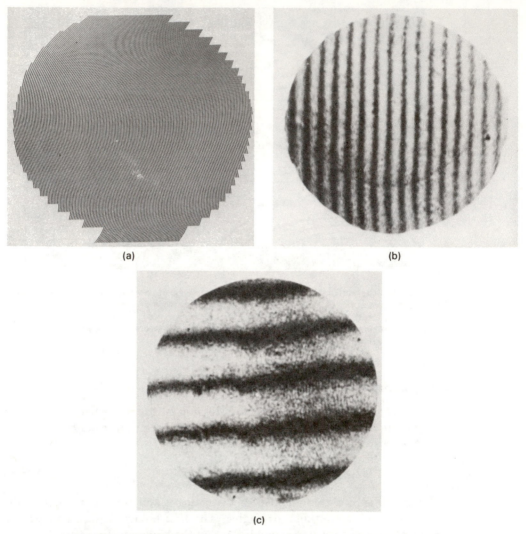

(a)

(b)

(c)

Fig. 26.20 (a) Master CGH plotted by a laser beam recorder, (b) CGH test of an aspheric wavefront having a maximum slope of 126 waves per radius and 64 waves departure, and (c) result of holographic test of lens having 50 waves of third- and fifth-order spherical aberration (from Ref. 8, p. 394).

plotted by a laser beam recorder. In (b), we show the CGH test of an aspheric surface, and in (c), we show the CGH test of a lens with 50 waves of third- and fifth-order spherical aberration.

26.11   CONCLUSIONS

In this chapter, various types of holographic interferometry have been discussed and shown to be qualitative extensions of our simple two-beam interferometer model. The reader interested in more detail regarding the analysis and use of the various types of holographic interferometry is referred to the books on the subject listed in the references. Additional techniques such as sandwich and heterodyne holographic interferometry have become practical in some applications. These techniques combine the principles of holographic interferometry described in this chapter with moiré techniques to create additional advantages as described in the literature.[3,6,7]

REFERENCES

1. R. L. Powell and K. A. Stetson, "Interferometric vibration analysis of three-dimensional objects by wavefront reconstruction," *J. Opt. Soc. Am.* 55(5), 612A (1965).
2. R. L. Powell and K. A. Stetson, "Interferometric vibration analysis by wavefront reconstruction," *J. Opt. Soc. Am.* 55(12), 1593–1598 (1965).
3. N. Abramson, *The Making and Evaluation of Holograms*, Academic Press, New York (1981).
4. *Handbook of Optical Holography*, H. J. Caulfield, ed., Academic Press, New York (1979).
5. R. J. Collier, C. B. Burckhardt, and L. H. Lin, *Optical Holography*, Academic Press, New York (1971).
6. R. Dändliker, *Heterodyne Holographic Interferometry, Progress in Optics*, E. Wolf, ed., Vol. 17. pp. 1–84, North Holland Publishing Co., Amsterdam (1969).
7. *Holographic Nondestructive Testing*, R. K. Erf, ed., Academic Press, New York (1974).
8. D. Malacara, *Optical Shop Testing*, John Wiley and Sons, Inc., New York (1978).
9. Y. I. Ostrovsky, N. N. Butvsov, and G. V. Ostrovskaya, *Interferometry by Holography, Springer Series in Optical Sciences*, Vol. 20, Springer-Verlag, New York (1980).
10. W. Schumann and N. Dubar, *Holographic Interferometry, Springer Series in Optical Sciences*, Vol. 16, Springer-Verlag, New York (1979).
11. C. M. Vest, *Holographic Interferometry*, John Wiley and Sons, Inc., New York (1978).
12. G. B. Brandt, "Holographic interferometry," in *Handbook of Optical Holography*, H. J. Caulfield, ed., pp. 463–502, Academic Press, New York (1979).
13. G. O. Reynolds, D. A. Servaes, L. Ramos-Izquierdo, J. B. DeVelis, D. C. Pierce, P. D. Hilton, and R. A. Mayville, "Holographic fringe linearization interferometry for defect detection," *Opt. Eng.* 24(5), 757–768 (1985).
14. R. F. Wuerker, L. O. Heflinger, and S. M. Zivi, "Holographic interferometry and pulsed laser holography," in *Holography*, B. G. Ponseggi, B. J. Thompson, eds., Proc. SPIE 15, 97–104 (1968).
15. H. M. Smith, *Principles of Holography*, 1st ed., p. 192, John Wiley and Sons, Inc., New York (1969).
16. C. R. Hazell, S. D. Liem, and M. D. Olson, "Real-time holographic vibration analysis of engineering structural components," in *Developments in Holography*, B. J. Thompson, J. B. DeVelis, eds., Proc. SPIE 25, 177–182 (1971).
17. R. W. Meier, "Magnification and third-order aberrations in holography," *J. Opt. Soc. Am.* 55(8), 987–992 (1965).
18. E. N. Leith, J. Upatnieks, and K. A. Haines, "Microscopy by wavefront reconstruction," *J. Opt. Soc. Am.* 55(8), 981–986 (1965).
19. E. N. Leith et al., "Correction of lens aberrations by means of holograms," *Appl. Opt.* 5(4), 589–593 (1966).
20. E. N. Leith and J. Upatnieks, "Holographic imagery through diffusing media," *J. Opt. Soc. Am.* 56(4), 523 (1966).
21. B. P. Hildebrand, "A general analysis of contour holography," Report 7421-35-T, Institute of Science and Technology, University of Michigan, Ann Arbor (1968).
22. B. P. Hildebrand and K. A. Haines, "Multiple-wavelength and multiple-source holography applied to contour generation," *J. Opt. Soc. Am.* 57(2), 155–162 (1967).

27 Applications of Holography*

27.1 INTRODUCTION

Since the first introduction of the basic concepts of the hologram and the holographic process by Gabor,[1-3] extensive advances have occurred in the techniques associated with the formation of the hologram, the materials used for recording the hologram, and the suggested applications of the various techniques. A hologram is, as explained in Chapters 25 and 26, the record of an interference pattern formed between a field (optical, acoustical, microwave, etc.) of interest and a known or reproducible background or reference field. When this record, the hologram, is illuminated with a beam equivalent to the reference field, the original field of interest can be recreated. Thus, both the amplitude and phase of the field of interest are stored and recreated, whereas normally, of course, if the field of interest is recorded alone, only the amplitude is stored and the field information is then restricted to the autocorrelation of that field.

Initially, despite the intriguing nature of the process itself, only limited positive results were obtained in the application of this unique invention. Nevertheless, some very important scientific results were obtained that provided the fundamental understanding of the process.

27.1.1 Early Developments, 1948-1962

The first period in the development of holography is clearly 1948-1962. During this time period, some extremely difficult problems were attacked.

27.1.1.1 Electron Microscopy

This technique was invented as a two-step coherent image forming process with the specific application of high-resolution electron microscopy in mind. The conceived process consisted of recording a hologram with an electron beam

---

*This is a reprint of selected sections of an article by B. J. Thompson published in *Rep. Prog. Phys.,* Vol. 41, pp. 633-674, 1978 (©1978 The Institute of Physics, London), and reprinted here with permission.

without the use of electron objectives, which limited the resolution to about 5 Å with the technology of 1948. The hologram was then magnified, for image formation with a light beam, by a factor that was the ratio of the wavelength of the light to that of the electrons, a factor of the order of 10^5. As Gabor pointed out in his original paper, the optical beam illuminating the hologram needed to match the original beam wavefront, including the aberrations. Since electron microscopy was the motivation for the development of holography, considerable attention was given to this problem[4] but without any outstanding success. The reason for this failure is usually stated as being a limitation in the coherence of the electron beam; however, it is more a matter of vibrational stability, because of long exposure times, and film resolution.

Early work in holography also suffered from the fact that two images are formed from the hologram and the light associated with these two images propagates in the same direction (see Chapter 25). Hence, any one of the images is modified by the presence of an out-of-focus contribution, which is actually a hologram of the other image. Considerable attention was given to the problem and a variety of solutions were suggested.[3,5,6] This problem was solved by the use of a separate off-axis reference beam.[7] In a sense, the solution was predicted by Gabor[1] for optical holography: "But it is very likely that in light optics, where beamsplitters are available, methods can be found for providing the coherent background which will allow better separation of object planes, and more effective elimination of the effect of the twin-wave than the simple arrangements which have been investigated."

27.1.1.2   X-Ray Microscopy

Interest in x-ray microscopy has been continuous since Bragg's original paper on a new x-ray microscope was published in 1939. This method was to produce an optical image of the atoms in a crystal from the diffracted x-ray data collected from that crystal. The idea consisted of recording the diffracted x rays from a particular crystal; this x-ray diffraction pattern comprises a discrete set of spots of various amplitudes and phases. A scaled mask is then made to represent the diffracted information and an image formed optically from that mask. Bragg recognized that under the correct experimental conditions the phase of the diffracted x rays could be constant and hence the field considered real and positive. To illustrate this idea, Bragg used one projection of the diffraction pattern of diopside, $CaMg(SiO_3)_2$, such that the calcium and the magnesium atoms overlapped, producing extremely strong scattering and hence eliminating the phase problem. In a sense, he had produced a hologram with the radiation scattered by the overlapping heavy atoms providing the background or reference wave. Subsequently, other workers[8-12] found ways to extend this method to structures where the phase was also required to produce an "image of the molecule." These methods are not holographic, since they involve making a physical mask with appropriate holes for amplitude control and mica phase plates for phase control. Stroke[13] suggested a modification of this technique which stores the various directions of diffraction as holograms and produces a genuine section of the molecule rather than a projection.

The original idea in x-ray microscopy using holography was to use a direct two-step imaging process.[14-17] In the final analysis, x-ray holography at the appropriate resolution might be possible if a significantly energetic source of x rays is available (an x-ray laser) and if extremely high-resolution recording materials are developed.

27.1.2   The Modern Phase of Holography, 1962–Present

The modern phase of holography started in the early 1960s with the implementation of the off-axis reference beam method.[7] This was closely followed by the use of the gas laser in holography,[18] which clearly allowed for considerable versatility in the experimental implementation and thus holograms could be made in reflected light,[19–21] leading to the so-called *three-dimensional photography*. (The typical hologram usually displayed today produces an image that appears like the real object in its parallax, and the hologram recording materials act as a window through which the object is viewed. It must be remembered that any image is three dimensional; it is only when that image is recorded that we have a two-dimensional representation. Indeed, it would be very disappointing if the holographic process did not produce an image of the original object equivalent to that produced by a lens.) High-resolution techniques were discussed[22] as a prelude to optical holographic microscopy. In a very different direction of development, the pulsed ruby laser was used in holographic recording,[23] and the concepts of far-field holography developed[24,25] and used for one of the first direct applications of holography—that of particle size analysis. Finally, it must be mentioned that the use of the hologram as an optical filtering element in coherent optical processing was first conceived and implemented by Vander Lugt.[26]

Starting in 1965, the subject of holography became one of considerable excitement and interest and many techniques and applications have been suggested. The literature from 1965 to 1978 contains almost 4000 references and hence a complete bibliography cannot be provided. Instead, the reader is referred to an excellent bibliography.[27] For more general reading, there are a number of books devoted to this field.[28–37] A series of review volumes also forms a standard part of the literature.[38] Major conferences have been held around the world on the subject of holography and its applications. A selected list of these published proceedings includes those by Thompson,[39] Robertson and Harvey,[40] Vienot et al.,[41] Barrekette et al.,[42] Camatini,[43] Thompson and DeVelis,[44] Aprahamian,[45] Erf,[46] Stroke et al.,[47] and Greguss.[48] A continuing series of specialized conferences on acoustical holography has resulted in many publications.[49–54]

Understandably, the major proportion of applications of holography has been in image formation and a significant part of this chapter is devoted to those applications dealing with optical, acoustical, x-ray, electron beam, and microwave holography. Perhaps the most important developments and applications of holography have been in interferometry and this continues to be an active and important area. Finally, holographic optical elements, such as filters, gratings, and lenses, deserve attention. Miscellaneous subjects, including 360-deg holography, local reference beam holography, incoherent holography, rainbow holography, and the principle properties of holography are discussed and referenced in the Appendix of Sec. 27.5.

27.2   IMAGE FORMATION

27.2.1   Optical Holography

As we have seen, the initial thrusts in holography were directed toward applications in electron and x-ray imaging, and optical experiments were only done for

evaluating ideas or illustrating principles. This situation changed rather dramatically in 1962 with the exploitation of the off-axis reference beam method, the
concept of being able to use reflected light and the obvious advantages of
illumination with laser light. Optical holography as a unique method of image
formation suddenly became a fascinating and exploitable technology. This
fascination caused a tremendous upsurge in work in holography—not all of it
with direction. Making holograms for the sake of making holograms became the
prime activity rather than careful evaluation of the need. It has often been stated
that holography should only be used if the information about the object cannot
be obtained in a conventional way. The public, too, became interested in this new
technology through a series of articles in the more popular journals.

27.2.1.1   *Photography*

Under the heading of photography, we include a variety of topics that use
holographic methods of image formation as an alternative to the conventional
photographic image (displays, information storage of three-dimensional objects,
stereo photography, etc.) or as an adjunct to conventional photography, e.g.,
television tape players.

Displays: One image-forming application that has come about as a result of
public interest is that of holographic displays. There is considerable impact in the
ability to look as if through a window and see a three-dimensional image on the
other side with all the parallax associated with the view through that window.
This particular effect catches the imagination much more than merely looking
through an equivalent window at the real object. Of course, the hologram can be
made from a model and the image formed can give the impression of being a
full-size object. (Benton[55] has described the status of holographic displays and
their future. He was optimistic about solving the problems of image quality at the
right price.) One of the most viewed holographic images was the Body by Fisher
display in the General Motors Building in New York City made by Conductron
Corporation. World Book Encyclopedia has distributed relatively large numbers
of simple reflection holograms.[a] For display purposes it is clear that artistic
ability plays an important role, and hence it is not surprising that artists have
been intrigued with holography as a new medium.[57]
Very clever displays have been made commercially using the 360-deg or
cylindrical holograms. Originally, these were made of a simple object so that it
could be viewed in all directions.[58] The Multiplex Company has combined the
basic idea into making a composite hologram from a series of frames of motion
picture film taken of an object placed on a rotating platform. A hologram, in the
form of a long thin strip, is made of each frame of the movie. The final composite
hologram is formed into an arc of a circle and the individual images have no
parallax or dimensionality, but since each eye is viewing a different image, a
stereoscope effect is produced. As the observer walks around the arc or as the
composite hologram is rotated, different images are seen.
Not all display techniques have been developed merely to intrigue the public
(or sell the product!), but considerable work has been done on applications that

---

[a]More recently, replication holography has been commercialized for use on credit cards to
discourage counterfeiting and has been used in mass production for display purposes on the cover of
various magazines: e.g., *National Geographic,* March 1984 and November 1985, *Lasers and
Applications,* May 1985, and *Optics News,* July 1988. This holographic display process is discussed
and illustrated in the literature.[56]

have a more serious goal. Asmus et al.[59] report on the use of holographic techniques to record accurate dimensions of statues so that if and when repair work is necessary, it can be accomplished correctly. Holograms can be useful educationally, especially for recording images of sculpture, statuary, and pottery so that students may have the ability to study these objects in a rather more real way than merely as flat two-dimensional photographs.

Holographic Movies: Holographic movies have intrigued people for some years, although only very limited results have been obtained. RCA has made a significant effort to reproduce conventional movies for home viewing using a holographic method.[62] In this system, the holograms were recorded as Fourier transform holograms because this gives an image location that is independent of lateral motion of the hologram. The required holograms were recorded initially on each frame of a motion picture film using a He-Cd laser at 4416 Å in photoresist coated on a cronar tape. (In fact, an array of holograms each 10 mm wide was produced for each input frame of photography.) When the photoresist had been developed, it was plated with nickel to produce a master hologram that could then be used to emboss a half-inch vinyl tape using heat and pressure. A large number of copies can then be produced from a master at a relatively low cost. To use the holographic tape, it is illuminated with a beam of light from a low-power (2-mW) He-Ne laser; the image is then picked up with a vidicon. For full-color images, the blue and red images are made up of overlapping modulated areas and the green information is recovered by subtracting the red and blue signals from the total luminance signal. This system worked well and was fully engineered; however, because of market considerations it was never released as a product. The techniques involved in this process are, however, quite important, particularly the use of an inexpensive embossed replica. The same technology of embossed holograms has also been used to produce a moving-map display.[63] In this application, full-color maps were stored as image plane phase holograms. Positional information to define the coordinate on the maps was stored separately as an off-axis Fresnel hologram. In the laboratory model, some 500 12.5- $\times$ 12.5-in.^{2} segments of air charts are stored on 1.2- $\times$ 1.2-cm^2 holograms. Other methods have also employed replica-type systems with simple white-light readout.[56,64,65]

More conventional recording techniques for movies have also been suggested[66] that use a two-dimensional storage format with scanning readout.

High-Speed Photography and Cinematography: Pulsed lasers were first used in dynamic particle size analysis, an application that will be described later. Clearly, the use of a pulsed laser to photograph fast-moving or transient events is important; however, if it is not known exactly where the event will occur within a given volume, then a pulse holographic method is an important solution to this problem. Brooks et al.[67] used a carefully matched path system to record off-axis holograms of a variety of objects. Photographs of bullets cutting a wire or cord, etc., are classic high-speed photographs; they have also become classic images produced holographically. Figure 27.1 illustrates rather well the importance of holographic photography of moving objects; it is an image formed from a hologram of two fruit flys in motion. The difficulty of carrying out ordinary photography on such a subject is perhaps too obvious to state.

Workers in conventional high-speed photography were quick to investigate and use holographic techniques. At the High-Speed Congress in Zurich in 1965,

Fig. 27.1 Image formed from a pulsed laser hologram of two fruit flies in flight. (Courtesy of R. F. Wuerker.)

Helwick[68] and Holland and Landré[69] described ultrahigh-speed laser framing cameras and suggested their use for holography; a pulsed argon laser was used in these systems. At the same conference, Ban and Greguss[70] discussed a method of combining high-speed cinematography with holography, using a ruby laser, to investigate bursting droplets. Perhaps a more important aspect of the same paper was the description of sonoholograms used for the same purpose. The Proceedings of the 11th International Congress also contains a series of useful papers.[71] Sessions on holography have been a part of the International Congress on High-Speed Photography since the Stockholm meeting[72] in 1968. Repetitive Q-switched laser systems were discussed as were superimposed holograms. Many papers at subsequent congresses combined holography with conventional rotating-mirror techniques (see, for example, Ref. 73).

Activity in this area of application continues. For example, Ebeling and Lauterborn[74] have used high-speed holocinematography with spatial multiplexing to produce eight images at a repetition rate of 20 kHz. The illumination was again a pulsed Q-switched ruby laser.

An advantage of holography in fluid mechanics is the fact that, once the hologram has been recorded, an image can be formed as if one or more of several diagnostic techniques had been used. A Schlieren photograph or shadow photograph can be obtained.[75] Flow visualization has become a reasonably important application of holography using imaging as well as interferometric methods.[76]

Figure 27.2 is an excellent illustration of this important area. The six photographs of the air flow over a wedge at Mach 8 are obtained from the same holographic recording. In this illustration, (a) is a shadowgraph, (b) a Schlieren, (c) an infinite fringe interferogram, (d) a finite fringe interferogram with horizontal fringes, (e) a finite fringe interferogram with fringe normal to bow wave, and (f) a finite fringe interferogram with diagonal fringes.

Finally, it might be mentioned that holographic portraiture has received some careful consideration and is usually done with a pulsed laser system.[77]

Stereo Holography: Holography has been considered for applications in mapping and photogrammetry.[78] The methods of extracting photogrammetric information from conventional stereo pairs of photography from the resulting

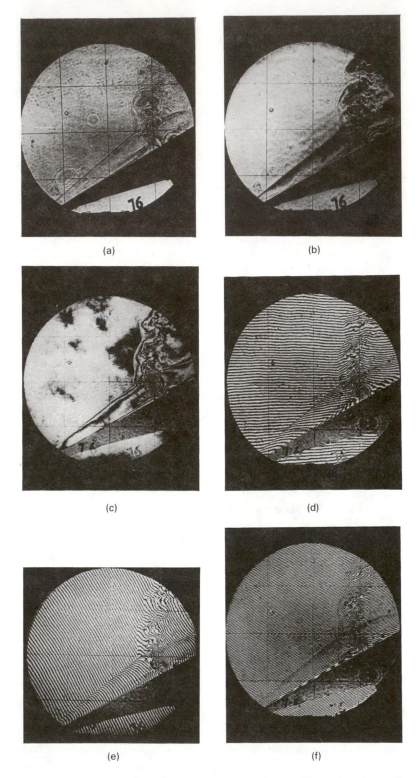

Fig. 27.2 Holography used in flow visualization studies. The six illustrations were all made from a single holographic record of the air flow over a wedge at Mach 8: (a) shadowgraph, (b) Schlieren, (c) zero fringe interferogram. Finite fringe interferograms are shown with (d) horizontal fringes, (e) fringes normal to bow wave, and (f) diagonal fringes (after Ref. 76). (Courtesy of J. E. O'Hare.)

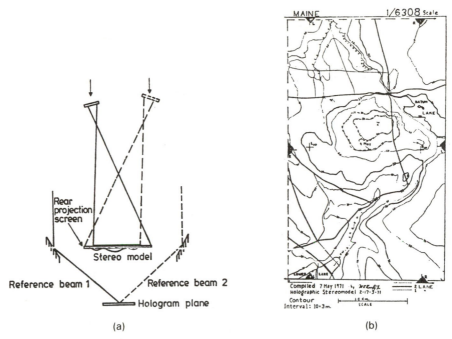

Fig. 27.3 Holographic stereo model: (a) the system for producing the holographic stereo model from a stereo pair and (b) a contour map made from a holographic stereo model (from Ref. 79).

stereo model are well developed technologically. Such stereo pairs produce subjective three-dimensional images. The virtual image produced from a hologram should have all the same information but, of course, represent a true three-dimensional image. This three-dimensional image can be used to produce contours by using a luminous dot connected to an *xyz* coordinate measuring device. This method works well for close-range photogrammetric applications since the required hologram can be recorded. When the hologram cannot be easily recorded, then some other intermediate technique has to be employed and then that intermediate record is converted into a hologram. One concept is to record on a single hologram the two stereo transparencies with the appropriate orientation; such a holographic record has been called a *holographic stereo model*.[79] Figure 27.3(a) shows the method used for recording the holographic stereo model as an off-axis Fresnel hologram. The two stereo transparencies are illuminated coherently and the stereo model produced on a rear projection screen. When the hologram is recorded, each projected transparency is recorded separately with its own reference beam. Image holograms have also been used for recording the stereo model. To produce contours from the holographic stereo model requires the same technique as that described earlier for the close-range application. Figure 27.3(b) shows a contour map of a region in Maine made from a holographic stereo model.

Holographic stereograms have been made by re-recording a series of ordinary photographs onto a single holographic plate. Pole[80] recorded a series of photographs of an object with the aid of a two-dimensional fly's-eye lens so that each lenslet records a slightly different view of the object. A synthesized real three-dimensional image is then produced by diffusely illuminating this photographic record and using the fly's-eye lens to project the individual images. If this is done

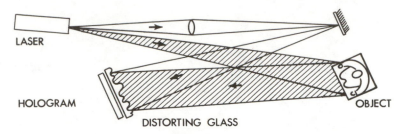

Fig. 27.4 Method of forming a lensless Fourier transform hologram in the presence of a distorting medium. The object and reference beam go through the same distorting glass (after Ref. 90).

coherently, the image may now be stored as a hologram. Basically, the same process can be achieved by recording sequential images (instead of the simultaneous recording that the fly's-eye lens accomplished).[81–83] These and other methods have been reviewed in Ref. 84.

Image Formation with Compensation for Wavefront Deformations:[b] Image formation in the presence of turbulence or other causes of distortion has always been a considerable problem. In principle, it appears that holographic recording under the correct circumstances can provide a useful solution. Kogelnik[85] and Leith and Upatnieks[86] showed that a hologram can be made of an object with a diffuser placed between the object and the holographic recording material; the reference beam does not go through the diffuser. If the hologram is reilluminated, the object will not be visible, but an image of the diffuser is formed. If the diffuser is retained in the system so that the image of the diffuser coincides with the actual diffuser, then an image of the object is formed.

If the reference beam also passes through the diffuser, then the reference beam and object beam have identical distortions and the hologram is formed of the field associated with the object, as if the distorting medium were not present.[87,88] This method was tested over rather long distances (up to 12 km) using a pulsed ruby laser and a horizontal air path.[89]

The basic principles of these methods are illustrated by the results of Ref. 90. Figure 27.4 shows the experimental arrangement, which uses an He-Ne laser to produce a lensless Fourier transform hologram of a small model. The distorting medium is placed directly in front of the recording plane, which ensures that the object beam and reference beam go through identical distortion. The results of such a method are quite good. Figure 27.5 shows a comparison between the image formed (a) normally through the distorting glass and (b) the image formed from the hologram.

The methods discussed above are one-step methods. It is possible to achieve a similar result using a two-step process in which a hologram of the diffuser or distorting glass is used as a corrector for direct imaging.[91] An image plane hologram is made of the diffuser; the object is then placed behind the diffuser and imaged through the diffuser and the hologram of the diffuser. In the conjugate image of the diffuser, the effect of the actual diffuser is removed. In this method, the distortion could be thought of as an aberration and so the holographic corrector plate is balancing the aberration. Upatnieks et al.[92] realized this and used a holographic corrector plate to balance the spherical aberration produced by a lens.

---

[b]Additional examples are discussed in Chapters 23 and 31.

(a) (b)

Fig. 27.5 Results obtained by the method illustrated in Fig. 27.4: (a) a direct photograph through the distorting glass and (b) the image formed from the hologram showing effective compensation for the distorting medium (after Ref. 90).

While all these techniques are interesting, very little use has been made of them for solving practical problems. The atmosphere's turbulence problem, particularly in astronomical imaging, has been attacked by the so-called speckle pattern method (see Ref. 93) as well as with adaptive optics (see Chapter 23).

27.2.1.2   *Microscopy*

In some sense, microscopy has been at the forefront of many efforts in the application of holography. We have seen that electron microscopy was at the very heart of Gabor's interest. In optical holography, this has also been true. There have been several driving forces for this effort: to obtain a larger field of view at high resolution, to examine transient events, and to increase depth of field. Initially, the idea of a lensless microscope was attractive, with the magnification obtained by the use of diverging beams and changes in wavelength between the formation of the hologram and the subsequent formation of the image.[14,15] By means of the lensless technique, Leith and Upatnieks[94] obtained a resolution of a few microns on objects such as a fly's wing. (It should be mentioned, however, that many of the suggested schemes actually used lenses to produce the required diverging beams.) These methods have not been pursued to any extent since they have been superseded by other holographic methods. The reader is referred to Ref. 30 for details on the magnification equations and effects of aberrations.

Holographic Microscopy with a Conventional Microscope: Holographic microscopy has been pursued most effectively by modification of a conventional microscope.[95-99] Some researchers limited their work to laboratory experimentation, while others, notably Van Ligten, pursued the method as far as a commercial holographic microscope that has now received some specialized use, for example in studying polymers.[100-102] Figure 27.6 shows a schematic diagram of the holographic microscope. One beam from the He-Ne laser passes through the sample on the microscope stage and then through the microscope itself in the usual way. The reference beam is brought along outside the body of the microscope and interferes with the object beam to produce the hologram. If we assume

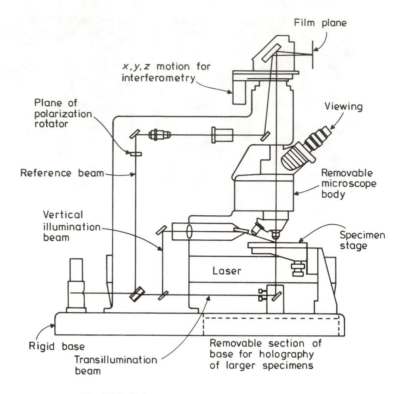

Fig. 27.6 Holographic microscope (after Ref. 96).

that the holographic recording process is not the limiting factor, then the resolution is that determined by the microscope itself; but it might be remembered that the microscope is being used coherently and hence speckle is a problem.

The advantages of holographic microscopy are well recognized. The focusing over a limited range can be done as the image is recovered from the hologram; hence, thicker samples can be studied. Furthermore, if the sample is changing with time, it may not be possible to study the sample in the detail required and through the focal ranges required by more conventional techniques. While these advantages are significant, they do not seem to have been significant enough to produce any widespread use of these methods.

In vivo studies using holographic microscopy are certainly attractive and some attention has been devoted to this area.[103] Holographic interference microscopy is discussed in the Appendix of Sec. 27.6.

Microscopy of Particulates: Microscopy has also been pursued in a very different way using the concepts of Fraunhofer holography. Indeed, the concepts of this type of holography were developed with the problems of particle size analysis in mind.[24,25] This particular field of endeavor was perhaps one of the first real applications of holography and was certainly the first use of pulsed laser holograms. A review,[104] published in 1974, gives a good overview of the subject up to that date. It is interesting that because of the use of this particular type of holography the process itself has been the focus of considerable attention. A reassessment of the Fraunhofer holographic process puts the methodology and mathematics on a firm foundation[105] and corrects some errors in earlier mathematical analyses of this process.

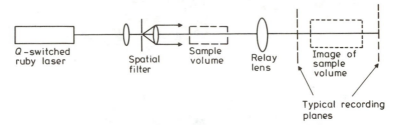

Fig. 27.7 Schematic diagram of a typical in-line Fraunhofer holographic system for particle size analysis.

The problem that this technique addresses may be clearly stated: It is the use of a two-step imaging process of a transient event occurring in a volume since the images of that event cannot be produced by conventional imaging methods. If a number of droplets or particles whose diameters are in the range of 2 to 200 μm exist in a small volume of, say, 1 cm^3, no normal imaging system can photograph all the particles in focus at the same time. (The depth of field of an imaging system that will resolve 2 μm is only about 20 μm!) If the particles are moving within the volume, then it is not possible to record a series of images as the volume is stepped through.

The methodology of Fraunhofer holography was initially developed to solve a problem just like the one stated in the last paragraph—the study of particle size distribution in the volume of naturally occurring fog with the possibility of following the time history of the particle size distribution. The initial instruments were referred to as laser fog disdrometers.[23] At the resolution of a few microns that is often required, the exposure times for recording holograms of moving particles becomes quite short and hence Q-switched pulsed ruby lasers were used to obtain exposure times down to a few tens of nanoseconds. Figure 27.7 shows a schematic of a typical in-line system. The source is a Q-switched ruby laser; the beam passes through a spatial filter and is then collimated (collimation is not essential) to pass through the sample volume, which is typically a few cubic centimeters for small particle analysis and correspondingly larger for larger size particles. To ease the resolution requirements on the recording material, the sample volume is often imaged with a magnification before the hologram is recorded. This has the added advantage of removing the recording plane from the immediate vicinity of the sample. Typical hologram recording planes are indicated. The actual configuration of these devices depends on the specific application.

Over the 12-yr history of this technique, a variety of problems have been successfully attacked. The original fog particle methods worked well, as did cloud chamber studies, explosively generated aerosol studies, marine plankton imaging, rocket engine exhaust studies, two-phase flow diagnostics, glass fiber measurements, etc. (see Ref. 104 for specific references). In addition, other problems were studied to extend the range of applications to smaller and larger particles and studies of the effect of index variations. Work in many of these fields is still continuing.

Figure 27.8 shows in (a) a hologram formed of a field of small pollen particles. The images formed from a portion of that hologram are shown in (b) and can be compared with (c) the conventional microscopic images of a similar set of particles. The diameter of these particles is 20 μm and some significant surface

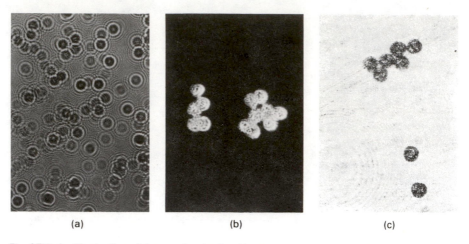

(a)   (b)   (c)

Fig. 27.8 An illustration of the results obtained by the in-line Fraunhofer hologram method: (a) hologram of a field of small pollen particles, (b) image formed from the hologram of a small section of the field, and (c) normal microscope image of similar particles.

detail can be detected. These particular results were preliminary results in a study to determine the applicability of the technique to blood cell analysis.

Trolinger[106] reports on the use of these in-line systems in wind tunnel and dust erosion studies. The off-axis method of holography has also found important applications in particle size analysis.[67] These systems have often been referred to as *holocameras* and have been applied to studies of rocket engine injectors and exhausts, smoke stacks, and limestone injection systems. Trolinger has used a hybrid system that combines off-axis and in-line systems. Studies carried out at the University of Wyoming Elk Mountain Weather Observatory simultaneously produced two in-line and two off-axis holograms of large volumes of ice and snow crystals. Two collimated beams with a 15-deg angle between them were used for illumination so that a stereo pair could be produced. Figures 27.9(a) and (b) show two images of snowflakes produced by this method. The quality of these images is certainly not as good as microphotographs of captured flakes but the conditions are quite different.

Finally, it should also be mentioned that similar systems have been flown in several aircraft to study cloud particulates. Figure 27.10 shows the schematic of such a system in the radome of a WBT7 weather reconnaissance aircraft. The system is folded and the hologram is recorded of a sample volume of 0.3 litre using a Q-switched ruby laser at 100 mJ per pulse. Resolution is 20 μm throughout the volume.

Quite clearly, these methods can produce a considerable amount of useful data that require analysis. Most workers have used variations of the basic reconstruction methods described by Thompson et al.[107] in which the hologram is mounted on a movable carriage and the image produced on a television screen. The operator can study the image in considerable detail as the volume is moved, though sample illustrations of these images when reduced to a mere two-dimensional photograph do not do justice to the method.

It is often important to obtain information about the velocity of the objects under study as well as their size and shape. This can be implemented by using a double- (or triple-) exposure hologram. A typical result is shown in Fig. 27.11

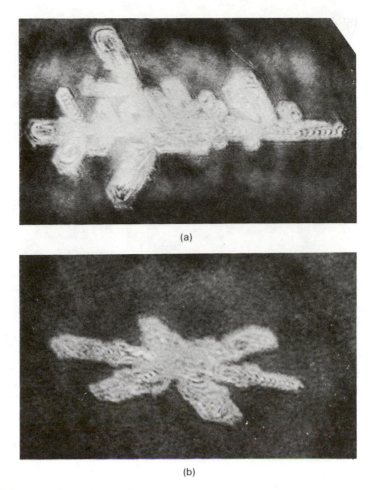

Fig. 27.9 Images of snowflakes. The photograph in (b) is a 0.2-mm-diam flake tilted at approximately 45 deg to the optical axis (after Ref. 76).

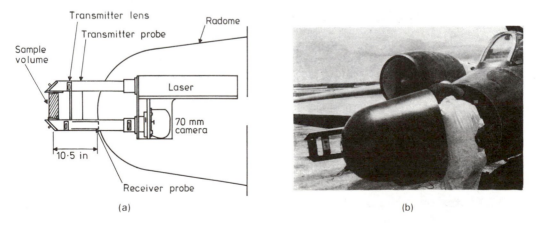

Fig. 27.10 Airborne holographic camera for cloud studies: (a) the schematic arrangement of the system and (b) the installation of the camera (after Ref. 76).

(a)   (b)

Fig. 27.11 Double-exposure method for velocity determination: (a) Double-exposure holo-gram of particulate matter of asbestos and tremolite in the respiratory range; the two expo-sures are 10 μs each and are 10 ms apart. (b) Image from the hologram of a 10-μm fiber. The fiber is in focus and the remainder of this group of matter is out of focus (after Ref. 108).

from a study on moving fibrous particulate matter in the respiratory range.[108] Figure 27.11(a) shows a portion of a double-exposure hologram of samples containing tremolite and asbestos; the two holograms were each made with a 10-μs exposure on the same recording area, 10 ms apart. Figure 27.11(b) shows an image of a 10-μm fiber in this field that is attached to a clump of material. In this image, only the fiber is in focus and clearly the clump is changing position as it falls and is moving at $15 \text{ cm} \cdot \text{s}^{-1}$.

The whole field of holographic studies of a wide variety of particulate-type problems is still extremely active and will continue to be so for some considerable time.

27.2.1.3   Image Storage and Holographic Memories

Since in a hologram the information about the field of interest is uniquely coded, it seems an interesting possibility to attempt to store multiple fields on the same holographic recording material. In a sense, this is an extension of the ideas of the multiple-exposure methods described in the last section, although historically it did not develop in that way. Leith and Upatnieks[21] illustrated this principle by recording two transparencies on the same holographic plate by placing the two transparencies at different distances from the plane in which the hologram was to be recorded. There are several variations on this theme; the holograms could be recorded with different angles for the reference beam or different orientations of the reference beam (for example, the hologram can be rotated between exposures). All these methods put down the various holograms on the same photographic recording material and essentially on top of each other. Obviously, this is but one implementation of optical multiplexing, which itself has had a long and interesting history. Mueller[109,110] reports on detailed nonholographic studies of multiple image storage of monochrome and color images that use spatial filtering techniques for separation.[c]

If amplitude holograms are being recorded, then the number of images that can be stored is limited by the dynamic range of the film and the density and noise buildup (a developed grain will scatter the incident radiation over the whole field and not just to the image that produced that particular grain). If phase recording is used, then considerably more images can be recorded in this way, but nevertheless, it is still a limited number.

The technology has moved away from overlapping holograms and has gone to the storage of small individual holograms arranged in a single holographic recording medium in an organized way. The advantage of holographic storage rather than merely storing small images is that the holographic recording is much less susceptible to the presence of local defects such as dust or scratches. It is the built-in redundancy in the holographic recording that achieves this, but only at the expense of higher resolution requirements on the recording medium.

The literature on holographic storage and memory devices is quite extensive. However, some review articles[111-113] provide an excellent source of detailed references. Memories that store analogue information, such as a series of images, seem to have been restricted to a limited set of applications that were described in the sections on displays and holographic movies. Current interest in optical memories is driven by the need for large computers and hence the information is more appropriately stored in a digital format.

Two types of memories have been identified by workers in the field to serve two different purposes. Memories that require a fast random access are called *block-organized* (also called *page-organized*) memories and require no moving mechanical parts. For slow access times, a different design philosophy can be used and the memory can be organized sequentially and mechanical components can be employed.

The concept in sequentially organized memories is to store the information in whatever holographic format is desired, one hologram at a time. Normally, the recording material is moved between exposures and the sequence of holograms laid down in a line format along the recording material. Two-dimensional formats can also be used. Sutherlin et al.[114] have developed a device called a *Holoscan,* which is a holographic read-only memory for credit verification. Individual holograms are recorded on 35-mm film; each memory measures 0.50×0.75 mm and produces as an image a readout that consists of 56 bits of data arranged in a 14×4 binary array. In an entirely different application, Nelson et al.[115] have combined in one film store both human-readable and machine-readable information. Figure 27.12 shows a schematic diagram of this system. The digital input is converted into a synthetic Fourier transform hologram (the

---

[c]See Chapter 34 for a discussion of modulated imagery.

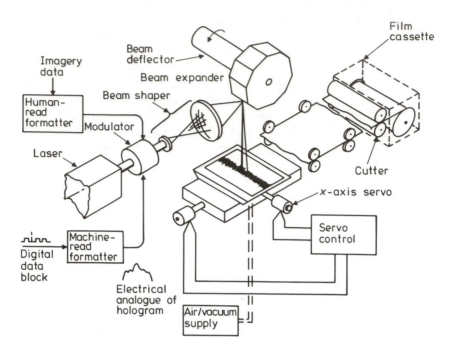

Fig. 27.12 Schematic diagram of a holographic data storage and retrieval system (after Ref. 115).

machine-readable part of the record), which is recorded on film by means of a modulator and laser scanner. This same scanning system can also be used to record images directly (the human-readable part of the record) at a $20\times$ reduction. In one mode of operation, five rows of 12 images can be recorded—this is a total of 2.5×10^6 bits, or a 4- $\times6$-in.^{2} microfiche format. In a separate mode, the entire area of the microfiche can be filled with digital information. The holographic method uses the same microfiche and puts down a series of synthetic Fourier transform holographic recordings, each 5 mm in diameter. The holograms are synthetic because they are not produced directly from an input object but are generated by writing with a modulated laser beam driven by a digital logic system. (This is but one application of many pieces of work using synthetic holograms generated by computers, often called *computer-generated holograms*. Some detailed discussion on computer-generated holograms can be found in a short review by Bryngdahl.[116]) The method described above has been developed into a complete system for the storage and retrieval of cartographic information, although it also has other potential applications. The advantages of the holographic system of recording are those discussed earlier, i.e., immunity to dust and scratches. In addition, several times greater reduction can be achieved over conventional microfilm storage; tolerances in the depth of focus are based both in recording and reading; and there is considerable tolerance in the lateral positioning of the hologram because of the invariant property of the Fourier transform hologram.

This and other applications of sequentially organized optical memories are used for archival storage where nonerasable storage media are appropriate.

The other approach to holographic storage is the block- or page-organized memory that uses no mechanically moving parts and has random access. In the

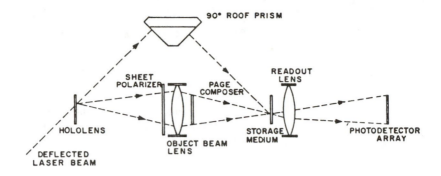

Fig. 27.13 Schematic diagram of an experimental block-organized read/write memory (after Ref. 117).

basic concept of such a memory, the holograms are arranged in a matrix so that it is possible in principle to erase specific information and rewrite; overlapping holograms would not allow this. The holograms are addressed by a laser beam directed to the appropriate hologram by an xy beam deflector; for example, an acousto-optic deflector. Each output falls on the same detector array for readout. The actual configuration of a system is, of course, more complicated than indicated by this brief statement, and a variety of designs have been considered (see, for example, Ref. 112). An experimental read/write memory, which has a capacity of 10^6 bits, has been described in some detail by Stewart et al.[117] This system is shown schematically in Fig. 27.13; looking down on the device, the system components include an argon-ion laser, a two-dimensional acousto-optic deflector, a liquid crystal page composer, a photothermoplastic recording material, and a silicon photodiode array. A polarized laser beam after deflection falls on the hololens, which consists of an array of permanent phase holograms that have been recorded so that one corresponds to each page or block of the memory. Light propagating from the hololens with the appropriate polarization falls on the page composer and hence onto the photothermoplastic storage medium, which is the image plane of the hololens. The hologram is thus written on a predetermined local area and the specific information is determined by the page composer, which is a liquid crystal light valve containing 2048 circular elements. The reference beam for the Fourier transform holographic recording is provided from the nondiffracted component of the original beam via the 90-deg roof prism. To read out the holograms, the plane of the polarization of the input illuminator is rotated through 90-deg and hence the reference beam illuminates the hologram to produce a virtual image, which is then relayed at unit magnification onto the photodetector array.

27.2.2   Nonoptical Holography

Holography was not developed as an optical technique but as a technique to improve electron microscopy. This aspect of the early days of holography was discussed in Sec. 27.1.1.1. Optical holography has, however, developed into the prime research and development area; most of the more significant developments have been in optical applications. This chapter would not be complete, however, without some comment on nonoptical methods of holography. These include predominantly acoustical holography, microwave holography, and x-ray and electron beam holography. The hologram is recorded with these radiations and then an optical image is produced from the holographic record.

27.2.2.1   Acoustical Holography

It is not possible to do complete justice to acoustical holography and those numerous workers in the field in the limited space available. For the reader who would like to obtain more detail, the information is readily available in reviews[118,119] and in detail in six volumes of international conference papers.[49-54]

Acoustic imaging and its applications is a relatively old art going back to the start of the century. Sonar systems are perhaps among the best known applications. Holography, then, fits within this body of scientific knowledge of technological applications. The same, perhaps, could have been said of optical holography except that there are very important differences. In acoustics, it is possible to record both the amplitude and phase of the acoustic field and so, in an important sense, holography is not required in the same way that it is with visible radiation. However, holography has played a very important role in revitalizing the area of acoustic imaging and bringing a new group of people into the field who consequently have approached the problem from a different viewpoint. The basic idea of acoustic holography is to form an optical image from an acoustic field produced by transmission or scattering of an acoustic beam falling on the object of interest.

The first hologram formed with sound waves appears to have been made by Greguss[120] using a sensitized photographic plate. The hologram was an interference pattern formed by two sound waves and as such is analogous to the conventional optical result described earlier. Clearly, to record a hologram similar to the holographic process described at optical frequencies means recording the intensity of the sound field produced by two interfering sound waves. One very interesting way to achieve this uses a surface deformation technique.[121] Figure 27.14 shows a schematic of this type of system. Two separate, but coherent, sources of ultrasound are used, one of which is used to illuminate (insonify) the object and the other to provide the required reference wave. Where the two beams of ultrasound overlap, an interference results which disturbs the surface of the water to produce a standing wave pattern. The water surface then acts as the "recording material" and the amplitude of the standing wave is proportional to the acoustic intensity at the surface. This surface deformation is a phase hologram that can be read out directly with a coherent light beam. Thus, in the schematic, light from a laser is used to illuminate the surface directly and an optical image of the object can be formed in real time. Variations on this basic method include placing a membrane on the liquid surface and covering this with a thin oil film. Systems based on this liquid surface deformation technique have been developed commercially by at least one company, particularly for medical diagnostics. (Hildebrand and Brenden[119] review the use of this technique and show some sample illustrations. More detailed discussions can be found in the proceedings of the international conferences referred to earlier.[49-54])

Other methods have been used for recording the intensity of the sound field. These include the use of radiation pressure on particles suspended in a liquid; the pressure can either result in a change in orientation of metal flakes that reflect the light or it can cause the particles to accumulate in the crests and troughs of the standing wave (see, for example, Cunningham and Quate[122]).

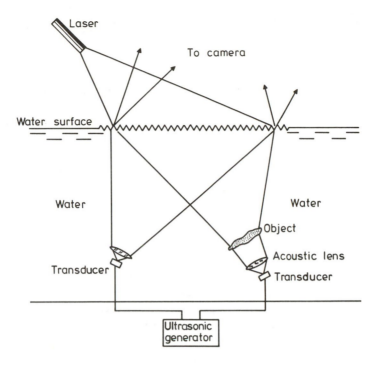

Fig. 27.14 Method of producing real-time images from an acoustic hologram using the surface deformation method (after Ref. 121).

An entirely different approach to detecting the sound field and turning it into an optical field relies essentially on modulating an optical beam with the complex amplitude of the sound wave. The first and perhaps least important method uses a linear surface interaction in which the surface deformation is proportional to the incident complex amplitude. This linear effect coexists with the usually stronger second-order effect that is proportional to the intensity. The linear effect is, however, larger than the second-order effect for solid/gas and liquid/solid interfaces and much smaller for liquid/liquid and liquid/gas interfaces. It is not appropriate to discuss the details of these effects here; Mueller[118] gives an excellent summary. This linear surface deformation is again read out directly with a laser beam, but because the surface deformations are small, heterodyne detection is often used.[123]

The second and important implementation of a linear effect is the so-called *Bragg diffraction method.[124]* When the sound passes through a liquid, a periodic variation of the index of refraction is produced. This variation in index is sensed by a coherent light beam passing through it since it acts very much like a three-dimensional phase grating. Thus, the familiar Bragg diffraction takes place and direct imagery can again be obtained. Figure 27.15 shows a schematic diagram of a Bragg imaging system. The object under test is immersed in the interactive cell and the laser beam passes through the cell at right angles to the sound wave. In this particular method,[125] the image is picked up by a vidicon and displayed on a monitor. It is a debatable point whether this is considered a holographic method since there is no reference wave involved. The complex amplitude of the sound wave is converted directly into a complex amplitude of a propagating optical wave.

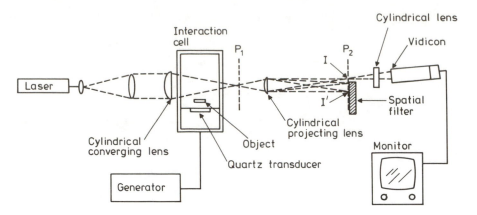

Fig. 27.15 Method of producing real-time images from the complex amplitude of a sound wave by the Bragg diffraction method (after Ref. 125).

Scanning methods in acoustic imaging have always been important and hence it is not surprising that they play a role in acoustical holography. The technique described in earlier paragraphs of this section used area detectors. All these techniques could, in fact, be used with a single detector that scans the deformed surface, for example. The laser beam can scan the surface or a single acoustic detector can detect the sound field. One of the first of these results was presented by Preston and Kreuzer.[126] It is important to note that holography is not really necessary in the conventional sense if the complex amplitude of the acoustic field is detected either directly or after some linear transducer has been employed. The linear surface deformation method has been adopted to be read out with a laser scanning system. Korpel and Desmares[127] and Auld et al.[128] have used a piezo-electric transducer. A scanning electron beam has also been used with a piezo-electric transducer.[129]

Acoustic imaging and acoustic holography provide a rich and fertile area of research and many interesting techniques exist. The dividing line between holography and normal imaging is fuzzy, but one thing is certain: holography has done a great deal for diagnostic methods using ultrasound even if a true ultrasonic hologram is never used. (In fact, of course, the intensity-sensitive surface deformation method described earlier is indeed a commercial product. Brendon[130] reviews the application of this process to medical diagnostics, characterization of geological features, industrial testing, and undersea imaging.)

27.2.2.2   X-Ray Holography

Some comments were made in Sec. 27.1.1.2 on attempts to build an x-ray microscope. The holographic approach to this problem encountered considerable difficulties.[14-17] Despite these difficulties, El-Sum and Kirkpatrick[131] did show that a photograph of an early record of an interference pattern produced with soft x rays by Kellstrom[132] could be used to form an optical image.

The hope that new sources of x rays with the correct combination of source size (so that the requisite spatial coherence can be achieved), spectral width (so that sufficient temporal coherence can be realized), and energy (so that short exposure times can be used) has spurred further effort in this research. The easiest type of hologram to record is the in-line Fraunhofer hologram. Giles[133] showed results on a 5.9-μm-diam glass fiber illuminated with radiation at 114 Å from a Be Kα source. Aoki and Kikuta[134] extended this method to two- and three-dimensional objects. They chose chemical fibers arranged in a three-dimensional sample. The fibers were 2 to 6 μm in diameter and were illuminated with a diverging 45-cm radius spherical wave of Al Kα radiation at 8.34 Å. The in-line hologram was recorded on a Sakura nuclear plate at a distance of 30 cm, and an exposure time of 80 min was required. Satisfactory results were obtained with the fibers and also with blood cells.

The use of a lensless Fourier transform hologram method was first suggested by Winthrop and Worthington[135] for x-ray holography. The advantage is that a lower resolution recording material is required, but the disadvantage is the need to provide a point reference source in the plane containing the object. Optical analogue experiments were carried out by Winthrop and Worthington[136] to illustrate the principle involved. Kikuta et al.[137] employed this method with an object consisting of three 2.8-μm-wide slits, 9 μm apart, that had been electro-formed in a 6-μm-thick nickel foil. A fourth slit in the same foil having a width of 2.5 μm, placed 30 μm from the nearest of the three slits, provides the required reference beam. The radiation was Cu Kα at 44.8 Å. Optical reconstructions using an He-Ne laser gave a recognizable image of the three slits. In a separate series of experiments, Aoki et al.[138] used the same object with monochromatic x rays at 60 Å with a half-width of 0.03 Å, which is considerably narrower than the Cu Kα radiation used in their previous experiments. These x rays were produced using synchrotron radiation and a spectroscope. A 1-h exposure was required on Fuji x-ray film. These experiments with a one-dimensional object showed that it could be done, but it is not easily modified for use with more general objects.

If a zone plate is used in the x-ray beam, one of the focused images of the source can be used for the reference beam and the remainder of the radiation from the zone plate used as the illuminating beam. This method was suggested by Röder et al.[139] and an optical simulation performed. Subsequently Reuter and Mahr[140] have attempted, with some limited success, to demonstrate the method using Cu Kα x rays at 4.48 nm. A zone plate was used that was made in gold foil 0.15 μm thick; the diameter of the central zone was 18.62 μm and the zone-plate focal length was 7.73 cm. Holograms were recorded on Kodak SWR film with 300-min exposures. Again, in these experiments, the combination of energy and coherence limited the performance.

27.2.2.3   *Electron Beam Holography*

Electron microscopy was the original driving force for the invention of holography. However, it has never materialized as an effective application despite considerable study in the early days of holography.[4] The progress made in electron microscopy has been such that there is very little interest or advantage to be gained in pursuing electron holography. High-quality electron optics are available and hence good images can be obtained. This is in considerable contrast to the x-ray situation where normal imaging methods are not available. There would be some advantages to electron holography if it could be carried out on actual objects without the need for making replicas. There are probably classes of objects where depth information is important. For example, Tonomura et al.[141] have used the in-line Fraunhofer holographic technique for electron beam holography; the electrons used had a wavelength of 0.037 Å. Opaque gold particles 100 Å in diameter were recorded and the image formed optically. In a second example, they produced an image of a small zinc oxide crystal.

27.2.2.4   Microwave Holography

The most famous microwave holograms were formed without the process being thought of as a holographic method. These holograms are those formed with side-looking, or synthetic aperture, radar. Doppler frequencies are generated as the aircraft, containing the microwave emitter, moves with respect to the ground, which results in a single point object on the ground producing an electrical signal in the form of a classical zone plate. This method was invented in the early 1950s but was not described in the literature until much later. Leith and Ingalls[142] discuss the optical processing portion of this method in terms of holographic processes in some detail and reviews by Leith[143,144] give an excellent overview of microwave analogies of holography as does the review by Tricoles and Farhat.[145]

The process of forming holograms at microwave frequencies was illustrated by Dooley[146] and by Stockman and Zarwyn[147] but little development was carried out with these techniques.

Farhat[148] has revitalized interest in microwave holography by examining scanning methods. The object is stationary and the transmitter and receiver are moved together or, conversely, the transmitter and receiver are fixed and the object is moved. The interest in this particular area of holography is for viewing objects that are transparent to microwaves but opaque to optical frequencies. In Farhat's experimental verification of the scanning method, the object is illuminated with a beam of coherent microwaves at 70 GHz, which is equivalent to a wavelength of 4 mm. The object was raster-scanned over a field 41.5×38 cm. The receiver detects the scattered field, which is mixed with a reference field from a local oscillator; this signal is then used to intensity-modulate a cathode-ray tube (CRT). A photographic record of the CRT display is the required hologram, which is then used to form an optical image. The resultant image shows better resolution than that obtained with the more conventional method in which only the receiver is scanned.[149]

27.3   HOLOGRAPHIC OPTICAL ELEMENTS

Thus far in this chapter, the hologram has been used to store information about some object field of interest so that an image can be formed. This image may be but one of several images used for a subsequent interferometric experiment. There was, however, one exception to that statement that was briefly described in Sec. 27.2.1.1. When attempting to image through a distorting glass or a diffuser, a hologram can be made of that distorting glass and the hologram used as a corrector plate since it produces a conjugate image to the distorting glass and hence cancels out the effect of the distortion.[91] Upatnieks et al.[92] in a related experiment, used a holographic corrector plate to balance out aberrations in an optical system. In these two examples, the hologram is being used in a different way—it is being used as an optical element in the system. Pastor[150] has used a synthetically generated hologram to replace a master reference surface for testing and controlling the manufacture of aspheric surfaces.

There are other ways in which a hologram can be used as an optical element. For example, the hologram of a point object is a zone lens and can be used as an imaging device in the same way that zone plates or conventional lenses are used. The holographic lens then merits some serious attention. The off-axis hologram of a point object can also be thought of as a grating with slightly varying periodicity. Hence, why not make gratings holographically?

The third type of holographic optical element is the complex filter for coherent optical processing applications that has contributed significantly to advances in that field since it solved a very difficult problem, as discussed in Chapter 33.

27.3.1   Lenses

The idea of a holographic lens (i.e., a zone lens or sine wave zone plate) was suggested by Kock[151] in 1966, although the idea was obviously apparent in many of Rogers' early papers.[152–155] The first demonstrated use of a hologram as an optical element appears to be in a paper by Schwar et al.[156] (We are considering lenses here; however, it is appropriate to mention that Denisyuk[19] had described a holographic mirror.) However, progress was not made until design procedures were available and ray tracing programs were developed, primarily by Latta.[157] A detailed understanding of the aberrations present in holograms of point objects was already available.[158,159] However, a general theory of aberrations was not available. Finally, in order to achieve a practical device, it was necessary to have high-efficiency recording materials in which to make the holographic lenses. Clearly, the phase holograms in materials like hardened dichromated gelatin are appropriate.

Close[160,161] has given an excellent review of the status of holographic lenses and this discussion is very much influenced by his writings. Close points out very correctly that: "In general, holographic optical elements will not replace conventional optical elements, but will be useful in special unique applications." Naturally, it should be borne in mind that the holographic lens is essentially monochromatic and that its properties vary considerably with a change in wavelength. Finally, the designer attempting to design a holographic lens system does not have the wealth of background of designs available to him that he would have if he were designing a conventional optical system.

One really successful use of holographic lenses has been in specialized display applications in which conventional optical systems could not be used. McCauley et al.[162] describe in considerable detail how optical elements can be formed in a dichromated gelatin layer on a variety of glass and plexiglass substrate shapes. These holograms were used for a helmut-mounted CRT display. The field of view of this device was ~5 deg; the limited field of view is, of course, a difficulty with holographic lenses. This problem has been partially overcome and a system has been built with a 30-deg field of view.[163] An extension[164] of this concept is head-up displays.[d]

27.3.2   Gratings

Research in holographic gratings, first described by Labeyrie and Flamand,[172] has led to fruitful commercial results and hence a variety of plane and concave gratings are now available. Up to 10,000 grooves per millimeter can be obtained with widths of 60 cm. The basic method of producing the gratings is, in principle, quite straightforward. A layer of photoresist or other photosensitive layer is coated onto an optical quality substrate; the required fringes are produced

---

[d] Holographic optical elements have been used for laser scanning applications such as supermarket bar code readers,[165,166] nonimpact printers,[167] and diode laser scanners.[168] In addition, HOEs have been used for multiple image generation,[169] narrow spectral wavelength filters,[161] and wavelength shifting applications.[170,171]

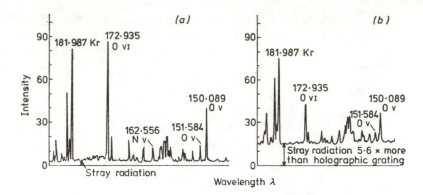

Fig. 27.16 A comparison of the use of (a) a holographically constructed concave grating and (b) an approximately similar conventional concave grating at grazing incidence (after Ref. 175).

interferometrically (i.e., as a hologram). After development, the relief surface is vacuum deposited with a reflective and a protective coating. The grating profile produced by this process is naturally sinusoidal. Square wave profiles can be obtained if the photosensitive layer is thin compared to the spacing of the incident interference fringes. Asymmetrical groove profiles can be obtained by a variety of methods including tilting the grating blank at an angle to the incident fringe pattern to produce a triangular profile (see, for example, Ref. 173). A Fourier synthesis method uses a multiple-beam interference technique to obtain controlled asymmetric groove profiles (see Ref. 174 for a further discussion).

The advantages of holographic gratings appear to lie in their efficiency and accuracy and the low level of scattered light, certainly below that of even the very best classically ruled gratings.

Figure 27.16 illustrates the comparison of the performance of a holographically ruled grating with an approximately similar conventional grating.[175] The gratings were concave with a radius of curvature of approximately 2 m and were used in grazing incidence; the holographic grating had 1800 grooves per millimeter and the conventional grating had 1200 grooves per millimeter. The point to notice is the stray radiation signal in the two cases—that from the holographic grating is much lower.

It is clear that the technology is here to stay and progress will continue to be made in this field and new types of analytical instruments may well be designed because of developments in holographic gratings (see, for example, Ref. 176).

27.3.3   Filters[c]

Holography has played an extremely important technological role in coherent optical processing. The basic method in coherent optical processing involves producing the Fourier transform of an input complex amplitude distribution by the analogue method (Fraunhofer diffraction as produced by a lens). This Fourier transform complex amplitude distribution is then multiplied by an appropriate filter function; the resultant complex amplitude is then retransformed to produce an "image" of the original input field. This image is, of course, a modified version of the input, modified in some predetermined and desired way by the filter function. Many linear processing methods can be designed to be

---

[c]Examples illustrating analog spatial filtering are discussed in Chapters 29 through 33.

carried out with this technique. The earliest of these methods was actually invented before the basic principles of coherent optical processing were expounded by Maréchal and Croce[177]—for example, the Foucault knife edge test, Schlieren systems, and the phase contrast microscope. Other applications include contrast enhancement, signal enhancement in the presence of additive noise, raster and half-tone removal, image replication, aberration balancing, matched and inverse filtering, etc. A discussion of these applications and the specifics of the methods employed are clearly beyond the scope of this chapter. However, it is probably sufficient to state that many of these applications require that the filter function be complex, i.e., that the function modify the phase and amplitude of the complex field in the Fourier plane in a continuous way. The amplitude portion of the filter can be fabricated in a variety of ways. However, the fabrication of the continuous phase variations is an entirely different matter. Furthermore, the required filter function has to be calculated first, which, while obviously possible, is still a restriction. Leith[178] has reviewed some of the history of the development of complex filters.

Vander Lugt[26] recognized that a matched filter could be fabricated as a Fourier transform hologram. A matched filter is one that will produce a delta function output in the coherent optical processing system. Thus, the filter is essentially the inverse of the Fourier transform of the input. If the object for which a matched filter has to be fabricated is the input to a coherent optical Fourier transform system, then the transform can be converted into a Fourier transform hologram by the addition of a separate off-axis reference beam. This record is then the required filter. The necessity of calculating the actual filter is hence removed and no continuous phase filter has to be fabricated since the hologram contains the necessary phase variations. If a new input is now chosen that contains in its field the object of interest, that object will be selected by the filter and a delta function output produced in one of the two resulting adjacent image planes. Thus, a cross correlation has been performed. This method has been used by Gara[179] for the real-time optical correlation of three-dimensional scenes. This is achieved with the use of a liquid crystal device acting as an incoherent-to-coherent converter so that the object may be illuminated incoherently and then imaged onto the converter. This converter is then read out coherently and provides the input to the coherent processor. A holographic filter is used to perform the necessary correlation. Figure 27.17 illustrates one of

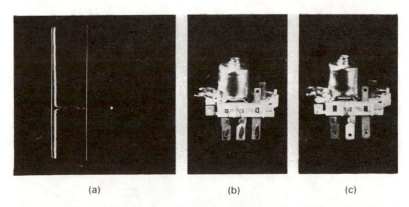

(a)   (b)   (c)

Fig. 27.17 Application of a matched holographic filter to perform a cross-correlation: (a) the correlation peak and its profile; the cross-correlation is between an electromechanical relay (b) with a similar relay (c) from which the filter is made (after Ref. 179).

Gara's results. A matched filter is made from one electromechanical relay shown in (c) in reflected light. This matched filter is then used to perform a cross correlation on a second relay (b). The output of the coherent processing part of the system is shown in (a). The optical output is sensed with a vidicon camera and then displayed on a television monitor; the cross-correlation peak and a profile of the peak can be shown simultaneously. The advances made in new devices such as the various incoherent-to-coherent light converters make hybrid optical processing systems an important reality (for a review, see Ref. 180).[f]

One of the very challenging problems in image enhancement is the correction of aberrated images. An incoherently recorded image is, of course, the result of a convolution of the actual object intensity distribution with the intensity impulse response of the aberrated imaging system. The necessary processing step requires, then, a deconvolution. Both digital and optical methods have been used for this purpose with some success. The fundamental step is to produce the Fourier transform of the recorded aberrated image, which is then the product of the transforms of the two functions that were originally convoluted. This transform is thus a product of the spatial frequency spectrum of the object intensity distribution and the transfer function of the system that recorded the image (i.e., the Fourier transform of the intensity impulse response of the aberrated system). A filter can then be used to attempt to divide out the effect of the transfer function. Naturally, this can only be achieved when the transfer function has a finite value above the noise level. Fortunately for early workers in the field, many optical systems have symmetric intensity impulse response functions and hence the transfer functions are real. Thus, the phase portions of the filter are only regions of constant 0 or π phase. In general, however, the filter function required is complex and hence, again, difficult to make optically. Holographic filters can once more provide an important solution. Following Vander Lugt's successful introduction of the holographic filter, Stroke and Zeck[185] applied the concept of the holographic filter to the deconvolution process discussed above. Stroke and his colleagues have pursued this area with some vigor and some success (see Ref. 186 for a review). Normally it is necessary to know the impulse response of the aberrated system, but this can often be inferred from the photographic record.

Image processing seems to have a useful future mainly because of the strong possibility that hybrid real-time systems will be realized. Holographic filters will play an integral role in this development.

27.4   CONCLUSIONS

The literature on holography is so large that it has been extremely difficult to review the entire spectrum of activity in this single chapter. Future reviews may have to treat individual applications or groups of applications and techniques separately. Naturally, a number of topics and groups of workers are not represented in this chapter but some choice had to be made. No discussion on computer-generated holograms or kinoforms is included; neither is the application of holographic spectroscopy in which spectral data are recorded holographically (see, for example, Ref. 187). Holography appears to be alive and healthy at the present time with workers in research and development settling

[f]More recently, it has been shown that digital phase-only holograms can be used to improve the signal-to-signal ratio of the correlation peak resulting from the use of Vander Lugt filters.[181-184]

down to the hard task of working on realistic problems that truly require solution.

27.5   APPENDIX: MISCELLANEOUS TERMINOLOGY

27.5.1   Introduction

There are several specialized pieces of terminology in holography that have become fairly standard and should be mentioned. The first of these is the 360-deg hologram,[58,188] which, as the name implies, is a hologram made so that a complete 360-deg view of the object is obtained. This type of hologram is usually recorded on a cylinder of film and makes an excellent and indeed dramatic illustration of the principles of holography.

Figure 27.18 shows the method used by Jeong[58] to produce cylindrical (360-deg) holograms. The light from a laser is focused by a high-power microscope objective so that part of the beam illuminates the cylindrical film directly and the remainder illuminates the object. The light scattered by the object also illuminates the film. When this cylindrical hologram is reilluminated, the image may be viewed from any direction.

Local reference beam holography refers to a method of forming the hologram where a portion of the light from the object is focused onto a small aperture and then the light from this aperture is used as a source for a reference wave.[189,190] As a specific method in holography it has had some application (for example, the study of laser-beam parameters[191]).

The term *incoherent hologram* appears to be a contradiction in terms since a hologram is an interference pattern and hence a coherent process. The terminology has been used in different ways in the literature. A hologram can, of course, be produced with light from an incoherent source since as the light propagates it gets more coherent. Synthetic holograms can be made if each point on an incoherent object can be made to produce its own zone-plate or zone-lens structure.[192] Young showed how this might be used for astronomy, but the subject has really been studied in detail for tomography and the term *coded aperture imaging* is usually used (for recent reviews, see Refs. 193 and 194).

A more direct way of producing a hologram of an incoherent object is to overlay two images of the object and, since there is point-to-point coherence between the two images, an interference will result.[195,196] The image is produced by illuminating the hologram with a coherent beam in the usual way. One paper[197] does discuss the "decoding of a noncoherent hologram using noncoherent light."

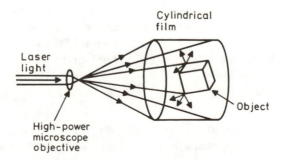

Fig. 27.18 Formation of a cylindrical (360-deg) hologram (after Ref. 58).

Courjon and Bulabois[198] have invented a different kind of incoherent process that uses an apodized aperture technique to modify the impulse response function of an incoherent system so that the photographic record of an input contains terms equivalent to those found in a Fourier transform hologram. Images are then formed coherently from this "incoherent hologram."

Benton[199] introduced a useful technique into holography; the holograms produced by this method have since been called *rainbow holograms* (see, for example, Refs. 143 and 144). In Benton's method, a hologram is first recorded in the usual way, a slit is then placed across the hologram, and a second hologram made from the real image produced from the first hologram. When this second hologram is viewed, the image is only seen if the eye is in the correct position. If the eye is placed at the location of the image of the slit, then the entire image will be seen but there is no parallax in the direction perpendicular to the length of the slit. This type of hologram can be used with white light and then, because of dispersion, the image of the slit is at different locations for different colors and as the eye moves in the direction perpendicular to the length of the slit, the image changes color.

27.5.2   Principal Properties of Holographic Systems[g]

27.5.2.1   *Resolution*

The resolution is determined in the recording step. If the reconstruction process limits the resolution, then something is seriously wrong with the technique. The major factors affecting resolution are the resolution and dismensions of the recording material, dimensions of the beam illuminating the hologram plane, and the temporal and spatial coherence of the illumination. Normally, the coherence is not a limitation in a well-designed system. Essentially, the size of the hologram is always the limiting factor, since that size will indeed be limited by the various factors listed above. This effective size then determines the resolution of the process. To a first approximation, the resolution is determined by the width of the impulse response function of the process, which in turn is directly determined by the effective aperture (size) of the hologram. Let the extent of the hologram be $2x$; then the half-width of the impulse response is $1.22z\lambda/2x$, where z is the distance at which the hologram is made. The resolution of the recording materials is, of course, a much more important factor in the off-axis and Fourier transform holograms.

27.5.2.2   *Field of View*

The field of view is also determined by the recording process. The extent of the hologram, regardless of whether it is limited by the format of the recording material, the spatial extent of the beam, the resolution of the recording material, or the spatial or temporal coherence of the illuminating beam, determines the field of view. The extent of the hologram acts as a window through which the object is viewed.

27.5.2.3   *Efficiency*

Diffraction efficiency is defined as the ratio of the power diffracted into one first-order wave to the power illuminating the hologram. For uniform illumina-

---

[g]Some of these properties were discussed in more detail in terms of the space bandwidth product in Chapter 25.

tion, the definition can be rewritten with intensity replacing power. When the holographic record is a normal photographic film and if linear recording is assumed, then the incident light will be divided into the three beams. The maximum amount of light associated with the image would be about 6% of the incident beam; however, in practice, it would be much less. Higher efficiency, up to 10%, can be obtained if linear recording is sacrificed, but then multiple images are formed.

Increased efficiency can be obtained by converting the density record into a phase record and thus producing a phase hologram. This can be achieved by bleaching the silver halide film. Various procedures have been given in the literature for this bleaching process.[200-202] The maximum efficiency that can be obtained is ~34% but, of course, this is not reached in practice.

Phase holograms can also be produced with other types of recording materials. Hardened dichromated gelatin has been used successfully.[203-205] The list of materials that have been used is quite lengthy, and it is not appropriate to discuss them here. A number of reviews do exist in the literature.[201,206,207]

Many of the comments made so far on efficiency have been concerned with thin holograms. However, there is considerable work on thick holograms and particularly thick-phase holograms since the latter can give diffraction efficiencies approaching 100%. For a detailed discussion of thick or volume holograms, see Ref. 35.

One important aspect of thick holograms is to obtain good full-color recording without the cross talk that occurs in simple thin holograms.[208] In addition, so-called *white-light* holograms can be produced[209] that use the Lippman-Bragg effect. The interference fringes that make up the thick hologram act as a filter to select the correct wavelength from the white-light illumination. More recently, holograms have been made and reconstructed with broadband white light by carefully equalizing the optical path lengths in the interferometer.[210-214]

27.6   APPENDIX: INTERFERENCE MICROSCOPY

The basic ideas of the application of holography to microscopy were discussed in Sec. 27.2.1.2. It is not surprising that these ideas were extended to the development of an interference microscope. The study of phase objects has always been an important challenge that, despite outstanding achievements, continues to attract the attention of active researchers. The holographic microscope developed by Van Ligten and Osterberg[95] was modified and the technique extended by Ellis[215] to illustrate phase contrast, dark field, and interference microscopy. A very significant advance was reported by Snow and Vandewarker[216] who developed a holographic interference microscope that alleviates some of the problems associated with conventional interference microscopes. In the conventional systems, very careful matching of the two optical paths and components is required. A schematic of the Snow and Vandewarker microscope for use with transmitted light is shown in Fig. 27.19. To use the system, a hologram is recorded with a blank microscope slide on the stage and the developed hologram replaced in position. An object of interest is then placed on the stage and an interference pattern occurs between the image of the blank slide and the image of the object. Thus, both beams pass through the same optics and hence a second set of matching optics is not required. Once the hologram of the blank slide is made, it becomes an integral permanent part of the system; a hologram must be made for each objective that is going to be used. Figure 27.20 illustrates the result of using

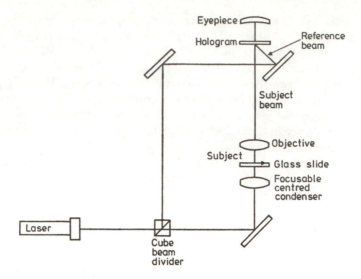

Fig. 27.19 Interference microscope for use in transmitted light (after Ref. 216).

this microscope on an etched glass specimen; in this particular example, the interference is a horizontal section interference.

Some variations in the use of the system are possible. For example, oblique section interference can be obtained by decentering the substage condenser. The process described in the previous paragraph can be reversed; a hologram of the object is made first, repositioned, and then a blank slide placed in position. Finally, it is possible to record both holograms in the same plate before viewing. Differential interference and total image doubling interference can also be achieved.

Fig. 27.20 Horizontal section interference of a sample of etched glass using the holographic microscope (after Ref. 216).

REFERENCES

1. D. Gabor, "A new microscopic principle," *Nature,* 161(4098), 777–779 (1948).
2. D. Gabor, "Microscopy by reconstructed wavefronts," *Proc. Royal Soc. A.* 197(1051), 454–487 (1949).
3. D. Gabor, "Microscopy by reconstructed wavefronts: II," *Proc. Phys. Soc. B* 64(part 6), 449–469 (1951).
4. M. E. Haine and T. Mulvey, "The formation of the diffraction image with electrons in the Gabor diffraction microscope," *J. Opt. Soc. Am.* 42(10), 763–773 (1952).
5. W. L. Bragg and G. L. Rogers, "Elimination of the unwanted image in diffraction microscopy," *Nature* 167(4240), 190–191 (1951).
6. P. Kirkpatrick and H. M. A. El-Sum, "Image formation by reconstructed wave fronts," *J. Opt. Soc. Am.* 46(10), 825–831 (1956).
7. E. N. Leith and J. Upatnieks, "Reconstructed wavefronts and communication theory," *J. Opt. Soc. Am.* 52(10), 1123–1130 (1962).
8. W. L. Bragg, "The x-ray microscope," *Nature* 149(3782), 470–471 (1942).
9. M. J. Buerger, "Generalized microscopy and the two-wave length microscope," *J. Appl. Phys.* 21(9), 909–916 (1950).
10. A. W. Hanson and H. Lipson, "Optical methods in x-ray analysis III. Fourier synthesis by optical interference," *Acta Crystallogr.* 5(3), 302–366 (1952).
11. B. D. Dunkerley and H. Lipson, "A simple version of Bragg's x-ray microscope," *Nature* 176(4471), 81–82 (1955).
12. G. Harburn and C. A. Taylor, "Three-dimensional optical transforms," *Proc. Royal Soc. A* 264(1318), 339–354 (1961).
13. G. W. Stroke, "A new assessment of optical and digital image processing for real-world applications," *Int. Opt. Comput. Conf. Digest,* pp. 1–9, IEEE, New York (1976).
14. A. V. Baez, "A study in diffraction microscopy with special reference to x rays," *J. Opt. Soc. Am.* 42(10), 756–762 (1952).
15. A V. Baez, "Resolving power in diffraction microscopy with special reference to x rays," *Nature* 169(4310), 963–964 (1952).
16. M. E. Haine and T. Mulvey, "Diffraction microscopy with x rays," *Nature* 170(4318), 202 (1952).
17. P. Kirkpatrick and H. H. Pattee, "X-ray microscopy," *Handb. des Phys.* 30, pp. 305-336, Springer Verlag (1957).
18. E. N. Leith and J. Upatnieks, "Wavefront reconstruction with continuous-tone objects," *J. Opt. Soc. Am.* 53(12), 1377–1381 (1963).
19. Y. N. Denisyuk, "Photographic reconstruction of the optical properties of an object in its own scattered radiation field," *Sov. Phys.-Dokl.* 7(6), 543–545 (1962).
20. Y. N. Denisyuk, "On the reproduction of the optical properties of an object by the wave field of its scattered radiation," *Opt. Spectrosc.* 15(4), 279–284 (1964).
21. E. N. Leith and J. Upatnieks, "Wavefront reconstruction with diffused illumination and three-dimensional objects," *J. Opt. Soc. Am.* 54(11), 12951301 (1964).
22. G. W. Stroke and D. G. Falconer, "Attainment of high resolutions in wavefront reconstruction imaging," *Phys. Lett.* 13(4), 306–309 (1964).
23. B. A. Silverman, B. J. Thompson, and J. H. Ward, "A laser fog disdrometer," *J. Appl. Meteorol.* 3(6), 792–801 (1964).
24. B. J. Thompson, "Diffraction by opaque and transparent particles," *J. Soc. Photo-Opt. Inst. Eng.* 2(2), 43–46 (1963).
25. G. B. Parrent and B. J. Thompson, "On the Fraunhofer (far field) diffraction patterns of opaque and transparent objects with coherent background," *Optica Acta* 11(3), 183–193 (1964).
26. A. Vander Lugt, "Signal detection by complex spatial filtering," *IEEE Trans. Inf. Theory* IT-10(1), 139–145 (1964).
27. S. Tanaka and T. Nakajima, *Bibliography on Holography,* Japan Optical Engineering Research Association, Tokyo (1968).
28. G. W. Stroke, *An Introduction to Coherent Optics and Holography,* 2nd ed., Academic Press, New York (1969).
29. J. B. DeVelis and G. O. Reynolds, *Theory and Applications of Holography,* Addison-Wesley Publishing Co., Reading, Mass. (1967).

30. H. M. Smith, *Principles of Holography,* 2nd ed., John Wiley and Sons, Inc., New York (1975).
31. W. E. Kock, *Lasers and Holography: An Introduction to Coherent Optics,* Doubleday Publishing Co., New York (1971).
32. H. J. Caulfield and S. Lu, *The Applications of Holography,* John Wiley and Sons, Inc., New York (1970).
33. M. Lehmann, *Holography: Technique and Practice,* Focal Press, Ltd., London (1970).
34. J.-C. Vienot, P. Smigielski, and H. Royer, *Holographie Optique, Developpements-Applications,* Dunod, Paris (1971).
35. R. J. Collier, C. B. Burckhardt, and L. H. Lin, *Optical Holography,* Academic Press, New York (1971).
36. J. N. Butters, *Holography and Its Technology,* Peter Peregrinus, Ltd., London (1972).
37. W. T. Cathey, *Optical Information Processing and Holography,* John Wiley and Sons, Inc., New York (1974).
38. N. H. Farhat, *Advances in Holography,* Vols. 1-3, Marcel Dekker, New York (1975/76).
39. *Holography,* B. G. Ponseggi, B. J. Thompson, eds., Proc. SPIE 15 (1968).
40. E. R. Robertson and J. M. Harvey, *The Engineering Uses of Holography,* Cambridge University Press, Cambridge (1970).
41. J.-C. Vienot, J. Bulabois, and J. Pasteur, *Applications of Holography,* Université de Besançon, Besançon, France (1970).
42. *Applications of Holography,* G. W. Stroke, ed., Plenum Press, New York (1971).
43. *Optical and Acoustical Holography,* E. Camatini, ed., Plenum Press, New York (1972).
44. *Developments in Holography,* B. J. Thompson, J. B. DeVelis, eds., Proc. SPIE 25 (1971).
45. *Proc. Engineering Applications of Holography Symp.,* R. Aprahamian, ed., February 16-17, 1972, Los Angeles, California, SPIE, Bellingham, Wash. (1972).
46. *Holographic Nondestructive Testing,* R. K. Erf, ed., Academic Press, New York (1974).
47. *Ultrasonic Imaging and Holography—Medical, Sonar, and Optical Applications,* G. W. Stroke, W. E. Kock, Y. Kikuchi, and J. Tsujiuchi, eds., Plenum Press, New York (1974).
48. *Holography in Medicine,* P. Greguss, ed., IPC Science and Technology Press, Ltd., Guildford, UK (1976).
49. *Acoustical Holography,* Vol. 1, A. F. Metherell, H. M. A. El-Sum, and L. Larmore, eds., Plenum Press, New York (1969).
50. *Acoustical Holography,* Vol. 2, A. F. Metherell and L. Larmore, eds., Plenum Press, New York (1970).
51. *Acoustical Holography,* Vol. 3, A. F. Metherell, ed., Plenum Press, New York (1971).
52. *Acoustical Holography,* Vol. 4, G. Wade, ed., Plenum Press, New York (1972).
53. *Acoustical Holography,* Vol. 5, P. S. Green, ed., Plenum Press, New York (1974).
54. *Acoustical Holography,* Vol. 6, N. Booth, ed., Plenum Press, New York (1975).
55. S. A. Benton, "Holographic Displays—A Review," *Opt. Eng.* 14(5), 402–407 (1975).
56. H. J. Caulfield, "The wonder of holography," *National Geographic* 165(3), 364–377 (1984).
57. M. Benyon, "Holography as an art medium," *Leonardo* 6(1), 1–9 (1973).
58. T. H. Jeong, "Cylindrical holography and some proposed applications," *J. Opt. Soc. Am.* 57(11), 1396–1398 (1967)
59. J. F. Asmus et al., "Holography in the conservation of statuary," *Stud. Conserv.* 18(2), 49–63 (1973).
60. E. N. Leith, D. B. Brumm, and S. S. H. Hsiao, "Holographic cinematography," *Appl. Opt.* 11(9), 2016–2023 (1972).
61. D. J. DeBitetto, "A front-lighted 3-D holographic movie," *Appl. Opt.* 9(2), 498–499 (1970).
62. R. Bartolini, W. Hannan, D. Karlsons, and M. Lurie, "Embossed hologram motion pictures for television playback," *Appl. Opt.* 9(10), 2283–2290 (1970).

63. B. R. Clay and D. A. Gore, "Holographic moving map display," *Opt. Eng.* 13(5), 435–439 (1974).

64. M. T. Gale, "Sinusoidal relief gratings for zero-order reconstruction of black-and-white images," *Opt. Commun.* 18(3), 292–297 (1976).

65. K. Knop, "Color pictures using the zero diffraction order of phase grating structures," *Opt. Commun.* 18(3), 298–303 (1976).

66. J. C. Palais, J. M. Watson, and S. A. Morrison, "Compact holographic storage and projection of two-dimensional movies," *Opt. Eng.* 15(2), 173–179 (1976).

67. R. E. Brooks, L. O. Heflinger, R. F. Wuerker, and R. A. Briones, "Holographic photography of high-speed phenomena with conventional and Q-switched ruby lasers," *Appl. Phys. Lett.* 7(4), 92–94 (1965).

68. *Kurzzeitphotographie, 7th Int. Kongr. für Kurzzeitphotographie,* O. Helwick, ed., Zurich, Switzerland, 1965, Springer Verlag, Darmstadt, West Germany (1967).

69. T. E. Holland and J. K. Landre, p. 105 in *Kurzzeitphotographie, 7th Int. Kongr. für Kurzzeitphotographie,* O. Helwick, ed., Zurich, Switzerland, 1965, Springer Verlag, Darmstadt, West Germany (1967).

70. T. Ban and P. Greguss, p. 426 in *Kurzzeitphotographie, 7th Int. Kongr. fur Kurzzeitphotographie,* O. Helwick, ed., Zurich, Switzerland, 1965, Springer Verlag, Darmstadt, West Germany (1967).

71. *Proc. 11th Int. Congress on High-Speed Photography,* P. J. Rolls, ed., Chapman and Hall, London (1975).

72. *High Speed Photography,* N. R. Nilsson and L. Hogberg, eds., John Wiley and Sons, Inc., New York (1968).

73. M. A. Lowe in *Proc. 9th Int. Congress on High-Speed Photography,* W. G. Hyzer and W. G. Chace, eds., SMPTE, p. 25 (1970).

74. K. J. Ebeling and W. Lauterborn, "High speed holocinematography using spatial multiplexing for image separation," *Opt. Commun.* 21(1), 67–71 (1977).

75. L. H. Tanner, "Some applications of holography in fluid mechanics," *J. Sci. Instrum.* 43(1), 81–85 (1966).

76. J. D. Trolinger, "Flow visualization holography," *Opt. Eng.* 14(5), 470–481 (1975).

77. D. A. Ansley and L. Siebert, "Coherent pulse laser holography," in *Holography,* B. G. Ponseggi, B. J. Thompson, eds., Proc. SPIE 15, 127–132 (1968).

78. N. Balasubramanian, "Holographic applications in photogrammetry," *Opt. Eng.* 14(5), 448–452 (1975).

79. E. M. Mikhail, "Holograms and holographic stereomodels: their mensuration and mapping," in *Coherent Optics in Mapping,* N. Balasubramanian, R. D. Leighty, eds., Proc. SPIE 45, 179–190 (1974).

80. R. V. Pole, "3-D imagery and holograms of objects illuminated in white light," *Appl. Phys. Lett.* 10(1), 20–22 (1967).

81. J. T. McCrickerd and N. George, "Holographic stereogram from sequential component photographs," *Appl. Phys. Lett.* 12(1), 10–12 (1968).

82. J. D. Redman, "The three-dimensional reconstruction of people and outdoor scenes using holographic multiplexing," in *Holography,* B. G. Ponseggi, B. J. Thompson, eds., Proc. SPIE 15, 117–122 (1968).

83. D. J. DeBitetto, "Holographic panoramic stereograms synthesized from white light recordings," *Appl. Opt.* 8(8), 1740–1741 (1969).

84. S. A. Benton in *Holography in Medicine,* P. Greguss, ed., pp. 69–77, IPC Science and Technology Press, Guildford, UK (1975).

85. H. Kogelnik, "Holographic image projection through inhomogeneous media," *Bell Syst. Tech. J.* 44(10), 2451–2455 (1965).

86. E. N. Leith and J. Upatnieks, "Holographic imagery through diffusing media," *J. Opt. Soc. Am.* 56(4), 523 (1966).

87. J. W. Goodman, W. H. Huntley, Jr., D. W. Jackson, and M. Lehmann, "Wavefront-reconstruction imaging through random media," *Appl. Phys. Lett.* 8(12), 311–313 (1966).

88. J. D. Gaskill, "Imaging through a randomly inhomogeneous medium by wavefront reconstruction," *J. Opt. Soc. Am.* 58(5), 600–608 (1968).

89. J. W. Goodman, D. W. Jackson, M. Lehmann, and J. Knotts, "Experiments in long-distance holographic imagery," *Appl. Opt.* 8(8), 1581–1586 (1969).

90. G. W. Stroke, "Image deblurring and aperture synthesis using *a posteriori* processing by Fourier-transform holography," *Optica Acta* 16(4), 401–422 (1969).

91. H. Kogelnik and K. S. Pennington, "Holographic imaging through a random medium," *J. Opt. Soc. Am.* 58(2), 273–274 (1968).
92. J. Upatnieks, A. Vander Lugt, and E. N. Leith, "Correction of lens aberrations by means of holograms," *Appl. Opt.* 5(4), 589–593 (1966).
93. A. E. Labeyrie, "High resolution techniques in optical astronomy," *Progress in Optics,* E. Wolf, ed., Vol. 14, pp. 47–87, North Holland Publishing Co., Amsterdam (1976).
94. E. N. Leith and J. Upatnieks, "Microscopy by wavefront reconstruction," *J. Opt. Soc. Am.* 55(5), 569–570 (1965).
95. R. F. Van Ligten and H. Osterberg, "Holographic microscopy," *Nature* 211(5046), 282 (1966).
96. R. F. Van Ligten, "A holographic microscope," *J. Opt. Soc. Am.* 57(4), 564A (1967).
97. R. F. Van Ligten, "Holographic microscopy," in *Holography,* B. G. Ponseggi, B. J. Thompson, eds., Proc. SPIE 15, 75–96 (1968).
98. W. H. Carter, P. D. Engeling, and A. A. Dougal, "Polarization selection for reconstructed wavefronts and application to polarizing microholography," *IEEE J. Quantum Electron.* JQE-2(2), 44–46 (1966).
99. W. H. Carter and A. A. Dougal, "Field range and resolution in holography," *J. Opt Soc. Am.* 56(12), 1754–1759 (1966).
100. M. B. Rhodes, "Holographic interferometric microscopy of polymer crystallization," *Appl. Opt.* 13(10), 2263–2267 (1974).
101. R. F. Cournoyer, M. B. Rhodes, and S. Siggia, "Application of holographic interferometric microscopy of the study of diffusion in the polyvinylpyrrolidone-iodine system," *J. Polymer Sci.* 13(5), 1023–1032 (1975).
102. R. F. Cournoyer, M. B. Rhodes, and S. Siggia, "Holographic interferometric microscopy," *Analyt. Chem.* 48(14), 2253–2258 (1976).
103. M. E. Cox, "*In vivo* applications of holographic microscopy," *Opt. Eng.* 14(3), 206–207 (1975).
104. B. J. Thompson, "Holographic particle sizing techniques," *J. Phys. E: Sci. Instrum.* 7(10), 781-788 (1974).
105. G. A. Tyler and B. J. Thompson, "Fraunhofer holography applied to particle size analysis: a reassessment," *Optica Acta* 23(9), 685– 700 (1976).
106. J. D. Trolinger, "Particle field holography," *Opt. Eng.* 14(5), 383–392 (1975).
107. B. J. Thompson, G. B. Parrent, J. M. Ward, and B. Justh, "A readout technique for the laser fog disdrometer," *J. Appl. Meteorol.* 5(3), 343–348 (1966).
108. E. A. Boettner and B. J. Thompson, "Multiple exposure holography of moving fibrous particulate matter in the respiratory range," *Opt. Eng.* 12(2), 56–59 (1973).
109. P. F. Mueller, "Linear multiple image storage," *Appl. Opt.* 8(2), 267–273 (1969).
110. P. F. Mueller, "Color image retrieval from monochrome transparencies," *Appl. Opt.* 8(10), 2051–2057 (1969).
111. G. R. Knight, "Holographic memories," *Opt. Eng.* 14(5), 453–459 (1975).
112. B. Hill, "Holographic memories and their future," in *Advances in Holography,* N. H. Farhat ed., Vol. 3, pp. 1–251, Marcel Dekker, New York (1976).
113. *Optical Mass Data Storage,* R. A. Sprague, ed., Proc. SPIE 529 (1985).
114. K. K. Sutherlin, J. P. Lauer, and R. W. Olenick, "Holoscan: a commercial holographic ROM," *Appl. Opt.* 13(6), 1345–1354 (1974).
115. R. H. Nelson, A. Vander Lugt, and R. G. Zech, "Holographic data storage and retrieval," *Opt. Eng.* 13(5), 429–434 (1974).
116. O. Bryngdahl, "Computer-generated holograms as generalized optical components," *Opt. Eng.* 14(5), 426–435 (1975).
117. W. C. Stewart et al., "An experimental read-write holographic memory," *RCA Rev.* 34(1), 3–44 (1973).
118. R. K. Mueller in *Advances in Holography,* N. H. Farhat, ed., Vol. 1, pp. 1-170, Marcel Dekker, New York (1975).
119. B. P. Hildebrand and B. B. Brenden, *An Introduction to Acoustical Holography,* Plenum Press, New York (1972).
120. P. Greguss, "Ultraschall hologramme," *Res. Film* 5, 330–337 (1965).
121. R. K. Mueller and N. K. Sheridan, "Sound holograms and optical reconstruction," *Appl. Phys. Lett.* 9(9), 328–329 (1966).

122. J. A. Cunningham and C. F. Quate, "Acoustic inteference in solids and holographic imaging," in *Acoustical Holography,* G. Wade, ed., Vol. 4, pp. 667–716, Plenum Press, New York (1972).

123. R. L. Whitman and A. Korpel, "Visualization of a coherent light field by heterodyning with a scanning laser beam," *Appl. Opt.* 8(8), 1567–1576 (1969).

124. A. Korpel, "Acoustical imaging in diffracted light-two-dimensional interaction," in *Acoustical Holography,* A. F. Metherell, H. M. A. El-Sum, L. Larmore, eds., Vol. 1, pp. 149–158, Plenum Press, New York (1969).

125. J. Landry and G. Wade, "Bragg-diffraction imaging and its application for nondestructive testing," in *Imaging Techniques for Testing and Inspection,* J. C. Urbach, B. B. Brenden, R. Aprahamian, eds., Proc. SPIE 29, 47–54 (1972).

126. K. Preston and J. L. Kreuzer, "Ultrasonic imaging using a synthetic holographic technique," *Appl. Phys. Lett.* 10(5), 150–152 (1967).

127. A. Korpel and P. Desmares, "Rapid sampling of acoustic holograms by laser-scanning techniques," *J. Acoust. Soc. Am.* 45(4), 881–884 (1969).

128. B. A. Auld, R. J. Gilbert, K. Hyllested, C. G. Roberts, and D. C. Webb, "A 1.1 GHz scanned acoustic microscopy," in *Acoustical Holography,* G. Wade, ed., Vol. 4, pp. 73–96, Plenum Press, New York (1972).

129. E. Marom, D. Fritzler, and R. K. Meuller, "Ultrasonic holography by electronic scanning of a piezoelectric crystal," *Appl. Phys. Lett.* 12(2), 26–28 (1968).

130. B. B. Brenden, "Acoustical holography," *Opt. Eng.* 14(5), 495–498 (1975).

131. H. M. A. El-Sum and P. Kirkpatrick, "Microscopy by reconstructed wavefronts," *Phys Rev.* 85(4), 763 (1952).

132. G. Kellstrom, "Experimental investigation of interferences and diffraction phenomena of long-wave x-rays," *Nova Acta Soc. Sci. Upsaliensis* 8(5), Series 4, 1–61 (1932).

133. J. W. Giles, "Image reconstruction from a Fraunhofer x-ray hologram with visible light," *J. Opt. Soc. Am.* 59(9), 1179–1188 (1969).

134. S. Aoki and S. Kikuta, "X-ray holographic microscopy," *Japan J. Appl. Phys.* 13(9), 1385–1392 (1974).

135. J. T. Winthrop and C. R. Worthington, "X-ray microscopy by successive Fourier transformation," *Phys. Lett.* 15(2), 124–126 (1965).

136. J. T. Winthrop and C. R. Worthington, "X-ray microscopy by successive Fourier transformation II. an optical analogue experiment," *Phys. Lett.* 21(4), 413–415 (1966).

137. S. Kikuta, S. Aoki, S. Kosaki, and K. Kohra, "X-ray holography of lensless Fourier-transform type," *Opt. Commun.* 5(2), 86–89 (1972).

138. S. Aoki, Y. Ichihara, and S. Kikuta, "X-ray hologram obtained by using synchrotron radiation," *Japan J. Appl. Phys.* 11(12), 1857 (1972).

139. U. Röder, D. Gutkowicz-Krusin, and H. Mahr, "Optical simulation experiment for x-ray holography," *Opt. Commun.* 9(3), 270–273 (1973).

140. B. Reuter and H. Mahr, "Experiments with Fourier transform holograms using 4.48 nm x-rays," *J. Phys. E: Sci. Instrum.* 9(9), 746–751 (1976).

141. A. Tonomura et al., "Optical reconstruction of image from Fraunhofer electron-hologram," *Japan J. Appl. Phys.* 7(3), 295 (1968).

142. E. N. Leith and A. L. Ingalls, "Synthetic antenna data processing by wavefront reconstruction," *Appl. Opt.* 7(3), 539–544 (1968).

143. E. N. Leith, "White-light holograms," *Sci. Am.* 235(4), 80–95 (1976).

144. E. N. Leith, "Some microwave analogs of holography," in *Advances in Holography,* N. H. Farhat, ed., Vol. 2, pp. 1–67, Marcel Dekker, New York (1976).

145. G. Tricoles and N. H. Farhat, "Microwave holography: applications and techniques," *Proc. IEEE* 65(1), 108–121 (1977).

146. R. P. Dooley, "X-band holography," *Proc. IEEE* 53(11), 1733–1735 (1965).

147. H. E. Stockman and B. Zarwyn, "Optical film sensors for RF holography," *Proc. IEEE* 56(4), 763 (1968).

148. N. H. Farhat, "High resolution microwave holography and the imaging of remote moving objects," *Opt. Eng.* 14(5), 499–505 (1975).

149. N. H. Farhat and P. C. Wang, "Holographic imaging with object synthesized apertures," *IEEE Trans. MTT* MTT-22(5), 531–535 (1974).

150. J. Pastor, "Hologram interferometry and optical technology," *Appl. Opt.* 8(3), 525–531 (1969).

151. W. E. Kock, "Three-color hologram zone-plates," *Proc. IEEE* 54(11), 1610–1612 (1966).

152. G. L. Rogers, "Gabor diffraction microscopy: the hologram as a generalized zone-plate," *Nature* 166(4214), 237 (1950).

153. G. L. Rogers, "Experiments in diffraction microscopy," *Proc. Royal Soc. Edin. A* 63(XIV), 193–221 (1950-51).

154. G. L. Rogers, "Artificial holograms and astigmatism," *Proc. Royal Soc. Edin. A* 63(XXII), 313–325 (1951).

155. G. L. Rogers, "Two hologram methods in diffraction microscopy," *Proc. Royal Soc. Edin. A* 64(XVI), 209–221 (1954).

156. M. J. R. Schwar, T. P. Pandya, and F. J. Weinberg, "Point holograms as optical elements," *Nature* 215(5098), 239–241 (1967).

157. J. N. Latta, "Computer-based analysis of hologram imagery and aberrations I. hologram types and their nonchromatic aberrations," *Appl. Opt.* 10(3), 599-608 (1971); see also same issue "II. aberrations induced by a wavelength shift," 609–618; "Fifth order hologram aberrations," 666–667; and "Computer-based analysis of holography using ray tracing," 2698–2710.

158. R. W. Meier, "Magnification and third-order aberrations in holography," *J. Opt. Soc. Am.* 55(8), 987–992 (1965).

159. E. G. Champagne, "Nonparaxial imaging, magnification, and aberration properties in holography," *J. Opt. Soc. Am.* 57(1), 51–55 (1967).

160. D. H. Close, "Holographic optical elements," *Opt. Eng.* 14(5), 408–419 (1975).

161. D. H. Close, "Optically recorded holographic optical elements," in *Handbook of Optical Holography,* H. J. Caulfield, ed., Chapter 8, pp. 573–585, Academic Press, New York (1979).

162. D. G. McCauley, C. E. Simpson, and W. J. Murbach, "Holographic optical element for visual display applications," *Appl. Opt.* 12(2), 232–242 (1973).

163. R. N. Winner and J. H. Brindle, "Holographic visor helmet-mounted display system," in *Proc. Conf. on Display and Device Systems,* pp. 43–52, Society for Information Display, New York (1975).

164. D. H. Close, "Hologram optics in head-up displays," *Soc. Inf. Display Int. Symp. Digest*, pp. 58–59, Society for Information Display, Los Angeles (1974).

165. P. S. Wu and J. C. Tandon, "Omnidirectional laser scanner for supermarkets," *Opt. Eng.* 20(1), 123–128 (1981).

166. L. D. Dickson and G. T. Sincerbox, "Optics and holography in the IBM super-market scanner," in *Advances in Laser Scanning Technology,* L. Beiser, ed., Proc. SPIE 299, 163–168 (1981).

167. C. J. Kramer, "Holographic laser scanners for nonimpact printing," *Laser Focus* 17(6), 70–82 (1981).

168. D. Doggett, S. Barasch, and M. J. Wegener, "Design trade-offs for a diode laser holographic scanner," in *Advances in Laser Scanning Technology,* L. Beiser, ed., Proc. SPIE 299, 151–156 (1981).

169. S. Lu, "Generating multiple images for integrated circuits by Fourier transform holograms," *Proc. IEEE* 56(1), 116–117 (1968).

170. M. Malin and H. E. Morrow, "Wavelength scaling holographic elements," *Opt. Eng.* 20(5), 756–758 (1981).

171. L. H. Lin and E. T. Doherty, "Efficient and aberration-free wavefront reconstruction from holograms illuminated at wavelengths differing from the forming wavelength," *Appl. Opt.* 10(6), 1314–1318 (1971).

172. A. Labeyrie and J. Flamand, "Spectrographic performance of holographically made diffraction gratings," *Opt. Commun.* 1(1), 5–8 (1969).

173. M. C. Hutley, "Blazed interference (holographic) diffraction gratings for the ultra-violet," in *Vacuum Ultraviolet Radiation Physics,* E. Koch, R. Hansel, and C. Kunz, eds., pp. 713–714, Pergamon Press, London (1974).

174. G. Schmahl and D. Rudolph, "Holographic diffraction gratings," *Progress in Optics,* E. Wolf, ed., Vol. 14, pp. 197–244, North Holland Publishing Co., Amsterdam (1976).

175. W. R. Hunter, "Diffraction gratings for the XUV-conventional VS holographics," *J. Spectrosc. Soc. Japan* 24, Vol. 28, Supplement No. 1, 37–54 (1975).

176. G. S. Hayat, F. Flamand, M. Lacroix, and A. Grillo, "Designing a new generation of analytical instruments around the new types of holographic diffraction grating," *Opt. Eng.* 14(5), 420–425 (1975).
177. A. Maréchal and P. Croce, "Un filtre de frequences spatiales pour l'amelioration du contraste de images optique," *C. R. Acad. Sci., Paris* 237(12), 607–609 (1953).
178. E. N. Leith, "Complex spatial filters for image deconvolution," *Proc. IEEE* 65(1), 18–28 (1977).
179. A. D. Gara, "Real-time optical correlation of 3-D scenes," *Appl. Opt.* 16(1), 149–153 (1977).
180. B. J. Thompson, "Hybrid processing systems—an assessment," *Proc. IEEE* 65(1), 62–76 (1977).
181. J. L. Horner, "Light utilization in optical correlators," *Appl. Opt.* 21(24), 4511–4514 (1982).
182. J. L. Horner and P. D. Gianino, "Phase-only matched filtering," *Appl. Opt.* 23(6), 812–816 (1984).
183. P. D. Gianino and J. L. Horner, "Additional properties of the phase-only correlation filter," *Opt. Eng.* 23(6), 695–697 (1984).
184. J. L. Horner and J. R. Leger, "Pattern recognition with binary phase-only filters," *Appl. Opt.* 24(5), 609–611 (1985).
185. G. W. Stroke and R. G. Zeck, "*A posteriori* image-correcting 'deconvolution' by holographic Fourier-transform division," *Phys. Lett.* 25A(2), 89–90 (1967).
186. G. W. Stroke et al., "Image improvement and three-dimensional reconstruction using holographic image processing," *Proc. IEEE* 65(1), 39–62 (1977).
187. H. J. Caulfield, "Holographic spectroscopy," in *Advances in Holography,* N. H. Farhat, ed., Vol. 2, pp. 139–184, Marcel Dekker, New York (1976).
188. T. H. Jeong, P. Rudolf, and A. Luckett, "360-deg holography," *J. Opt. Soc. Am.* 56(9), 1263–1264 (1966).
189. W. T. Cathey in *Holography: State of the Art Review,* J. Kallard, ed., p. 96, Optosonic Press, New York (1969) (US Patent 3415 587).
190. H. J. Caulfield, J. L. Harris, H. W. Hemstreet, Jr., and J. G. Cobb, "Local reference beam generation in holography," *Proc. IEEE* 55(10), 1758 (1967).
191. C. Roychoudhuri and B. J. Thompson, "Application of local reference beam holography to the study of laser beam parameters," *Opt. Eng.* 13(4), 347–361 (1974).
192. L. Mertz and N. O. Young, "Fresnel transformations of images," in *Proc. ICO Conf. on Optical Instrumentation,* K. J. Habell, ed., pp. 305–312, Chapman and Hall, London (1962).
193. H. H. Barrett et al., "Coded apertures derived from the Fresnel zone plates," *Opt. Eng.* 13(6), 539–549 (1974).
194. H. Weiss et al., "Coded aperture imaging with x-rays (flashing tomosynthesis)," *Optica Acta* 24(4), 305–325 (1977).
195. A. W. Lohmann, "Wavefront reconstruction for incoherent objects," *J. Opt. Soc. Am.* 55(11), 1555–1556 (1965).
196. G. W. Stroke and R. C. Restrick, "Holography with spatially noncoherent light," *Appl. Phys. Lett.* 7(9), 229–231 (1965).
197. R. Silva and G. L. Rogers, "Decoding of a noncoherent hologram using noncoherent light," *J. Opt. Soc. Am.* 65(12), 1448–1450 (1975).
198. D. Courjon and J. Bulabois, "Holographic recording with a spatially incoherent imaging process," *Opt. Commun.* 21(1), 96–101 (1977).
199. S. A. Benton, "Hologram reconstructions with extended incoherent sources," *J. Opt. Soc. Am.* 59(11), 2283 (1969).
200. R. L. Lamberts and C. N. Kurtz, "Reversal bleaching for low flare light in holograms," *Appl. Opt.* 10(6), 1342–1347 (1971).
201. R. G. Zech, "Data storage in volume holograms," pp. 198-200, PhD Thesis, Univ. of Michigan (1974).
202. A. Graube, "Advances in bleaching methods for photographically recorded holograms," *Appl. Opt.* 13(12), 2942–2946 (1974).
203. T. A. Shankoff, "Phase holograms in dichromated gelatin," *Appl. Opt.* 7(10), 2101–2105 (1968).
204. L. H. Lin, "Hologram formation in hardened dichromated gelatin films," *Appl. Opt.* 8(5), 963–966 (1969).

205. D. Meyerhofer, "Phase holograms in dichromated gelatin," *RCA Rev.* 33(1), 110–131 (1972).
206. J. C. Urbach, "Advances in hologram recording materials," in *Developments in Holography,* B. J. Thompson, J. B. DeVelis, eds., Proc. SPIE 25, 17–42 (1971).
207. R. L. Kurtz and R. B. Owen, "Holographic recording materials—a review," *Opt. Eng.* 14(5), 393–401 (1975).
208. K. S. Pennington and L. H. Lin, "Multicolor wavefront reconstruction," *Appl. Phys. Lett.* 7(3), 56–57 (1965).
209. G. W. Stroke and A. E. Labeyrie, "White-light reconstruction of holographic images using the Lippmann-Bragg diffraction effect," *Phys. Lett.* 20(4), 368–370 (1966).
210. E. N. Leith and G. J. Swanson, "Achromatic interferometers for white light optical processing and holography," *Appl. Opt.* 19(4), 638–644 (1980).
211. E. N. Leith and G. J. Swanson, "Recording of phase-amplitude images," *Appl. Opt.* 20(17), 3081–3084 (1981).
212. G. D. Collins, "Achromatic Fourier transform holography," *Appl. Opt.* 20(18), 3109–3119 (1981).
213. G. J. Swanson, "Recording of one-dimensionally dispersed holograms in white light," *Appl. Opt.* 20(24), 4267–4270 (1981).
214. E. N. Leith, "Some unsolved (and partially solved) problems in holography," in *Holographic Data Nondestructive Testing,* D. Vukicević, ed., Proc. SPIE 370, 2–6 (1982).
215. G. W. Ellis, "Holomicrography: transformation of image during reconstruction *a posteriori,*" *Science* 154(3753), 1195–1197 (1966).
216. K. A. Snow and R. Vandewarker, "An application of holography to interference microscopy," *Appl. Opt.* 7(3), 549–554 (1968).

28   Communication Theory
Techniques in Optics*

28.1   INTRODUCTION

In this chapter, we discuss some further methods that are useful in the analysis of linear optical systems. We will assume that the optical system is a stationary linear black box whose input/output relationship is described by the convolution process in the space domain. The stationary linear optical system has the property that it is completely characterized by either its point spread function in the space domain or by the Fourier transform of its point spread function—the optical transfer function—in the frequency domain.

We first discuss the sampling process whereby a given band-limited function is represented in terms of its values at a series of discrete sample points. The technique is described in both the space and spatial frequency domains. A space-bandwidth product is also discussed in terms of the sampled function, and a number of examples are given.

Methods for describing statistical processes are introduced. These include ensemble and spatial averages, correlation functions, and the concept of spectral density. The use of these methods in linear system analysis is demonstrated by examples.

28.2   SAMPLING THEOREM

28.2.1   Space Domain

The sampling theorem of Refs. 1 and 2 is a curve-fitting device for representing functions with a finite spectrum centered around zero frequency. If the function to be sampled is truly band-limited, then the sampling theorem will introduce no error into the functional representation.

Theorem: If a one-dimensional function $a(x)$ contains no frequencies higher than μ_0 line pairs/mm, the function is completely determined by giving its

---

*This is a reprint of an article originally titled "Communication Theory" by John B. DeVelis and George O. Reynolds published in *Handbook of Optical Holography,* H. J. Caulfield, ed. (© 1979 Academic Press, Inc.) and reprinted here with permission.

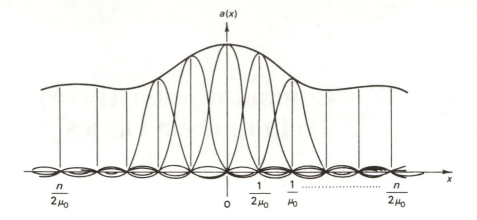

Fig. 28.1 Graphical representation of the sampling theorem in the space domain.

ordinates at a series of points extending throughout the space domain and spaced $(\tfrac{1}{2}\mu_0)$ mm apart. For example, in optics, the function $a(x)$ could be an aerial image, i.e., an object scene limited by the lens transfer function; or it could be a photographic transparency, which is band-limited by the film transfer function, which has a finite cutoff frequency.

In one dimension, the sampling theorem is

$$a(x) = \sum_{n=-\infty}^{\infty} a\left(\frac{n}{2\mu_0}\right) \operatorname{sinc} 2\pi\mu_0 \left(x - \frac{n}{2\mu_0}\right) , \qquad (28.1)$$

where the interpolation function is the sinc function. Equation (28.1) states that to reconstruct the function $a(x)$ we must find its value at the sampled points $x_n = (n/2\mu_0)$ and at each of these points multiply the value of $a(x)$ by the interpolation function $\operatorname{sinc} 2\pi\mu_0[x - (n/2\mu_0)]$. Graphically, this corresponds to plotting sinc functions at each of the sampled ordinates such that the magnitude of the sinc function at the specific ordinate is the value of the function at that point, as shown in Fig. 28.1. Due to the periodicity of the sinc function, it does not contribute to the function at the other sampled ordinates.

28.2.2   Frequency Domain

The function $a(x)$ in Eq. (28.1) is represented by a summation over all sampled points of the sampled values multiplied by the interpolation function $\operatorname{sinc} 2\pi\mu_0(x - n/2\mu_0)$. Taking the Fourier transform on both sides of Eq. (28.1) yields

$$\tilde{a}(\mu) = \operatorname{rect}(\mu|\mu_0) \sum a\left(\frac{n}{2\mu_0}\right) \exp\left(\frac{2\pi i\mu n}{2\mu_0}\right) .$$

The significance of the two factors is clear. The rect function, $\operatorname{rect}(\mu|\mu_0)$, is equal to zero for all frequencies whose magnitude is greater than μ_0. That is, this factor indicates that the function is band limited. The second factor, the summation, indicates that each sampled point gives rise to a periodic variation in the spectrum, the magnitude of which is proportional to the sampled value. In the development of the sampling theorem, the sampled signal is obtained by convolv-

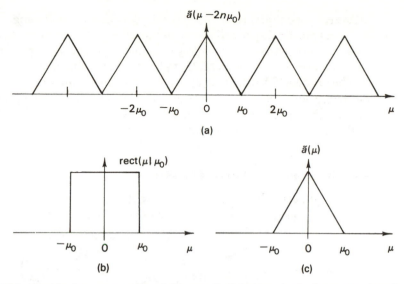

Fig. 28.2 Graphical representation of the sampling theorem in the frequency domain: the product of (a) the sampled function spectrum $\tilde{a}(\mu - 2n\mu_0)$ and (b) the spectrum of the interpolation function yields (c) the spectrum $\tilde{a}(\mu)$.

ing the original function $a(x)$ with a Dirac comb. The band-limited Fourier spectrum is a precise representation of the function $a(x)$.

An alternate, but closely related representation, is frequently useful in the description of stationary linear systems, particularly in optical systems. Physically, an optical field may be sampled by passing it through a grating. This has the effect of multiplying the field $a(x)$ by a periodic function. If the grating is composed of sufficiently narrow slits, it can be approximated by a Dirac comb. Denoting the sampled function by $b(x)$, we may write

$$b(x) = a(x) \circledast \sum \delta(x - n/2\mu_0) \ .$$

Taking the Fourier transform of $b(x)$ and using the convolution theorem, we can see that its spectrum is simply the spectrum of $a(x)$ replicated at intervals of $n/2\mu_0$. As long as the frequency of the grating exceeds the band limit of the signal by at least a factor of 2 these replicated spectra will not overlap and it is possible to unambiguously reconstruct the original signal $a(x)$. This method of sampling is frequently used in optical systems and is illustrated graphically in Fig. 28.2.

28.2.3   Sampling Criteria and Space Bandwidth Product

To recover the function $a(x)$ successfully in Eq. (28.1), the sampled ordinates must satisfy the condition that

$$x_0 \geq 1/2\mu_0 \ , \tag{28.2}$$

where x_0 is the separation of the sampled ordinates in the space domain. This ensures that the replicated spectra in Fig. 28.2(a) do not overlap. If the sampling interval x_0 is greater than $\frac{1}{2}\mu_0$, the spectra are more widely separated in Fig. 28.2(a) and the function is still recovered even though more sampling points have been used. If the sampling interval x_0 is less than $\frac{1}{2}\mu_0$, the spectra in Fig. 28.2(a) overlap and the function is not completely recovered. This degradation of the recovered function, due to undersampling, is known as *aliasing*.

In any practical application, the summation in Eq. (28.1) will extend from $-N$ to N rather than from $-\infty$ to ∞, resulting in a total of $2N$ sample points. Further, the condition given in Eq. (28.2) applies to each sampled point.

The product $(2N)(2\mu_0)$, resulting from the total number of sample points and the bandwidth requirement of the function (which determines the sampling interval), is called the *one-dimensional space bandwidth product* of that portion of the function being considered.

28.2.4   Two-Dimensional Sampling Theorem

The sampling theorem can be readily extended to two dimensions. For the two-dimensional rectangular case, we obtain the extension of Eq. (28.1) in the form

$$a(x,y) = \sum_{n=-\infty}^{\infty} \sum_{m=-\infty}^{\infty} a\left(\frac{n}{2\mu_0}, \frac{m}{2\nu_0}\right) \text{sinc} 2\pi\mu_0 \left(x - \frac{n}{2\mu_0}\right)$$

$$\times \text{sinc} 2\pi\nu_0 \left(y - \frac{m}{2\nu_0}\right), \tag{28.3}$$

where μ_0 and ν_0 are the cutoff frequencies associated with the μ and ν axes for the Fourier spectrum of $a(x,y)$. For the rotationally symmetric two-dimensional case, one uses the Bessel function of the first kind for sampling, and the result is given by[2]

$$a(x,y) = \sum_{n=-\infty}^{\infty} \sum_{m=-\infty}^{\infty} a\left(\frac{n}{2\nu_0}, \frac{m}{2\nu_0}\right)$$

$$\times \frac{2\pi\nu_0^2 J_1\left(2\pi\nu_0\{[x-(n/2\nu_0)]^2 + [y-(m/2\nu_0)]^2\}^{\frac{1}{2}}\right)}{2\pi\nu_0\{[x-(n/2\nu_0)]^2 + [y-(m/2\nu_0)]^2\}^{\frac{1}{2}}}, \tag{28.4}$$

where $\mu_0 = \nu_0$ for this case.

28.2.5   Examples

Three well-known processes that utilize the sampling principle are halftone reproduction of photography, facsimile transmission and display, and the display of television images. These applications combine electronic and optical principles to define and implement the bandwidth limitation necessary to obtain an optimum display.

Other applications that utilize sampling as a fundamental process are electro-optical scanning systems.[3,4] In this case, the input information on the film is further band-limited by the f-number of a diffraction-limited optical scanning system. The sampling interval in the space domain, which is given by the reciprocal of the system bandwidth (as determined from a square lens aperture), is

$$x_0 = \lambda(\text{f-number})/2. \tag{28.5}$$

In such systems, aperture sizes are usually variable, while the scanning speed is

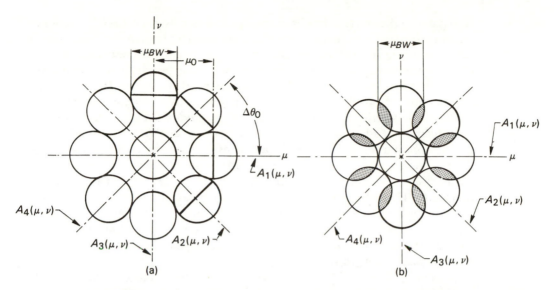

Fig. 28.3 (a) Angular separation of four properly sampled image spectra in frequency space. (b) Example of undersampling in frequency space showing spectra overlap, which results in aliasing $\mu_0 = f_1 \lambda/p$, μ_{BW} = bandwidth (after Ref. 5).

constant; thus, the rate of information is set by the aperture size. The electronic sampling rate is determined by the sampling interval of Eq. (28.5) and the system speed.

Another optical example of the use of the sampling theorem is modulated imagery (see Chapters 3 and 4).[5] In this process the image is band-limited by a lens and sampled with a diffraction grating having the desired fundamental frequency for sampling.

The principle of angular modulation can be used to store more than one image on the film. The grating must be rotated between exposures by the amount

$$\Delta\theta_0 = 2\sin^{-1}[\mu_{BW}/(2\mu_0)] \, , \tag{28.6}$$

where $\Delta\theta_0$ is the amount of rotation, μ_{BW} is the bandwidth of the individual images, and μ_0 is the sampling frequency. The angle defined in Eq. (28.6) ensures that the image spectra are spatially separated as shown in Fig. 28.3(a). The multiply stored images on the developed film can be individually retrieved in a coherent optical processing system by placing an aperture over the appropriate image spectrum in the transform plane of the processing system.

This process results in aliasing when the sampling interval is too large. This results in overlapping spectra as shown in Fig. 28.3(b).

In the retrieved image plane, the filtered image is degraded by the presence of high-frequency components from adjacent image spectra.

28.3   STATISTICAL DESCRIPTION OF RANDOM SAMPLES

28.3.1   Ensemble and Coordinate Averaging Descriptions of Random Processes

The basic entity of any communication channel (system) is *information,* and the fundamental entity that most characterizes such systems is the *information capacity.* Whether the system is electrical, optical, or electro-optical, it processes signal information which can best be classified as completely deterministic or statistical. In the deterministic case, the signal is usually given a Fourier series or integral representation, i.e., it is a periodic or transient waveform whose value is completely determined for all values of the independent variable (time or space). On the other hand, the magnitude of the statistical signals is not completely determined for any particular value of the independent variable (time or space) that are not completely determined, i.e., they are only known in a probabilistic sense. These statistical signals, usually called *random signals,* are treated by introducing statistical or probabilistic methods for analyzing and synthesizing the information content of such signals. In essence, for random signals over an infinite limit, a Fourier representation does not exist, and one is forced to consider a statistical representation. These resulting statistical methods can be applied to the deterministic case; however, they have found wider application and interest in the analysis of random processes. In the optical case, such methods are used as the basic tool in the formulation of the classical theory of partial coherence, the analysis of film grain noise, and the analysis of coherent optical noise usually called *speckle.*

A random signal (or process) $F(\mathbf{x}, t)$ may be defined as one that does not depend on the independent variable (either space, time, or both) in a completely deterministic manner. Generally, we work with random signals that obey a simplifying constraint, stationarity. If we have a physical process that gives rise to a random signal, such a signal will be considered stationary with respect to the time (space) coordinate if the impulse response depends only on the coordinate difference. The underlying physical process that gives rise to stationary random signals is described by statistics which are not time dependent.

In principle, we have two ways of handling these problems. In the first case, we can assume we know the function over a long period of time (space) from which we determine the probability distribution functions used to determine both time and space averages. In the second case, we have an ensemble of similar functions from which we determine the probability distribution functions by an examination of all members of the ensemble. These distribution functions are then used in the determination of ensemble averages. The ergodic hypothesis is that coordinate and ensemble averages should yield the same results. Hence, as we now define our correlation functions, we shall assume we have ergodic stationary signals and only define the averages over spatial coordinates.

28.3.2 Correlation Functions

The correlation integral given by Eq. (28.7) can be used to correlate the same two functions, $a(x)$ and $s(x)$, shown in Fig. 28.6(a). Figure 28.6(b) illustrates the non-folded shift indicated by the cross-correlation integral and (c) shows the cross-correlation:

$$c(x) = <a^*(x_1)s(x_1 + x)> \equiv \lim_{L \to \infty} \frac{1}{2L} \int_{-L}^{L} a^*(x_1)s(x_1 + x) \, dx_1 \; . \quad (28.7)$$

It should be noted that changing the sense of the correlation by putting the shift variable in the second function in Eq. (28.7) yields $c(-x)$. In developing general theorems, we will shift the second function in the negative direction. However, in the physical examples, the physics of the experiment will determine which function should be shifted in which direction.

The autocorrelation function of a complex function $a(x_1)$ is defined by

$$c(x) = <a^*(x_1)a(x_1 + x)> . \qquad (28.8)$$

The following properties of the autocorrelation function are useful:

1. It is an even function of the delay variable:

$$c(x) = c(-x) , \qquad (28.9)$$

2. It is a maximum at the origin:

$$c(0) \geq |c(x)| \quad \text{for } x \neq 0 . \qquad (28.10)$$

3. For $x = 0$, we get the average of the square of the function, which in many physical cases is the energy of the system:

$$c(0) = <|a(x_1)|^2> . \qquad (28.11)$$

28.3.2.1 Example: Convolution Versus Correlation

The form of the integrals in Eqs. (28.7) and (28.8) should not be confused with the form of the convolution integral. The convolution process involves folding, shifting, and summing procedures, whereas the correlation procedure involves a shifting and summing without folding. This is not merely a semantic distinction. Unless the functions involved are even functions, the differences between their convolution and correlation are dramatic.

Symmetric Functions: If we consider the function $a(x_1)$ to be an even function, e.g., a rectangular function of width $2a$, the folding process yields the same function since $a(x_1) = a(-x_1)$. Thus, correlation of the rectangular function with itself (autocorrelation) and convolution of the rectangular function with itself both yield the same triangular function as shown in Fig. 28.4.

Nonsymmetric Functions: As an example of the difference between convolution and correlation, consider the two functions shown in Fig. 28.5(a) and given by

$$a(x_1) = \begin{cases} 1 & \text{for} \quad 0 \leq x_1 \leq 1 , \\ 0 & \text{for} \quad 0 > x_1 > 1 , \end{cases} \qquad (28.12)$$

$$s(x_1) = \begin{cases} \delta(x_1) - e^{-x_1} & \text{for} \quad x_1 \geq 0 , \\ 0 & \text{for} \quad x_1 < 0 . \end{cases} \qquad (28.13)$$

Use of the convolution integral, shown schematically in Fig. 28.5(b), yields

$$b(x) = \begin{cases} \displaystyle\int_0^x [\delta(x - x_1) - e^{x_1 - x}] dx_1 = e^{-x} & \text{for} \quad x < 1 , \\ \\ -\displaystyle\int_0^1 e^{x_1 - x} dx_1 = e^{-x}(1 - e) & \text{for} \quad x \geq 1 . \end{cases} \qquad (28.14)$$

The resulting function $b(x)$ is plotted in Fig. 28.5(c).

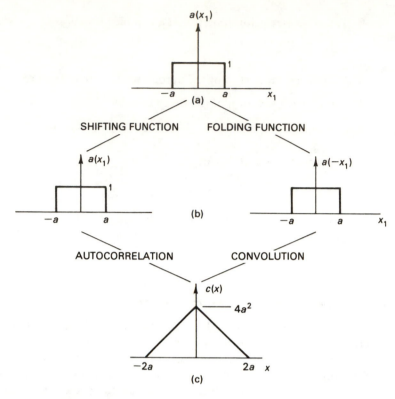

Fig. 28.4 Comparison of convolution and autocorrelation of symmetric functions: (a) the symmetric rectangular function, (b) the symmetric shifting and folding functions, and (c) the resulting triangular function.

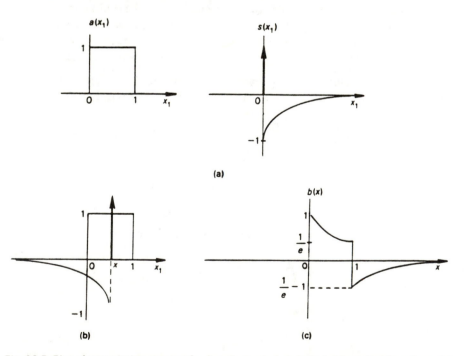

Fig. 28.5 Plot of convolution process for functions $a(x_1)$ and $s(x_1)$: (a) original functions, (b) illustration of the folded shifted function $s(x - x_1)$ relative to the unshifted function $a(x_1)$, and (c) result of the convolution process.

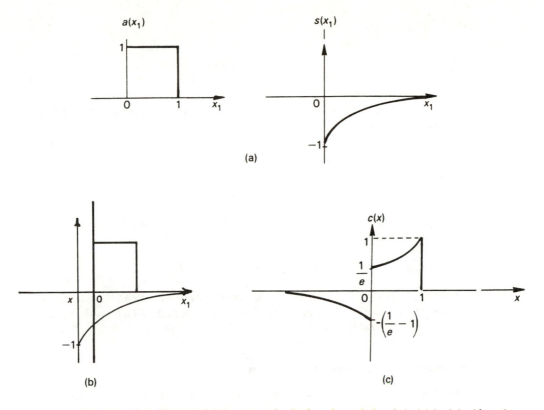

Fig. 28.6 Plot of the correlation process for the functions $a(x_1)$ and $s(x_1)$: (a) original functions, (b) illustration of the unfolded shifted function $s(x + x_1)$ relative to the unshifted function $a(x)$, and (c) result of the correlation process.

The autocorrelation integral given by Eq. (28.8) can be used to autocorrelate the same two functions, $a(x_1)$ and $s(x_1)$, shown in Fig. 28.6(a). This autocorrelation shown schematically in Fig. 28.6(b) yields

$$c(x) = \begin{cases} \displaystyle\int_x^1 [\delta(x_1 + x) - e^{-(x_1+x)}] \, dx_1 = 1 - e^{-2x} + e^{-(1+x)} & \text{for } 0 < x < 1 , \\[4mm] \displaystyle -\int_0^1 e^{-(x_1+x)} \, dx_1 = e^{-x}(1/e - 1) & \text{for } x \le 0 . \end{cases} \tag{28.15}$$

The resulting function is plotted in Fig. 28.6(c). Comparison of Figs. 28.5(c) and 28.6(c) shows that the processes of convolution and correlation yield quite different results when the functions are nonsymmetric.

28.3.3   Spectral Density

The spectral density of a stationary random signal $a(x)$ is extremely useful in the analysis of random signals because of its measurability and its relationship to the autocorrelation function. The spectral density is sometimes referred to as the *power spectral density* or the *power spectrum*. It is defined by

$$C(\mu) = \lim_{L \to \infty} \frac{1}{L} |\tilde{a}_T(\mu)|^2 , \tag{28.16}$$

where $\tilde{a}_T(\mu)$ is the Fourier transform of a truncated form of $a_T(x)$.

A very important relationship exists between the spectral density defined by Eq. (28.16) and the autocorrelation defined by Eq. (28.8). This relationship is the Wiener-Khintchine theorem and states that the spectral density and autocorrelation functions are Fourier transform pairs, i.e.,

$$C(\mu) \cong \tilde{c}(\mu) = \int_{-\infty}^{\infty} c(x) \exp(-2\pi i \mu x) \, dx . \tag{28.17}$$

If the random signals are the input to a linear system, the statistical description of the system output is

$$C_{oo}(\mu) = |S(\mu)|^2 C_{ii}(\mu) , \tag{28.18}$$

where $C_{oo}(\mu)$ is the spectral density of the output of the linear system, $C_{ii}(\mu)$ is the spectral density of the input to the linear system, and $|S(\mu)|$ is the modulus of the system transfer function.

28.3.4   Examples of Statistical Techniques

28.3.4.1   *Example 1: Stationary Linear System With a Gaussian Impulse Response*

As an example illustrating these concepts, consider a linear system with a Gaussian mathematical spread function given by

$$s(x) = \exp(-\pi x^2/\alpha^2) , \tag{28.19}$$

where α is a real number defining the width of the Gaussian. Equation (28.19) is plotted in normalized form in Fig. 28.7(a). The transfer function or frequency response of the system is obtained by taking the Fourier transform of Eq. (28.19) to obtain

$$S(\mu) = \alpha \exp(-\pi \alpha^2 \mu^2) , \tag{28.20}$$

which is plotted in normalized form in Fig. 28.7(b). If the input to the linear system is white noise, its autocorrelation function is given by

$$c_{ii}(x) = \overline{I^2} \delta(x) , \tag{28.21}$$

where $\overline{I^2}$ is the mean square brightness of the object (input). The spectral density (power spectrum) is given by

$$C_{ii}(\mu) = \overline{I^2} . \tag{28.22}$$

The output power spectrum can be determined from Eq. (28.18) to be

$$C_{oo}(\mu) = \overline{I^2} \alpha^2 \exp(-2\pi \alpha^2 \mu^2) . \tag{28.23}$$

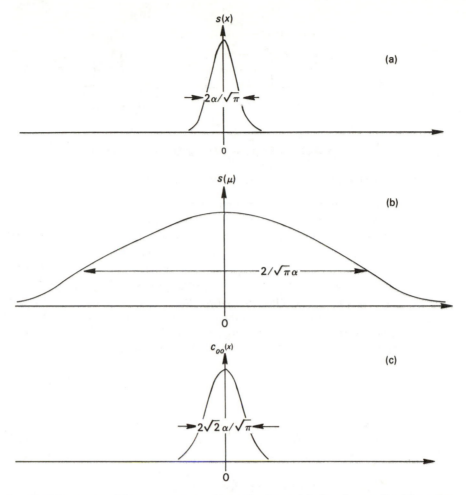

Fig. 28.7 Response of linear system to white-noise input: (a) plot of normalized Gaussian spread function given by Eq. (28.19), (b) plot of the normalized Gaussian transfer function of the linear system given by Eq. (28.20), and (c) plot of the normalized output correlation function given by Eq. (28.24) resulting from the white-noise input to the linear system.

Taking the Fourier transform of this equation, we obtain the output autocorrelation function:

$$c_{oo}(x) = (\overline{I^2}\alpha/\sqrt{2})\exp(-\pi x^2/2\alpha^2) \ . \tag{28.24}$$

Equation (28.24) is plotted in Fig. 28.7(c) in normalized form.

This example illustrates that when a random function (white noise in this example) is the input to a linear system the output correlation function [Eq. (28.24)] is broader than the input correlation function [Eq. (28.21)]. The degree of the spread is determined by the width of the spread function α, which is determined from the system bandwidth. A comparison of Fig. 28.7(c) with (a) shows that the width of the output correlation function is greater than that of the system spread function.

To determine the output correlation function when the correlation function of the input differs from white noise, one must convolve the correlation function of the input with the impulse response given in Eq. (28.24).

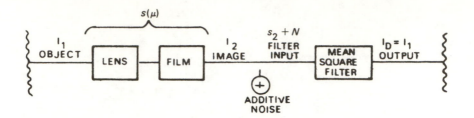

Fig. 28.8 Block diagram of imaging and mean square filtering system.

28.3.4.2   *Example 2: Filtering Signals in the Presence of Additive Noise*

Wiener Filter: One of the fundamental problems in the application of optical spatial filtering techniques[1,2,4,6] to actual photographs is the presence of photographic grain noise, which appears as a spatially irregular structure deteriorating the image of interest (see Chapters 20 and 21). Since this irregular structure is representative of a random process, we require statistical methods for minimizing its effect. An approach for filtering signals in the presence of additive noise has been developed and widely applied in both electrical and optical systems.[7-14]

In this approach, the filter minimizes the mean square error between the desired signal and the input signal to the filtering system. The general form of the transfer function of this optimum filtering system is[7,13]

$$H(\mu) = C_{id}(\mu)/C_{ii}(\mu) , \tag{28.25}$$

where $C_{id}(\mu)$ is the cross power spectrum of the input signal (object) and the desired signal, and $C_{ii}(\mu)$ is the power spectrum of the input signal. To apply this technique to the filtering of photographs having granular structure, we refer to the system diagram shown in Fig. 28.8. The purpose of the filtering system is to extract from the degraded noisy image $I_2 + N$ the original object I_1 as exactly as possible. The transfer function of the optical spatial filtering system that performs this task in the least-mean-square sense is[1,7]

$$H(\mu) = [C_{I_2 I_1}(\mu) + C_{I_1 N}(\mu)]/[C_{I_2 I_2}(\mu) + C_{NN}(\mu) + 2C_{I_2 N}(\mu)] , \tag{28.26}$$

where

$$C_{I_2 I_1}(\mu) = C_{I_1 I_1}(\mu)S^*(\mu) = \text{cross power spectrum of the object and the image}$$

$$C_{I_1 N}(\mu) = \text{cross power spectrum of the object and the noise}$$

$$C_{I_2 I_2}(\mu) = C_{I_1 I_1}(\mu)|S(\mu)|^2 = \text{power spectrum of the image}$$

$$C_{I_1 I_1}(\mu) = \text{power spectrum of the object}$$

$$C_{NN}(\mu) = \text{noise power spectrum}$$

$$C_{I_2 N}(\mu) = \text{cross power spectrum of the image and the noise}$$

$$S(\mu) = \text{lens/film system transfer function.}$$

In general, the cross power spectrum of the object and the noise requires knowledge of the Fourier spectrum of the noise rather than its power spectrum. Therefore, in optical applications involving noise, it is assumed that the cross power spectral terms in Eq. (28.26) are negligibly small, i.e., the object and the noise are not correlated. In general photographic situations, the noise is directly dependent on the object; however, the lack of correlation is a reasonable approximation when the image contrast is low, such as in aerial photography or grain-limited imagery. Assuming that the object and the noise are not correlated, Eq. (28.26) becomes

$$H(\mu) = C_{I_1 I_1}(\mu) S^*(\mu) / [C_{I_1 I_1}(\mu) | S(\mu) |^2 + C_{NN}(\mu)] \ . \tag{28.27}$$

Examination of this equation leads to two cases of interest which lend some physical insight into the filtering processes.

The first case occurs when the noise is vanishingly small so that $C_{NN}(\mu) \cong 0$. In this case, Eq. (28.27) reduces to

$$H(\mu) \cong 1/S(\mu) \ . \tag{28.28}$$

This is the inverse filter for the system and is the noise-free least-mean-square filter (see Chapters 31 and 33).

The second case of interest occurs when the noise is small but not negligible, i.e.,

$$\frac{C_{NN}(\mu)}{C_{I_1 I_1}(\mu) | S(\mu) |^2} \ll 1 \quad \text{or} \quad \frac{C_{I_1 I_1}(\mu)}{C_{NN}(\mu)} \gg \frac{1}{| S(\mu) |^2} \ . \tag{28.29}$$

For this approximation, Eq. (28.27) reduces to

$$H(\mu) \cong \frac{1}{S(\mu)} \left[1 - \frac{C_{NN}(\mu)}{C_{I_1 I_1}(\mu) | S(\mu) |^2} \right] \ . \tag{28.30}$$

In this approximation, where the noise is not negligible, the value of the optimum filter is reduced from the corresponding value of the inverse filter by a factor dependent on the signal-to-noise ratio at each frequency.

For the case of a Gaussian transfer function reaching its $1/e$ value at 22.5 line pairs/mm, Eq. (28.27) is plotted in Fig. 28.9 for various constant values of the signal-to-noise ratio. For decreasing values of signal-to-noise ratios in Fig. 28.9, we observe that the cutoff frequency of the optimum filter, as determined by the maximum ordinate of the curve, decreases. This implies that frequencies higher than the filter cutoff frequency only contribute to noise and hence are rejected by the filter. Experimental evidence showing the superior performance of the optimum filter compared to the inverse filter for the case of imaging through a turbulent medium has appeared in the literature.[12]

Matched Filter: Another linear filter useful for detecting known signals from additive random background noise is the matched filter.[8] This filter, which maximizes the ratio of peak signal to rms noise, is given by

$$H(\mu) = (\text{const}) O^*(\mu) / C_{NN}(\mu) \ , \tag{28.31}$$

where $O^*(\mu)$ is the complex conjugate of the signal (object) Fourier spectrum and $C_{NN}(\mu)$ is the noise power spectrum (see Chapters 31 and 33).

28.3.4.3   *Example 3: Speckle Photography*

This is an example of ensemble averaging. When imaging through the atmosphere with a lens, the principal effect of the turbulent medium is to distort the

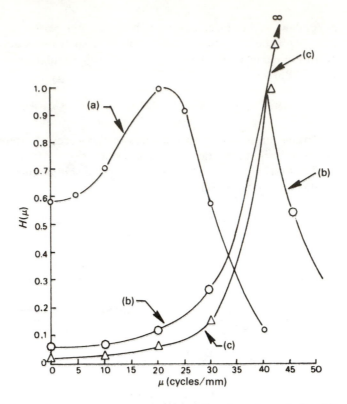

Fig. 28.9 Calculated optimum least-mean-square filter for assumed signal-to-noise ratio $[C_{I_1I_1}(\mu)/C_{NN}(\mu)]$ of (a) 10, (b) 100, and (c) infinity (after Ref. 12).

transmitted wavefront from each point in the object so that the wavefront reaching the optical system contains random structure in both amplitude and phase. The thermal variations present in the atmosphere cause density inhomogeneities which, in turn, impress phase distortions on the wavefront. After sufficient propagation, the phase variations produce random amplitude variations in the entrance pupil of the lens. Phase shifts can originate near to or far from the entrance pupil of the imaging system and cause the following types of image distortions:

1. scintillation, intensity fluctuations

2. distortion, shifting of all or parts of the image

3. blur, broadening of the instantaneous point-spread function.

A technique for obtaining increased resolution for turbulence-degraded photographic images has been described in the literature.[15] The underlying principle is that on any fast exposure through the atmosphere, many points of the image are distorted, but at some points in the image the full-aperture resolution is present in one or more directions. Adding a series of short exposures can build up the resolution in all directions over the image format. Thus, an ensemble of very short exposure images of an object having a center of symmetry (double star) are recorded on film. The exposures must be fast enough to "stop" the atmospheric motion. The Fourier transforms of the individual ensemble members are added

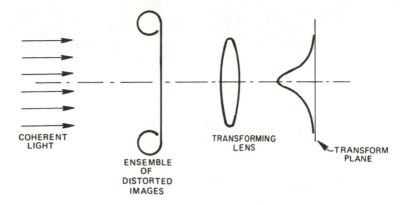

COHERENT
LIGHT

ENSEMBLE
OF
DISTORTED
IMAGES

TRANSFORMING
LENS

TRANSFORM
PLANE

Fig. 28.10 Optical system for adding Fourier transforms of an ensemble of distorted images.

sequentially in a coherent system, such as that shown in Fig. 28.10. The shift theorem of Fourier analysis ensures that the Fourier transform of the object is centered on the optical axis for each ensemble member, whereas the Fourier transform of the noise is randomly distributed. A sequential recording of the intensities in the Fourier transform plane results in an addition of the signal power spectra from the ensemble members and the simultaneous averaging of their noise contributions. This addition (ensemble averaging) increases the signal-to-noise ratio of the signal by retaining all signal information, in register, from each member of the ensemble. Since the technique records the intensity of the Fourier transform, the object phase is lost and only objects with a center of symmetry can be treated successfully by this technique.

Upon development of the film in the transform plane of Fig. 28.10 to a gamma of one half, the enhanced image is obtained by Fourier transforming the developed film. The concept has been extended to include objects without a center of symmetry by determining the relative phase of the autocorrelation function by the use of a computer algorithm.[16]

REFERENCES

1. E. L. O'Neill, *Introduction to Statistical Optics*, Addison-Wesley Publishing Co., Reading, Mass. (1963).
2. J. W. Goodman, *Introduction to Fourier Optics*, McGraw-Hill Book Co., New York (1968).
3. L. Beiser, "Laser scanning systems," in *Laser Applications*, M. Ross, ed., Vol. 2, pp. 53–159, Academic Press, New York (1974).
4. J. C. Dainty and R. Shaw, *Image Science*, Academic Press, New York (1974).
5. P. F. Mueller, "Linear multiple image storage," *Appl. Opt.* 8(2), 267–273 (1969).
6. F. T. S. Yu, *Introduction to Diffraction, Information Processing and Holography*, Massachusetts Institute of Technology Press, Cambridge, Mass. (1973).
7. R. J. Becherer and J. D. Geller, "Optimum shading apertures for reducing photographic grain noise," in *Image Information Recovery*, A. Derr, ed., Proc. SPIE 16, 89–96 (1968).
8. W. M. Brown, *Analysis of Time Invariant Systems*, McGraw-Hill Book Co., New York (1963).

9. W. B. Davenport and W. L. Root, *Introduction to Theory of Random Signals and Noise*, Chapter 11, McGraw-Hill Book Co., New York (1958).
10. S. Goldman, *Information Theory*, Prentice Hall, Englewood Cliffs, N.J. (1953).
11. C. W. Helstrom, "Image restoration by the method of least squares," *J. Opt. Soc. Am.* 57(3), 297–303 (1967).
12. J. L. Horner, "Optical restoration of images blurred by atmospheric turbulence using optimum filter theory," *Appl. Opt.* 9(1), 167–171 (1970).
13. Y. W. Lee, *Statistical Theory of Communication*, John Wiley and Sons, Inc., New York (1960).
14. D. W. Slepian, "Linear least-squares filtering of distorted images," *J. Opt. Soc. Am.* 57(7), 918–922 (1967).
15. A. Labeyrie, "Attainment of diffraction limited resolution in large telescopes by Fourier analysis speckle patterns in star images," *Astronom. and Astrophys.* 6(1), 85–87 (1970).
16. K. T. Knox and B. J. Thompson, "Recovery of images from atmospherically degraded short-exposure photographs," *Astrophys. J.* 193(1, Part 2), L45–L48 (1974).

29 Analog Optical Computing: Experimental Fourier Analysis

29.1 INTRODUCTION

In previous chapters, the important role of the Fourier transformation in modern physical optics has been demonstrated in diffraction and interference phenomena and in the description of imaging systems, with emphasis on spatial spectral analysis. By solving the wave equation in Chapter 1, it was shown that the optical wavefront in the far field of the diffracting aperture (Fraunhofer zone) or in the focal plane of a lens is the Fourier transform of the optical wavefront in the diffracting aperture. In Chapter 3, the Fourier transform was used to develop the array theorem and the convolution theorem was derived mathematically using an integral representation of the delta function. Chapters 4 through 8 applied Fourier theory to imaging systems in order to treat them as linear. These systems are described equally well by either an optical impulse response or its Fourier transform, the optical transfer function. Fourier analysis was also used in Chapters 11, 22, 23, 25, and 26 to develop the principles of interferometry, coherence, and holography.

In this chapter and the next four chapters, we will develop the application of Fourier transforms to the analysis and synthesis of optical signals using both optical analog methods as well as hybrid methods. In particular, we show how an analog optical system can be used to perform optical harmonic analysis resulting in many useful mathematical operations. As has been the case in these chapters, no attempt is made to include references to all the original work; rather, useful supplemental reference material is cited to support the particular approach being taken.

29.2 OPTICAL FOURIER TRANSFORMS

We now present some of the basic ideas of Fourier harmonic analysis. As illustrated in Fig. 29.1, diffraction theory shows that in the far-field (Fraunhofer zone) of the object (i.e., $z \gg d^2/\lambda$)

$$\tilde{\psi}(\mathbf{x}) = K \int \psi(\boldsymbol{\xi}) \exp[-2\pi i(\boldsymbol{\xi} \cdot \mathbf{x}/\lambda z)] \, d\boldsymbol{\xi} \,, \tag{29.1}$$

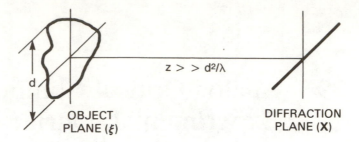

Fig. 29.1 Schematic for Fraunhofer diffraction.

where $\tilde{\psi}(\mathbf{x})$ denotes the Fourier transform of the function $\psi(\boldsymbol{\xi})$; $d\boldsymbol{\xi} = d\xi d\eta$; K is a complex multiplication factor containing a quadratic phase term in $\mathbf{x}$; $\boldsymbol{\xi} = \mathbf{i}\xi + \mathbf{j}\eta$; $\mathbf{x} = \mathbf{i}x + \mathbf{j}y$; $\mathbf{i}$ and $\mathbf{j}$ are unit vectors directed along the coordinate axes; and the integral is assumed to have infinite limits unless otherwise noted. Thus, the amplitude distribution in the diffraction plane is seen to be proportional to the Fourier transform of the amplitude distribution in the object plane.

A similar situation exists between the front and back focal planes of a lens when illuminated with coherent radiation. For the convenience of analysis, we will use the coherent optical system shown in Fig. 29.2. For this system,

$$\tilde{\psi}(\mathbf{x}) = A \int \psi(\boldsymbol{\xi}) \exp\left[\frac{-2\pi i(\boldsymbol{\xi} \cdot \mathbf{x})}{\lambda f}\right] d\boldsymbol{\xi}, \tag{29.2}$$

which shows that the amplitude distribution in the back focal plane of the lens is the two-dimensional spatial Fourier transform of the amplitude distribution in the front focal plane apart from a multiplicative complex constant A. Using this result, we can demonstrate some of the interesting Fourier transform relationships discussed in previous chapters.

29.3   SLIT APERTURE

One simple example is to place a one-dimensional, uniformly illuminated slit of width $2a$ in the front focal plane of the lens in Fig. 29.2. In this one-dimensional case, Eq. (29.2) yields

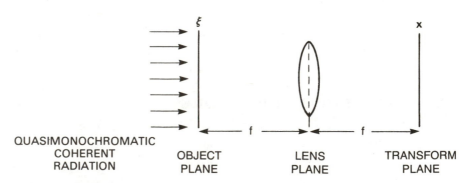

Fig. 29.2 Schematic for a coherent optical Fourier transform system.

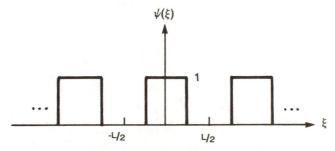

Fig. 29.3 Diffraction pattern of a slit: (a) normalized plot of Eq. (29.4) with $x' = 2\pi ax/\lambda f$ and (b) experimental result showing central and side order maxima and minima (from Ref. 1).

$$\tilde{\psi}(x) = \int \text{rect}(\xi | a) \exp\left[-2\pi i\left(\frac{x\xi}{\lambda f}\right)\right] d\xi = \int_{-a}^{a} \exp\left[-2\pi i\left(\frac{x\xi}{\lambda f}\right)\right] d\xi$$

$$= 2a \, \text{sinc}\left(\frac{2\pi ax}{\lambda f}\right), \tag{29.3}$$

where we have set the constant A equal to unity for convenience. The resulting intensity distribution is given by

$$I(x) = |\tilde{\psi}(x)|^2 = 4a^2 \, \text{sinc}^2\left(\frac{2\pi ax}{\lambda f}\right), \tag{29.4}$$

which is sketched in Fig. 29.3(a) and, when compared to the experimental result in Fig. 29.3(b), illustrates the agreement between theory and experiment. The side lobes of the sinc function appear as secondary maxima in Fig. 29.3(b).

29.4   PERIODIC RECTANGULAR APERTURES

The one-dimensional mathematical square wave shown in Fig. 29.4 possesses a Fourier series representation given by

$$\psi(\xi) = \frac{1}{2} + \sum_{n=1}^{\infty} \left(\frac{2}{n\pi}\right) \sin\left(\frac{n\pi}{2}\right) \cos\left(\frac{n\pi\xi}{L}\right). \tag{29.5}$$

Note that $\psi(\xi)$ as expressed by Eq. (29.5) and represented in Fig. 29.4 corresponds to the transmission of an optical grating; that is, the transmission is unity

Fig. 29.4 Schematic diagram of a square wave.

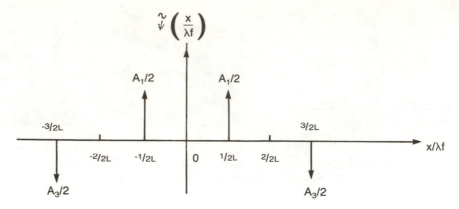

Fig. 29.5 Fourier decomposition of a square wave.

in the ruled portions of the grating and zero in the unruled portions (the reader is referred to Chapter 8 for a description and discussion of Ronchi rulings). If we place such a grating in the front focal plane of the lens system in Fig. 29.2 and illuminate it with a collimated quasimonochromatic beam, then in the back focal plane, we obtain, aside from a multiplicative constant, the amplitude Fourier transform given by

$$\tilde{\psi}(x) = \frac{1}{2}\delta(x) + \sum_{n=1}^{\infty} \frac{A_n}{2}\left[\delta\left(\frac{x}{\lambda f} + \frac{n}{2L}\right) + \delta\left(\frac{x}{\lambda f} - \frac{n}{2L}\right)\right], \qquad (29.6)$$

where $A_n = (2/n\pi)\sin(n\pi/2)$. In Fig. 29.5, we show a schematic diagram of the discrete Fourier spectrum given by Eq. (29.6), which is an amplitude-modulated Dirac comb function.

In Fig. 29.6, we see schematically that the Fourier coefficients given by Eq. (29.6) are just the values of the sinc function envelope at the harmonic frequencies. An alternative method of performing this calculation is given in the appendix in Sec. 29.8.

Note that for a square wave the even harmonics coincide exactly with the zeros of the sinc function. Figure 29.7 shows two experimental Fourier spectra created from a rectangular (wave) grating and a square (wave) grating, respectively.

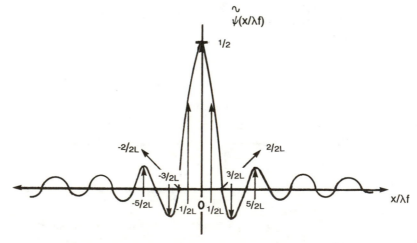

Fig. 29.6 Schematic diagram of a Fourier transform of a square wave.

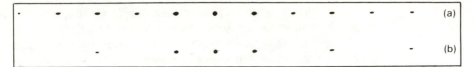

Fig. 29.7 Experimental Fourier transforms of (a) rectangular (wave) and (b) square (wave) gratings demonstrating the missing even orders. (Courtesy of P. F. Mueller.)

These results were obtained using the optical Fourier transform system shown in Fig. 29.2. It should be noted that the even harmonics are present in the spectrum of the rectangular (wave) grating. This is easily seen in Eq. (29.6) if the half-width of the transmitting portions of the rectangular wave are changed from $L/2$ to an arbitrary size b. However, if the ratio $2L/b =$ (period/half-width) is an integral number N, then all the spectral orders that are multiples of $N/4L$ will be exactly zero. This experiment also illustrates nonlinear film processing as discussed in Chapter 21.

29.5   OPTICAL ADDITION

If two rectangular waves in perpendicular directions are added by subsequent exposures onto photographic film which is then linearly procesed, the resulting amplitude transmission of the film placed in the ξ plane of Fig. 29.2 is given by

$$\psi(\xi) = S(\xi) + S(\eta) \ , \tag{29.7}$$

where $S(\xi)$ and $S(\eta)$ are the representations of the rectangular wave functions in the ξ and η directions, respectively. If we use the fact that the two-dimensional Fourier kernel is separable, the normalized Fourier spectrum in the **x** plane of Fig. 29.2, except for a multiplicative factor, is given by

$$\tilde{\psi}(x,y) = \delta(y) \sum_{n=0}^{\infty} B_n \delta \left(x + \frac{n}{2L} \right) + \delta(x) \sum_{n=0}^{\infty} B_n \delta \left(y + \frac{n}{2L} \right) \ , \tag{29.8}$$

where the B_n are the appropriate Fourier coefficients for a rectangular wave.

Figure 29.8 shows the result obtained using the system of Fig. 29.2 to optically

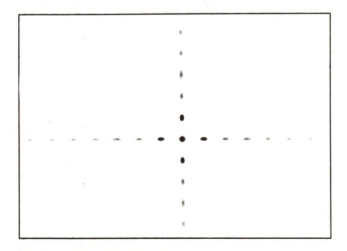

Fig. 29.8 Fourier spectrum demonstrating the addition of two rectangular (wave) gratings in perpendicular directions. (Courtesy of P. F. Mueller.)

Fourier transform a photographic transparency consisting of the sum of two gratings. The sum of the gratings, represented by Eq. (29.8) was experimentally created by first exposing a rectangular grating in one direction and then exposing the same grating, oriented in a perpendicular direction, onto photographic film. The film was subsequently linearly processed and placed in the ξ plane of Fig. 29.2.

29.6   OPTICAL CONVOLUTION

If two rectangular (wave) gratings are placed together at right angles, their transmissions multiply, i.e.,

$$\psi(\xi) = S(\xi) S(\eta) .$$

(29.9)

The Fourier transform of this transmission function, apart from a multiplicative factor, is

$$\tilde{\psi}(\mathbf{x}) \cong \delta(y) \sum_{n=0}^{\infty} B_n \delta\left(x - \frac{n}{2L}, y\right) \circledast \delta(x) \sum_{n'=0}^{\infty} B_n \delta\left(y - \frac{n'}{2L}\right) ,$$

(29.10)

where $\circledast$ denotes the convolution process and the B_n are the Fourier coefficients for a rectangular wave. The experimental spectrum obtained by optically Fourier transforming two such multiplicative gratings is shown in Fig. 29.9. It is interesting to note that the double-exposure photograph of two separate gratings and the single-exposure photograph of two crossed gratings are nearly indistinguishable to the eye. However, as shown by Eqs. (29.8) and (29.10) and Figs. 29.8 and 29.9, there is a definite difference between addition and multiplication with optical systems, as one intuitively expects.

We have invoked the convolution theorem in Eq. (29.10), which says that the Fourier transform of the product of two functions is the convolution of their respective transforms. Using $\mathscr{F}$ to denote a Fourier transform, the convolution theorem may be expressed as

$$b(x) = \mathscr{F}[f_1(\xi) f_2(\xi)] = \int \mathscr{F}[f_1(x - x_1)] \mathscr{F}[f_2(x_1)] dx_1$$
$$= \mathscr{F}[f_1(x)] \circledast \mathscr{F}[f_2(x)] .$$

(29.11)

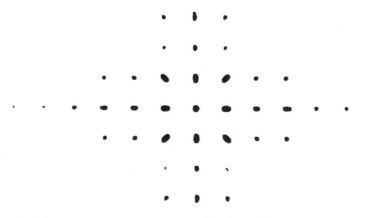

Fig. 29.9 Fourier transforms of multiplicative gratings in perpendicular directions demonstrating the convolution process. (Courtesy of P. F. Mueller.)

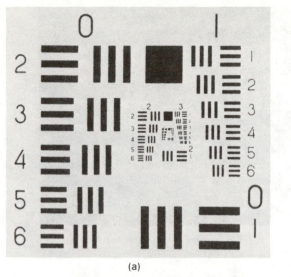

(a)   (b)

Fig. 29.10 (a) Three-bar target and (b) the intensity distribution corresponding to the Fourier transform of a photographic transparency of the three-bar target.

In Chapter 28, we used the result expressed in Eq. (29.11) to calculate the convolution of symmetric and nonsymmetric functions in order to compare the results to correlations of the same functions.

29.7 OPTICAL SPECTRUM REPLICATION BY MULTIPLICATION

If a band-limited photographic copy of a two-dimensional three-bar target, as shown in Fig. 29.10(a), is optically Fourier transformed, its transform will have the characteristic appearance shown in Fig. 29.10(b). If the bar target $\psi(\xi)$ is multiplied by a rectangular grating and then Fourier transformed, we obtain

$$b(\mathbf{x}) = \bar{\psi}(\mathbf{x}) \circledast \delta(y) \sum_{n=0}^{\infty} B_n \delta\left(x - \frac{n}{2L}\right) , \tag{29.12}$$

where the B_n are the appropriate Fourier coefficients for a rectangular wave. The optical experimental result corresponding to Eq. (29.12) is demonstrated in Fig. 29.11. This operation is referred to as *spectrum replication* (i.e., identical spectra of varying intensity are located at each diffraction order of the grating). We note, once again, that the energy distribution of the various orders is determined by the Fourier coefficients described previously in Eqs. (29.5) and (29.6). Such a technique has application in multichannel analysis, and has been used to advantage in seismic data analysis[2] and in multiple-image storage techniques.[3]

29.8 APPENDIX: FOURIER TRANSFORM OF A RECTANGULAR WAVE

An alternative way to express the square wave transmission given by Eq. (29.5) is

$$\psi(\xi) = \text{rect}(\xi|L/2) \circledast \sum_{n=-\infty}^{\infty} \delta(\xi - 2nL) , \tag{29.13}$$

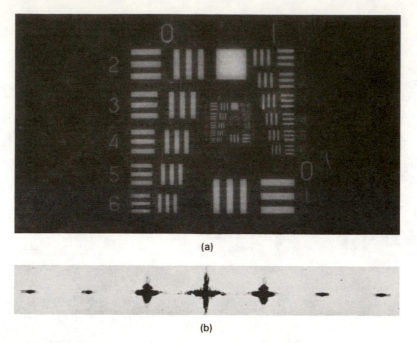

(a)

(b)

Fig. 29.11 (a) Modulated three-bar target and (b) convolution of bar target spectrum with Dirac comb.

where the rect function represents the central region of transmission (pulse) in Fig. 29.4. Fourier transforming Eq. (29.13) yields

$$\tilde{\psi}(x) = L \operatorname{sinc}\left[\frac{2\pi(L/2)x}{\lambda f}\right]\left[\frac{1}{2L}\sum_{n=-\infty}^{\infty}\delta\left(\frac{x}{\lambda f}-\frac{n}{2L}\right)\right] , \tag{29.14}$$

where we have utilized the fact that the Fourier transform of an array of delta functions is given by[4]

$$\mathscr{F}\left[\sum_{n=-\infty}^{\infty}\delta(\xi-2nL)\right] = \frac{1}{2L}\mathscr{F}\left[\operatorname{comb}\left(\frac{\xi}{2L}\right)\right]$$

$$= \frac{1}{2L}\sum_{n=-\infty}^{\infty}\delta\left(\frac{x}{\lambda f}-\frac{n}{2L}\right) . \tag{29.15}$$

The sinc function in Eq. (29.14) has a zero when

$$2\pi(L/2)(x/\lambda f) = \pi \tag{29.16}$$

or

$$x/\lambda f = 1/L , \tag{29.17}$$

indicating that the first zero, which occurs at $1/L$, coincides with the second harmonic of the square wave. When $x/\lambda f$ equals $1/2L$, i.e., the position of the first harmonic in Eq. (29.14), the value of the sinc function in Eq. (29.14) is

$$\operatorname{sinc} 2\pi(L/2)(1/2L) = 2/\pi . \tag{29.18}$$

The value of the Fourier transform $\tilde{\psi}(x)$ of Eq. (29.14) at the fundamental is $1/\pi$, which agrees with Eq. (29.6).

In general, any rectangular wave of period $2L$ can be represented by Eq. (29.13) by changing the width of the rect function so that its Fourier coefficients are proportional to the values of the sinc function at the harmonic frequencies, as illustrated in Fig. 29.6 for a square wave.

REFERENCES

1. E. Hecht and A. Zajac, *Optics*, p. 345, Addison-Wesley Publishing Co., Reading, Mass. (1979).
2. P. Jackson, "Diffractive processing of geophysical data," *Appl. Opt.* 4(4), 419 (1965).
3. P. F. Mueller, "Linear multiple image storage," *Appl. Opt.* 8(2), 267–273 (1969).
4. J. D. Gaskill, *Linear Systems, Fourier Transforms and Optics*, Chapter 7, John Wiley and Sons, Inc., New York (1978).

30   Analog Optical Computing: Fourier Synthesis Utilizing Amplitude Filters

30.1   GENERALIZED OPTICAL SYSTEM FOR FOURIER FILTERING

In Chapter 29 we demonstrated that a lens can be used to perform optical Fourier transformations. As discussed previously in Chapter 19, a second lens can be used to perform a second Fourier transform to produce an image, and controlled manipulations can be performed in the first transform plane. This system, shown in Fig. 30.1, is often used for optical spatial filtering. This system is also referred to as a *one-to-one imaging system* since a nonmagnified image of the object appears in the back focal plane of the second lens. Magnification can be introduced by varying the ratio of the focal lengths of lenses L_2 and L_3. It should be noted that a one-lens system could also be used to perform this filtering function [1] with magnification; however, we have chosen the system in Fig. 30.1 for simplicity.

30.2   MULTIPLICATION WITH BINARY FILTER FUNCTIONS

Optical filtering of imagery can be achieved by using multiplicative binary masks (blocking filters) in the **x** plane of Fig. 30.1. One example of an experiment (from the Abbe-Porter experiments) in which this was done was discussed in Chapter

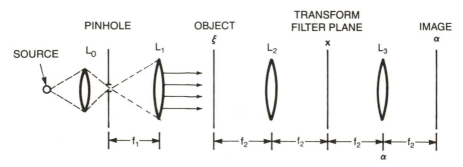

Fig. 30.1 Generalized optical filtering system: L_0, source imaging lens; L_1, collimating lens; L_2, first transform lens; and L_3, second transform lens.

19 where dots were changed to lines by altering the position of the multiplicative blocking filter in the Fourier plane of the optical system. If the entire spectrum of the object was not passed by the multiplicative filter, then the form of the image was altered.

This concept is further illustrated in Fig. 30.2 where the duck, seen in Fig. 30.2(b), is placed in the object plane of Fig. 30.1. The filtered images—passed through multiplicative blocking filters having various diameters—are illustrated in Figs. 30.2(b), (d), (f), (h), (j), and (l), where the corresponding filters are shown in Figs. 30.2(a), (c), (e), (g), (i), and (k). As the diameter of the blocking filter is reduced, the higher frequency information in the image is lost until finally the image of the duck is not even recognizable. Figure 30.2(l) is the image formed when an off-axis part of the transform, shown in (k), is used.

The experimental arrangement in Fig. 30.1 can be used to perform a number of other operations using multiplicative blocking filters. Some of these are

1. removal of halftone

2. raster removal

3. edge sharpening

4. enhancement of a periodic signal in additive noise

5. filtering of seismic data.

In Fig. 30.1, the coherent image of the object amplitude distribution appears in the image plane. If a filter mask is placed in the transform plane, then the convolution of the Fourier transform of the filter with the object amplitude distribution appears in the back focal plane of the second transform lens. In *symbolic notation,* if the object amplitude distribution is $f(\xi)$, and its Fourier transform is represented by $\tilde{f}(\mathbf{x})$ and the filter by $M(\mathbf{x})$, the image amplitude distribution is given by

$$\psi(\alpha) = f(\alpha) \circledast \tilde{M}(\alpha) \ . \tag{30.1}$$

30.2.1   Halftone Removal

A simple, but illustrative, example using the system shown in Fig. 30.1 demonstrates the use of spatial filtering to remove halftoning. For a halftone transparency [see Fig. 30.3(a)], the object distribution may be represented by

$$f(\xi) = O(\xi) S(\xi) \ , \tag{30.2}$$

where $O(\xi)$ is the band-limited object of interest, and $S(\xi)$ is a two-dimensional periodic binary function that properly samples the object. Since the Fourier transform of a two-dimensional periodic function is an amplitude-modulated Dirac comb function, the transform plane consists of the Fourier transform of the object convolved with this amplitude-modulated Dirac comb function [see Fig. 30.3(c)]. If a mask consisting of an aperture in an opaque screen having a diameter equal to the object bandwidth is used as a filter and centered on one of the spectral orders of the Dirac comb (any convenient order may be chosen), the intensity distribution in the image plane becomes $|O(\alpha)|^2$. Thus, the halftoning has been removed by a spatial filter as shown in Fig. 30.3(b). This illustrates one use of binary filters in the system.

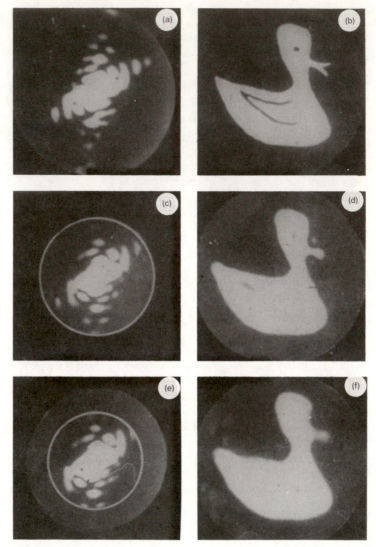

Fig. 30.2 (a–f) Example of blocking filters. Reconstructed duck images and their corresponding filter plane distributions as limited by circular holes in an otherwise opaque background (from Taylor and Lipson).

30.2.2 Raster Removal

Another example of filtering a periodicity in the transform plane is shown in Fig. 30.4. In Fig. 30.4(a), the original object has a raster since it was photographed from a television display. The enlargement of the ear in the upper left-hand corner shows the raster lines. The filtered image (raster has been removed) with the same insert is shown in Fig. 30.4(b). In the insert, the removal is quite dramatic. In this case, the filter was a band-limiting aperture placed in the transform plane and centered at one of the orders of the diffraction pattern.

30.2.3 Edge Enhancement

Figure 30.5 shows the results of high-pass filtering, which demonstrates edge enhancement. The filter, which was designed to entirely block the low spatial

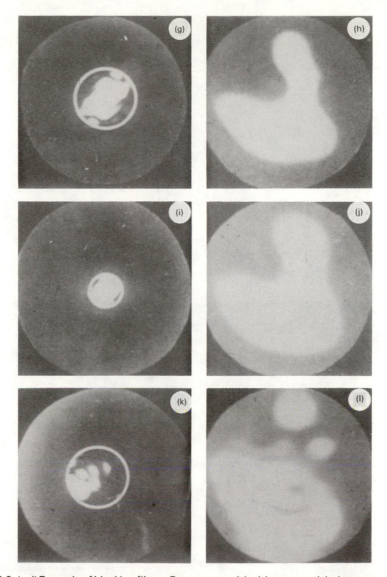

Fig. 30.2 (g–l) Example of blocking filters. Reconstructed duck images and their corresponding filter plane distributions as limited by circular holes in an otherwise opaque background (from Taylor and Lipson).

frequencies and pass the high spatial frequencies, consisted of a black, absorbing circular mask centered on the optical axis in the transform plane. The high-pass filtered image shown in Fig. 30.5(b) illustrates the absence of dc (on-axis undiffracted energy in the transform plane) in that the larger bars are dark in the middle and have sharp edges. To further emphasize this edge sharpening, the magnified center section of the original object is shown in Figs. 30.5(c) and its high-pass filter image in Fig. 30.5(d). This enlarged filtered image has sharp edges but the centers of the bars start to fill in, which is indicative of a high-frequency structure at these locations in the object. Notice that the filter acts similar to a differentiating filter in that the edges are shown as lines. Figures 30.5(c) and (d) represent frequency information from 4 to 30 line pairs/mm. The filter starts passing an image at the 0 to 3 group, which corresponds to 1.25 line pairs/mm.

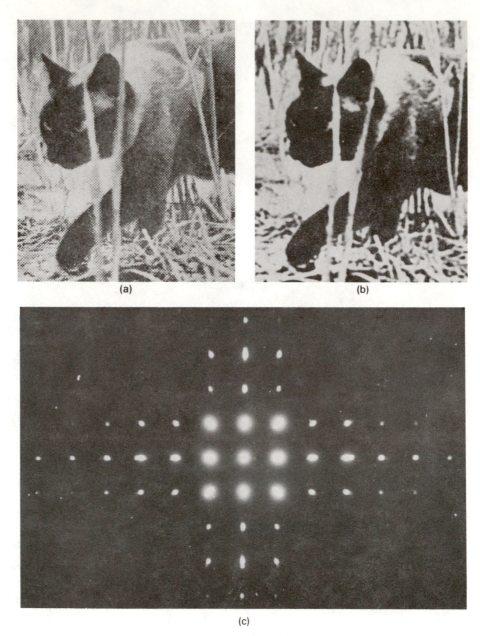

Fig. 30.3 Removal of halftone: (a) halftone original, (b) filtered image with halftone removed, and (c) Fourier spectrum demonstrating convolution (after Ref. 2).

30.2.4   Enhancement of a Periodic Signal in Additive Random Noise

In this experiment,[3] a periodic object (a diffraction grating in Fig. 30.6) is added to random noise [Fig. 30.6(b)] by double exposure onto photographic film [Fig. 30.6(c)] to form the transparency placed in the object plane of Fig. 30.1. The random noise pattern was created by sprinkling white circular dots onto a black background. A periodic array of pinholes in an opaque mask, each centered at the harmonic frequencies of the grating, was used as a multiplicative filter to

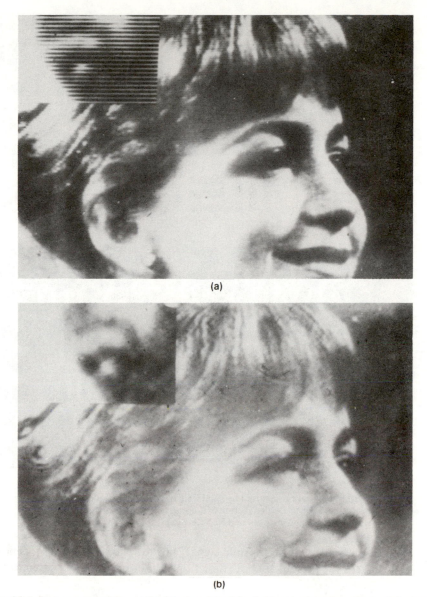

(a)

(b)

Fig. 30.4 Raster removal by spatial filtering: (a) original object (enlargement at top left shows raster) and (b) filtered image with raster removed. (Courtesy of P. F. Kellen.)

remove the noise and increase the signal-to-noise ratio of the periodic function in the filtered image, as shown in Fig. 30.6(d).

30.2.5 Filtering of Seismic Data

Blocking filters can be used to filter a seismic signal (consisting of variations about the mean path) from the periodic traces. The seismic data shown in Fig. 30.7(a) are placed in the object plane of Fig. 30.1 and a horizontal slit in the transform plane is used to reject the periodic traces and pass the information

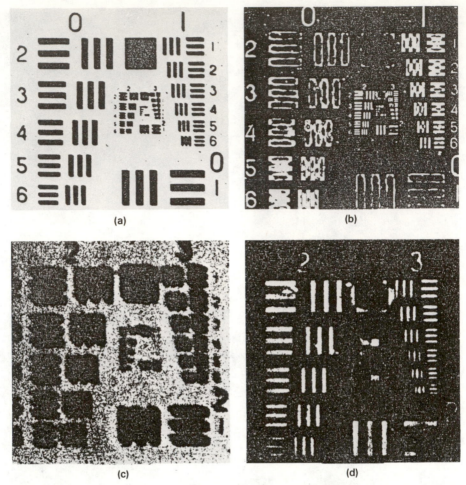

Fig. 30.5 Edge enhancement by spatial filtering: (a) original object, (b) filtered image showing enhanced edges, (c) center section of original object, and (d) its image showing enhanced edges. (Courtesy of P. S. Considine and R. A. Profio.)

associated with the seismic signal, as shown in Fig. 30.7(b). Since the timing marks are periodic in both directions, they are partially passed by the slit filter.

30.3   OBJECT REPLICATION AS AN EXAMPLE OF MULTIPLICATION WITH A PERIODIC BINARY FILTER

The multiplication concept can also be used to produce object replication by multiplying the spectrum of the object with a diffraction grating in the transform plane of Fig. 30.1, and then taking the optical Fourier transform of the product. The resulting convolution produces replicated images of the object in the image plane of Fig. 30.1. By using crossed gratings, one can utilize the full two-dimensional nature of the Fourier transform plane for object replication as seen in Fig. 30.8. This principle has also been used to build a pulsed laser sensitometer.[4]

30.4   OPTICAL SUBTRACTION BY MULTIPLICATION WITH A PERIODIC AMPLITUDE FILTER

Analog optical subtraction can be accomplished by using the generalized optical filtering system shown in Fig. 30.1. Many techniques have been introduced in the

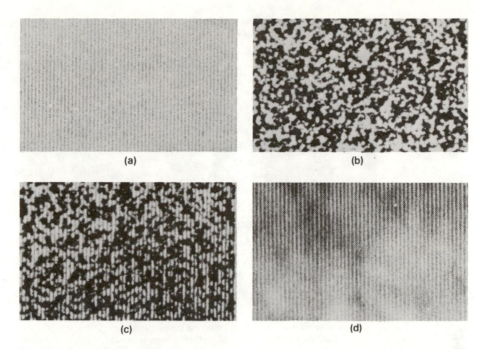

Fig. 30.6 Spatial filtering experiment illustrating the suppression of random noise on a periodic signal with a Dirac comb filter: (a) periodic signal, (b) random noise, (c) signal plus noise (object), and (d) filtered image of object (after Ref. 3).

literature[5,6] to accomplish optical subtraction, and we shall utilize one method[7] that can be used readily with the optical system of Fig. 30.1.

If we place the two functions $f_1(\xi)$ and $f_2(\xi)$ to be subtracted in the object plane of Fig. 30.1 and separate them by a distance of $2b$, then the spatial Fourier transform of this object scene appears in the $\mathbf{x}$ plane. For simplicity, the following analysis is done in one dimension. However, experimental subtraction of two-dimensional images can be performed using this system.

A filter whose amplitude transmission is given by

$$T_A(x) = A + B\sin\omega x \,, \tag{30.3}$$

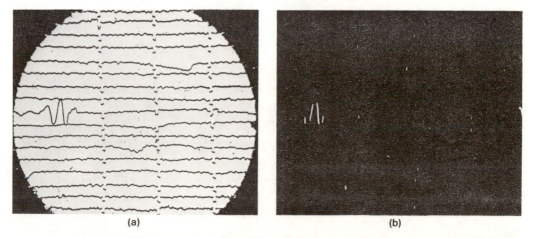

Fig. 30.7 (a) Seismic data and (b) filtered data from a slit filter illustrating the removal of periodic clutter. (Courtesy of P. S. Considine and R. A. Profio.)

Fig. 30.8 Replicated images resulting from one three-bar target in the object plane of Fig. 30.1 and crossed diffraction gratings in the transform plane. (Courtesy of P. F. Kellen.)

where $\omega = 2\pi b / \lambda f$ is placed in the **x** plane. The subtraction property of the filter is created by linearly shifting an interferometrically produced cosine grating through a distance of three quarters of a period in the transform plane. The amplitude transmission of this shifted cosine grating in the transform plane is

$$T_A(x) = A + B\cos\beta\left(x + \frac{3\pi}{2\omega}\right) ,\tag{30.4}$$

which is equivalent to the desired filter given in Eq. (30.3).

The Fourier transform of Eq. (30.3) gives the amplitude impulse response of the filter as

$$h(\alpha,\beta) = \tilde{T}_A(\alpha,\beta) = A\delta(\alpha)\,\delta(\beta) - \frac{i}{2}\,\delta(\beta)\,B\delta(\alpha + b)$$

$$+ \frac{i}{2}\,\delta(\beta)\,B\delta(\alpha - b) ,\tag{30.5}$$

where the two off-axis delta functions are 180 deg out of phase with respect to each other, and both are 90 deg out of phase with the delta function at the origin.

The transform of the two separated objects in the **x** plane is

$$\tilde{f}(\mathbf{x}) = \tilde{f}_1(\mathbf{x})\exp\left(\frac{ikb\mathbf{x}}{f}\right) + \tilde{f}_2(\mathbf{x})\exp\left(\frac{-ikb\mathbf{x}}{f}\right) .\tag{30.6}$$

When placed in the transform plane, the grating of Eq. (30.4) multiplies the object spectrum given by Eq. (30.6). The Fourier transform of this product is performed by the second transforming lens in Fig. 30.1, resulting in object

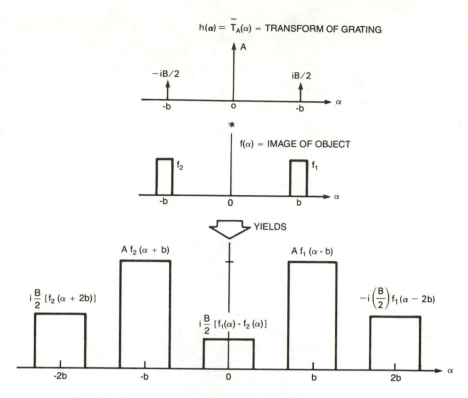

Fig. 30.9 Schematic illustrating subtraction of two on-axis images.

replication. Since $\omega = 2\pi b/\lambda f$, the lower side order image from the $+1$ diffraction order, $f_1(\alpha - b)$, will exactly overlap the upper side order image from the -1 diffraction order, $f_2(\alpha + b)$, as shown in Fig. 30.9. This results in image subtraction between the two overlapping images centered on axis because of the 180-deg phase shift between the two off-axis delta functions in Eq. (30.5).

Mathematically, convolution of the object

$$f(\xi) = f_1(\xi - b) + f_2(\xi + b) \tag{30.7}$$

with the impulse response of the filter of Eq. (30.5) gives

$$F'(\alpha) = f(\xi) \circledast h(\alpha - \xi, \beta)$$

$$= Af_1(\alpha - b) + Af_2(\alpha + b) + \frac{iB}{2} f_1(\alpha) - \frac{iB}{2} f_2(\alpha - 2b)$$

$$+ \frac{iB}{2} f_1(\alpha + 2b) - \frac{iB}{2} f_2(\alpha) . \tag{30.8}$$

This result is shown schematically in Fig. 30.9.

The third and sixth terms in Eq. (30.8) are centered on axis and have opposite signs, resulting in optical subtraction of the two functions f_1 and f_2 as shown experimentally in Fig. 30.10. The separation of the two objects, $2b$, must be chosen large enough so that the various terms in Eq. (30.8) do not overlap.

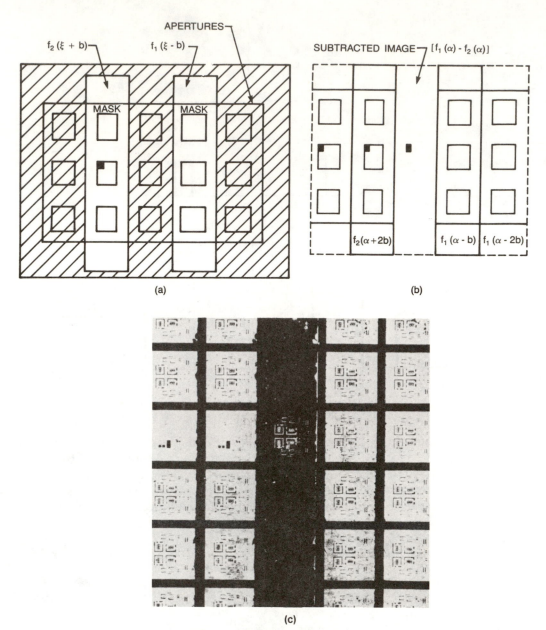

Fig. 30.10 Example illustrating optical subtraction: (a) schematic of the photomask array shown with the aperture mask superimposed, $f_1(\xi - b) + f_2(\xi + b)$; (b) image that would be observed at the output showing the subtracted image and the display of an error; and (c) experimental result of subtracting two columns of features. Aperture width is equal to one column and the separation is equal to two columns (from Ref. 7).

In the experiment of Fig. 30.10, two columns of images from an integrated circuit photomask were used as two separated objects $f_1(\xi - b)$ and $f_2(\xi + b)$. One of the images in $f_2(\xi + b)$ had a defect making it different from $f_1(\xi - b)$. Fabrication of a filter satisfying Eq. (30.3) and having an impulse response given by Eq. (30.5) resulted in image subtraction so that the defect is observed, as shown in Fig. 30.10(c).

REFERENCES

1. J. W. Goodman, *Introduction to Fourier Optics,* Chapter 5, McGraw-Hill Book Co., New York (1968).
2. P. S. Considine and R. A. Profio, "Image processing with in-line optical systems," in *Image Information Recovery,* A. Derr, ed., SPIE Proc. 16, 113–119 (1968).
3. E. L. O'Neill, *Introduction to Statistical Optics,* p. 102, Addison-Wesley Publishing Co., Reading, Mass. (1963).
4. J. P. Fallon and P. F. Kellen, "Film sensitometry with laser sources," *Opt. Eng.*12(2), 75–79 (1973).
5. J. F. Ebersole, "Optical image subtraction," *Opt. Eng.* 14(5), 436–447 (1975).
6. S. R. Dashell, A. W. Lohmann, and J. D. Michael, "Real time optical-electronic image subtraction," *Opt. Comm.* 8(2), 105–108 (1973).
7. L. S. Watkins, "Application of spatial filtering subtraction to thin film and integrated circuit mask inspection," *Appl. Opt.* 12(8), 1880–1884 (1973).

31 Analog Optical Computing: Fourier Synthesis Utilizing Amplitude and/or Phase Filters

31.1   OPTICAL DIVISION

In many spatial filtering systems, inverse filters, which can be real, purely imaginary, or complex, are used to remove image degradations. These systems assume that a degraded image was created in a linear and stationary (incoherent) imaging system having a known or measurable impulse response. Linear storage of the degraded image yields

$$I_{im}^{deg}(\alpha,\beta) = I_{ob}(\xi,\eta) \circledast \mathscr{I}_{deg}(\alpha,\beta|\xi,\eta) \ , \tag{31.1}$$

where

$$I_{im}^{deg}(\alpha,\beta) = \text{degraded image intensity}$$

$$\mathscr{I}_{deg}(\alpha,\beta|\xi,\eta) = \text{intensity impulse response of the degrading imaging system assumed to be linear and stationary}$$

$$I_{ob}(\xi,\eta) = \text{intensity distribution of the object}$$

$$\circledast \quad \text{denotes convolution.}$$

Fourier transformation of Eq. (31.1) yields

$$\tilde{I}_{im}^{deg}(\mu,\nu) = \tilde{I}_{ob}(\mu,\nu)\,\tau_{deg}(\mu,\nu) \ , \tag{31.2}$$

where

$$\tilde{I}_{im}^{deg}(\mu,\nu) = \text{Fourier spectrum of the degraded image}$$

$$\tilde{I}_{ob}(\mu,\nu) = \text{Fourier spectrum of the object}$$

$$\tau_{deg}(\mu,\nu) = \text{transfer function of the degraded imaging system and}$$

$\mu = x/\lambda z_2$ and $\nu = y/\lambda z_2$ where z_2 is the image distance.

(a)   (b)

Fig. 31.1 Contrast enhancement by inverse spatial filtering: (a) original object and (b) contrast-enhanced image (after Ref. 1).

If $\tau(\mu) \neq 0$, then dividing Eq. (31.2) by the transfer function of the degrading system, which is obtained in an independent measurement, gives the Fourier spectrum of the original object. A subsequent Fourier transform gives an image that is no longer degraded. Several examples illustrating this optical division procedure in which the inverse filters are real, purely imaginary, and complex are given in the next sections.

31.2   CASE I: REAL FILTERS

31.2.1   Example 1: Contrast Enhancement

The results of a contrast enhancement experiment using the system shown in the previous chapter (Fig. 30.1) are shown in Fig. 31.1. The original object is shown in Fig. 31.1(a) and the contrast-enhanced image is shown in (b). The filter used in this experiment was an inverse real filter whose transmission varied approximately as an inverse Gaussian function. This filter is centered on the optical axis in the transform plane and strongly absorbs light associated with low spatial frequencies while transmitting most of the energy associated with high spatial frequencies. The intermediate frequencies are subject to inverse Gaussian suppression. This suppression of low spatial frequencies, while simultaneously emphasizing the presence of higher spatial frequencies, causes contrast enhancement in the filtered image as illustrated in Fig. 31.1.

31.2.2   Example 2: The Cosine Experiment

Another example of an optical division experiment using real filters is the cosine experiment, which was used as a system calibration procedure for filtering the effects of random image degradation due to a random medium.[2]

The procedure was accomplished by first degrading the image of a three-bar target by photographing it through a diffraction-limited lens masked so that its amplitude transmission was proportional to one plus the $\cos(x)$. Next the impulse response was obtained by suitably photographing a point source through the same system. From this image, the inverse filter was obtained by recording its Fourier transform with a photographic gamma of 2.

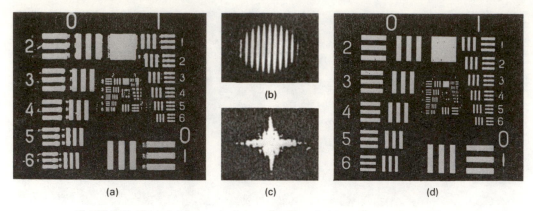

Fig. 31.2 Cosine division experiment: (a) degraded bar target image, (b) inverse transfer function filter, (c) spectrum of degraded bar target, and (d) retrieved bar target (after Ref. 2).

This inverse real filter appears as a cosine fringe pattern with a π phase shift, i.e., dark fringes where the transfer function has light fringes. In addition, the inverse transfer function has the same frequency as the distorting transfer function but with a different modulation.

The inverse filter and the degraded image spectrum are shown in Figs. 31.2(b) and (c), respectively. If we now place the degraded object shown in Fig. 31.2(a) in the object plane of Fig. 30.1, and place the inverse filter shown in Fig. 31.2(b) in the x plane of Fig. 30.1, we obtain the retrieved object three-bar target with the cosine degradation divided out as shown in Fig. 31.2(d). The fabrication of these filters requires precise sensitometry, as described in detail in the literature.[2]

31.2.3   Example 3: Removal of Random Media Distortions[2]

A turbulent medium such as the air above an open-coil electric hot plate placed in front of the lens of an incoherent imaging system distorts the image. Similar effects occur in optical telescopes due to clear air turbulence. The distorted image obtained with an exposure time that is short compared to the characteristic correlation time of the medium is an example of a linear, nonstationary system, i.e., the system impulse response varies as a function of position in the image plane. Small regions in the image plane over which the impulse response is constant are called *isoplanatic patches*. As discussed in Chapter 23, adaptive optics technology can be utilized to remove the atmospheric distortions over the field of view corresponding to a given isoplanatic patch.

The imaging system can be made stationary over its field of view by making the exposure time much longer than the characteristic correlation time of the medium. This time was determined empirically to be 10 to 15 s by placing point sources at the target extremities and increasing the exposure time until the impulse responses were similar, as shown in Fig. 31.3(a).

An example of a distorted image of a printed page target made through a severe turbulence with a 15-s exposure time is shown in Fig. 31.3(b). A point source was placed in the center of the target to diagnose the effect of the time-averaged turbulent medium. The resulting impulse response is seen in the center of Fig. 31.3(b). This impulse was optically Fourier transformed to create the inverse transfer function of the time-averaged random medium. This inverse transfer function filter is shown in Fig. 31.3(c). Placing the distorted image in the

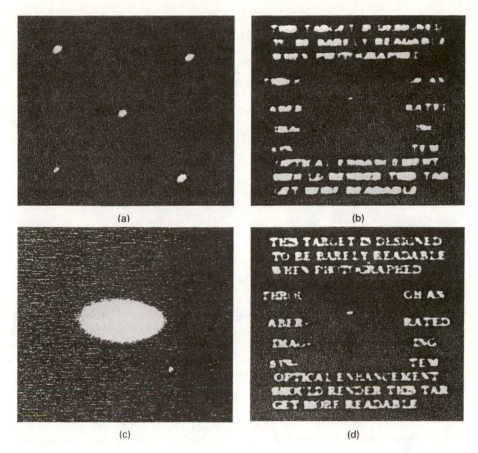

Fig. 31.3 Experimental results showing the removal of random media distortions: (a) time-averaged impulse responses over the object field of view illustrating system stationarity, (b) distorted image of object with diagnostic impulse response in the center for a 15-s exposure time, (c) inverse transfer function made from the diagnostic impulse response, and (d) enhanced image obtained from the analog optical division process (after Ref. 2).

object plane and the inverse filter in the filter plane of Fig. 30.1 results in the optically enhanced image shown in Fig. 31.3(d). Sensitometry measurements on the filter showed it to be an inverse Gaussian function. Subsequent measurements made with a hard-clipped, non–Gaussian filter caused additional image degradations, illustrating that the process of optical division is quite sensitive to the functional form of the filter.

31.3   CASE II: PURELY IMAGINARY INVERSE FILTERS

31.3.1   Example: Phase Correction of Human Cataracts[3,4]

For this case, we illustrate phase-only inverse filtering, which is equivalent to phase conjugation. This is the only example in this chapter in which the degraded object is created using coherent light.

The process of phase conjugation, which results from multiplying a complex wavefront by its complex conjugate, can be illustrated by referring to the schematic diagram shown in Fig. 31.4. The random phase wavefront can arise from propagation of a plane wave through a phase distorting plate having random pits and ripples on its surface. The resulting wavefront then consists of random phase

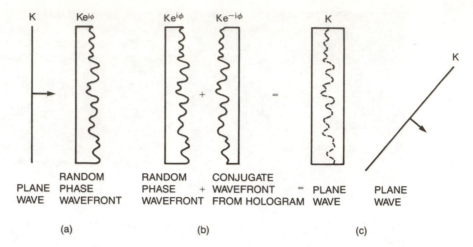

Fig. 31.4 Schematic demonstrating phase conjugation: (a) random wavefront, (b) conjugation process, and (c) resulting plane wave (after Ref. 3).

delays, as shown schematically in Fig. 31.4(a). The conjugate wavefront is realized as one of two holographic images. Its random pits and ripples are positioned such that a peak on the conjugate exactly matches the complementary valley on the phase plate, as shown in Fig. 31.4(b). Light propagating through both elements results in phase conjugation so that the original plane wave propagates in the direction of the holographic side order image as shown in Fig. 31.4(c).

The technique of correcting imagery taken through severe aberrations (*in vitro* human cataracts) with holographic phase conjugation over reasonable fields of view has been discussed in the literature.[3,4] In this work, an extracted cataract from a human patient was suitably chambered and mounted in a Mach-Zehnder type interferometer [Fig. 31.5(a)] and an image hologram was formed. Since the cataract is a pure phase function, its conjugate (inverse) is stored in the hologram as one of the side order images.

In the reconstruction process, the hologram was placed in the same position as it was recorded using an (x, y) stage and the reference beam was blocked [Fig. 31.5(b)]. The resulting target was viewed through the hologram in the direction of the conjugate cataract image. The positional tolerance of registering the hologram to the cataract was ~ 50 μm in both the x and y directions.

For this experiment, a standard Air Force three-bar resolution chart served as the target in Fig. 31.5(b). The target was placed 25 mm from the cataract and the conjugated image photographed in the image plane of lens L_5. This system was applied to 30 excised cataracts of varying degrees of severity. In all cases, improvement of resolution was obtained.

A typical result is shown in Fig. 31.6. Figure 31.6(a) represents the view through a 20/200 cataract without correction while (b) represents the view recorded through the same cataract, now conjugated by its correction hologram (inverse filter). Figure 31.6(c) is a magnified view of the corrected imagery. Imagery through this cataract was improved from the Snellen reading of 20/200 to an equivalent better than 20/15. These results, which were previously discussed in Chapter 21 as an example of noise reduction in holography, also demonstrate the phase-only inverse filtering property of holographic filters.

These results demonstrate that holographic phase conjugation techniques can be used to view a target through badly aberrated optical systems when laser

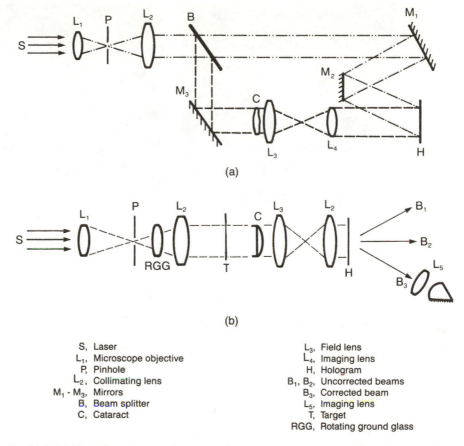

(a)

(b)

S, Laser	L₃, Field lens

S, Laser
L₁, Microscope objective
P, Pinhole
L₂, Collimating lens
M₁ - M₃, Mirrors
B, Beam splitter
C, Cataract

L₃, Field lens
L₄, Imaging lens
H, Hologram
B₁, B₂, Uncorrected beams
B₃, Corrected beam
L₅, Imaging lens
T, Target
RGG, Rotating ground glass

Fig. 31.5 (a) Mach-Zehnder interferometer for holographic recording. The cataract (C) is imaged onto the hologram plane. (b) Hologram reconstruction system. The three-bar resolution target (T) is imaged through the cataract along the B₃ direction. The rotating ground glass (RGG) eliminates coherent speckle noise (from Ref. 3).

illumination is used. Holographic phase conjugation systems (phase-only inverse filters) for coding and decoding images have also been discussed,[5] but these systems had difficult registration problems due to the severe scattering characteristic of the ground glass code plate. Additional examples of phase filtering are contained in Chapter 35 on phase contrast imaging methods.

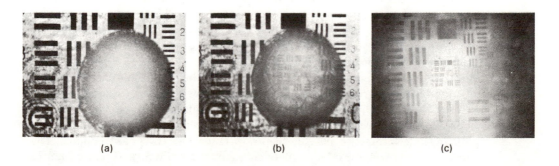

(a)   (b)   (c)

Fig. 31.6 Holographic correction of imagery through a cataract: (a) degraded image of resolution target placed directly behind cataract, (b) resolution target observed through a holographic correction filter, and (c) magnified portion of (b) showing high-resolution imagery observed through cataract and holographic filter (after Ref. 3).

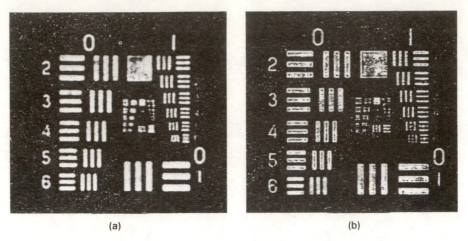

Fig. 31.7 (a) Out-of-focus image and (b) the image restored by complex filtering (after Ref. 1).

31.4   CASE III: COMPLEX INVERSE FILTERS

31.4.1   Example 1: Correction of Defocused Images

The third case to be considered for analog optical division is the general case of a complex inverse filter.[6,7] We consider a photograph taken by an out-of-focus camera of an object distribution $I_0(\xi, \eta)$. The camera has an intensity impulse response given by $\mathscr{I}_{deg}(\xi, \eta)$. The image distribution is given by Eq. (31.1).

The inverse filter to remove the effect of defocusing has an amplitude transmission of the form:

$$T_A(\mu, \nu) = \frac{D}{\tau_{deg}(\mu, \nu)} \text{ and } \tau(\mu, \nu) \neq 0 , \tag{31.3}$$

where $\tau_{deg}(\mu, \nu)$, the Fourier transform of the system intensity impulse response, is the transfer function of the defocused optical system and, in general, is a complex function. D is a constant and μ and ν are defined by Eq. (31.2).

A typical result illustrating the correction of a defocused image is shown[1] in Fig. 31.7. In Fig. 31.7(a), a defocused image made with a camera having five waves of defocus aberration is shown. The spurious resolution indicating the negative regions of the transfer function is seen near group 2–3 where three bars shift to two bars; i.e., there has been a contrast reversal of these bars. The impulse response for this experiment may be approximated by a circ function[8] and the corresponding defocused transfer function to be corrected is a besinc function. An inverse filter for this experiment is fabricated by utilizing a photographic positive of the besinc function (processed such that $\gamma = 1$) as the amplitude portion of the filter. The phase portion of the filter consists of a series of rings having alternating phases of zero and π. The phase filter can be made by deposition techniques or by controlling exposures on photoresist material and utilizing interferometric measurements to determine the phase heights. The corrected image obtained by using the composite amplitude and phase filter in the Fourier plane to remove the effect of defocus is described by Eq. (31.2) and shown in Fig. 31.7(b). Notice that the areas of spurious resolution have been removed; i.e., the contrast has been returned to the correct sense with the black bar in its proper location.

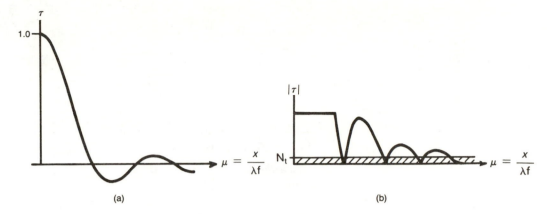

Fig. 31.8 Defocused and corrected transfer function: (a) transfer function to be corrected, normalized besinc function, and (b) modulus of corrected transfer function where N_t is the noise threshold (from Ref. 9).

The operations performed by the analog computer in this experiment are shown schematically in Fig. 31.8. Figure 31.8(a) shows a radial representation of the circularly symmetric defocused transfer function. The modulus of the transfer function after correction is shown in Fig. 31.8(b) indicating the regions in frequency space not corrected because of the zeros in the transfer function of (a) and the noise inherent in the photographic inverse amplitude filter.

31.4.2   Example 2: Correction for Image Motion

As a second example of complex inverse filtering, consider the case of a photographic image taken in a camera having linear image motion. In this case, the ideal impulse response, shown schematically in Fig. 31.9(a), is smeared into the rect function shown in (b). The transfer function (i.e., the Fourier transform of the impulse response) is the sinc function shown in Fig. 31.9(c). This transfer function has negative values which require that phase filtering be used. The inverse filter consists of two parts as shown in (d)—an amplitude portion and a phase portion. The amplitude portion is realized by recording the intensity in the transfer plane on film and developing to $\gamma = 1$. The phase filter will have π phase steps in those regions where the transfer function is negative and zero phase otherwise. This filter is made with deposition techniques as discussed in the last example. The results obtained using such a filter are shown in Figs. 31.9(e) and (f). The bar target, which is smeared in one direction from linear image motion in Fig. 31.9(e), has enhanced resolution after filtering, as seen in (f) where a few of the spurious resolution lobes of the complex transfer function have been corrected.

31.4.3   Example 3: Correction of Arbitrary Image Motion

As a third example of complex inverse filtering, we consider the case of a blurred photographic image taken with a camera moving in a Z pattern during the exposure. The original object shown in Fig. 31.10(a) was a printed page containing a diagnostic point that measures the impulse response of the system for the Z motion. The transfer function of the Z motion is obtained by optically Fourier transforming this impulse response.

The inverse filter in this case was made holographically using a Mach-Zehnder interferometer with the photograph of the system impulse response in one path.

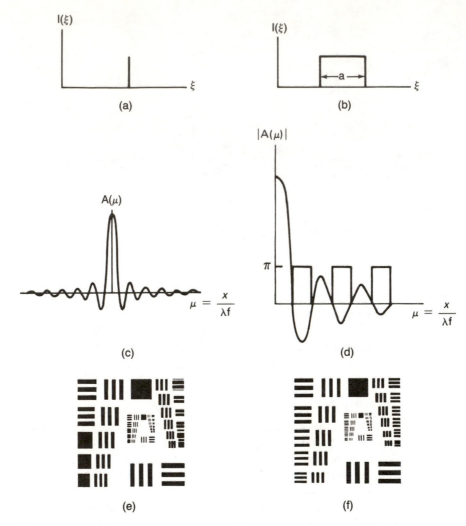

Fig. 31.9 Linear image motion compensation: (a) ideal impulse response; (b) smeared impulse response; (c) transfer function that corresponds to (b); (d) the modulus and phase of the transfer function in (c); (e) the smeared image; and (f) the filtered image with image motion aberration removed. (Courtesy of P. S. Considine.)

The holographic process is used because the asymmetry in the impulse response resulted in a complex transfer function that is asymmetric in both amplitude and phase (note that in the previous two examples the phase was binary).

The filter is made in two steps. First we fabricate the inverse squared magnitude of the transfer function of the Z motion, $1/|\tau_{deg}(\mu, \nu)|^2$, by taking the Fourier transform of the impulse response and recording the result on film as a positive transparency. We then holographically record the complex transfer function in amplitude and phase using the Mach-Zehnder interferometer. The positive transparency is placed in front of the hologram before recording so that the conjugate term of the holographic filter has the form

$$T_A(\mu, \nu) = \frac{1}{|\tau_{deg}(\mu, \nu)|} [\exp -i\phi(\mu, \nu)] \text{ and } \tau(\mu, \nu) \neq 0 , \qquad (31.4)$$

where $\phi(\mu, \nu)$ is the phase of the transfer function corresponding to the Z motion. Equation (31.4) describes the desired complex inverse filter.

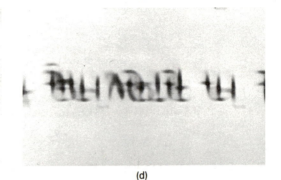

(a)

(b)

(c)

(d)

Fig. 31.10 Compensation for Z image motion by complex inverse filtering: (a) original printed page object and diagnostic point source; (b) blurred Z motion image of printed page object in (a); (c) image of a printed page corrected for the Z motion blur shown in (b); and (d) enlargement of portion of (c). (Courtesy of P. S. Considine.)

If we now place the original smeared photograph shown in Fig. 31.10(b) in the object plane of the filtering system and place the filter given by Eq. (31.4) in the filter plane, then we obtain the corrected image shown in Fig. 31.10(c) in the conjugate image direction of the holographic filter. An enlargement of a portion of the corrected image is shown in Fig. 31.10(d).

31.4.4   Example 4: Detection of Objects by Complex Inverse Filtering

As a fourth example of an inverse filter, we consider the important case of detection filtering. Two methods for fabricating these complex inverse detection filters are discussed and illustrated. In this application, an inverse filter is designed to locate (detect) a particular object in the presence of other objects (clutter).

Method 1: As a simple example, consider the geometrical objects shown in Fig. 31.11(a) and assume that we desire to detect the presence of the three large circles. Mathematically, we can represent the amplitude transmission of this object as

$$F(\xi, \eta) = \sum_{n=1}^{3} \text{circ}(|\boldsymbol{\xi} - \boldsymbol{\xi}_n| \,|\, a) + f(\xi, \eta) \ , \tag{31.5}$$

where $\sum_{n=1}^{3} \text{circ}(|\boldsymbol{\xi} - \boldsymbol{\xi}_n| \,|\, a)$ represents the three large circles and $f(\xi, \eta)$ represents the remainder of the objects in Fig. 31.11(a).

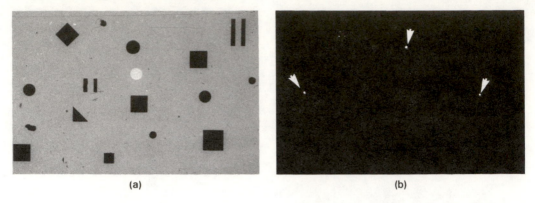

(a)   (b)

Fig. 31.11 Complex inverse detection filtering: (a) original object showing the three large circles of interest and other geometrical shapes and (b) output of detection filtering system showing the locations of the three large circles. (Courtesy of J. Ward.)

In the transform plane, the amplitude distribution corresponding to the object described by Eq. (31.5) using the result given in Chapter 2 is

$$\tilde{F}(x,y) = \tilde{f}(\mu, \nu) + (\pi a^2) \sum_{n=1}^{3}$$

$$\times \exp[2\pi i(\xi_n \mu + \eta_n \nu)] \operatorname{besinc}(2\pi a|\mu|) . \tag{31.6}$$

We fabricate a complex filter, having an amplitude transmission inversely proportional to the besinc function in Eq. (31.6), by the same techniques used in the previous examples. If this inverse complex filter is centered at the origin in the filter plane, the output amplitude in the image plane is given by

$$F(\alpha, \beta) = (\pi a^2) \sum_{n=1}^{3} \delta(|\alpha - \xi_n|) + f(\alpha, \beta) \circledast \mathscr{F}\left(\frac{1}{\operatorname{besinc}}\right), \tag{31.7}$$

where besinc $\neq 0$ and undefined otherwise. Equation (31.7) shows that the energy density at the locations of the centers of the three large circles in the original object is much greater than the energy density associated with the other objects, since they get convolved with the Fourier transform of the inverse filter. Hard clipping the energy in the output plane gives the result shown in Fig. 31.11(b), illustrating the detection and location of the presence of the circles of interest in the object.

This filter has a size discrimination capability of 10 to 15% due to the symmetry of the object used in the experiment. In this experiment, the objects of interest were positive definite and symmetric. Therefore, the only phase values necessary to create a filter representing the inverse Fourier spectrum of the object were zero and π. For objects having spatial asymmetries, the phase portions of their Fourier spectra are complicated and require more sophisticated methods for fabricating phase filters. Two methods for fabricating such complicated complex filters are holography and computer-generated hybrid filters. The holographic method is discussed in Chapter 33.

Method 2: Complex inverse filters can be designed in a computer and subsequently used in an analog optical filtering system. We illustrate this method by detection filtering of the object shown[1] in Fig. 31.12(a). The object consists of a

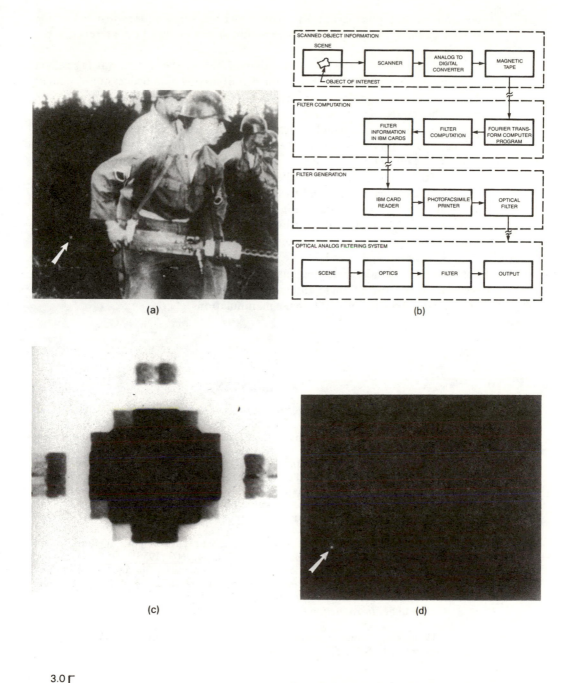

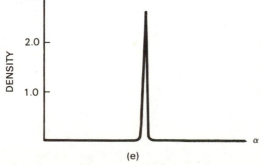

Fig. 31.12 Detection filtering using computer-generated filter fabrication method: (a) original continuous-tone soldier and square (courtesy of P. S. Considine), (b) block diagram of digital-analog data processing system (after Ref. 1), (c) facsimile array of the amplitude distribution of the inverse sinc function (after Ref. 1), (d) filtered image (courtesy of P. S. Considine), and (e) microdensitometer trace of detected square. (Courtesy of P. S. Considine.)

continuous-tone photograph of a soldier with a small square just below the soldier's elbow. The purpose of the experiment was to detect the presence of the square.

In this experiment, the inverse detection filter for the square was fabricated using the methodology described in the digital-analog data processing system of Fig. 31.12(b). A scanning microdensitometer was used to digitize information about the square object to be detected for input to the computer. The Fourier transform was computed and the inverse sinc filter needed to detect the presence of the square was fabricated. The inverse sinc filter consists of the inverse magnitude [Fig. 31.12(c)] and the corresponding π phase steps for the negative lobes of the sinc function. The filter thus generated was placed into the filter plane of the optical system and used to detect the location of the square in the scene of Fig. 31.12(a). The detection of the square is shown in Fig. 31.12(d) by the spot of light indicated by the arrow and the corresponding microdensitometer trace across the detection spot is seen in (e).

In Chapters 29, 30, and 31, we have demonstrated the various forms that spatial filters can take: real, imaginary, and complex. In the following chapters, we discuss the use of these filters in two classes of applications: additional mathematical operations and image manipulations.

REFERENCES

1. P. S. Considine and R. A. Profio, "Image processing with in-line optical systems," in *Image Information Recovery,* A. Derr, ed., Proc. SPIE 16, 113–120 (1968).
2. P. F. Mueller and G. O. Reynolds, "Image restoration by removal of random-media degradations," *J. Opt. Soc. Am.* 57(11), 1338–1344 (1967).
3. G. O. Reynolds, J. L. Zuckerman, D. Miller, and W. A. Dyes, "Holographic phase compensation techniques applied to human cataracts," *Opt. Eng.* 12(1), 23–35 (1973).
4. G. O. Reynolds, J. L. Zuckerman, W. A. Dyes, and D. Miller, "Phase aberration balancing of simulated cataracts in the reflection mode," *Opt. Eng.* 12(2), 80–82 (1973).
5. E. N. Leith and J. Upatnieks, "Holographic imagery through diffusing media," *J. Opt. Soc. Am.* 56(4), 523 (1966).
6. A. Marechal and P. Croce, "Un filtre de fréquences spatiales pour amélioration du contrast des images optique," *Compt. Rend.* 237, 607–609 (1953).
7. J. Tsujiuchi, "Correction of optical images by compensation of aberrations and by spatial frequency filtering," *Progress in Optics,* E. Wolf, ed., Vol. 2, pp. 133–180, North Holland Publishing Co., Amsterdam (1963).
8. J. W. Goodman, *Introduction to Fourier Optics,* McGraw-Hill Book Co., New York (1968).
9. B. J. Thompson, "Coherent optical processing: a tutorial review," in *Coherent Optical Processing,* Proc. SPIE 52, 1–22 (1974).

32 Analog Optical Computing: Additional Mathematical Operations

In this chapter, we demonstrate that the generalized optical system of Fig. 30.1 can also be utilized to display Fresnel and Mellin transforms. We also demonstrate that the same system can be utilized to display the differentiation and integration of an optical signal.

32.1 FRESNEL TRANSFORM

Two one-dimensional functions which are Fresnel transform pairs are given by

$$F(\xi') = \int_{-\infty}^{\infty} G(x') \exp[i\pi(\xi' - x')^2] \, dx' \, , \tag{32.1}$$

$$G(x') = \int_{-\infty}^{\infty} F(\xi') \exp[-i\pi(\xi' - x')^2] \, d\xi' \, . \tag{32.2}$$

To demonstrate the Fresnel transformations given by Eqs. (32.1) and (32.2), we rewrite Eq. (32.1) in the form

$$[F(\xi') \exp(-i\pi\xi'^2)] = \int_{-\infty}^{\infty} [G(x') \exp(i\pi x'^2)] \exp(-2\pi i\xi' x') \, dx' \, . \tag{32.3}$$

Equation (32.3) indicates that the Fourier transform of the function $[G(x') \exp(i\pi x'^2)]$ is given by $[F(\xi') \exp(-i\pi\xi'^2)]$. Taking the inverse Fourier transform of Eq. (32.3), we have

$$[G(x') \exp(i\pi x'^2)] = \int_{-\infty}^{\infty} [F(\xi') \exp(-i\pi\xi'^2)] \exp(2\pi i\xi' x') \, d\xi' \, . \tag{32.4}$$

Rewriting Eq. (32.4), we get Eq. (32.2), the inverse Fresnel transform of $F(\xi')$, which completes the proof. Thus, Eqs. (32.1) and (32.2) form a Fresnel transform pair.

Physically, we point out that, apart from the obliquity factor K, Eq. (32.1) simply describes paraxial light propagation in the Fresnel diffraction region as described in Eq. (9.8) of Chapter 9, where $\xi' = \xi/\sqrt{\lambda z}$ and $x' = x\sqrt{\lambda z}$. Special cases of this propagation condition occur when the field distance z satisfies the inequality

$$z \gg \frac{(d_{max})^2}{\lambda} ,$$
(32.5)

where d_{max} is the maximum dimension of the diffracting aperture. This condition [Eq. (32.5)] reduces Eq. (32.1) to a Fourier transformation multiplied by a quadratic phase factor. As we have indicated many times, the same result (i.e., Fraunhofer diffraction) occurs if we go to the focal plane of a lens used behind the diffracting aperture.

Fresnel diffraction is the basis of Fresnel holography where the complex Fresnel diffraction pattern is mixed with a coaxial reference wave and stored with a square law detector. The subsequent reconstruction process is an inverse Fresnel transform. To retrieve a replica of the object, a focusing condition is utilized to scale the Fresnel transform properly as discussed in Chapter 25.

Equation (32.1) also arises in a defocused lens and is usually removed by focusing as discussed previously in Chapter 8. The Fresnel transform is not very useful in spatial filtering in the Fourier plane because each filter is only unique for one value of defocus, Δz.

32.2   MELLIN TRANSFORM

Mellin transforms can be displayed using the system illustrated in Fig. 30.1. The one-dimensional Mellin transform is given by

$$M(\mu) = \int_0^\infty f(x)x^{-(1+i\mu)}\,dx .$$
(32.6)

We can write Eq. (32.6) as a Fourier transform by making the following change of variable:

$$x = e^\xi ,$$
(32.7)

or

$$\xi = \ln x .$$
(32.8)

Substituting Eq. (32.7) into (32.6) yields:

$$M(\mu) = \int_{-\infty}^\infty f(e^\xi)\,e^{-i\xi\mu}\,e^{-\xi}(e^\xi\,d\xi) ,$$

or

$$M(\mu) = \int_{-\infty}^\infty f(e^\xi)\,e^{-i\xi\mu}\,d\xi ,$$
(32.9)

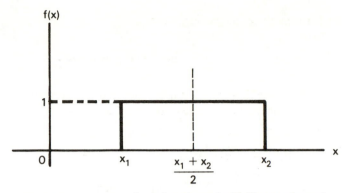

Fig. 32.1 Schematic of rect function to be Mellin transformed.

which indicates that the Mellin transform of $f(x)$ is the same as the Fourier transform of $f(e^\xi)$. Thus, we can obtain a Mellin transform of a function $f(x)$ by taking the Fourier transform of the modified function $f(e^\xi)$, where $x = e^\xi$.

A useful property of the Mellin transform is its scale invariance, i.e., the transform does not change due to a change in the size of the object. This property is obtained from Eq. (32.6) by replacing $f(x)$ with $f(ax)$, where a is a scaling parameter, to give

$$M'(\mu) = \int_0^\infty f(ax)\, x^{-i\mu-1}\, dx \ . \tag{32.10}$$

The change of variable, $y = ax$ in Eq. (32.10), yields

$$M'(\mu) = a^{i\mu} M(\mu) = \exp[i\ln(a)]\, M(\mu) \ , \tag{32.11}$$

so that $|M'(\mu)| = |M(\mu)|$. Equation (32.11) shows that the scale factor a in object space appears as a linear phase factor multiplying the unscaled Mellin transform in transform space. When used in scale invariant correlators,[1,2] a subsequent Fourier transform of Eq. (32.11) will shift the correlation peak by an amount proportional to $\ln(a)$.

We illustrate this scale invariance of the Mellin transform optically by choosing a rect function for $f(x)$ as shown in Fig. 32.1. Mathematically, this rect function is given by

$$f(x) = \text{rect}\left[x - \left(\frac{x_2 + x_1}{2}\right)\bigg|\frac{(x_2 - x_1)}{2}\right] \ . \tag{32.12}$$

Substituting Eq. (32.12) into Eq. (32.6), we get

$$M(\mu) = \int_{x_1}^{x_2} \frac{dx}{x x^{i\mu}} \ . \tag{32.13}$$

Transforming the variable x by using Eq. (32.7), we can write Eq. (32.13) as

$$M(\mu) = \int_{\ln x_1}^{\ln x_2} e^{-i\mu\xi}\, d\xi \ . \tag{32.14}$$

Performing the ξ integration yields

$$M(\mu) = \frac{1}{-i\mu} \left[\exp(-i\mu\ln x_2) - \exp(-i\mu\ln x_1) \right] . \tag{32.15}$$

To illustrate that the Mellin transform of the rect function is scale invariant, we determine the magnitude of the Mellin transform given by Eq. (32.15). This yields

$$|M(\mu)| = \sqrt{M(\mu) M^*(\mu)} ,$$

$$|M(\mu)| = \frac{1}{\mu} \sqrt{2 - 2\cos\left[\mu\ln\left(\frac{x_2}{x_1}\right)\right]} ,$$

or

$$|M(\mu)| = \left| \frac{2}{\mu} \sin\left[\frac{\mu}{2} \ln\left(\frac{x_2}{x_1}\right)\right] \right| . \tag{32.16}$$

In Eq. (32.16), the scale invariance of the Mellin transform is shown by the fact that its magnitude depends only on the ratio (x_2/x_1) in agreement with Eq. (32.11). For example, a rect function two times wider than the one in Fig. 32.1 would have its limiting values at $2x_2$ and $2x_1$, respectively, leaving its Mellin transform [Eq. (32.16)] scale invariant. Because of this scale invariance property, the Mellin transform is useful in pattern recognition.[1]

To illustrate the scale invariance of a Mellin transform, two optical transparencies containing squares differing in scale by a factor of 2 were used. A closed circuit television camera is focused on the square inputs. The outputs of the vertical driver circuits are then level shifted, amplified, and passed through log modules before entering the modulated input of a CRT.[1] Transparencies of the CRT, which represent $f(e^\xi)$, are then placed in the object plane of Fig. 30.1 from Chapter 30. The Fourier transform appears in the x plane of the system. Results demonstrating the scale invariance of two squares having dimensions W and $2W$ appear in Fig. 32.2. In Figs. 32.2(a) and (b), we show a square of dimension W and its Mellin transform; in (c) and (d), we show a square of dimension $2W$ and its Mellin transform. These results illustrate the scale invariance property of the Mellin transform because the transforms have the same dimensions, as predicted by Eqs. (32.11) and (32.16). This Mellin transformation process can be accomplished in real time by utilizing optical light modulators.[1]

32.3   DIFFERENTIATION AND INTEGRATION OF OPTICAL SIGNALS

It is possible to perform differentiation and integration of optical signals by the use of appropriate optical filters in the transform plane if the signal to be differentiated or integrated is in the object plane of a two-lens Fourier transforming system.

We will show that, from the amplitude transmission as a function of position on a piece of film $f(\xi)$, we can generate either $f^{(n)}(\xi)$, where (n) means the n'th derivative with respect to ξ, or its indefinite integral $\int f(\xi)\, d\xi$.

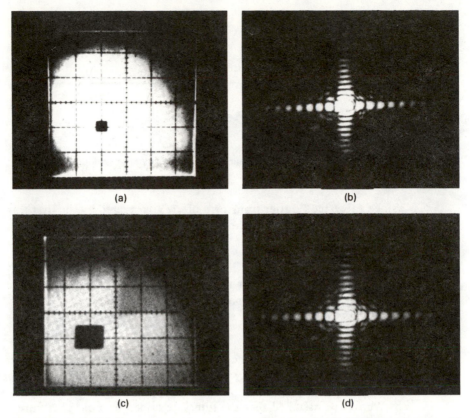

Fig. 32.2 Optical Mellin transform of a square input function of width W: (a) input function, (b) Mellin transform, (c) square input function of width $2W$, and (d) Mellin transform (after Ref. 2).

32.3.1 Differentiation

If the object to be differentiated is placed in the ξ plane of Fig. 30.1, its Fourier transform is spatially displayed in the x plane. (For mathematical simplicity, we restrict ourselves to the one-dimensional problem since the extension to two dimensions is straightforward.)

The filter function in the x plane necessary to produce the n'th-order derivative of the object $f(\xi)$ in the α plane is now derived. To obtain this n'th-order derivative, the complex amplitude distribution in the filter plane, with the filter in place, must be

$$T_A(x) = \int \frac{d^n f(\xi)}{d\xi^n} \exp(-2\pi i \xi x) \, d\xi \, . \tag{32.17}$$

Since $f(\xi) = \int F(x') \exp(2\pi i \xi x') \, dx'$, Eq. (32.17) can be written as

$$T_A(x) = \int \left[\int F(x') \, (2\pi i x')^n \exp(2\pi i \xi x') \, dx' \right] \exp(-2\pi i \xi x) \, d\xi \, , \tag{32.18}$$

where the ξ integration gives a delta function, and its sifting property reduces Eq. (32.18) to

$$T_A(x) = F(x) \, (2\pi i x)^n \, . \tag{32.19}$$

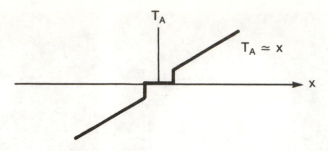

Fig. 32.3 Approximation of a differentiation filter.

Since $F(x)$, the Fourier transform of $f(\xi)$, is created by the first transforming lens in Fig. 30.1, a filter of the form

$$M(x) = (2\pi i x)^n \tag{32.20}$$

is needed in the **x** plane to create the desired derivative of the function in the α plane. The intensity at the image plane will be given by

$$I(\alpha) = |f^{(n)}(\alpha)|^2 . \tag{32.21}$$

Although Eq. (32.21) gives the square of the derivative, we show in the following discussion that the intensity in the image plane can be made proportional to $f^{(n)}(\alpha)$.

Above, we assumed that the filters described by Eq. (32.20) are realizable. Actually, the necessary transmission filters for differentiation cannot be achieved at the origin in the transform plane since this requires infinite density in the filter material. A blocking filter inserted over the low-frequency components of the spectrum in the **x** plane has been utilized to approximate the differentiation filter. Then, except for those frequencies that the blocking filter covers, differentiation occurs. The amplitude transmission of this approximate filter for achieving the *first* derivative of the object is shown in Fig. 32.3. Negative values of x are realized by placing a π phase step over the negative half of the **x** plane.

Since the low-frequency information in the object is not differentiated with this filter, other realizable filters that avoid this difficulty have been fabricated. A spatial filter whose amplitude transmittance is of the form

$$T_A = T_0 + Cx , \tag{32.22}$$

where C and T_0 are experimentally controllable constants, overcomes this difficulty by placing the nonrealizable portion of the filter (point of zero transmission) outside the field of view of the optical system.

The intensity image created in the α plane by the filter of Eq. (32.22) in the **x** plane is

$$I(\alpha) \cong \left| C \frac{df(\alpha)}{d\alpha} + T_0\, f(\alpha) \right|^2 , \tag{32.23}$$

which shows that the true derivative is present but that it is superimposed on the image of the object.

A filter for creating a true derivative, which separates the differentiated image in

Eq. (32.23) from the image itself, has been realized by modulating the filter function onto a carrier.[3] This filter, which has an amplitude transmission given by

$$T_A(x) = 1 + \epsilon x \sin px \; , \tag{32.24}$$

can be realized by combining shifted cosine gratings to create the modulation. Thus, the filter function given by

$$M(x) = \lim_{\epsilon \to 0} [1 + \cos px - \cos(px + \epsilon x)] \tag{32.25}$$

is equivalent to Eq. (32.24) in the limit as $\epsilon \to 0$. This is easily seen by using the cosine identity for the sum of two angles. In Eq. (32.25), p is $2\pi/$ period; the minus sign indicates a half-period shift; and ϵ represents an additional phase shift between the cosine gratings. Such a filter can be experimentally realized by computer generation techniques, by Talbot interferometry,[3] or by more standard interference construction techniques (holographic gratings with phase shifters between exposures).[4]

The amplitude impulse response of the filter described by Eq. (32.25) is

$$h(\alpha) = \lim_{\epsilon \to 0} \left\{ \delta(\alpha) + \frac{1}{2} \left[\delta \left(\alpha \pm \frac{1}{p} \right) - \delta \left(\alpha \pm \frac{1}{p + \epsilon} \right) \right] \right\} \; , \tag{32.26}$$

which is shown schematically in Fig. 32.4. The two side order terms shown in Fig. 32.4 are mathematically true first derivatives of positive and negative delta functions, respectively, providing that the optical system can resolve the phase shift introduced in the grating, i.e., ϵx in Eq. (32.25). In other words, the impulse response diameter of the second transform lens must be less than the separation between the side order delta functions $\delta(\alpha \pm 1/p) - \delta[\alpha \pm 1/(p + \epsilon)]$ of Eq. (32.26), as shown in Fig. 32.4(b). Therefore, this filter will yield true first derivatives of any object transparency placed in Fig. 30.1 by convolving the amplitude impulse response of the filter [Fig. 30.4(b)] with the amplitude distribution of the object. In a manner similar to the problem of image replication discussed in Chapter 30, the frequency of the carrier in the filter must be large enough to prevent overlap of the on-axis image from the side order differentiated images.

An example of an object, differentiated with the filter of Eq. (32.25) constructed in a Talbot interferometer, is shown in Fig. 32.5. Derivatives in both the x and y directions were realized by rotating the filter 90 deg. Second derivatives (i.e., d^2/dy^2), cross derivatives (i.e., $d^2/dxdy$), and higher order derivatives can also be achieved by these or similar modulation techniques.[3,4,6]

32.3.2   Integration

The process of optical analog integration can also be achieved with the generalized optical filtering system shown in Fig. 30.1. The filter necessary to perform optical integration of an arbitrary function has an amplitude transmission varying inversely with distance in the transform plane.[7]

This is most easily demonstrated by letting the one-dimensional initial distribution in the ξ plane itself be a derivative, i.e.,

$$\psi(\xi) = \frac{df(\xi)}{d\xi} = \int 2\pi i \eta' \bar{f}(\eta') \exp(2\pi i \xi \eta') \, d\eta' \; , \tag{32.27}$$

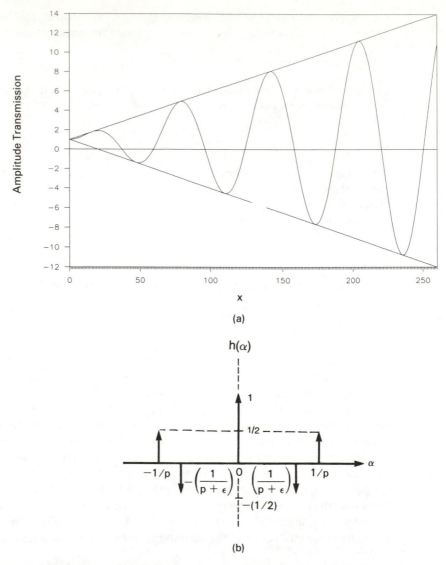

Fig. 32.4 (a) Plot of modulated derivative filter of Eq. (32.24) for $p = 1$ and $\epsilon = 0.05$ and (b) its impulse response given by Eq. (32.26).

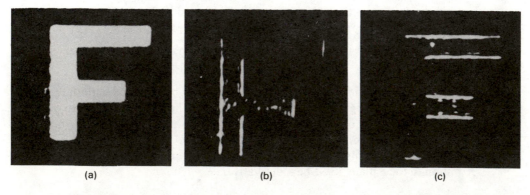

Fig. 32.5 First-order differentiation: (a) the object, (b) x derivative, and (c) y derivative (after Ref. 5).

where $\eta' = x/\lambda f$ and $\tilde{f}(\eta')$ denotes the Fourier transform of $f(\xi)$. Insertion of the integration mask (varying inversely with distance) in the transform plane gives

$$\bar{\psi}(\eta') = \int \frac{1}{\eta'}\, \psi(\xi)\exp(2\pi i\xi\eta')\,d\xi \; , \tag{32.28}$$

which, using Eq. (32.27), can be written in the form

$$\bar{\psi}(\eta') = \int \frac{1}{\eta'} \int 2\pi i\eta''\, \tilde{f}(\eta'')\exp(2\pi i\xi\eta'')\exp(2\pi i\xi\eta')\,d\eta''\,d\xi \; , \tag{32.29}$$

where η'' is a dummy variable. Performing the ξ integration gives $\delta(\eta'' + \eta')$, so that

$$\bar{\psi}(\eta') = 2\pi i\tilde{f}(\eta') \; . \tag{32.30}$$

Fourier transformation gives

$$\psi(\alpha) = 2\pi if(\alpha) \; , \tag{32.31}$$

as the amplitude distribution, which shows that $f(\alpha)$ is recovered by integration.

In general, the $1/x$ integration filter is difficult to fabricate using transmission masks whose values are limited to between zero and unity. One way to circumvent this difficulty is to utilize a mask whose transmission is given by[4]

$$T_A(x) = 2a\,\mathrm{sinc}\left(\frac{2\pi ax}{\lambda f}\right) . \tag{32.32}$$

This mask is readily fabricated by optically Fourier transforming a rect function (slit). The negative lobes of the sinc function can be realized with π phase filters or by making a Fourier transform hologram integration mask in a Mach-Zehnder interferometer.

If the function $f(\xi)$ is placed in the object plane of Fig. 30.1, and the mask given by Eq. (32.32) is placed in the filter plane, the resulting amplitude transmission in the filter plane is given by

$$T'_A(x) = 2a\,\tilde{f}\left(\frac{x}{\lambda f}\right)\sin\left(\frac{2\pi ax}{\lambda f}\right)\left(\frac{1}{x}\right)\left(\frac{\lambda f}{2\pi a}\right) . \tag{32.33}$$

The amplitude distribution in the image plane α resulting from this filter is obtained by Fourier transforming Eq. (32.33). In symbolic notation, this yields

$$A(\alpha) = \left(\frac{1}{\pi}\right)\mathscr{F}\left[\left(\frac{\lambda f}{x}\right)\tilde{f}\left(\frac{x}{\lambda f}\right)\right] \circledast \mathscr{F}\left[\sin\left(\frac{2\pi ax}{\lambda f}\right)\right], \tag{32.34}$$

where the $\mathscr{F}$ represents the Fourier transform operation. The first Fourier transform term appearing in Eq. (32.34) is the Fourier transform of the product of two functions, which is a convolution of their respective transforms. The transform of the term $(\lambda f/x)$ is given by[8]

$$\mathscr{F}\left(\frac{\lambda f}{x}\right) = \frac{2\pi i}{\lambda f}\left[H(\alpha) - \frac{C}{2}\right], \tag{32.35}$$

where $H(\alpha)$ is the unit step function and C is a constant. Using Eq. (32.35), we can express the first Fourier transform in Eq. (32.34) as

$$\mathscr{F}\left[\left(\frac{\lambda f}{x}\right)\tilde{f}\left(\frac{x}{\lambda f}\right)\right] = \frac{2\pi i}{\lambda f}\left[H(x) - \frac{C}{2}\right] \circledast f(\alpha)$$

$$= -K + \frac{2\pi i}{\lambda f}\int_{-\infty}^{\alpha} f(\xi)\,d\xi\ , \tag{32.36}$$

where K is the constant resulting from convolving $f(\alpha)$ with the constant $\pi i C/\lambda f$. Equation (32.36) gives the desired result, the integral of the original function $f(\xi)$ riding on a bias.

The second Fourier transform term in Eq. (32.34), which is present because of the method used in fabricating the filter, can be written as

$$\mathscr{F}\left[\sin\left(\frac{2\pi a x}{\lambda f}\right)\right] = \frac{1}{2i}[\delta(\alpha - a) - \delta(\alpha + a)]\ . \tag{32.37}$$

Substitution of Eqs. (32.36) and (32.37) into Eq. (32.34) gives the image amplitude distribution in the α plane as

$$A(\alpha) = \frac{\pi}{\lambda f}\int_{-\infty}^{\alpha+a} f(\xi)\,d\xi - \frac{\pi}{\lambda f}\int_{-\infty}^{\alpha-a} f(\xi)\,d\xi\ . \tag{32.38}$$

The constant term in Eq. (32.36) drops out of the analysis because the two of them are equal with opposite signs. With this assumption, we see that the image displays both the integral of the object and a displaced negative integral of the object, separated by the width of the rect function used to fabricate the filter of Eq. (32.32). This rect function width must be chosen to guarantee that the two filtered images created by the sinc filter of Eq. (32.32) do not overlap, i.e., a field-of-view limitation is placed on the experiment.

An example illustrating one-dimensional integration by a holographic filter satisfying Eq. (32.32) is shown in Fig. 32.6. The two objects used in the experiments are shown in Figs. 32.6(a) and (c). The results of integrating in the horizontal direction in the analog optical system (microdensitometer traces) are shown in Figs. (b) and (d), respectively. Since the objects are rect functions (constants) the results of integrating give straight lines with positive slopes on the left side of the microdensitometer traces and straight lines with negative slopes on the right side of the microdensitometer traces in Figs. 32.6(b) and (d), respectively, which is in agreement with Eq. (32.38).

Another way of interpreting the results of Fig. 32.6 is to Fourier transform Eq. (32.33) directly. In symbolic notation, this gives

$$A(\alpha) = f(\alpha) \circledast \text{rect}(\alpha|a)\ . \tag{32.39}$$

Since the convolution of a function with the unit step function is an integral, Eq. (32.39) can be rewritten as

$$A(\alpha) = \int_{-\infty}^{\alpha+a} f(\xi)\,d\xi - \int_{-\infty}^{\alpha-a} f(\xi)\,d\xi\ , \tag{32.40}$$

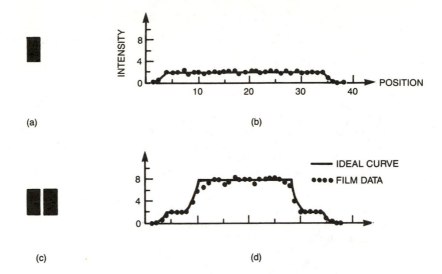

Fig. 32.6 Example of optical integration: (a) single-bar object and (b) corresponding microden-sitometer trace of integrated output; (c) double-bar object and (d) corresponding microdensi-tometer trace of the integrated output (after Ref. 4).

where

$$\text{rect}(\alpha|a) = H(\alpha + a) - H(\alpha - a) ,$$ (32.41)

and

$$H(\alpha) = \begin{cases} 1 & \alpha > 0 \\ 0 & \alpha < 0 \end{cases}$$ (32.42)

is the unit step function.

For the rect function objects used in the Fig. 32.6 experiment, this analysis would give the results shown in Fig. 32.7 for $a \gg b$. Comparison of the convolution results in Fig. 32.7 with the microdensitometer traces in Fig. 32.6 shows the qualitative agreement beween the analysis and the experiment.

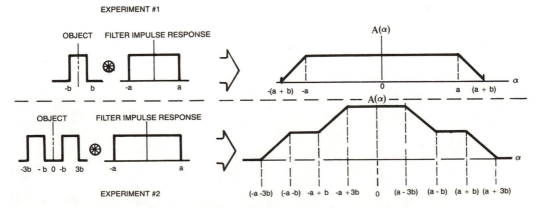

Fig. 32.7 Schematic of integration described by Eq. (32.39) corresponding to the experiment in Fig. 32.6.

REFERENCES

1. D. Casasent and D. Psaltis, "New optical transforms for pattern recognition," *Proc. IEEE* 65(1), 77–81 (1977).
2. D. Casasent and D. Psaltis, "Scale invariant optical transform," *Opt. Eng.* 15(3), 258–261 (1976).
3. J. K. T. Eu, C. Y. C. Liu, and A. W. Lohmann, "Spatial filters for differentiation," *Opt. Comm.* 9(2), 168–171 (1973).
4. S. K. Yao and S. H. Lee, "Spatial differentiation and integration by coherent optical-correlation method," *J. Opt. Soc. Am.* 61(4), 474–477 (1971).
5. J. K. T. Eu, C. Y. C. Liu, and A. W. Lohmann, "Spatial filters for differentiation—Erratum," *Opt. Comm.* 10(2), 211 (1974).
6. S. H. Lee, "Image Processing," in *Handbook of Optical Holography,* H. J. Caulfield, ed., pp. 537–560, Academic Press, New York (1979).
7. J. B. DeVelis and G. O. Reynolds, *Theory and Applications of Holography,* Chapter 8, p. 157, Addison-Wesley Publishing Co., Reading, Mass. (1967).
8. J. D. Gaskill, *Linear Systems, Fourier Transforms and Optics,* p. 201, John Wiley and Sons, Inc., New York (1978).

33   Analog Optical Computing: Optical Correlation Techniques

33.1   INTRODUCTION

Optical correlation techniques have been used successfully in applications involving the detection of signals in noise, in multichannel processing, and in pattern and character recognition.[1,2] The systems are advantageous in that signals having a large space-bandwidth product can be processed rapidly. In this chapter, we discuss some of the analog optical correlation techniques and give an example of the use of each type of system discussed. No attempt is made to discuss all correlator systems or applications as this would be well beyond the scope of this chapter. The subjects of correlation and convolution and their comparisons were treated previously in Chapter 28.

33.2   INCOHERENT LIGHT CORRELATION

If two transparencies, emulsion facing emulsion, are shifted with respect to each other, then the measurement of the total light output as a function of the shift variable is the *autocorrelation function.*

    For purposes of demonstration, two negative transparencies of the distribution shown in Fig. 33.1 will exhibit an autocorrelation function when slid across each other in front of an incoherent light source and viewed with an unaided eye. This figure may be used to demonstrate that the autocorrelation length of the distribution is determined primarily by the larger sized particles.

33.3   COHERENT LIGHT CORRELATION

Optical correlation can also be accomplished using coherent light and the generalized system of Fig. 30.1 of Chapter 30. In this case, one of the functions $f(\xi, \eta)$ is imaged with unit magnification by an additional optical system onto the object plane in which the transparency of another function $g(\xi, \eta)$ is shifted to avoid scratching of the emulsion (this could also have been done in the incoherent light case). Thus, in the object plane we have constructed the function, $f(\xi, \eta)g(\xi - \xi_0, \eta)$ by shifting only in the positive ξ direction. The Fourier

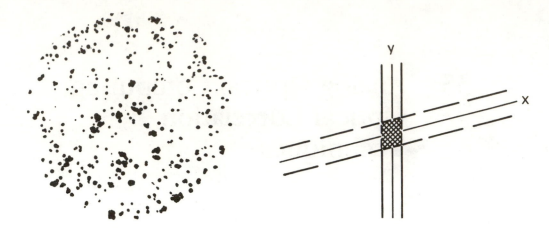

Fig. 33.1 A random distribution of particles useful for demonstrating incoherent autocorrelation.

Fig. 33.2 Schematic of correlator output plane.

transform of this product will appear in the (x, y) plane in the form

$$\int\!\!\!\int_{-\infty}^{\infty} f(\xi, \eta) g(\xi - \xi_0, \eta) \exp[-2\pi i(\xi x + \eta y)] \, d\xi \, d\eta \ . \tag{33.1}$$

If a filter consisting of two crossed slits, rect$(x|a)$ and rect$(y|a)$, is placed in the (x, y) plane (see Fig. 33.2), then the transmission in that plane is given by

$$\text{transmission}(x, y) = \text{rect}(x|a) \, \text{rect}(y|a) \ ;$$

and if a photomultiplier is inserted in the system so as to capture all light that passes through the crossed slits, then the output of the photomultiplier will be proportional to $f(\xi_0)$ given by

$$f(\xi_0) = \int\!\!\!\int\!\!\!\int\!\!\!\int_{-\infty}^{\infty} \text{rect}(x|a) \, \text{rect}(y|a) \, f(\xi, \eta) \, g(\xi - \xi_0, \eta)$$

$$\times \exp[2\pi i(\xi x + \eta y)] \, dx \, dy \, d\xi \, d\eta \ .$$

This equation may be readily integrated over x to yield

$$f(\xi_0) = \int\!\!\!\int_{-\infty}^{\infty} \text{sinc}\left(\frac{2\pi a\xi}{\lambda f}\right) \text{sinc}\left(\frac{2\pi a\eta}{\lambda f}\right) f(\xi, \eta) \, g(\xi - \xi_0, \eta) \, d\xi \, d\eta \ .$$

If a is suitably small, the central lobe of the two sinc functions will exceed the maximum dimension of the two functions $f(\xi, \eta)$ and $g(\xi - \xi_0, \eta)$ and the integral can be further evaluated. Denoting the maximum dimensions of $f(\xi, \eta)$ and $g(\xi - \xi_0, \eta)$ by ξ_{max} and η_{max}, respectively, and requiring that $\xi_{max} \ll f/2a$ and $\eta_{max} \ll f/2a$, the sinc functions can be considered constant over the nonzero regions of $f(\xi, \eta)$ and $g(\xi - \xi_0, \eta)$, and the equation reduces to the square of the cross correlation and the square root of the output produces the magnitude of the desired result, i.e.,

$$\iint_{-\infty}^{\infty} f(\xi, \eta) g(\xi - \xi_0, \eta)\, d\xi\, d\eta \ .$$

(33.2)

This condition is a significant constraint on the applicability of the method. For instance, if the functions to be correlated fill the aperture of the final or condensing lens, then the two rect functions must be narrow compared to the diffraction pattern of the lens. This method also suffers from the fact that most of the energy is lost to the measurement. This method, when applicable, yields the magnitude of a one-dimensional cross correlation of 2 two-dimensional functions with one direction, namely, η, averaged out (i.e., the output is an average of the channels created by the resolution elements of the film in the η direction).

When used to measure autocorrelation functions, this system has the advantage that the photocell, which measures the light, sees the same geometrical area throughout the experiment and only the amount of light inside it changes. This procedure of measurement is most desirable from a practical point of view and is used because of its experimental simplicity.

As an example of coherent light correlation of a spatially random function,[3] consider the following. A coherent light correlator has been used to measure the autocorrelation function of a random medium. The medium consisted of dyed gelatin particles embedded in a clear gelatin matrix. The particles were 1.4 to 5 mm in size and represented 0.5% by volume of the matrix cell and were generated by first dying a sample of particles. The dyed particles were then randomly placed in a container of clear gelatin and two photographs were made with high-contrast copy film. The two photographs were optically autocorrelated using the optical system illustrated in Fig. 33.3. One of the photographs is shown in Fig. 33.1. The first photograph is imaged at a 1:1 magnification onto the second, which is scanned incrementally in the ξ direction. All the energy passing through the second photograph is collected by a lens that transforms it onto the crossed slits. The energy passing through the slits is then collected by the photomultiplier tube and the correlation function is read directly. The data and the corresponding best fit curve for the optically measured correlation function are shown in Fig. 33.4.

In obtaining the correlation function by observing the dc term of the Fourier transform of the product of the two negatives with a photodetector as described

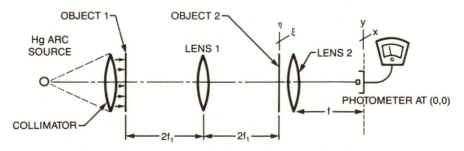

Fig. 33.3 Coherent optical system for measuring the correlation function. (Courtesy of A. E. Smith.)

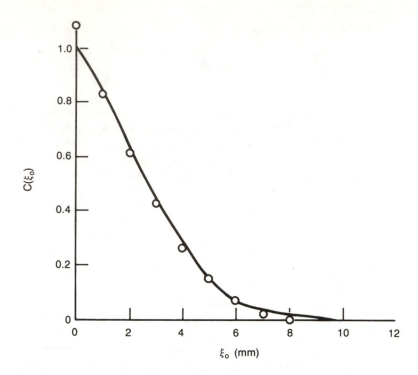

Fig. 33.4 Average correlation curve $C(\xi_0)$ for a medium containing 1- to 5-mm particles. The dots represent experimental data points. (Courtesy of A. E. Smith.)

in Eq. (33.2) (see Fig. 33.3), we immediately get the square of the average of the correlation function over the η direction for each value of ξ. By plotting the square root of the measured intensity as a function of ξ_0 (with a suitable normalizing factor), the average correlation function is obtained, as shown in Fig. 33.4.

33.4   TRUE ONE-DIMENSIONAL, MULTICHANNEL CORRELATION SYSTEM

To realize a multichannel correlation system,[4] we use the system shown in Fig. 33.5. One of the functions to be correlated is placed in the P_1 plane and imaged 1:1 onto the other function, which is placed in the P_3 plane (ξ, η). The function in the P_3 plane can be shifted in the positive ξ direction. Both the spherical and cylindrical lenses between the P_3 and P_4 planes have the same focal length f. The combination (two thin lenses in contact) behaves like a lens of focal length $f/2$ for the η axis; however, for the ξ axis, it behaves like a lens of focal length f. Thus, the η axis is imaged with unit magnification, while the ξ axis is Fourier transformed. Let $h(\xi, \eta)$ represent the product of the two functions in the P_3 plane of Fig. 33.5; then, in the P_4 plane, we have the optical amplitude distribution given by

$$F(x,y) = \exp\left(\frac{iky^2}{f}\right) \int_{-\infty}^{\infty} h(\xi,-y) \exp\left(\frac{-ik\xi x}{f}\right) d\xi . \tag{33.3}$$

The quadratic phase factor arising from the coherent image in the y direction

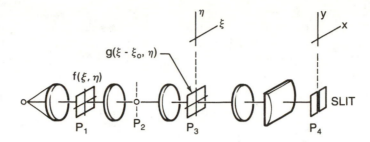

Fig. 33.5 Multichannel cross-correlator configuration (from Ref. 5).

simply indicates that the image of the y axis is multiplied by a cylindrical wavefront.

Equation (33.3) can be rewritten in terms of the functions $f(\xi,\eta)$ and $g(\xi - \xi_0,\eta)$ as

$$F(x,y) = \exp\left(\frac{iky^2}{f}\right) \int_{-\infty}^{\infty} f(\xi,-y)g(\xi-\xi_0,-y)\exp\left(\frac{-ik\xi x}{f}\right) d\xi \ . \quad (33.4)$$

If we now introduce the slit along the y axis (i.e., $x=0$) in the back focal plane P_4, we obtain

$$F(0,y) = \exp\left(\frac{iky^2}{f}\right) \int_{-\infty}^{\infty} f(\xi,-y)g(\xi-\xi_0,-y)\,d\xi \ . \quad (33.5)$$

Measuring the intensity along the slit [$x=0$ in the (x,y) plane] and then taking its square root yields

$$|F(0,y)| = \left| \int_{-\infty}^{\infty} f(\xi,-y)g(\xi-\xi_0,-y)\,d\xi \right| \ . \quad (33.6)$$

Thus, at each position y_n along the line $x=0$, a detector having a finite pixel size sees the magnitude of the one-dimensional cross-correlation of the two functions f and g, i.e.,

$$|F(0,y_n)| = \left| \int_{-\infty}^{\infty} f(\xi,-y_n)g(\xi-\xi_0,-y_n)\,d\xi \right| \ . \quad (33.7)$$

Each value of y_n and the detector defines a channel.

In practice, the number of channels is restricted, for example, by the resolution of the film. By measuring the total light output from the slit, we obtain an average of all the channels in the η direction; by measuring the light output from various parts of the slit, we obtain the magnitude of the true one-dimensional cross-correlation of the two functions f and g. This method is best suited for photometric measurement because the geometrical area seen by the photocell is

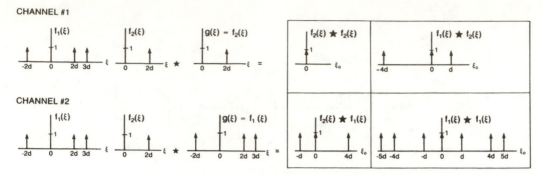

Fig. 33.6 Theoretical example of a two-channel correlator where $f_1(\xi)$ and $f_2(\xi)$ are the inputs into each channel. In channel 1, the input is correlated with $f_2(\xi)$, and in channel 2, the input is correlated with $f_1(\xi)$. The four-element output having autocorrelation functions placed on the diagonal is shown as the output of the two-channel correlator.

fixed in size throughout the experiment. However, the total light output changes with the cross-correlation, and the quadratic phase factor limits the number of useful channels if it is to be used as an input to a future experiment.

33.4.1   Example 1: Two-Channel Correlation

A simple theoretical example illustrating the operation of multichannel correlation is shown in Fig. 33.6. A two-channel correlator operating on a two-signal input function is used to create 2×2 matrix output from the correlator. The two input functions are

$$f_1(\xi) = \delta(\xi \pm 2d) + \delta(\xi - 3d) \ , \tag{33.8}$$

$$f_2(\xi) = \delta(\xi - 2d) \ . \tag{33.9}$$

The functions in the two separate channels to be correlated separately with the two input functions are

$$g_1(\xi) = \delta(\xi - 2d) \ , \tag{33.10}$$

$$g_2(\xi) = \delta(\xi \pm 2d) + \delta(\xi - 3d) \ . \tag{33.11}$$

The output from the two-channel correlator gives a matrix-like 2×2 array of correlations consisting of two autocorrelation (symmetric) functions, which we write as the diagonal elements in Fig. 33.6, and two cross-correlation terms, which we write as the off-diagonal elements. The off-diagonal elements are given by

$$C_{12}(\xi_0) = f_1(\xi) \star f_2(\xi) \ , \tag{33.12}$$

$$C_{21}(\xi_0) = f_2(\xi) \star f_1(\xi) = C_{12}(-\xi_0) \ . \tag{33.13}$$

Here, the star denotes correlation. In general, if this technique were extended to n channels in a square matrix-like format, then

$$C_{nm}(\xi_0) = C^*_{mn}(-\xi_0) \ . \tag{33.14}$$

To implement a multichannel correlator experimentally, the one-dimensional input functions $f_n(\xi)$ would be placed in the P_1 plane of Fig. 33.5 as parallel channels and imaged by the one-to-one imaging system onto the P_3 plane, which contains the $g_n(\xi - \xi_0)$ functions as parallel channels.

If a film is moved past the output slit, synchronously with the motion of the g functions, then the multichannel correlator output consisting of both cross and autocorrelation functions would be recorded on the film in an $n \times n$ matrix-like format. Examples of cross and autocorrelation functions using these multichannel correlators have been discussed in the literature.[4,5]

33.4.2   Example 2: Three-Channel Correlation

An example of correlating pseudorandom binary codes is illustrated here. The one-dimensional input functions $f(\xi)$ to be correlated are shown in Fig. 33.7(a). The film sprocket holes are used for reference. This same input function will be used in all three channels of a three-channel system utilizing Fig. 33.5.

The three functions $g_n(\xi - \xi_0)$ to be correlated are each located at a different height y_n from the reference sprocket holes in plane P_3 of Fig. 33.5, as shown in Figs. 33.7(b), (c), and (d). This difference in height utilizes the multichannel capability of the system. The correlation outputs from the three-channel system are shown in Fig. 33.7(e) where the three autocorrelation functions appear in the right-hand column rather than on the diagonal.

33.4.3   Example 3: Matched Filtering [6]

A hologram technique for producing an optimum filter that can detect a weak signal from background noise has been developed.[4,7] It is assumed that the input to the linear system is

$$f(\xi) = s(\xi) + n(\xi) \; , \tag{33.15}$$

where $s(\xi)$ denotes the signal of interest and $n(\xi)$ denotes a stationary, random noise function having spectral density $\tilde{N}(\mathbf{x})$. Then the optimum linear filter that maximizes the ratio of peak signal-to-rms noise is[8]

$$G(\mathbf{x}) = c \frac{[\tilde{s}(\mathbf{x})]^*}{\tilde{N}(\mathbf{x})} \; , \tag{33.16}$$

where c is a constant and $[\tilde{s}(\mathbf{x})]^*$ is the complex conjugate of the Fourier spectrum of the signal.

Note that this matched filter is a special case of the more general optimum (in the least-mean-square sense) estimating filter. This case arises when the average signal power spectrum is small compared to the noise for all spatial frequencies (i.e., the filter is matched to the signal). To construct this filter using holographic techniques, the numerator and denominator of Eq. (33.16) are produced by separate experiments.[7] Since $\tilde{N}(\mathbf{x})$ is a real, positive function, its inverse is easily realizable photographically. However, to produce $(\tilde{s})^*$, which is a complex quantity, a Mach-Zehnder interferometer having an optical Fourier transformer in one arm is used. If $s(\xi)$ is placed in the object plane in one arm of the interferometer, the intensity distribution in the back focal plane of the lens is

$$I(\mathbf{x}) = |K(\mathbf{x}) + \tilde{s}(\mathbf{x})|^2 \; , \tag{33.17}$$

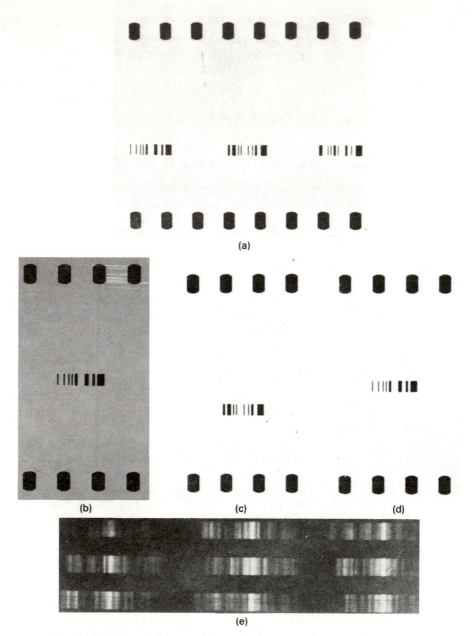

Fig. 33.7 Three-channel correlation of pseudorandom signals: (a) correlation input signals, (b)–(d) the three function $g_n(\xi - \xi_0)$ to be correlated, and (e) cross-correlation of the signal with itself and two other signals (from Ref. 5).

where $K(\mathbf{x})$ represents the reference beam, which is slightly off axis. Thus, a Fourier transform hologram has been formed. Dividing Eq. (33.17) by the noise spectrum, which was separately realized, we obtain

$$M(\mathbf{x}) = \frac{I(\mathbf{x})}{\tilde{N}(\mathbf{x})} = \frac{|K(\mathbf{x})|^2}{\tilde{N}(\mathbf{x})} + \frac{|\bar{s}(\mathbf{x})|^2}{\tilde{N}(\mathbf{x})} + \frac{K(\mathbf{x})[\bar{s}(\mathbf{x})]^*}{\tilde{N}(\mathbf{x})} + \frac{K^*(\mathbf{x})\bar{s}(\mathbf{x})}{\tilde{N}(\mathbf{x})} \cdot \quad (33.18)$$

Comparing Eqs. (33.18) and (33.16), we see that the third term in Eq. (33.18) is

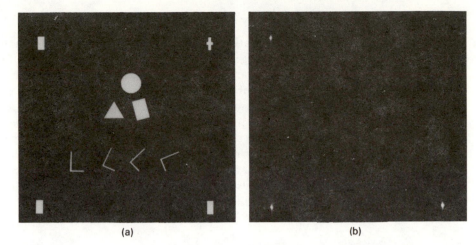

(a)   (b)

Fig. 33.8 Optical detection filter for geometrical shapes. An optical matched filter was made as defined by Eq. (33.16): (a) the object consisting of various geometric shapes and (b) the detection of the small rectangles emphasizing that such a complex holographic filter is sensitive to geometry, size, and orientation of the object to be detected (from Ref. 7, ©1964 IEEE).

the desired optimum filter. Its output will appear at one of the side orders in the image plane of the reconstructed hologram. The background intensity must be adjusted so that $|K(\mathbf{x})|$ is a constant. For the case of white noise, when $M(\mathbf{x})$ is placed in the filter plane of Fig. 30.1 with $s(\boldsymbol{\xi})$ as an object, the output at the side order corresponding to the optimum filter is the expected result {that is, $s(\boldsymbol{\alpha})$ convolved with $[s(\boldsymbol{\alpha})^*]$ has maximum energy}. It should be emphasized that the desired filter is strongly dependent on the shape and orientation of the object.

In Fig. 33.8, a filter was constructed to detect the small rectangles in the corners of the object [Fig. 33.8(a)]. The image using the matched filter is shown in (b). Figure 33.9 illustrates a case where the noise density is not constant. This technique has been used for isolating a desired binary signal from additive noise; however, in this experiment, the optimum property of linear filters (yielding maximum peak signal power-to-rms noise) is reduced.[9]

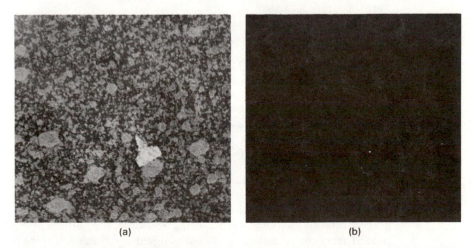

(a)   (b)

Fig. 33.9 Optical detection filter for detection of a signal in random noise. An optical matched filter was made to detect a signal in random noise: (a) the signal and random noise and (b) the signal as detected by the appropriate complex holographic filter (from Ref. 7).

33.4.4   Example 4: Ambiguity Function Processing

In radar and sonar applications, known waveforms are used to search a volume of space to locate point targets of interest.[10] The received signal is confusing when multiple targets having various ranges, sizes, and velocities are present. The multiple signals cause ambiguity in determining the range and velocity of a specific target in its temporal return because the phase changes caused by range and Doppler shifts are combined as a single phase change in the time waveform. As a result, the same target strength can be observed for multiple targets, making it difficult to determine the range and velocity of each target independently. The ambiguity function associated with a temporal waveform[11,12] can be utilized to resolve this problem because it separates the phase shifts associated with range and Doppler onto a two-dimensional display.

The ambiguity function is defined as[2,11,12]

$$\psi(\tau,\omega) = \int_{-\infty}^{\infty} f(t)\, f(t+\tau)\, e^{-i\omega t}\, dt \;, \tag{33.19}$$

where

$$f(t) = \text{transmitted signal at frequency } \nu$$
$$f(t+\tau) = \text{received signal having a Doppler shift } \nu_d$$
$$\tau = \text{time delay between sending and receiving the signals}$$
$$e^{-i\omega t} = \text{Fourier kernel}$$
$$\omega = 2\pi\nu \;.$$

Following Casasent,[2] Eq. (33.19) lends itself readily to optical processing by using the multichannel correlator shown in Fig. 33.10. Prior to using the correlator, the desired transmitted signal $f(t)$ of Eq. (33.19) is transduced to a spatial signal $f(\xi,\eta)$ and placed into each of the N channels in plane P_3. An array of matched filter holograms of the desired signal is recorded in plane P_4, in which the signal has been transformed in the x direction and imaged in the y direction. Thus, each of the N channels in plane P_4 has an identical filter matched to the desired transmitted signal spectrum and represented by a transmission of the form

$$T_A(x,y) = F^*(\mu, y - nd) \exp(-2\pi i \mu \xi_r) \exp(-2\pi i \alpha_0 x) \;, \tag{33.20}$$

where

$$n = n\text{'th channel}$$
$$F = \text{Fourier transform of the desired transmitted signal } f(\xi,\eta)$$
$$d = \text{separation between channels in the } P_4 \text{ plane}$$
$$\mu = \frac{x}{\lambda f} = \text{spatial frequency conjugate to the } x \text{ direction in the } P_4 \text{ plane}$$
$$\xi_r = \text{reference location for the beginning of the signal}$$

and $\alpha_0 = \sin\theta/\lambda$ = spatial frequency associated with the angle θ of the holographic off-axis reference wave. During operation, the outputs of N antennas in a phased array are transduced and used as signal inputs to each of the N channels in plane P_3 of the correlator (see Fig. 33.10). These signals have amplitude transmissions represented by

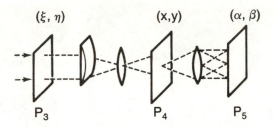

(ξ, η) (x, y) (α, β)

P_3 P_4 P_5

Fig. 33.10 Schematic representation of a multichannel correlator (after Ref. 2).

$$F(\xi, \eta) = \sum_{m=1}^{N} f(\xi - \xi_s + m\Delta\xi_\phi, y - md) , \qquad (33.21)$$

where ξ_s designates the beginning location for the return signal out of the n'th antenna and $\Delta\xi_\phi$ is proportional to the target azimuth because of the locations of the elements in the antenna receiver array. Note that $\xi_s - \xi_r$ is proportional to the target range.

In the P_4 plane of Fig. 33.10, the one-dimensional Fourier transform of the N channel input function described by Eq. (33.21) multiplies the array of stored matched filters described by Eq. (33.20). This multiplication occurs in a channel-by-channel sequence because of the one-dimensional imaging property of the two lenses in the y direction. These multiplications yield

$$T_A(x, y) = \sum_{m=1}^{N} F(\mu, y - md) \exp(2\pi i \mu \xi_s) \exp(2\pi i \mu m \Delta\xi_\phi)$$

$$\times F^*(\mu, y - md) \exp(-2\pi i \mu \xi_r)$$

$$\times \exp(-2\pi i \alpha_0 x) . \qquad (33.22)$$

Before taking the two-dimensional Fourier transform of Eq. (33.22) (i.e., plane P_4 in Fig. 33.10), we will discuss the unique characteristic of the azimuthal phase terms in Eq. (33.22) given by

$$\psi(\mu, y) = \exp(2\pi i \mu m \Delta\xi_\phi) . \qquad (33.23)$$

These azimuthal phase terms have a different slope in each channel. This occurs because the path difference from the target to each individual antenna gives rise to different spatial shifts in the starting position of the signal recorded in the ξ direction of each channel. This is indicated by the term $m\Delta\xi_\phi$ in Eq. (33.21). Thus, the summation of phase factors in the μ variable described in Eq. (33.23) actually has the characteristic of a linear phase factor in the y direction because of the geometry of the process.[2] Under these conditions, Eq. (33.23) can be rewritten as

$$\psi(\mu, y) = \exp(2\pi i \mu y \Delta\xi_\phi) . \qquad (33.24)$$

Using the spherical lens between planes P_4 and P_5 of Fig. 33.10 to take the Fourier transform of Eq. (33.22) modified by Eq. (33.24), we obtain

$$F(\alpha, \beta) = [f(\alpha) \star f^*(\alpha)] \circledast \delta[\alpha - \alpha_0 \lambda f - (\xi_s - \xi_r), \beta - \Delta\xi_\phi] , \qquad (33.25)$$

(a)   (b)

Fig. 33.11 Real-time multichannel correlator output for a linear phased array and a bore-sighted target for two different values of range (from Ref. 12).

which shows that the ambiguity in phase due to range and azimuth of the target is removed because the range information ξ_r determines the location of the correlation peak of the transmitted signal in the α direction, and the azimuth information $\Delta\xi_\phi$ determines the location of the correlation peak in the β direction. A more detailed analysis also shows a spreading of the correlation peak in both directions due to the geometry of the antenna array relative to the target.[12] An example illustrating the correlation output for two different ranges is shown in Fig. 33.11.

REFERENCES

1. *Transformations in Optical Signal Processing,* W. T. Rhodes, J. R. Fienup, B. E. A. Saleh, eds., SPIE Proc. 373 (1983).
2. D. Casasent, "Pattern and character recognition," in *Handbook of Optical Holography,* H. J. Caulfield, ed., pp. 503–536, Academic Press, New York (1979).
3. A. E. Smith, L. Liederman, and M. J. Beran, "Laboratory simulation of a random media," *J. Opt. Soc. Am.* 60(5), 728A (1970).
4. L. J. Cutrona, E. N. Leith, C. J. Palermo, and L. J. Porcello, "Optical data processing and filtering systems," *IRE Trans. Info. Th.*, IT-6(3), 386–400 (1960).
5. L. J. Cutrona, "Recent developments in coherent optical technology," in *Optical and Electro Optical Information Processing,* J. T. Tippett, D. A. Berkowitz, L. C. Clapp, C. J. Koester, and A. Vanderburgh, Jr., eds., pp. 83–124, Massachusetts Institute of Technology Press, Cambridge, Mass. (1965).
6. J. B. DeVelis and G. O. Reynolds, *Theory and Applications of Holography,* Chapter 8, Addison-Wesley Publishing Co., Reading, Mass. (1967).
7. A. Vander Lugt, "Signal detection by complex spatial filtering," *Trans. IEEE* IT-10, 139–145 (1964).
8. W. M. Brown, *Analysis of linear time invariant systems,* McGraw-Hill Book Co., New York (1963).
9. A. Kozma and D. L. Kelly, "Spatial filtering for detection of signals submerged in noise," *Appl. Opt.* 4(4), 387–392 (1965).
10. E. Brookner, *Radar Technology,* Chapter 7, Artech House Inc., Dedham, Mass. (1977).
11. P. M. Woodward, *Probability and Information Theory with Application to Radar,* Pergamon Press, New York (1960).
12. D. Casasent and E. Klimas, "Multichannel correlation for radar signal processing," *Appl. Opt.* 17, 2058 (1978).

34   Optically Modulated Imagery

34.1   INTRODUCTION

In Chapter 28, we showed that the theoretical communication techniques that are so common in electrical engineering are also very useful in optics because of the linear, spatially invariant nature of many optical systems. In a similar manner, the concept of modulating signals on unique carriers, which is a common technique in radio and television, has also found its way into optics in the form of modulated imagery and holography.

We have already seen that the image is retrieved from the hologram by looking in the angular direction, which was unique to the holographic reference beam. In fact, different images have been stored and retrieved from a single hologram by using different carrier frequencies[1] and/or recording distances.[2]

In this chapter, we discuss storage and retrieval by carrier-modulated imaging techniques. In this process, different images are stored on a single medium (film) with unique carriers and retrieved by tuning the retrieval system to the specific carrier of interest. Different implementations of the concept are given and the capability of carrier-modulated imagery is demonstrated.

34.2   THE CONCEPT OF CARRIER-MODULATED IMAGING

We have already given a simple example of carrier-modulated imagery in Chapter 30 where we discussed halftone and raster removal. Since these images are derived from a transmitted signal where the sampling theorem was invoked at the sending site (see Chapter 28), it is not surprising that the periodic raster overlaid on the image could be removed by spatial filtering techniques. It is interesting to note in these examples that the information content of the image is not increased by spatial filtering; rather, the periodicity in the image is removed. We now extend the concept of a rastered image to illustrate the usefulness of carrier-modulated imagery.

The rastered image is an example of a carrier-modulated image. A carrier-modulated image could also be created by placing a transmission diffraction grating in front of the film plane of a camera as shown in Fig. 34.1. If the fundamental frequency of the diffraction grating is chosen to be at least twice the bandwidth of the imaging system, then the sampling theorem is obeyed and all of the information content of the scene passed by the imaging system is properly sampled and stored on the linear portion of the film's H and D curve. The image

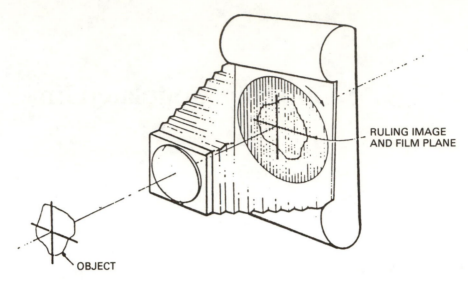

Fig. 34.1 Multiple image camera (from Ref. 3).

will have a rastered appearance. We know from the results already shown in Chapter 30 that a raster-free image of the object can be obtained by spatial filtering with an aperture in a blocking mask in the transform plane, which passes only one of the diffraction orders of the modulated grating. Mathematically, we represent the carrier-modulated image as the product

$$I_{im}(\xi) = I_{ob}^{BL}(\xi) P(\xi) \ , \tag{34.1}$$

where

$I_{im}(\xi)$ = recorded image (assumed to be linear with a $\gamma = -2$ film developing process)[3]

$I_{ob}^{BL}(\xi)$ = band-limited object, which has been low pass filtered by the transfer function of the imaging lens

$P(\xi)$ = transmission function of the periodic diffraction grating.

Using the system shown in Fig. 30.1 of Chapter 30, we optically Fourier transform Eq. (34.1) and then filter the spectrum by passing only one order of the grating with an aperture having a diameter equal to the object bandwidth in frequency space. Upon re–Fourier transforming the filtered image with the second lens, we obtain

$$I_{im}^{filtered}(\alpha) = I_{ob}^{BL}(\alpha) \ , \tag{34.2}$$

i.e., a raster-free image of the band-limited object.

34.3   MULTIPLE IMAGE STORAGE WITH ANGULARLY DEPENDENT CARRIERS[3,4]

We now consider using the camera in Fig. 34.1 to store many band-limited images N on one film with a multiple exposure process. Between each exposure the grating will be moved through an angle θ_n such that $N\theta_n = \pi$. The recorded multiply exposed image is mathematically represented by

$$I_{im}(\xi) = I_{ob}^{(1)}(\xi) P^{\theta_1}(\xi) + I_{ob}^{(2)}(\xi) P^{\theta_2}(\xi) + \ldots + I_{ob}^{(n)}(\xi) P^{\theta_n}(\xi)$$

$$+ \ldots + I_{ob}^{(N)}(\xi) P^{\theta_N}(\xi) , \tag{34.3}$$

where $I_{ob}^{(N)}(\xi)$ represents the n'th object used in the experiment and $P^{\theta_N}(\xi)$ represents the orientation of the grating in the n'th exposure.

The film containing the multiple images described by Eq. (34.3) is linearly processed ($\gamma = -2$) so that the amplitude transmittance of the film is linearly proportional to the intensity distribution in the image plane of the camera, as discussed in Chapter 21. This film is placed in the object plane of a Fourier processing system, and the aperture filter used in the raster removal experiment is allowed to rotate and be centered sequentially over the first side orders of the diffraction patterns of the various modulated gratings. The various images, caused by the multiple exposures, will appear sequentially in the image plane. This demonstrates the principle of multiple image storage and retrieval by using *theta modulation*, i.e., angular modulation. An example illustrating this multiple image procedure experimentally is shown in Fig. 34.2.

In Fig. 34.2(a), we show the four separate objects used in the experiment and in (b) we see the multiply stored images. In (c) we show the corresponding Fourier spectrum of the multiple images. The retrieved images from a linear storage experiment are shown in Fig. 34.2(d), demonstrating the importance of film processing on system linearity when contrasted with (e), where the output images from a nonlinear storage process ($\gamma = +2$) are shown.

As discussed in Chapter 28, if the image bandwidth is much less than the carrier frequency, then the number of images that can be stored on the film by this technique can be determined by first calculating the circumference of the circle in frequency space whose radius is the maximum carrier frequency that can be stored on the film at a modulation of 50%. Only half of this circumference can be used to store different carrier-modulated images because of the diffraction symmetry in the frequency plane. Dividing this available circumference by the image bandwidth yields the number of stored images as

$$N = \pi\omega_0/\omega_{BW} , \tag{34.4}$$

where ω_0 is the maximum carrier frequency that can be stored on the film at a modulation of 50% and ω_{BW} is the image bandwidth. Twenty encoded map images stored on one film are shown in Fig. 34.3(a) and three retrieved images are shown in (b), (c), and (d).

The number of images stored on the film can be increased by utilizing noncommensurate carriers to fill in the frequency plane as shown in Fig. 34.4. In this application, however, because of the large number of images involved, one must be careful not to exceed the dynamic range of the recording medium (film); otherwise, the process becomes nonlinear.

In linear multiple image storage with single-frequency angular carriers, the number of low-frequency images that can be stored is given by Eq. (34.4). With reference to Fig. 34.4 for noncommensurate carriers, this number can be increased to

$$N \cong \frac{1}{2} \left[\frac{\pi\omega_0^2 - \pi(\omega_0/2)^2}{\pi(\omega_{BW}/2)^2} \right] = \frac{3}{2} \frac{\omega_0^2}{\omega_{BW}^2} , \tag{34.5}$$

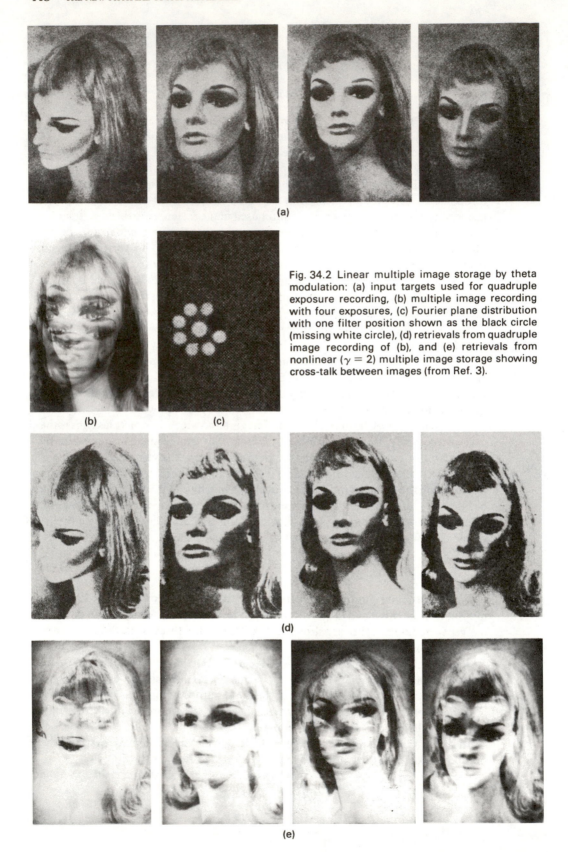

Fig. 34.2 Linear multiple image storage by theta modulation: (a) input targets used for quadruple exposure recording, (b) multiple image recording with four exposures, (c) Fourier plane distribution with one filter position shown as the black circle (missing white circle), (d) retrievals from quadruple image recording of (b), and (e) retrievals from nonlinear ($\gamma = 2$) multiple image storage showing cross-talk between images (from Ref. 3).

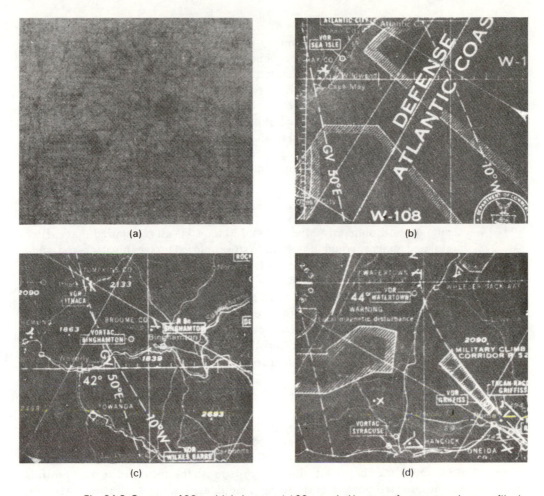

Fig. 34.3 Storage of 20 multiple images: (a) 20 encoded images of maps stored on one film by the carrier-modulated technique and (b), (c), and (d) 3 of the 20 retrieved images. (Courtesy of P. F. Mueller and D. A. Servaes.)

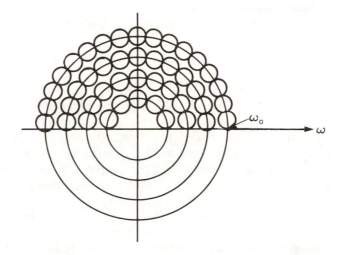

Fig. 34.4 Schematic of noncommensurate carriers used to fill frequency space.

where $\omega_0 \gg \omega_{BW}$. Equation (34.5) shows a sizable increase in the number of images that can be stored with noncommensurate carriers compared to the limit given by Eq. (34.4) for single-frequency angular modulation.

As shown previously in Chapter 28, the penalty paid for undersampling carrier-modulated imagery is an overlap of the image spectra, commonly called *aliasing*. This disturbing effect is occasionally seen on television as a moiré phenomenon when striped moving objects are present in the scene (referee's shirts in sporting events, striped clothing on dancers, etc.). The television industry is well aware of this phenomenon and avoids the problem by using plain or nonperiodic patterns on clothing and background whenever possible.

34.4   ENCODING COLOR IMAGES ON BLACK-AND-WHITE FILM

Armitage and Lohmann,[4] Mueller,[5] and Macovski[6] all recognized that carrier-modulated imagery could be utilized to store colored images on black-and-white film and avoid the time-consuming task of processing color film. However, in practice, the need for special Fourier projectors to play back the image in full color has limited the usefulness of the technique. Because of its inherent simplicity, we include it here as yet another example of carrier-modulated imagery.

In the Lohmann method,[4] each primary color in the object had a grating painted on it in a unique characteristic direction. Mueller's technique is more useful because the information is encoded in the camera just in front of the film plane with a specialized color filter.

34.4.1   Sequential Exposures

Mueller's method[5] is easily understood by referring to Fig. 34.1. Suppose that a red filter is placed over the camera lens and a carrier-modulated image on black-and-white achromatic film is recorded with the grating in angular position θ_1. Thus, the red portion of the scene is modulated in the θ_1 direction. After the first exposure, the red filter is replaced with a green filter and the grating is rotated to position θ_2. The second exposure on the same film is a registered modulated image containing the green portion of the scene. For the third exposure, a blue filter is used and the grating is placed in the θ_3 orientation so that the blue portions of the scene image have been encoded by a carrier in the θ_3 direction. Mathematically, this image is represented by

$$I_{im}(\xi) = I_R(\xi) P^{\theta_1}(\xi) + I_G(\xi) P^{\theta_2}(\xi) + I_B(\xi) P^{\theta_3}(\xi) , \qquad (34.6)$$

where $I_R(\xi)$, $I_G(\xi)$, and $I_B(\xi)$ represent the red, green, and blue portions of the scene, respectively, and θ_1, θ_2, and θ_3 represent the three orientations of the gratings used in the three exposures.

The film is then linearly processed, assuming that the three exposures are controlled to be within the straight-line portion of the D–log E curve of the film. The resulting transparency is used as the input to the system shown in Fig. 30.1. A color filter is designed so that a red filter is placed over the first diffraction orders in the θ_1 direction, a green filter in the θ_2 direction, and a blue filter in the θ_3 direction. Use of a collimated white-light illuminator in the f-f system yields a full-color image in the output plane of the processor.

Images retrieved by this technique are shown in Fig. 34.5. (Of course, false color images can be created by interchanging the positions of the color filters in

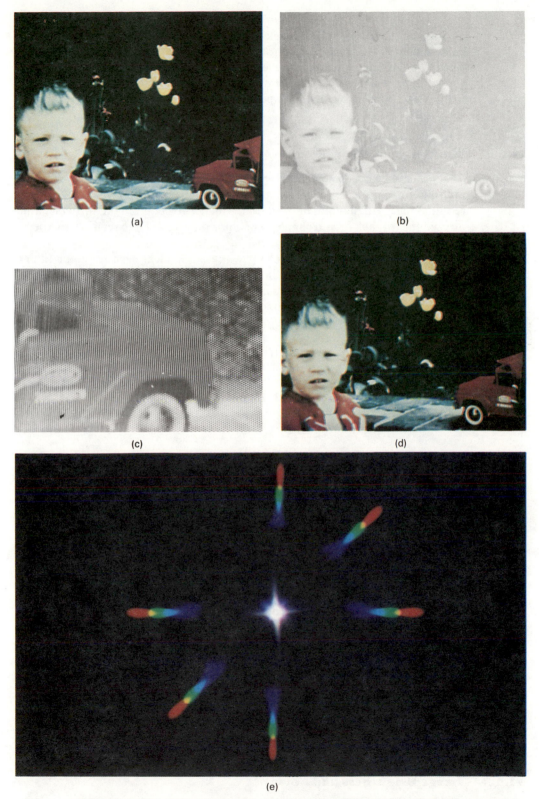

Fig. 34.5 Stored and retrieved color images from black-and-white film: (a) original color object, (b) black-and-white modulated image, (c) enlargement of (b) showing modulation, (d) retrieved color image, and (e) the Fourier plane. (Courtesy of P. F. Mueller.)

the Fourier plane.) The original object used in the experiment is shown in Fig. 34.5(a) and the color-encoded black-and-white transparency is shown in (b). An enlargement of the truck wheel showing the halftone-like pattern is seen in (c). The full-color filtered image retrieved from the black-and-white transparency is shown in (d) and the Fourier spectrum of the encoded transparency with the white-light source showing the dispersive property of the diffraction gratings is seen in (e).

34.4.2   Simultaneous Exposures

This process—while simple and easy to understand as a direct extension of the multiple imaging technique—is not practical because of the three exposures required to record the color information. Mueller[5] recognized this difficulty and utilized the tri-pack filter technology of color photography to realize his experimental goal.

Simplifying Mueller's analysis,[5] we will assume ideal negative color filters having the transmissions shown in Fig. 34.6. If these filters are made into a tri-pack with each layer being ruled in a different direction with a square wave, all having the same fundamental frequency, then the transmittance of the tri-pack filter is

$$T_A(\xi, \lambda) = [(1 - R) P^{\theta_1}(\xi)] [(1 - G) P^{\theta_2}(\xi)] [(1 - B) P^{\theta_3}(\xi)] \, , \qquad (34.7)$$

where R, G, and B represent the idealized positive red, green, and blue filters, respectively, as shown in Fig. 34.7. Equation (34.7) may be written

$$T_A(\xi, \lambda) = 1 - R\, P^{\theta_1}(\xi) - G\, P^{\theta_2}(\xi) - B\, P^{\theta_3}(\xi) \, , \qquad (34.8)$$

where the fact that $RG = GB = BR = 0$ has been used in going from Eq. (34.7) to (34.8) (the transmission through a red-green combined filter is zero since the red filter passes only red and the positive green filter absorbs it; similarly, the same effect holds for the GB and BR combinations). A photograph of a tri-level carrier-modulated filter is shown in Fig. 34.8.

When the carrier-modulated filter described by Eq. (34.8) is placed in front of the film plane in Fig. 34.1, the recorded image is

$$I_{im}(\xi) = I_{ob}(\xi) - I_{ob}^R(\xi) P^{\theta_1}(\xi) - I_{ob}^G(\xi) P^{\theta_2}(\xi) - I_{ob}^B(\xi) P^{\theta_3}(\xi) \, , \qquad (34.9)$$

where $I_{ob}^R(\xi) = R I_{ob}(\xi)$ = red portions of the object, with similar expressions for green and blue. Except for the minus sign common to all three color-modulated image terms, the terms in Eq. (34.9) are similar to (34.6). In fact, in the diffraction orders, Eqs. (34.6) and (34.9) are identical except for the minus sign. Filtering the encoded image of Eq. (34.9) in an *f-f* white-light Fourier processing system with color blocking filters placed over the first diffraction orders yields a positive color image of the original object, since the minus sign is common to all three amplitude terms in the image plane. An example of an image retrieved in this manner from a black-and-white encoded transparency is shown in Fig. 34.9.

34.4.3   Image Brightness and Noise

The Fourier color-encoding process just described has a few inherent limitations: diffraction efficiency, speckle noise, and the film resolution needed to record the

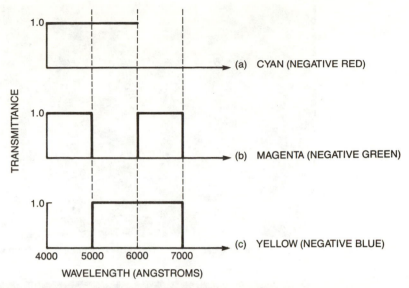

Fig. 34.6 Spectral transmittance of idealized negative color filters.

grating with a high modulation. Since the Fourier projector (the *f-f* imaging system) is inherently a diffraction process, the radiation in the object plane of the projector must have sufficient spatial coherence to enable the grating to diffract. As discussed in Chapters 11 and 28, this means that the grating must have a very high frequency or the source must be extremely small. This limitation is best overcome by utilizing high-frequency gratings and relatively large, broadband, bright sources to reduce speckle noise and increase image brightness. The high-frequency grating requirement limits the range of applications of this system since high-resolution (slow-speed) film is needed to record the high-frequency grating with a modulation index (visibility) of 50% or greater. In applications such as archival copying of color microfilm, 200 cycles/mm grating frequencies

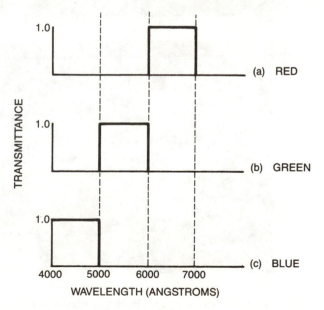

Fig. 34.7 Transmittance of idealized positive color filters.

Fig. 34.8 Section of a carrier-modulated tri-pack color filter (from Ref. 5).

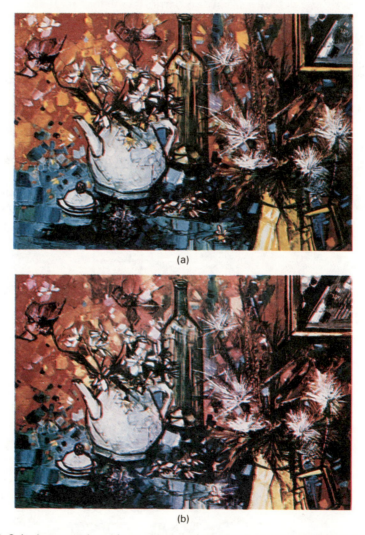

(a)

(b)

Fig. 34.9 Color image retrieved from a black-and-white transparency encoded with a single exposure and the tri-pack angular encoding color filter shown in Fig. 34.8: (a) original scene and (b) retrieved image. (Courtesy of P. F. Mueller.)

were found to be adequate in this trade-off.[7] These large sources coupled with broad-bandwidth color filters have low spatial and temporal coherence. These factors help reduce speckle noise, as illustrated earlier in Chapter 21. Spatially encoded sources are also utilized to reduce noise.[7,8]

In addition to providing more brightness to a system by utilizing high-frequency gratings with a high-modulation index, additional brightness was gained by utilizing an encoded source consisting of multiple off-axis sources. These configurations are useful because the Fourier plane is an image of the source plane. Six off-axis sources are placed in a configuration equivalent to the first-order diffraction pattern geometry in the Fourier plane due to the encoded image. Each off-axis source has one of its first diffraction orders at the dc location when used to illuminate the same encoded image. A blocking filter having a diameter equivalent to the image bandwidth is used on axis in the Fourier plane. The color filtering is accomplished by placing color filters over the appropriate pairs of sources in the collimator as shown in Fig. 34.10. The multiple source system shown here increases the retrieved image brightness by a factor of 6 compared to that of a system having a single on-axis source. This process of adding additional sources can be repeated for the higher diffraction orders to further increase the image brightness. However, these brightness gains are smaller due to diffraction losses.

Image brightness can also be increased by bleaching the color-encoded transparency to create a modulated phase image. Since phase gratings can have very high diffraction efficiencies and no absorption, factors of 10 to 20 in image brightness were realized by utilizing a modulated phase image.[7] While diffraction efficiencies of 1 to 10% are common with amplitude diffraction gratings, diffractive efficiencies of 30 to 60% can be realized with modulated phase images. Further increases in image brightness may be obtained by suitable design of the condenser optics.

34.5   PHASE-MODULATED IMAGES

As just described, carrier-phase-modulated images may be created by bleaching an intensity-modulated image on a photographic emulsion. The bleaching is

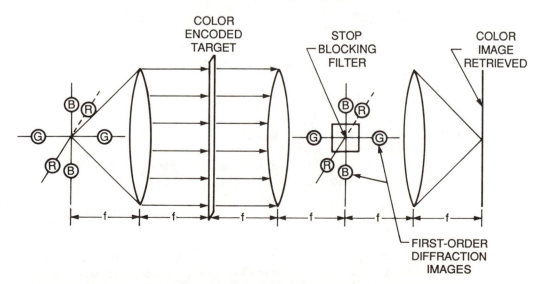

Fig. 34.10 Schematic of color retrieval system with six off-axis sources.

linear in that phase height is proportional to photographic density over a wide dynamic range as shown by Altman[9] and Smith.[10]

In frequency space, the modulation centers the image spectrum about the carrier frequency, which avoids the low-frequency modulation transfer function (MTF) fall-off characteristic of photographic phase images, as illustrated in Fig. 34.11.

In a phase-modulated image, the phase is referenced within each period by the nonmodulated portion of the rectangular (square) wave carrier. This gives the system an excellent continuous-tone capability. This system is also an example of a phase contrast imaging system and is analyzed in detail in Chapter 35.

An example of a continuous-tone image retrieved from a phase-modulated image using an off-axis source is shown in Fig. 34.12. This image was duplicated from an original negative by utilizing an optical contact duplicating method in which the image is phase modulated onto a carrier in photoresist and subsequently transferred into various plastic materials.[11] A conventional photographic duplicate of the same original negative is shown in Fig. 34.12(b) for purposes of comparison. Resolution measurements on the carrier-modulated phase image duplicates showed that the resolution transfer was better in the phase images than it was in the density images.[11] This result was expected since optical contact reduces the distance between the original negative and the duplication medium. This avoids diffraction losses and improves the system MTF as shown in Fig. 34.13.

34.6   THE SQUARE-ARRAY-MODULATED IMAGE CONCEPT

The Lohmann principle of detour phase[13,14] gives the path difference at the n'th diffraction order due to a diffraction grating and a shifted grating. As shown in Fig. 34.14, the grating shift p creates a path difference of $\Delta_m = p \sin\theta_m$ relative to the m'th diffraction order at the point A in the observation plane. As p increases up to a period, the path difference Δ_m increases linearly with angle—up to a wavelength—in the small-angle approximation. From the grating equation $m\lambda = d \sin\theta$, the small-angle approximation means that the period d must be a few wavelengths or greater. Under these conditions, we see from Fig. 34.14 that

$$\frac{\Delta_m}{\Delta} = \frac{p \sin\theta_m}{d \sin\theta} = \frac{\Delta_m}{m\lambda} \cong \frac{p}{d} , \tag{34.10}$$

where $\theta \cong \theta_m$. Thus, the detour phase is

$$\Delta_m = \frac{m\lambda p}{d} , \tag{34.11}$$

where p is the amount of lateral shift, d is the grating period, and m is the diffraction order being considered.

Mueller[15,16] realized that if the grating shift is implemented by interleaving modulated images, then $p = d/2$ and Eq. (34.11) becomes

$$\Delta_m = \frac{m\lambda}{2} . \tag{34.12}$$

Thus, an effective phase shift of π is created at the odd diffraction orders. Therefore, image subtraction results when the interleaved modulated images are

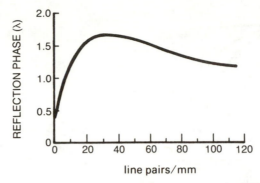

Fig. 34.11 Relative phase relief as a function of spatial frequency for 3404 film (from Ref. 11).

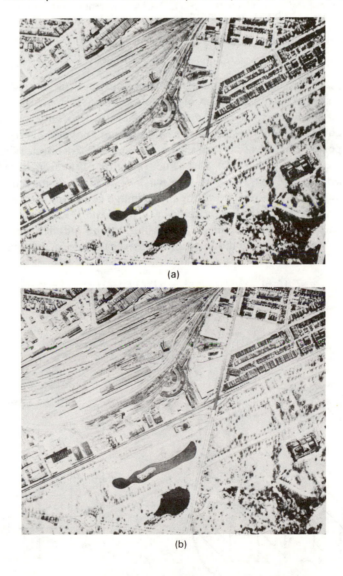

(a)

(b)

Fig. 34.12 (a) Image retrieved from a phase-modulated parylene duplicate of an original negative and (b) image from a conventional photographic duplicate of the same negative. This result shows the excellent continuous-tone capability of a phase-modulated image (from Ref. 11).

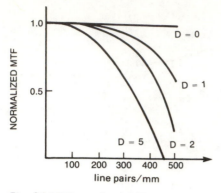

Fig. 34.13 Normalized MTF of contact printing as a function of the air gap D (in microns) and the spatial frequency (in line pairs/mm). Curves plotted using data from Ref. 12.

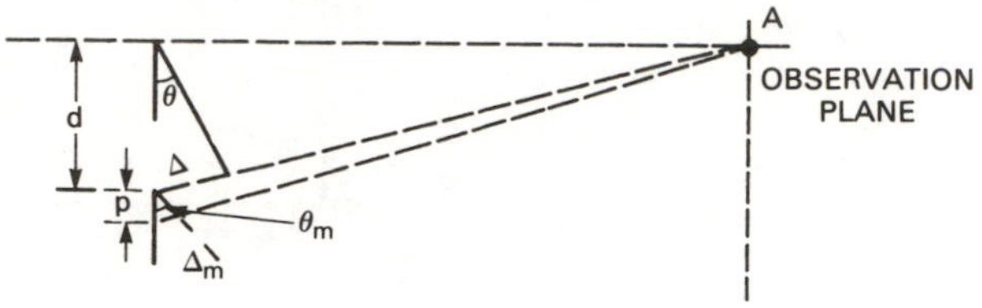

Fig. 34.14 Schematic diagram illustrating the concept of detour phase Δ_m.

placed in the object plane of a coherent optical Fourier processing system having a spatial filter that passes only the first diffraction order.

Mueller[15] then expanded the concept to two dimensions and considered a checkerboard or a square-array-modulated image geometry as a shifted series of periodic masks of the type shown in Fig. 34.15. An interleaving of four sequential messages (A, B, C, D) on the linear portion of the film is made by replacing the grating of Fig. 34.1 with this mask. The mask is moved by a distance $1/\omega_0$ between exposures, resulting in the image geometry shown in Fig. 34.16, where $1/\omega_0$ describes the period of the basic pixel and $1/2\omega_0$ represents the period of the checkerboard. After linear processing, the amplitude transmittance of the film containing the four modulated (interleaved) images in the square array geometry is

$$T_A(\xi, \eta) = AP(\xi)\,P(\eta) + BP(\xi)\,P\left(\eta - \frac{1}{2\omega_0}\right) + CP\left(\xi - \frac{1}{2\omega_0}\right)P(\eta)$$

$$+ DP\left(\xi - \frac{1}{2\omega_0}\right)P\left(\eta - \frac{1}{2\omega_0}\right), \tag{34.13}$$

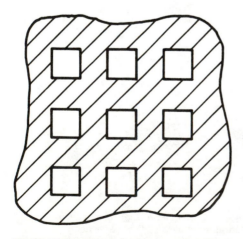

Fig. 34.15 Basic mask used to create a square array of four interleaved modulated images.

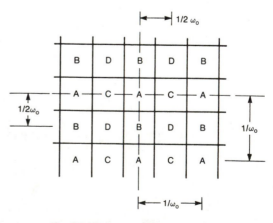

Fig. 34.16 Square array geometry.

where $P(\xi)$ and $P(\eta)$ represent high-contrast square wave gratings in the ξ and η directions, respectively, and A, B, C, and D are functions of ξ and η.

The Fourier transform of a product of period functions of different variables is given by

$$\tilde{P}(\mu)\,\tilde{P}(\nu) = \sum_{n}\ \sum_{m} a_n a_m \delta(\mu - n\omega_0, \nu - m\omega_0) \ , \tag{34.14}$$

where $\mu = x/\lambda f$ and $\nu = y/\lambda f$.

The Fourier transform of the square array of modulated images described by Eq. (34.13) is

$$\tilde{T}_A(\mu,\nu) = \sum_{n}\ \sum_{m}\ \{a_n a_m [\tilde{A}(\mu - n\omega_0, \nu - m\omega_0)] + e^{in\pi}\,\tilde{B}(\mu - n\omega_0, \nu - m\omega_0)$$

$$\times e^{im\pi}\,\tilde{C}(\mu - n\omega_0, \nu - m\omega_0)$$

$$+ e^{i(m+n)\pi}\,\tilde{D}(\mu - n\omega_0, \nu - m\omega_0)\} \ . \tag{34.15}$$

This Fourier transform has square symmetry and we can analyze the content of the characteristic orders by looking at the specific makeup of one quadrant of the Fourier transform plane, as shown in Fig. 34.17. Although $A(\xi,\eta)$, $B(\xi,\eta)$, $C(\xi,\eta)$, and $D(\xi,\eta)$ can, in general, be spatial functions or images, we will assume them to be constants in the present analysis for simplicity.

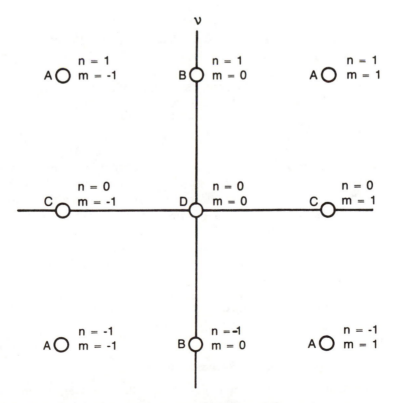

Fig. 34.17 Square array Fourier spectrum geometry for the algorithm for $A + B + C = D$ or the algorithm $A = B + C$, $B = C + A$, $C = B + A$, and $D = 0$.

The values of the first-order Fourier spectra of Eq. (34.15) are given by

$$\tilde{T}_A(\mu,\nu)\Big|_{\substack{n=0\\m=0}} = a_0^2\delta(\mu,\nu)(A+B+C+D) \,. \tag{34.16a}$$

$$\tilde{T}_A(\mu,\nu)\Big|_{\substack{n=\pm1\\m=0}} = a_1 a_0 \delta(\mu\mp\omega_0 f_2\lambda,\nu)(A-B+C-D) \,. \tag{34.16b}$$

$$\tilde{T}_A(\mu,\nu)\Big|_{\substack{n=0\\m=\pm1}} = a_0 a_1 \delta(\mu,\nu\mp\omega_0 f_2\lambda)(A+B-C-D) \,. \tag{34.16c}$$

$$\tilde{T}_A(\mu,\nu)\Big|_{\substack{n=\pm1\\m=\pm1}} = a_1^2\delta(\mu\mp\omega_0 f_2\lambda,\nu\mp\omega_0 f_2\lambda)(A-B-C+D) \,. \tag{34.16d}$$

These results suggest a powerful encoding scheme; namely, if we let $D = A + B + C$ and let A, B, and C be separate functions (images), then we have the following terms as located by the appropriate delta function:

$$\tilde{T}_A\Big|_{\substack{n=0\\m=0}} = 2a_0^2 D; \qquad \tilde{T}_A\Big|_{\substack{n=1\\m=0}} = -2a_1 a_0 B;$$

$$\tilde{T}_A\Big|_{\substack{n=1\\m=1}} = 2a_1^2 A; \qquad \tilde{T}_A\Big|_{\substack{n=0\\m=1}} = -2a_1 a_0 C. \tag{34.17}$$

This suggests that the Fourier spectra of the various images separate, meaning that the various images can be reconstructed by a Fourier processing system. An identical distribution in the transform is obtained by the algorithm $A = B + C$, $B = C + A$, $C = B + A$, and $D = 0$.

These algorithms have been demonstrated with colored imagery stored on black-and-white film so that registration of the colors was preserved and colored information could be retrieved.[17] The algorithm, $D = A + B + C$ is referred to as *additive* when encoding color images on black-and-white film because for normal color imagery A, B, and C would be the additive primary separations (red, green, and blue) and D would be the integrated full-color image as recorded on a panchromatic black-and-white film. The second scheme, $A = B + C$, $B = C + A$, $C = B + A$, and $D = 0$, is *subtractive* since the subsample exposures would be equivalent to cyan (white minus red), magenta (white minus green), and yellow (white minus blue). However, whether or not the three zones to be recorded are in the visible, ultraviolet, or infrared regions does not matter. As long as the logic of one of these two schemes is followed, we can store the information in registration and subsequently obtain spatially separated zonal spectra in the transform plane of an optical processing system. By inserting color filters over the various zonal spectra, all channels can be projected simultaneously in perfect registration to produce true or false color images.

This technique has been demonstrated experimentally with both analog and digital approaches, illustrating that the coding algorithm is unique and devoid of cross-talk.

34.6.1   Example 1: Image Subtraction [15]

A binary object consisting of the three words *Fourier, optical,* and *subtraction,* as shown in Fig. 34.18(a), was encoded as a modulated image. The square array periodic encoding mask of Fig. 34.15 was used in the image plane of the camera

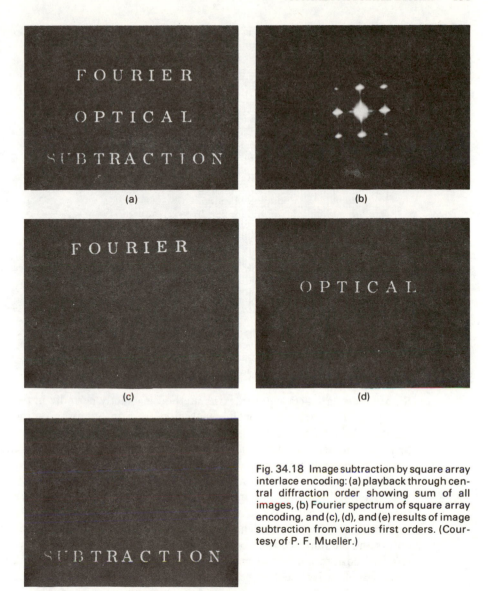

Fig. 34.18 Image subtraction by square array interlace encoding: (a) playback through central diffraction order showing sum of all images, (b) Fourier spectrum of square array encoding, and (c), (d), and (e) results of image subtraction from various first orders. (Courtesy of P. F. Mueller.)

shown in Fig. 34.1. The entire object of Fig. 34.18(a) was exposed through the square array onto the linear portion of the films D–log E curve. With the square array moved one-half period in the x direction, the words *optical* and *subtraction* were blocked and the word *Fourier* was imaged; a subsequent one-half period motion of the square array in the y direction was used to image the word *optical* after blocking the words *Fourier* and *subtraction* in the object plane. Finally, a one-half period movement of the array back in the x direction (this is equivalent to a single one-half period shift in the y direction) was used to encode the word *subtraction* as a modulated image with the words *Fourier* and *optical* blocked in the object plane. After linear processing, the playback of this square-array-encoded image through a Fourier optical projector gives the results shown in Fig. 34.18. Retrieval of the central or dc order shows all three words of the original object whereas retrievals from the "first" diffraction orders [Figs. 34.18(c), (d), and (e)] show that the various words are subtracted as predicted by Eq. (34.17).

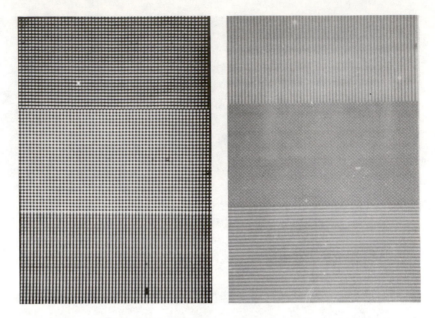

Fig. 34.19 Digitally encoded color patches.

34.6.2   Example 2: Digital Encoding of Color [17]

Since this square array multiplexing method is exactly periodic in two dimensions, the square array approach is well suited for printout by an x-y mechanical scanner. A digital encoding algorithm (see Fig. 34.17) was used with the scanner printout geometry to create a synthetic color target that is equivalent to an exposure through a subtractive square array color filter. The synthetic target graphically represented the amplitude transmittances of a color bar chart (blue, red, green, yellow, cyan, and magenta patches). The color patch was encoded in the square array geometry in a computer and the output data were used as the input for an x-y film writer. The black-and-white encoded image for the primary colors and the subtractive colors are shown in Fig. 34.19. Here the patches in the left half correspond to the negative colors and those in the right half to the primary colors. Notice that the *red* patch has *vertical* lines and the *blue* patch has *horizontal* lines since these correspond to symmetry in the x and y directions. The

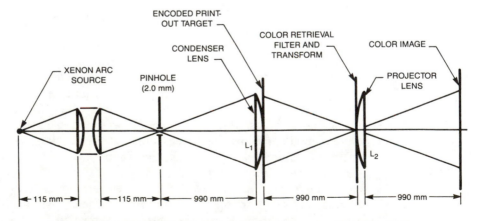

Fig. 34.20 Schematic of Fourier projector. (Courtesy of P. F. Mueller).

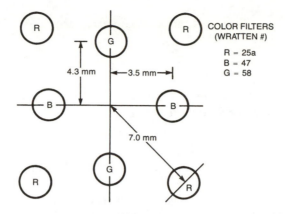

Fig. 34.21 Color retrieval filter geometry. (Courtesy of P. F. Mueller.)

green patch consists of cross-hatched lines since two directions of symmetry are involved in the 45-deg direction. The secondary colors are made up as a combination of two primaries, resulting in a cross-hatched type of grating geometry.

The linearly processed image was placed in a white-light source Fourier projector (Fig. 34.20) where red, green, and blue color filters were introduced in the appropriate diffraction orders. The geometry of the diffraction pattern in the filter plane is shown in Fig. 34.21. The retrieved image showing the color test patch is shown in Fig. 34.22. This simple experiment demonstrates that informa-

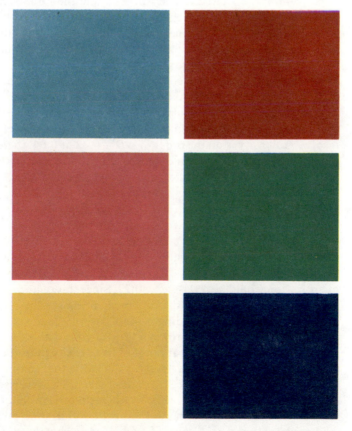

Fig. 34.22 Retrieval of color patches. (Courtesy of P. F. Mueller.)

tion can be encoded in the square array geometry and retrieved in an optical projector.

34.6.3   Example 3: Square Array Digital Picture Storage of Wide Dynamic Range Data [18]

We now discuss an algorithm whereby the digital information of a pixel can be encoded in the square array geometry so that either the image can be viewed or the digital information can be retrieved from the image for subsequent processing.

Digital imagery is made up of resolution elements (called *pixels*) having various shades of gray arranged in a two-dimensional geometry. The basic concept concerns storing the N-bit digital word associated with each pixel on film. The square array geometry, shown in Fig. 34.23 and consisting of four channels (subpixels) within the pixel area, will be used to encode the digital word for each pixel. Three of the channels are used to store the digital word code and the fourth channel is a reference channel. This enables the image to be viewed visually. The film resolution must be good enough to resolve the individual channels (subpixels) well. The viewer is designed to filter the square array geometry so that it is viewed in a raster-free configuration. This is easily accomplished with independent off-axis sources in the viewer analogous to those discussed previously for color viewing.

Therefore, the viewed image will resemble the nonrastered image. We will show that proper attenuation of the various sources in the viewer, or use of different colors in each channel, will enable one to view the full dynamic range of the image associated with the digital data. In addition, the data can be retrieved in digital format if the image area of interest is scanned and digitized.

The encoding technique consists of density slicing the data and storing it in each of the three channels. As an example, if we consider an eight-bit word ($N = 8$), there are $(2)^8 = 256$ possible levels within any one pixel. Since most films have a linear working range of approximately two density levels and since the just resolvable density is on the order of $\Delta D = 0.02$ or 0.04, there are only 50 or 100 distinguishable levels available for recording the data on the linear portion of the film. Thus, normal film cannot record 256 levels of digital data in its gray scale with any hope of accurate retrieval of the data.

The square array code will associate each level with a distinguishable density level on the film. Thus, a uniformly distributed code over three channels for an eight-bit word would require seven levels/channel since $(7)^3 = 343$, and since $(6)^3 = 216$ is less than the desired number of 256 levels, for a 20-dB range on the linear portion of the film's H and D curve (two density units), the density quantization would be $\Delta D = 2/7 \cong 0.29$. Since this quantization is much greater than the just detectable density difference of most films, the repeatability and reliability of the uniform coding process is very good.

This uniform coding algorithm is illustrated in Table 34.I. The logic assumes that the film has a usable linear range of two density units (between a D_{max} of 2.5 and a D_{min} of 0.5). The value/level for the individual channels is determined by building up the desired number in units of seven. The seven values of 0 through 6 are placed in channel 1. The values 7 to 48 are placed in channel 2, and the values 49 to 342 are placed in channel 3. Channel 1 will have seven levels with a ΔD of 0.29 per level, but with a value of 1 assigned to each level. Channel 2 will have

Fig. 34.23 Sampled image showing subpixel code for encoding wide dynamic range data. (From *Scientific Honeyweller* July 1985 cover.)

seven levels with a ΔD of 0.29 per level and a value of 7 assigned to each level. Channel 3 will also have seven levels with a ΔD of 0.29 per level and a value of 49 assigned to each level. This logic has equal probability for errors in all channels, but the code is very reliable since the values of 0.29 per level are significantly greater than the granularity values of most films, provided the pixel size is not too small.

To illustrate the uniform coding logic in Table 34.I, the number 222 is chosen as a number to code. As seen in Table 34.II, this number would have a density value of 1.34 in channel 3, a value of 1.63 in channel 2, and a value of 1.05 in channel 3. The reference channel (channel 4) would have a density value corresponding to the sum of the transmittances of channels 1, 2, and 3, which for this case would be a value of 0.80. The formula for the reference density D_R is

$$D_R = -\log_{10}(10^{-D_3} + 10^{-D_2} + 10^{-D_1}) .\qquad(34.18)$$

TABLE 34.I
Example of a Uniform Code for an Eight-Bit Word

Channel	Number of Levels	Value/Level	D/Level
3	7 levels at	49/level	0.29
2	7 levels at	7/level	0.29
1	7 levels at	1/level	0.29
4	Reference	Sum of transmittance of channels 1, 2, and 3	

TABLE 34.II
Encoding of the Number 222

Channel	$\left(\dfrac{\text{Number or Remainder}}{\text{Value/Level}}\right)$	Numerical Value	Remainder	Density Calculation	Density Value for Channel (Positive Retrieval)[a]
3	$\dfrac{222}{49} = 4$	$49 \times 4 = 196$	26	$D_3 = 0.29 \times 4$	1.34
2	$\dfrac{26}{7} = 3$	$7 \times 3 = 21$	5	$D_2 = 0.29 \times 3$	1.63
1	$\dfrac{5}{1} = 5$	$5 \times 1 = 5$	0	$D_1 = 0.29 \times 5$	1.05
4	—	Reference	—	D_R[Eq.(34.18)]	0.80

[a]Density values for positive retrieval obtained by subtracting D_3, D_2, or D_1 from D_{max}.

The density value in a given channel, for a positive retrieval, is calculated by subtracting the density value (D_3, D_2, D_1) from D_{max} ($D_{max} = 2.5$ in the example). This method ensures that when the image is processed as a positive it will have positive polarity when viewed.

The second example chosen was the coding for the number 41 as shown in Table 34.III. In this case, the value of channel 3 is 0 so the maximum density (D_{max}) is printed in the channel. Channel 2 has a density value of 1.05 and channel 1 has a density value of 0.76. Reference channel 4 has a density value equaling the sum of the transmittances of channels 1, 2, and 3, which for this case would be a density value of 0.57. This example illustrates the concept of positive retrieval. The value of 0 in channel 3 means that no light should be transmitted through that subpixel when the film is viewed. The printed value of D_{max} in the subpixel means that minimal light (the threshold level for the system) will be transmitted through the subpixel when actually viewing the film.

TABLE 34.III
Encoding of the Number 41

Channel	$\left(\dfrac{\text{Number or Remainder}}{\text{Value/Level}}\right)$	Numerical Value	Remainder	Density Calculation	Density Value for Channel (Positive Retrieval)[a]
3	$\dfrac{41}{49} = 0$	0	41	$D_3 = 0 \times 0.29$	D_{max}
2	$\dfrac{41}{7} = 5$	$5 \times 7 = 35$	6	$D_2 = 5 \times 0.29$	1.05
1	$\dfrac{6}{1} = 6$	$6 \times 1 = 6$	0	$D_1 = 6 \times 0.29$	0.76
4	—	Reference	—	D_R[Eq.(34.18)]	0.57

[a]Density values for positive retrieval obtained by subtracting D_3, D_2, or D_1 from D_{max}.

34.6.4   Viewing Considerations

34.6.4.1   Attenuation

To display the image with its original gray scale, one could attenuate the various orders by appropriate amounts or, more practically, use off-axis sources in the conjugate image positions of the diffraction orders and control the brightness of these sources separately by the relative weighting levels of the Fourier coefficients.

In addition, for uniform coding of eight-bit imagery, if the channel 3 source ($m = 0, n = 1$) has a normalized brightness level, then the channel 2 source ($m = 1$, $n = 0$) should have a relative brightness of $1/49$, while the channel 1 ($m = 1, n = 1$) source should be attenuated by a value of $1/343$.

34.6.4.2   Display

Following the above procedure might not be the best way to display the image for the human observer because very little contribution to the image would be detected from the attenuated zones. A more striking presentation would be observed if all three zones were displayed at the same brightness level (sequentially or simultaneously) or if a green source were used for channel 1, a red source for channel 2, and a blue source for channel 3. This would be similar to the manner in which color TV monitors are used to display density sliced images.

However, the true nature of the image is only observed in this mode when channel 3 is viewed alone, since channel 3 best approximates the full value of a given pixel.

34.6.4.3   Simulated Density Slicing Experiments

To demonstrate the advantages of presenting digital imagery in a density sliced mode, a special viewer is utilized. The viewer uses multiplexed, multispectral earth resources photographs to simulate density slices. The carrier modulation principles of this viewer are similar to those of the square array code and the results are intended to illustrate the display characteristics typical of square array storage.

The photograph in Fig. 34.24 simulates the print of a multichannel digital signal. The highlight whites are extremely overexposed leading to blooming and distortion, while the low-level shadow details are still muddy. There are actually three brightness channels in Fig. 34.24 with relative maximum values of 1.0, 0.01, and 0.001.

By amplifying the dimmer channels and attenuating the brighter channel so that their visual brightnesses were matched, the result shown in Fig. 34.25 was obtained. This operation is equivalent to density slicing and normalizing intensity as explained in the square array scheme. Blooming and distortion have been eliminated, but the proper brightness scale has also been lost in this black-and-white print. To retain the advantages achieved by density slicing and to avoid a loss of brightness, the same information of Fig. 34.25 can be viewed in color as shown in Fig. 34.26 where the highest signal levels appear orange, the intermediate levels yellow and blue, and the low levels cyan and red. The choice of colors for each density level is arbitrary, but the use of color in conjunction with the large dynamic range encoded image of Fig. 34.24 does provide an additional display parameter.

Fig. 34.24 Black-and-white superposition of three Landsat multispectral scenes. Note over-exposure at shadow. (Courtesy of P. F. Mueller.)

Fig. 34.25 Black-and-white retrieval from encoded image with density slicing achieved by illuminating sources in viewers. Note information recovery at arrow. (Courtesy of P. F. Mueller.)

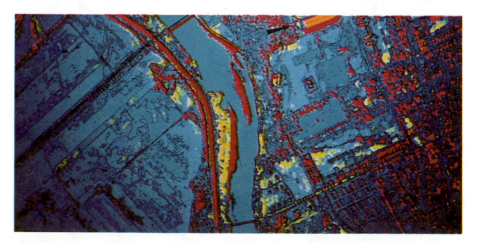

Fig. 34.26 Color view of wide dynamic range encoded scene. Note information at arrow. (Courtesy of P. F. Mueller.)

A clear advantage of the density sliced display can be seen in the large orange object toward the middle of the left side of Fig. 34.26. The evenly spaced dots show up immediately in Figs. 34.25 and 34.26 while they have been washed out in the nonsliced display of Fig. 34.24. Also, the effluent formations along the river banks are much easier to detect in the colored picture.

The prime advantage offered by this technique is that the user interacts with the visual image, which is in register with the digital information. The technique could be useful for the storage, dissemination, and analysis of digital imagery such as Landsat since the user would have access to large volumes of preregistered data with which he can interact either visually or digitally.

34.7   IMAGE HOLOGRAPHY: THREE-DIMENSIONAL IMAGE MODULATION

Image holography differs from conventional holography in that the object field is an image rather than a diffraction pattern of the object. The reference wave usually consists of a plane or spherical wave that is derived from a separate path in a two-beam interferometer (see Fig. 34.27). An image hologram is capable of storing a three-dimensional image. One advantage of image holography over conventional holography is that the speckle phenomenon normally associated with holography can be removed by spinning a ground glass in the reconstructing laser beam. This effect is illustrated dramatically in Fig. 34.28, which shows the reconstructed image from an image hologram with and without rotating ground glass in the reconstruction laser beam. This noise-averaging effect was also used to advantage when phase compensating human cataracts with image holography.[20]

Even though image holography is capable of storing a three-dimensional image, we consider it to be a kind of modulated imagery since the image is amplitude modulated onto a carrier, i.e.,

$$I(x) = |\exp(ik\alpha_0 x) + A(x)|^2$$
$$= 1 + |A(x)|^2 + A(x)\cos k\alpha_0 x , \qquad (34.19)$$

where $I(x)$ = intensity in the hologram, $A(x)$ = the image amplitude, and α_0 = reference beam angle. As seen in Eq. (34.19), the image hologram differs from

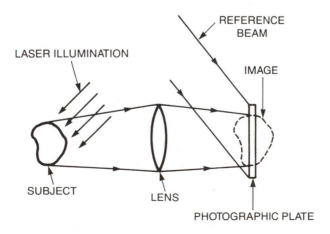

REFERENCE BEAM

IMAGE

LASER ILLUMINATION

SUBJECT

LENS

PHOTOGRAPHIC PLATE

Fig. 34.27 Schematic for formation of an image hologram (from Ref. 19).

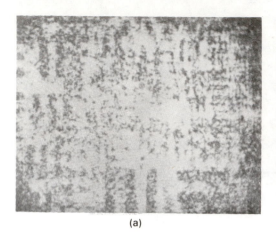

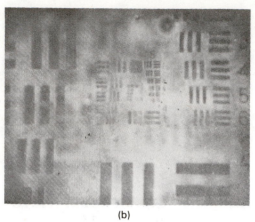

(a)   (b)

Fig. 34.28 Noise averaging in image holography: (a) image reconstructed with collimated laser illumination and (b) image reconstructed with a rotating diffuser near the laser source, illustrating the speckle average in image holography. (Courtesy of P. F. Mueller.)

carrier-modulated imagery in that the image is not amplitude modulated onto the bias, i.e., the zero-order spectrum of the image hologram is the self-convolution of the image spectrum and not the spectrum itself.

34.7.1   Example 1: Experimental Demonstration of High-Resolution Duplication Using Image Holography

To demonstrate the depth-of-focus capability of image holography in duplicating high-resolution photographs, an image hologram system was set up in the configuration shown in Fig. 34.27. An f/1.9 lens was used in a unit magnification imaging configuration. The reference beam was split off and recombined with the image field in the vicinity of the image plane at an angle of 180 deg, corresponding to a carrier frequency of ~600 cycles/mm. A black-and-white tri-bar target (228 cycles/mm) was used as the object and out-of-focus holograms were recorded. A typical result is shown in Fig. 34.29(a), (b), and (c).

The distribution in the plane of the hologram as seen in normal incoherent illumination is seen in Fig. 34.29(a). From this picture, it is clear that the image is quite far out of focus (~0.25 mm) since the image is quite blurred. An out-of-focus image hologram recorded in this plane as seen in coherent light is shown in Fig. 34.29(b). (Coherent light was used to photograph the hologram because it had such low contrast that it was difficult to obtain a good picture with incoherent photography.) The hologram was placed in a reconstruction system using rotating ground glass in the illumination beam to both sharpen focal definition and average coherent artifacts. The image retrieved from this system is shown in Fig. 34.29(c). The resolution seen in (c) is group 7-6 (228 cycles/mm), which is all that was present on the original resolution target.

The results show that out-of-focus information may indeed be recovered when reconstructed by focusing through the hologram.

34.7.2   Example 2: Contact Printing Using Holography

In applications requiring both high-resolution and large format films, a contact-type printer would be needed since the high-quality lens required for use in Fig.

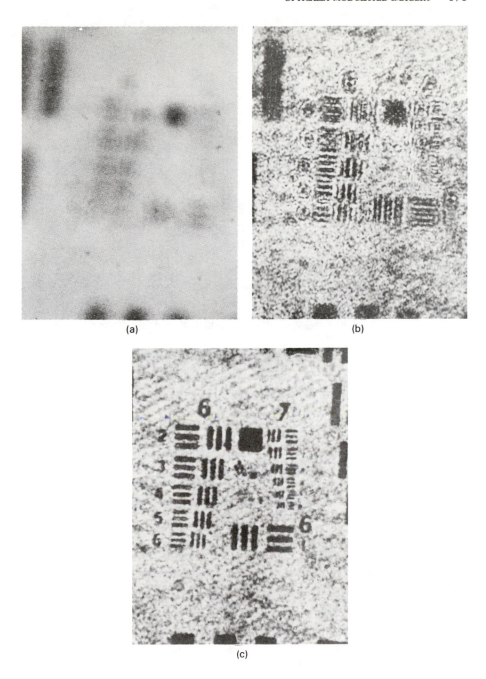

Fig. 34.29 High-resolution duplication by image holography: (a) out-of-focus image at hologram recording position (incoherently illuminated), (b) the image hologram as seen with
coherent light (speckle deteriorates image resolution), and (c) the retrieved image from the
image hologram with speckle noise averaged. (Courtesy of P. F. Mueller.)

34.27 does not exist. One such technique for accomplishing this task is total
internal reflection or "evanescent wave" holography, which is shown[19,21,22]
diagrammatically in Fig. 34.30. In this type of hologram, a surface wave reference beam is coupled into the hologram through the back of the film by utilizing a
prism liquid gated to the film and using a reference wave at the critical angle.

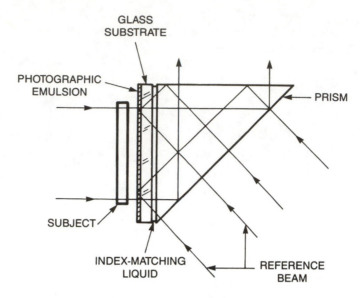

Fig. 34.30 Hologram formation using total internal reflection (from Ref. 19).

Phase gratings may also be used for this purpose.[23] Holograms having image resolutions in excess of 500 cycles/mm have been recorded using these techniques in the laboratory.[21]

The three-dimensional nature of the image hologram provides the advantage that the observer can focus through the image on reconstruction and not suffer losses due to the diffusion of light through the depth of the emulsion, as illustrated in Fig. 34.29. This information, which is usually lost, is stored on the carrier and retrieved in the reconstruction step.

If holographic techniques are considered for the duplication problem, one must bear in mind these distinct differences between holographic duplication and conventional contact prints:

1. Holography is imagery on a carrier and the film resolution must be high enough to resolve the carrier well. This will influence other film characteristics such as dynamic range, gray scale, gamma, etc.

2. The speckle problem normally found in reconstructing holograms must be avoided in order to obtain "aesthetically pleasing pictures."

3. To realize the three-dimensional nature of the holographic system, the *reference beam* must reach the hologram plane *without passing through the object*.

REFERENCES

1. J. B. DeVelis and G. O. Reynolds, *Theory and Applications of Holography,* p. 160, Addison-Wesley Publishing Co., Reading, Mass. (1967).
2. E. N. Leith and J. Upatnieks, "Wavefront reconstruction with diffused illumination and three-dimensional objects," *J. Opt. Soc. Am.* 54(11), 1295–1301 (1964).
3. P. F. Mueller, "Linear multiple image storage," *Appl. Opt.* 8(2), 267–273 (1969).
4. J. D. Armitage and A. W. Lohmann, "Theta modulation in optics," *Appl. Opt.* 4(4), 399–403 (1965).
5. P. F. Mueller, "Color image retrieval from monochrome transparencies," *Appl. Opt.* 8(10), 2051–2057 (1969).
6. A. Macovski, "Encoding and decoding of color information," *Appl. Opt.* 11(2), 416–420 (1972).
7. P. F. Mueller, "Standard microfilm for recording color and multiple images," in *Proc. Natl. Microfilm Assn. Annual Convention,* May 1969.
8. F. T. S. Yu, *Optical Information Processing,* John Wiley and Sons, Inc., New York (1983).
9. J. Altman, "Pure relief images on type 649-F plates," *Appl. Opt.* 5(11), 1689L (1968).
10. H. M. Smith, "Photographic relief images," *J. Opt. Soc. Am.* 58(4), 533–539 (1968).
11. P. F. Mueller, D. J. Cronin, and G. O. Reynolds, "Novel image duplication technique utilizing Fourier optics," in *Processing of Images and Data from Optical Sensors,* W. H. Carter, ed., Proc. SPIE 292, 47–58 (1981).
12. F. G. Kasper, "Computation of light transmitted by a thick grating for application to contact printing," *J. Opt. Soc. Am.* 64(12), 1623–1630 (1974).
13. B. R. Brown and A. W. Lohmann, "Complex spatial filtering with binary masks," *Appl. Opt.* 5(6), 967–969 (1966).
14. A. W. Lohmann and D. P. Paris, "Computer generated spatial filters for coherent optical data processing," *Appl. Opt.* 7(4), 651–655 (1968).
15. P. F. Mueller, "Optical processing of information including synthesis by complex amplitude addition of diffraction spectra," U.S. Patent No. 3,664,248, May 23, 1972 (filed May 3, 1968).
16. P. F. Mueller, "Optical reconstruction of phase images," in *Processing of Images and Data from Optical Sensors,* W. H. Carter, ed., Proc. SPIE 292, 39–46 (1981).
17. P. F. Mueller, "Multiple image storage overlay methods," Final Technical Report, Contract NA S5-21301, prepared for NASA Goddard Space Flight Center (June 1971).
18. P. F. Mueller and G. O. Reynolds, "Storage of digital information on film in register with the image," *J. Opt. Soc. Am.* 67(10), 1407A (1977).
19. R. J. Collier, C. B. Burckhardt, and L. H. Lin, *Optical Holography,* Academic Press, New York (1971).
20. G. O. Reynolds, J. L. Zuckerman, and W. A. Dyes, "Holographic phase compensation techniques applied to human cataracts," *Opt. Eng.* 12(1), 23–35 (1973).
21. K. A. Stetson, "Holography with total internally reflected light," *Appl. Phys. Lett.* 11(7), 225–226 (1967); see also "Improved resolution and signal-to-noise ratios in total internal reflection holograms," *Appl. Phys. Lett.* 12(11), 362–364 (1968).
22. O. Bryngdahl, "Holography with evanescent waves," *J. Opt. Soc. Am.* 59(12), 1645–1650 (1969).
23. P. K. Tien, "Light waves in thin films and integrated optics," *Appl. Opt.* 10(11), 2395–2413 (1971); see also P. K. Tien, G. Smolinsky, and R. J. Martin, "Thin organosilicon films for integrated circuits," *Appl. Opt.* 11(3), 637–642 (1972).

35   Phase Contrast Imaging

35.1   INTRODUCTION

In Chapters 29 through 33 on analog optical computing, we considered filters that modify the amplitude and/or phase of the optical wave field in the Fourier transform plane of the system shown in Fig. 35.1. The most general filter function is one that modifies both the amplitude and the phase of the optical wave field in the Fourier transform plane. These filters are known as *complex filters* and, as pointed out previously, are quite difficult to fabricate. A special case of the complex filter is a filter that modifies only the optical path difference or phase of the optical wave field in the Fourier transform plane. A pure phase filter satisfies the condition

$$|\tilde{f}(\mathbf{x})| = 1 \ , \tag{35.1a}$$

so that

$$\tilde{f}(\mathbf{x}) = \exp[i\psi(\mathbf{x})] \ , \tag{35.1b}$$

where the tilde, ~, represents a spatial Fourier transformation.

Both phase and complex filters have been successfully utilized in phase contrast imaging for purposes of making phase objects (which are invisible with normal incoherent light) observable. In this chapter, discussions of these various techniques are given as additional examples of complex optical filtering.

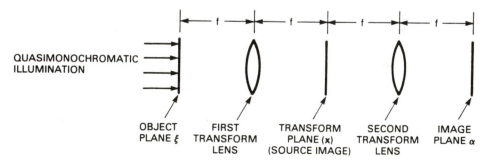

Fig. 35.1 Generalized optical filtering system. (For a detailed discussion of this system, see Chapter 30.)

35.2   PHASE CONTRAST VIEWING METHODS

35.2.1   Zernike Techniques

As an example of complex filtering using a phase filter, we describe the phase contrast microscope, a fundamental experiment by Zernike[1,2] that had tremendous impact on the scientific community. Using the system shown in Fig. 35.1, let us consider a pure phase object of the form

$$f(\xi) = \exp[i\phi(\xi)] , \tag{35.2}$$

where $\phi(\xi)$ is a function describing the optical phase in the object as a function of position. If the variations of $\phi(\xi)$ are small compared to unity (1), Eq. (35.2) may be written as

$$f(\xi) \cong 1 + i\phi(\xi) , \tag{35.3}$$

which is a real constant added to the spatially varying object in phase quadrature with the constant. The Fourier transform of this distribution is given by

$$\tilde{f}(\mathbf{x}) = \delta(\mathbf{x}) + i\tilde{\phi}(\mathbf{x}) , \tag{35.4}$$

which is a real delta function at the origin added to the spectrum of the phase quadrature signal.

At this point, a number of specific filtering operations can be performed in the Fourier transform plane to make the two terms of Eq. (35.4) interact to form a visible image. These filtering operations are schematically represented throughout this chapter and discussed below. For simplicity, the schematics of the filter functions shown in these figures are one-dimensional representations of the two-dimensional functions described by the appropriate equations.

The first sketch, Fig. 35.2(a), represents the unfiltered distribution where the background term $\delta(\mathbf{x})$ and the signal spectrum $\tilde{\phi}(\mathbf{x})$ are 90 deg out of phase

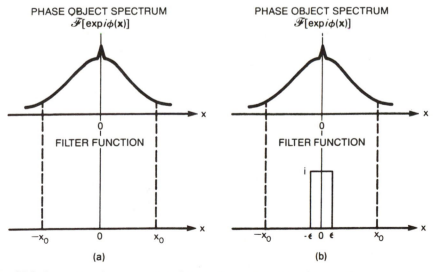

Fig. 35.2 Frequency plane operations for two phase contrast techniques. The filter function is $M(\mathbf{x})$ as used in the text, and $\mathcal{F}[\exp i\phi(\xi)]$ is the spectrum of the phase object, where the $\mathcal{F}$ represents the Fourier transform operation: (a) unfiltered spectrum using the system bandwidth $\mathrm{rect}(x|x_0)$ as the filter function and (b) positive phase contrast using the filter function of Eq. (35.5).

(phase quadrature) and $2|\mathbf{x}_0|$ is the bandwidth of the system centered at the center of the filter plane. If this unfiltered distribution is observed in the image plane (the α plane) of Fig. 35.1, a uniform intensity distribution is all that appears, i.e., the original phase distribution which is not visible.

35.2.1.1   Example 1: Positive Phase Contrast

Consider a filter function, whose amplitude transmission is given by

$$M(\mathbf{x}) = \begin{cases} \exp\left(\dfrac{i\pi}{2}\right) & |\mathbf{x}| \leq \epsilon \\ 1 & |\mathbf{x}| > \epsilon \end{cases} , \tag{35.5}$$

where ϵ is a very small distance from the origin and is used in the transform plane of Fig. 35.1 to shift the delta function. Using the filter function shown in Fig. 35.2(b), the intensity distribution in the image plane of Fig. 35.1 is given by

$$|\psi(\alpha)|^2 = |1 + \phi(\alpha)|^2 \cong 1 + 2\phi(\alpha) , \tag{35.6}$$

where $|\phi(\alpha)|^2 \ll 1$. Thus, the original phase distribution becomes visible as an intensity distribution added to a bright background (positive phase contrast). This technique is illustrated in Fig. 35.3 with the phase image retrieved using a positive phase contrast filter.

35.2.1.2   Example 2: Negative Phase Contrast

If the filter function

$$M(\mathbf{x}) = \begin{cases} \exp\left(\dfrac{i3\pi}{2}\right) & |\mathbf{x}| \leq \epsilon \\ 1 & |\mathbf{x}| > \epsilon \end{cases} \tag{35.7}$$

is used in the transform plane of Fig. 35.1 to shift the phase of the delta function, as seen in Fig. 35.4(a), then the intensity distribution in the image plane is given by

$$|\psi(\alpha)|^2 = |1 - \phi(\alpha)|^2 \cong 1 - 2\phi(\alpha) , \tag{35.8}$$

where $|\phi(\alpha)|^2 \ll 1$. The original phase distribution now becomes visible as an

Fig. 35.3 Positive phase contrast image. (Courtesy of P. F. Mueller and D. J. Cronin.)

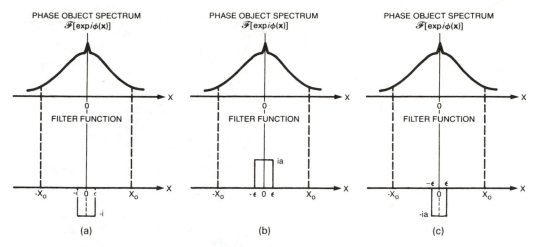

Fig. 35.4 Frequency plane operations for various phase contrast techniques: (a) negative phase contrast using the filter function of Eq. (35.7), (b) enhanced positive phase contrast using the filter function of Eq. (35.9) and (c) enhanced negative phase contrast using the filter function of Eq. (35.9).

intensity distribution subtracted from the bright background (negative phase contrast image).

35.2.1.3   *Example 3: Enhanced Positive and Negative Phase Contrast*

In the previous two examples, the filters have been pure phase filters; however, as discussed in Chapter 31, it is possible to fabricate filters of the form

$$
M(\mathbf{x}) = \begin{cases} a \exp(i\gamma) & |\mathbf{x}| \le \epsilon \; ; \; \gamma = \dfrac{\pi}{2} \, , \, \dfrac{3\pi}{2} \\ 1 & |\mathbf{x}| > \epsilon \end{cases} , \tag{35.9}
$$

where a allows for partial absorption of the background illumination [Figs. 35.4(b) and (c)]. This filter will enhance the contrast of the viewed image in the ratio $1/a$. Thus, for example, if we reduce the strength of the background intensity by a factor of 4, we double the contrast of the observed image.

35.2.2   Dark Field Technique

A special case of the filter function of Eq. (35.9) occurs when $a = 0$ such that

$$
M(\mathbf{x}) = \begin{cases} 0 & |\mathbf{x}| \le \epsilon \\ 1 & |\mathbf{x}| > \epsilon \end{cases} . \tag{35.10}
$$

When the filter of Eq. (35.10) is used in the transform plane of Fig. 35.1 to remove the effect of the background illumination shown in Fig. 35.5, dark field illumination results. The observed intensity distribution in this image is given by

$$
|\psi(\boldsymbol{\alpha})|^2 = \left| \phi(\boldsymbol{\alpha}) - \int \text{rect}(\mathbf{x}|\epsilon)\tilde{\phi}(\mathbf{x}) \exp\left(\frac{-2\pi i \mathbf{x} \cdot \boldsymbol{\alpha}}{\lambda f}\right) d\mathbf{x} \right|^2
$$

$$
\cong A^2 - 2A\phi(\boldsymbol{\alpha}) + |\phi(\boldsymbol{\alpha})|^2 \, , \tag{35.11}
$$

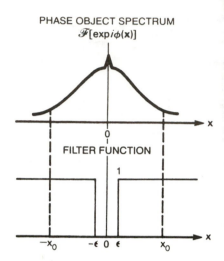

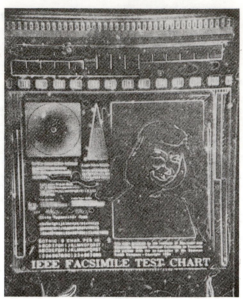

Fig. 35.5 Frequency plane operation for dark field method of phase contrast using the filter function of Eq. (35.10).

Fig. 35.6 Example illustrating the dark field method of phase contrast imaging. (Courtesy of P. F. Mueller and D. J. Cronin.)

where

$$A \cong 2\tilde{\phi}(0) \int\limits_{-\epsilon}^{\epsilon} \exp\left(\frac{-2\pi i x \cdot \alpha}{\lambda f}\right) d\mathbf{x} \; . \tag{35.12}$$

For small values of ϵ, A is approximately constant over the region of interest. Since A is not large enough to justify dropping the $|\phi(\alpha)|^2$ term in Eq. (35.11), the dark field method of phase contrast results in a nonlinear relationship between the observed phase image and the original phase distribution. Thus, this method of observation, while simple experimentally and useful for the detection of information, cannot be used to measure the thickness of a phase object accurately.

An example of a phase image made visible by the dark field imaging technique is shown in Fig. 35.6.

35.2.3 Oblique Illumination Techniques

35.2.3.1 Example 1: Dark Field Single-Sideband Method

The dark field method of phase contrast imaging may also be achieved by using oblique illumination in the optical system shown in Fig. 35.1. In this case (see Fig. 35.7), an off-axis point source is used to produce a physical shift of the spectrum in the Fourier plane. Since the zero-order of the spectrum occurs at the point of the source image, dark field illumination is created by imaging the point source outside the collection angle of the second transform lens in Fig. 35.1. This case is further complicated in that only part of the Fourier spectrum is passed; however, in one direction, the resolution limit of the spectrum which is passed is nearly doubled compared to the on-axis case. Again, this technique is highly nonlinear

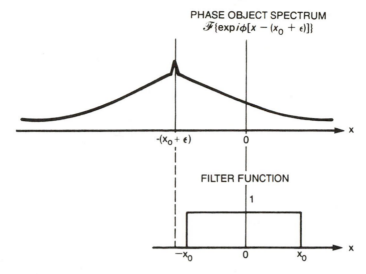

Fig. 35.7 Frequency plane operation for oblique dark field single-sideband method using the filter function of Eq. (35.13).

but very useful for visualizing phase images because of its experimental simplicity.

The mask (filter function) to be used in the transform plane of Fig. 35.1 is

$$M(\mathbf{x}) = \text{rect}(x|x_0)\,\text{rect}(y|y_0) \ . \tag{35.13}$$

The shifting of the spectrum causes an asymmetry in the transform plane because only one side of the spectrum is passed by the mask (filter function). It is easier to describe the operation of such systems by introducing the concept of an equivalent system filter to determine the spatial frequency distribution of the image. This filter utilizes the attenuation and phase shift characteristics of the mask and is centered about the location of the dc portion of the object spectrum. The equivalent filter, defined in this manner, includes the source configuration as part of the filter. Identification of the equivalent filter with the transfer function of the system later in this chapter means that the source configuration is considered to be an integral part of the system in the experiment of interest.

The equivalent filter function of the system for the dark field single-sideband method of Fig. 35.7 is

$$F_e(\mathbf{x}) = \text{rect}(x - x_0 + \epsilon|x_0)\,\text{rect}(y|y_0) \ . \tag{35.14}$$

This equivalent filter function for this dark field single-sideband system is shown in Fig. 35.8.

35.2.3.2 Example 2: Oblique Positive and Negative Single-Sideband Phase Contrast Imaging

The oblique dark field single-sideband method illustrated in Fig. 35.7 can be converted to a phase contrast method by incorporating Zernike phase quadrature delays into the mask (filter function), as illustrated in Figs. 35.9(a) and (b). The angle of the illumination beam must be chosen so that the image of the source is located just inside the edge of the system's bandpass in the transform

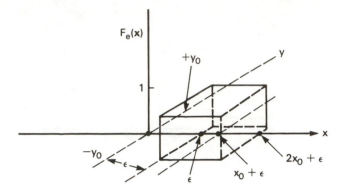

Fig. 35.8 Schematic of equivalent filter function for an oblique dark field single-sideband phase viewing system.

plane of Fig. 35.1. In addition, a phase quadrature filter is placed at the inside edge of the system bandpass so that the source image (dc) can be phase shifted by $+\pi/2$ or $-\pi/2$ to create positive or negative phase contrast images, respectively. The masks (filter function) for achieving these results are given by

$$M(\mathbf{x}) = \begin{cases} 1 & -y_0 \leq y \leq y_0; \ -x_0 \leq x \leq x_0 - 2\epsilon \\ a\exp(i\gamma) & x_0 - 2\epsilon \leq x \leq x_0; \ -\epsilon \leq y \leq \epsilon \\ 0 & \text{otherwise} \end{cases}, \qquad (35.15)$$

where $\epsilon \ll x_0$ and y_0, $\gamma = \pi/2$ or $3\pi/2$, and $a \leq 1$. The $-\epsilon \leq y_0 \leq \epsilon$ dimension is included in Eq. (35.15) because of the finite size of the phase filter in the y direction. This was not necessary in Eq. (35.13) because it was dark field; hence, the dc was outside the passband of the optical system.

The equivalent system filter function for the system is given by making the origin of the mask function coincident with the dc of the spectrum to yield

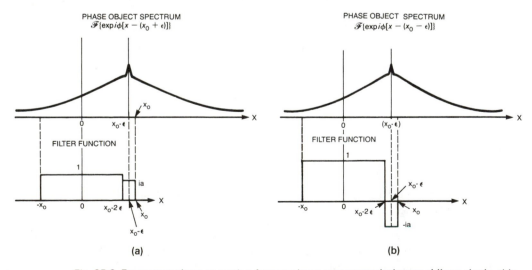

Fig. 35.9 Frequency plane operation for two phase contrast techniques: oblique single-sideband (a) positive and (b) negative phase contrast.

$$F_e(\mathbf{x}) = \begin{cases} 1 & -2\epsilon \geq x > -2x_0; \ -y_0 \leq y < y_0 \\ a\exp(i\gamma) & 0 > x > -2\epsilon; \ -\epsilon \leq y \leq \epsilon \\ 0 & \text{otherwise} \end{cases} , \qquad (35.16)$$

where $\epsilon \ll x_0$ and y_0, $\gamma = \pi/2$ or $3\pi/2$, and $a \leq 1$. Images through these systems would appear similar to those of the system described by Eqs. (35.13) and (35.14) except that they would be of lower contrast and have a continuous-tone characteristic due to the presence of the low spatial frequencies.

35.2.3.3 *Example 3: Double-Sideband Phase Contrast Imaging*

We have already shown that oblique illumination increases the single-sideband resolution of dark field imaging, as shown in Fig. 35.8. The same phenomenon also occurs in phase contrast imaging by controlling the illumination such that the dc is just commensurate with a phase quadrature delay filter located at the edge of the passband in the transform plane [see Figs. 35.9(a) and (b)].

The resolution asymmetry that is characteristic of all single-sideband systems can be removed by introducing a second equally spaced source on the opposite side of the optical axis. The dc portion of the shifted spectrum (which will be shifted in phase by $\pi/2$ or $3\pi/2$ to create positive or negative phase contrast) behaves as a common reference source for each single sideband (see Fig. 35.10). The two sidebands maintain their relative phase with the dc reference, resulting in a coherent addition to create a double-sideband phase contrast image having a

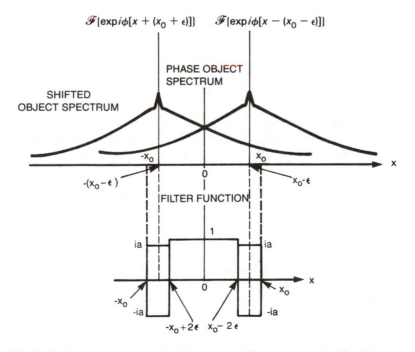

Fig. 35.10 Frequency plane operation for a symmetric two-source double-sideband system for positive (*ia*) or negative (*−ia*) phase contrast. (Note: The coherent transfer function is shown in Fig. 35.11).

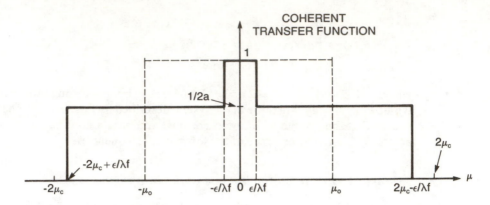

Fig. 35.11 One-dimensional coherent transfer function of a symmetrical two-source double-sideband phase contrast system showing increased coherent bandwidth. The dotted line represents the coherent system transfer function of the lens used in this experiment (from Ref. 3).

coherent bandwidth twice that of the coherent imaging system. The mask for this system is

$$M(\mathbf{x}) = \begin{cases} 1 & -x_0 + 2\epsilon \le x \le x_0 - 2\epsilon;\ -y_0 \le y \le y_0 \\ a\exp(i\gamma) & x_0 - 2\epsilon \le x \le x_0;\ -x_0 \le x \le -x_0 + 2\epsilon;\ -\epsilon \le y \le \epsilon, \\ 0 & \text{otherwise} \end{cases} \quad (35.17)$$

where $\epsilon \ll x_0$ and y_0, $\gamma = \pi/2$ or $3\pi/2$, and $a \le 1$. The equivalent one-dimensional system filter function of the system is

$$F_e(x) = \begin{cases} a\exp(i\gamma) & -\epsilon \le x \le \epsilon \\ 1 & \epsilon \le x \le 2x_0 + \epsilon \\ 1 & -(2x_0 + \epsilon) \le x \le -\epsilon \\ 0 & \text{otherwise} \end{cases} \quad (35.18)$$

The equivalent system filter function is the coherent transfer function of the system if the oblique illumination is considered to be an integral part of the system, i.e., all phase objects see the same system transfer function when being imaged by the oblique illumination system. The one-dimensional equivalent system filter function corresponding to Eq. (35.18) is shown schematically in Fig. 35.11 in spatial frequency coordinates. The corresponding coherent transfer function of the lens used without oblique illumination is also shown for purposes of comparison.

In the Appendix, Sec. 35.6, we show how a phase image having bandwidth doubling is obtained with an obliquely illuminated double-sideband system.

35.2.3.4   *Example 4: Two-Dimensional Double-Sideband Phase Contrast Imaging*

The double-sideband concept of a dynamic coherent imaging system has been extended[4] to two dimensions by using a rotating prism in the object illumination beam shown in Fig. 35.12. The mask for this system is a thin annular phase ring centered on axis and having a $\pi/2$ phase shift. The radius of the ring is chosen

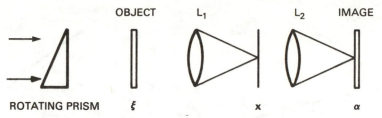

Fig. 35.12 Optical schematic of dynamic coherent system (from Ref. 4).

such that the phase step occurs at the inside edge of the bandpass in the transform plane of Fig. 35.12.

The equivalent filter function is determined from the fractional arc length that each spatial frequency subtends within the mask aperture during each revolution of the prism, as illustrated in Fig. 35.13. The dc is at the pupil edge so that its arc length is the circumference of the mask.

The equivalent filter function (coherent transfer function) for this system is

$$F_e(\mathbf{x}) = \frac{1}{2}\,\delta(0) + \frac{1}{\pi}\arccos\left(\frac{|\mathbf{x}|}{|\mathbf{x}_0|}\right)\cdot \tag{35.19}$$

A comparison of the filter function of the two-dimensional dynamic coherent system with that of the coherent transfer function of the lens is shown in Fig. 35.14. In Fig. 35.14(b), contrast enhancement is achieved by utilizing an additional 0.3 ND absorbing filter in the annular ring to renormalize Eq. (35.19).

A comparison of phase contrast images taken through the two systems having the transfer functions shown in Fig. 35.14 are shown in Fig. 35.15. This illustrates the resolution doubling and noise reduction capabilities of the dynamic coherent system. Equivalent results have been obtained by replacing the spinning prism illumination system of Fig. 35.12 with a uniformly illuminated incoherent narrow bandwidth annular ring source in the focal plane of the collimating lens of the illumination system.[3]

35.2.3.5   *Example 5: Dark Field Double-Sideband Phase Imaging*

To remove the restriction concerning the one-sided nature of oblique illumination in dark field phase viewing systems, annular sources have been utilized.

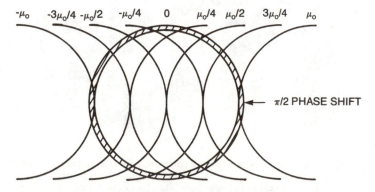

Fig. 35.13 Trajectories of various spatial frequencies. The aperture stop coincides with the 0 trajectory; μ_0 is the maximum spatial frequency transmitted (from Ref. 4).

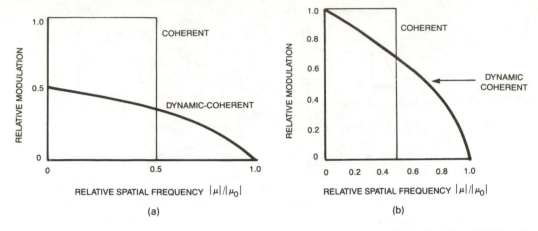

Fig. 35.14 Transfer function of the two-dimensional (a) unfiltered (from Ref. 3) and (b) filtered dynamic coherent (annular source) phase contrast system (from Ref. 4).

These have been realized with both passive[5] (annular ring source) and active[4] sources (spinning prism). In principle, these two techniques are equivalent[3] in that they double the coherent resolution limit set by the lens aperture and simultaneously average the coherent noise of the system.

This resolution increase is achieved due to the coherent mixing of the various Fourier components of the image spectrum passed by the available bandwidth of the system. Physically, this process is very complicated because the reference wave is not controlled as is the usual practice in interferometry and holography. The result is a nonlinear relationship between phase height and image intensity and, like all dark field techniques, is good for visualization of phase objects but not useful for precise mensuration.

The one-dimensional mask (filter function) for this system is $M(x) = \text{rect}(x|x_0)$, as seen in Fig. 35.16. The equivalent one-dimensional system filter function (coherent transfer function) for this system (see Fig. 35.17) is given by

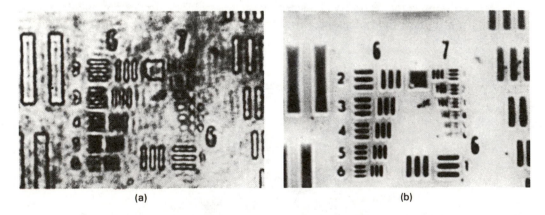

Fig. 35.15 Comparison between coherent phase visualization and dynamic coherent systems: (a) Coherent image of target without prism in system. The phase contrast is achieved because the lens acts as a low-pass filter. Resolution cutoff is at the 6-3 or 6-4 group, and the image is noisy. (b) Dynamic coherent phase imaging system. Resolution cutoff is at the 7-3 or 7-4 group, and the coherent noise is averaged (from Ref. 3).

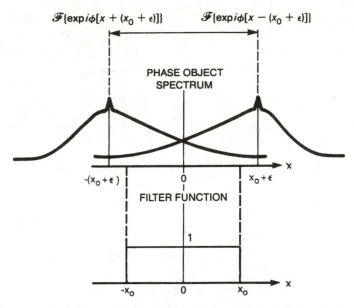

$\mathscr{F}\{\exp i\phi[x + (x_0 + \epsilon)]\}$ $\mathscr{F}\{\exp i\phi[x - (x_0 + \epsilon)]\}$

PHASE OBJECT
SPECTRUM

$-(x_0 + \epsilon)$ 0 $x_0 + \epsilon$

FILTER FUNCTION

1

$-x_0$ 0 x_0

Fig. 35.16 Frequency plane operation for annular source or dynamic coherent dark field system as given by Eq. (35.20).

$$F_e(x) = \begin{cases} 1 & \epsilon \le x \le 2x_0 + \epsilon \\ 1 & -(2x_0 + \epsilon) \le x \le -\epsilon \\ 0 & \text{otherwise} \end{cases} \tag{35.20}$$

For the two-dimensional annular source, the equivalent filter function (see Fig. 35.18) becomes[4]

$$F_e(|\mathbf{x}|) = \frac{1}{\pi} \arccos\left(\frac{|\mathbf{x}|}{|2\mathbf{x}_o + \epsilon|}\right) \quad |\epsilon| \le |\mathbf{x}| \le |2\mathbf{x}_o + \epsilon| . \tag{35.21}$$

Since this dark field system is conceptually different from the previous phase contrast systems, an example illustrating the coherent mixing principles is discussed.

If a phase grating is the object to be viewed by the dynamic coherent dark field system, a comparison of the object spectrum with the coherent transfer function of the system immediately illustrates how the coherent mixing of the Fourier components creates a visible image. In Fig. 35.19, the coherent transfer functions

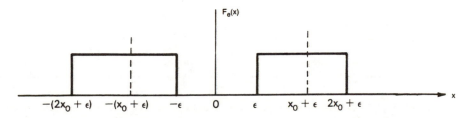

$F_e(x)$

$-(2x_0 + \epsilon)$ $-(x_0 + \epsilon)$ $-\epsilon$ 0 ϵ $x_0 + \epsilon$ $2x_0 + \epsilon$ x

Fig. 35.17 Equivalent system filter function of a one-dimensional dynamic coherent dark field viewing system.

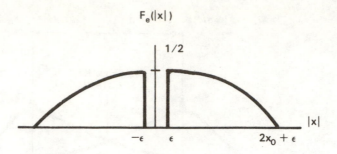

Fig. 35.18 Equivalent filter function of two-dimensional dynamic coherent dark field viewing systems as given by Eq. (35.21).

of the on-axis dark field system, the one-dimensional dynamic coherent system, and the object spectrum are shown to the same scale. The fundamental frequency of the phase grating ω_0^- is chosen to be slightly less than the coherent bandwidth of the oblique illumination optical system ω_0.

A simple comparison of Figs. 35.19(a) and (c) shows that the phase grating is never visualized with the dark field system since none of the Fourier components are within the system passband. Since the two sidebands of the dynamic coherent system [Fig. 35.19(b)] do not occur simultaneously within the coherent passband, it appears that this system will not image the phase grating either. However, if the prism angle of the dynamic coherent system is increased to correspond to a frequency slightly greater than $2\omega_0^-$, then the fundamental and the second order are passed by the lens simultaneously and beat together to create a periodic sine wave, which becomes observable. When the prism is rotated to the 180-deg

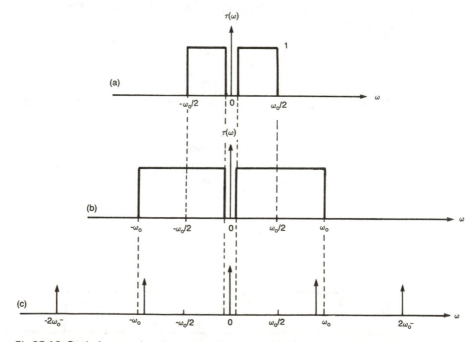

Fig 35.19 Scaled comparison between coherent transfer function of (a) the on-axis dark field, (b) the dynamic coherent dark field, and (c) the Fourier spectrum of a periodic phase grating having fundamental frequency, ω_0^-, where ω_0^- is chosen to be slightly less than the coherent bandwidth ω_0 of the system.

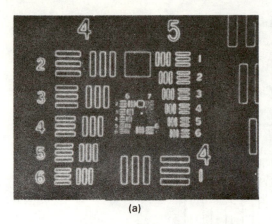 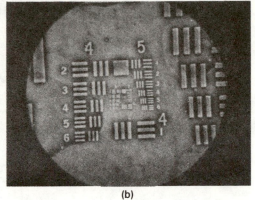

(a)   (b)

Fig. 35.20 Comparison of phase images obtained in (a) dynamic coherent dark field and (b) dynamic coherent phase contrast imaging systems. In (a) the dark field image of a resolution target is produced by stopping down the entrance pupil of the imaging lens to just exclude the annular trajectory of the source image. In (b) the phase image is reproduced by the same dynamic coherent system with a $\lambda/4$ phase annulus over the dc trajectory (from Ref. 4).

position, the two orders are passed by the other side of the coherent transfer function. These two interference images, which add intensity incoherently, have the same reference source (i.e., the fundamental frequency of the phase grating) which behaves as the dc reference for the dynamic coherent image. This reference preserves that portion of the coherent spectrum of the object, $\tilde{O}(\omega)$, which falls within the bandwidth of the dynamic coherent system shown in Fig. 35.19(b).

If the phase grating is replaced by a phase object having a continuous Fourier spectrum and viewed with the dark field dynamic coherent system, coherent mixing between all possible pairs of Fourier components falling within the lens bandwidth occurs for both the 0- and 180-deg positions of the prism. Each Fourier component acts as a coherent reference for all the others and, thus, preserves the Fourier spectrum of the phase object. Due to the multiple interference present in this system, the output image is nonlinearly related to the object phase even though the effective bandwidth of the coherent viewing system has doubled. An experimental example showing the output of the dark field double-sideband and the double-sideband phase contrast systems appears in Fig. 35.20.

35.3   PHASE VISUALIZATION BY DEFOCUS AND SCHLIEREN TECHNIQUES: NONLINEAR METHODS

35.3.1   Defocus

One of the simplest and most commonly used techniques for visualizing phase objects in a nonlinear mode is to view the image with the optical system slightly out of focus (Fig. 35.21). This technique is extremely difficult to analyze mathematically in closed form. Physically, this defocusing of the optical system changes the coherent transfer function from the diffraction-limited rect function to a product of a rect function times a quadratic phase factor.[6,7] The quadratic phase factor introduces a complex phase shift between the various Fourier components of the phase object, rendering the phase distribution visible in the image plane of the optical system. The visualization of the phase image by this technique is associated with a resolution loss depending on the amount of

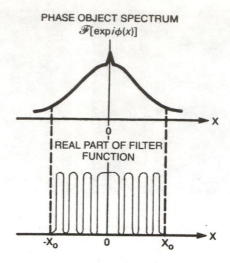

Fig. 35.21 Frequency plane operation for defocused ring structure of the filter function demonstrating dependence on the degree of defocusing.

defocus introduced into the experiment. This resolution loss is usually tolerable because of the extreme simplicity of performing phase contrast by this method. An example of phase contrasting by this method is shown in Fig. 35.22.

35.3.2   Schlieren Knife Edge

Another simple experimental method for visualizing phase objects is the well-known Foucault or Schlieren knife edge test first introduced by Foucault in 1859. In this test, a knife edge filter is used to block out half of the Fourier spectrum of the phase object but the dc is passed (see Fig. 35.23 for the one-dimensional schematic representation of this filter). The test is used for detecting large phase errors in lens or mirror surfaces and for detecting rates of change of the refractive index of a medium.[9] The method produces a visual image whose intensity is

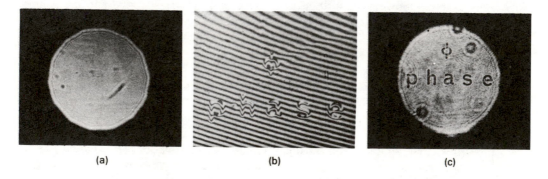

(a)   (b)   (c)

Fig. 35.22 Phase visualization of a pure phase object by defocusing of a lens: (a) shows a direct image of a pure phase object obtained by suitably bleaching a Kodak 649-F emulsion. The object is the ϕ and the word *phase* (20 mm in length). (b) This is an interferogram, using laser light of 6328 Å, of the phase object. (c) The reconstructed image of the phase object. Phase contrast enhancement methods on the image were used to obtain this photograph. The actual enhancement procedure in this case was defocusing of the image by an amount $-f/4$ (from Ref. 8).

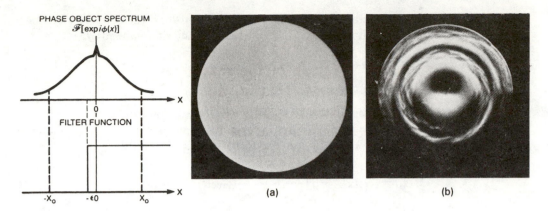

Fig. 35.23 Frequency plane operation for a Schlieren knife edge.

Fig. 35.24 Foucault graphs of an irregular mirror: (a) before and (b)
after introducing the knife edge (from Ref. 7).

proportional to the derivative of the phase object. The method may be represented by the filter function

$$M(\mathbf{x}) = \begin{cases} 1 & x \geq -\epsilon; \text{ all } y \\[2mm] 0 & x < -\epsilon; \text{ all } y \end{cases} , \qquad (35.22)$$

which gives rise to an image intensity of the form[10]

$$I(\alpha, \beta) = c + b\, d\phi(\alpha, \beta)/d\alpha , \qquad (35.23)$$

where c and b are constants. An example of an irregular mirror whose phase
variations were visualized by this method is shown in Fig. 35.24.

35.4   PHASE CONTRAST IMAGING WITH EXTENDED LINEARITY

35.4.1   The Problem of Phase Redundancy

The phase contrast techniques described thus far are limited in their region of
linearity to phase functions obeying the Zernike condition given by Eq. (35.3).
Optical phase functions are measured in terms of wavelength because

$$\phi(x,y) = \frac{2\pi}{\lambda} \phi'(x,y) , \qquad (35.24)$$

where $\phi'(x,y)$ is the optical path difference (OPD) of the phase object. To
preserve a linear relationship between phase height and output intensity for
phase objects not obeying the Zernike condition, various methods have been
introduced for extending the region of linearity of phase viewing devices. One
basic problem encountered when extending the region of linearity is phase
redundancy; i.e., phase information is a multivalued function of period 2π. To
measure phase heights greater than a wavelength, a mechanism is needed to
overcome the multivalued nature of the function. We now discuss systems that

extend linearity without removing redundancy[11,12] and then consider systems that extend linearity and remove phase redundancy simultaneously.[13,14]

35.4.2   Extended Linearity with Redundancy

35.4.2.1   Method 1: Henning Method

This technique, which extends linearity without removing redundancy,[11] achieves its goal by replacing the $\pi/2$ filter in the Zernike positive phase contrast microscope [Eq. (35.5)] with a filter of the form [see Fig. 35.25(a)]:

$$M(\mathbf{x}) = \text{rect}(\mathbf{x}|\epsilon) \exp\left(\frac{i\pi}{4}\right) + \frac{1}{\sqrt{2}} [\text{rect}(\mathbf{x}|x_0) - \text{rect}(\mathbf{x}|\epsilon)] \ , \tag{35.25}$$

where 2ϵ describes the width of the dc response and $2x_0$ is the bandwidth of the imaging system.

A negative phase contrast filter having extended linearity is realized by using a $7\pi/4$ phase step [see Fig. 35.25(b)]. When this filter is used in the Fourier transform plane of Fig. 35.1, the Fourier transform of the phase object, $\mathscr{F}\{\exp[i\phi(x,y)]\}$, multiplies the filter given in Eq. (35.25) to give

$$T(\mathbf{x}) = T(x,y) = \mathscr{F}\{\exp[i\phi(x,y)]\}$$

$$\times \left[\frac{i}{\sqrt{2}} \text{rect}(\mathbf{x}|\epsilon) + \frac{1}{\sqrt{2}} \text{rect}(\mathbf{x}|x_0)\right] \ . \tag{35.26}$$

Fourier transformation of Eq. (35.26) by the second lens in Fig. 35.1 gives the output intensity in the form

$$I(\alpha,\beta) \cong \left(\frac{1}{\sqrt{2}}\right)^2 |iK + \cos\phi(\alpha,\beta) + i\sin\phi(\alpha,\beta)|^2 \ , \tag{35.27}$$

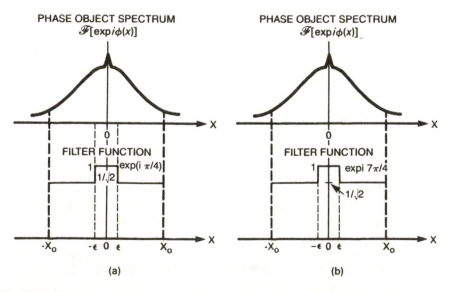

Fig. 35.25 Frequency plane operations for (a) positive and (b) negative phase contrast with extended linearity.

where K is the result of convolving the phase image with the sinc functions $2\epsilon\,\mathrm{sinc}(2\pi\epsilon\alpha/\lambda f)$ and $2x_0\,\mathrm{sinc}(2\pi\alpha x_0/\lambda f)$, and the impulse response of the phase contrast imaging system has been approximated by a delta function in deriving Eq. (35.27), which reduces to

$$I(\alpha,\beta) \cong \frac{1}{2}\,[1 + K^2 + 2K\sin\phi(\alpha,\beta)] \;. \tag{35.28}$$

When $K = 1$, Eq. (35.28) reduces to

$$I(\alpha,\beta) = 1 + \sin\phi(\alpha,\beta) \;. \tag{35.29}$$

To understand the significance of the extended linearity for the phase distribution in Eq. (35.29), we must write it in a form similar to Eq. (35.6).

Using the series form for the sine term in Eq. (35.29) and choosing only the linear term, we have

$$I(\alpha,\beta) \cong 1 + \phi(\alpha,\beta) \;, \tag{35.30}$$

which is a result similar to that of Eq. (35.6). In essence, we are using the linear portion of the sine function, which is linear over a wider range of phase heights than that obtainable in the small phase approximation which restricts Eq. (35.6). This can be seen from the following argument. If $\phi^2 \ll 1$, then the phase difference must be much less than the wavelength λ; hence, if we choose to assume the limit of the OPD to be one or two orders of magnitude, then we have the range of phase difference given by

$$\mathrm{OPD} = \text{optical path difference} < \lambda/10 \quad \text{or} \quad \lambda/100 \;, \tag{35.31}$$

which corresponds to $\pi/5$ or $\pi/50$ radians (36 or 3.6 deg).

On the other hand, the sine function $\sin\phi(\alpha,\beta)$ is nearly linear to values of $\pi/3$ radians or 60 deg. Since we have

$$\phi = (2\pi/\lambda)\,(\mathrm{OPD}) < \pi/3 \quad \text{or} \quad \mathrm{OPD} < \lambda/6 \;, \tag{35.32}$$

we can see that the allowable OPD for linearity is greater for this case than for the small phase approximation. This is shown schematically in Fig. 35.26.

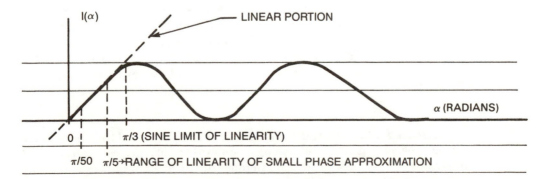

Fig. 35.26 Schematic comparison of the linearity of extended range and small-phase approximation phase contrast methods.

35.4.2.2 *Method 2: Carrier-Modulated Phase Technique*

Phase objects with large phase variations can also be visualized by utilizing a screening function and then spatially filtering the frequency spectrum.[12] This approach is particularly useful for continuous-tone objects and its implementation is straightforward. The continuous-tone capability arises because the image is referenced within each period of the carrier. Either sinusoidal or square wave carriers can be used. The phase-modulated image is prepared by using a bleaching process from a prebleached modulated density image. Carrier-modulated imagery was covered in Chapter 34.

Sine Wave Carriers: It is well known[15] that the amplitude transmittance of a sine wave phase grating having modulation index $m/2$ is

$$T(\xi,\eta) = \exp[i(m/2)\sin 2\pi\omega_0\xi] = \sum_{n=-\infty}^{\infty} J_n(m/2)\exp(2\pi i n\omega_0\xi) , \qquad (35.33)$$

where ω_0 is the grating frequency, m is expressed in radians, and $J_n(m/2)$ is the n'th-order Bessel function plotted in Fig. 35.27(a). In Eq. (35.33), the η dependence is constant.

If this amplitude distribution appears in the object position (front focal plane) of Fig. 35.1, the Fourier transform appears in the back focal plane of the lens and is given by

$$T(x,y) = \tilde{T}(\xi,\eta) = \sum_{n=-\infty}^{\infty} J_n(m/2)\,\delta(x - n\lambda f\omega_0, y) , \qquad (35.34)$$

where the phase modulation m has been treated as a constant. If we now make a blocking filter which passes only the n'th order, and place it in the filter plane of Fig. 35.1, then in the back focal plane of the second Fourier transforming lens we obtain the intensity distribution given by

$$I(\alpha,\beta) = J_n^2(m/2) . \qquad (35.35)$$

If we now allow m to be spatially variable, then each diffraction order n gives rise to the corresponding Bessel function $J_n(m/2)$, which describes the polarity and gray-scale characteristics of the image as seen in Fig. 35.27. In Figs. 35.27(b), (c), and (d), a linear step wedge was exposed through a grating in contact with a Kodak high-resolution plate.[12] Various plates were processed with different photographic gammas and then bleached. After spatial filtering the first order, the retrieved images in (b), (c), and (d) were obtained. This illustrates the sensitivity of output dynamic range with input density.

The polarity characteristics for various ranges of phase heights m for the image intensities resulting from the zeroth and first diffraction orders are given in Table 35.I.

Modulated Phase Images with Square Wave Carriers: Modulated phase images[12] may be created by bleaching modulated density images[12] or by contact duplicating density images with interferometric fringes onto surface relief materials such as photoresist.[16] Modulated phase images with square wave carriers can be modeled as an on/off amplitude grating and a phase-modulated square

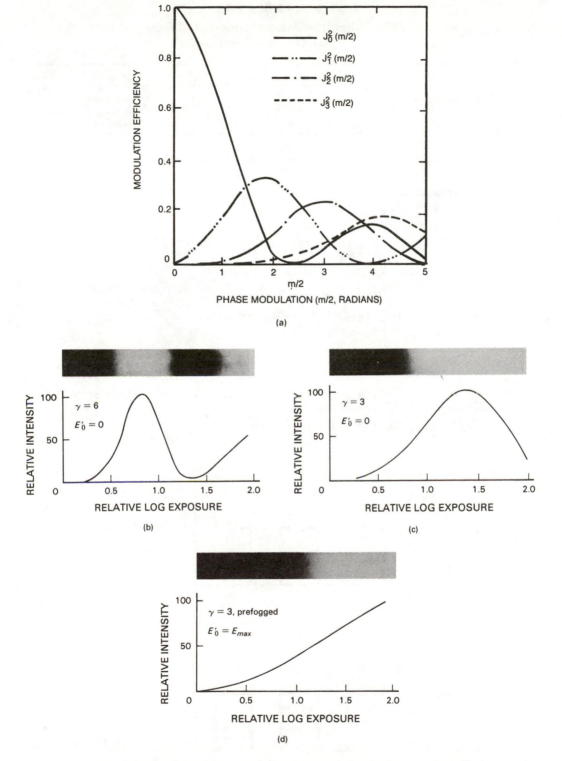

Fig. 35.27 Dynamic range and polarity of modulated phase images where E'_0 denotes the prefog exposure: (a) Plot of $J_n^2(m/2)$ for $n = 0, 1, 2,$ and 3. Phase retrieval of (b) very high contrast ($\gamma = 6$), (c) high contrast ($\gamma = 3$), and (d) moderate contrast ($\gamma = 3$ prefogged) density wedge targets. The top portions of (b), (c), and (d) show photographs of the images and the bottom portions show the corresponding relative intensity image traces (from Ref. 12).

TABLE 35.I
Polarity Characteristics of Modulated Phase Images for Zero and First Order

n	Range of $m/2$	Polarity
0	0 to 2.4	Negative
±1	0 to 1.8	Positive
	1.8 to 3.9	Negative

wave of the same fundamental frequency shifted by a half-period, as shown in Fig. 35.28.

The carrier-modulated phase image represented schematically in Fig. 35.28 is mathematically given by[12]

$$\text{phase object} = T(\xi) + T\left(\xi - \frac{L}{2}\right)\{\exp[ikhD(\xi, \eta)]\} \ , \tag{35.36}$$

where

$$T(\xi) = \sum_{n=-\infty}^{\infty} \text{rect}\left(\xi \left| \frac{L}{4}\right.\right) \odot \delta(\xi - nL) \ , \tag{35.37}$$

and

h_0 = reference relief height of the phase grating (see Fig. 35.28)

h = relief height frequency response of the phase grating[17] assumed to be constant

$D(\xi, \eta)$ = prebleached density image.

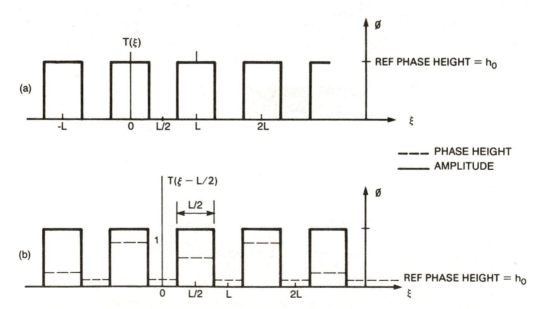

Fig. 35.28 Representation of carrier-modulated phase image: (a) amplitude grating (reference phase height is h_0) and (b) half-period shifted phase-modulated square wave of same period relative to reference phase height h_0.

In Eq. (35.37), it is tacitly assumed that the modulated phase image height is proportional to the density image. The Fourier transform of the phase object in Eq. (35.36) is given by

$$f(x,y) = \sum_{n=-\infty}^{\infty} \frac{L}{2} \operatorname{sinc} \frac{n\pi}{2} \left[\delta\left(x - \frac{n}{L}, y\right) \right.$$

$$\left. + \exp(in\pi)\, \tilde{\psi}\left(x - \frac{n}{L}, y\right) \right], \tag{35.38}$$

where $\tilde{\psi}(x,y)$ is the Fourier transform of the term $\exp[ikhD(\xi,\eta)]$ in Eq. (35.36) and $\tilde{\psi}(x - n/L, y)$ resulted from the convolution of $\tilde{\psi}(x,y)$ with the Dirac comb (sum of delta functions).

If we place a spatial filter having a bandwidth of $1/2L$ over any of the diffraction orders and retransform to the image plane (α, β) of Fig. 35.1, the image intensity is

$$I_{\text{filtered}}(\alpha,\beta) = \left| \exp\left(-2\pi i\, \frac{n\alpha}{L}\right) a_n \{1 + \exp(in\pi) \exp[ikhD(\alpha,\beta)]\} \right|^2, \tag{35.39}$$

where n represents the order filtered, and $a_n = (L/2)\operatorname{sinc}(n\pi/2)$ so that the even orders are missing due to the square wave nature of the carrier.

In obtaining Eq. (35.39), the filtering of a single order removes the summation in Eq. (35.38) and the minus sign in the linear phase factor arises from the Fourier kernel, which is always negative in optical propagation problems.

For $n = 1$, Eq. (35.39) becomes

$$I_{\text{filtered}}(\alpha,\beta) = 4\left(\frac{L}{\pi}\right)^2 \sin^2\left[\frac{1}{2} khD(\alpha,\beta)\right]. \tag{35.40}$$

This result again shows extended linearity in this phase contrast technique since the sine squared term is nearly linear up to angles of 50 to 60 deg, as shown previously in Fig. 35.26.

For $n = 0$, Eq. (35.39) becomes

$$I_{\text{filtered}}(\alpha,\beta) = L^2 \cos^2\left[\frac{1}{2} khD(\alpha,\beta)\right], \tag{35.41}$$

which indicates that the filtered image is polarity reversed relative to Eq. (35.40) and that its range of linearity is less than that of Eq. (35.40) because of its quadratic (rather than linear) dependence in this small-angle approximation.

An example of such imagery is shown in Fig. 35.29. Figure 35.29(a) shows the continuous-tone density object used in the experiment and (b) is a photomicrograph of the modulated density target before the bleaching process has been undertaken. In Fig. 35.29(c), we show the spatial Fourier transform of the bleached modulated continuous-tone object shown in (b). A positive retrieval from the first order is shown in (d) and a negative image retrieved from the zeroth order is shown in (e).

This method has a capability for high resolution over large fields of view because the reference occurs within each period of the grating. This advantage also makes the image relatively speckle-free since retrieval can be achieved with a

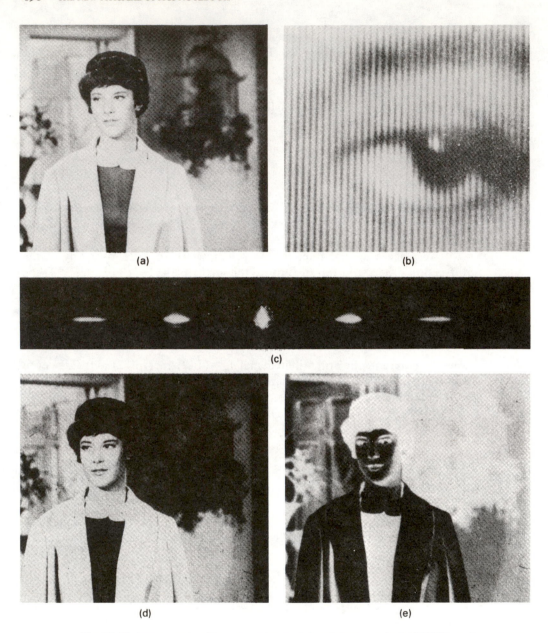

Fig. 35.29 Image retrieval from a carrier-modulated phase angle: (a) SMPTE continuous-tone object, (b) photomicrograph of modulated target ($\omega_0 = 40$ line pairs/mm), (c) frequency spectrum of modulated target, (d) positive image retrieval from first order, and (e) negative image retrieval from zero order (from Ref. 12).

viewer having a small degree of spatial coherence. Noise-free continuous-tone images having nearly 300 line pairs/mm resolution have been retrieved with this technique.[16]

35.4.3 Extended Linearity Without Redundancy

35.4.3.1 *Method 1: Differentiation and Integration Technique*

The visualization of large variation phase objects without phase redundancy can be achieved by means of coherent object differentiation and integration.[10,13] In this procedure, an intermediate photographic step is required; however, it results in no restriction on the maximum slope of the phase variation. The phase object of interest is placed in the object plane of Fig. 35.1 and a derivative filter (see Chapter 32) is placed in the filter plane. The filter has a one-dimensional amplitude transmittance given by

$$T_A(x) = C_1(x_0 - x) , \qquad (35.42)$$

where C_1 and x_0 are constants. Equation (35.42) is a linear function of x. The resulting intensity distribution in the α plane is given by

$$I(\alpha) = \left\{ \left[\frac{x_0}{f} + ik \frac{d\phi(-\alpha)}{d\alpha} \right] \exp[ik\phi(-\alpha)] \right\}^2 . \qquad (35.43)$$

The image distribution described by Eq. (35.43) is recorded on film and then a positive transparency is made. The film is processed such that $\gamma_n \gamma_p = 1$. The value of x_0/f is chosen such that its amplitude transmittance is the positive definite function (assuming immersion in an index matching liquid to remove phase variations) given by

$$T_A'(\xi) = \left(\frac{x_0}{f} \right) + ik \frac{d\phi(-\xi)}{d\xi} . \qquad (35.44)$$

The liquid gated film whose amplitude transmittance is given by Eq. (35.44) (i.e., the derivative of the phase object plus a constant) is placed back in the object plane of Fig. 35.1. An integration filter (see Chapter 32) is placed in the filter plane. The amplitude transmittance of this integration filter is given by

$$T_A'(x) = \begin{cases} iC_2 & \text{for } |x| < x_1 \\ C_3/x & \text{for } x > x_1 \\ -C_3/x & \text{for } x < -x_1 \end{cases} , \qquad (35.45)$$

where C_2 and C_3 are constants. Equation (35.45) describes a complex filter that has a π phase shift over the negative half plane and a small $\pi/2$ phase shift about the origin in addition to the $1/x$ amplitude filter. The resulting amplitude image distribution in the image plane is given by

$$A'(\alpha) = i \frac{C_4 x_0}{f} + ik C_3 \phi(\alpha) , \qquad (35.46)$$

where C_4 is a new constant. By choosing C_4 to be sufficiently large, Eq. (35.46) yields the following intensity distribution:

$$I(x) \cong \frac{C_4^2 x_0^2}{f^2} + \left(\frac{2C_4 x_0 k}{f} C_3 \right) \phi(\alpha) , \qquad (35.47)$$

which is the desired phase distribution $\phi(\alpha)$, biased on a constant without any restrictions on its magnitude.

It is interesting to note that the constant term in the differentiated image [Eq. (35.45)] does not get integrated because it transforms to a delta function in the filter plane. In the filter plane, the delta function passes through the central

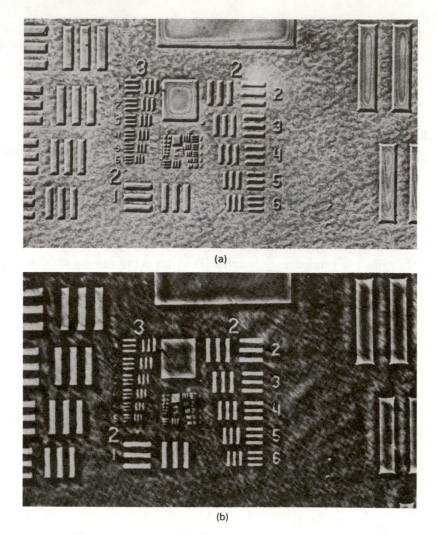

(a)

(b)

Fig. 35.30(a) Derivative of the phase resolution target. The x axis is oriented at 45 deg to the vertical. The random pattern in the broad clear areas is the actual derivative of the phase target surface, which was slightly irregular in these areas. The fringes inside the bars are due to multiple reflections between the front and back surfaces of the phase target. (b) Integral of the recorded derivative of a phase resolution target. The two-dimensional numbers that label the individual bar groups have been almost completely reproduced, despite the one-dimensional differentiation and integration. The larger bars exhibit a distinct edge sharpening (from Ref. 13).

portion of the integrating filter [Eq. (35.45)] and thereby is multiplied by a complex number (iC_2) and retransformed to a new constant.

In Fig. 35.30, the experimental results of this phase contrast imaging method are shown. In Fig. 35.30(a), we show the derivative of a phase resolution target and in (b), we see the corresponding integral of the derivative of the phase resolution target. Also, in (b), note the loss of information in the large square area which resembles edge sharpening. This occurs because of the poor frequency response of the film phase relief image at low spatial frequencies.[17]

The experimental resolution limit obtained with the system [see Fig. 35.30(b)] is between groups 4-3 and 4-4, which corresponds to a three-bar resolution of

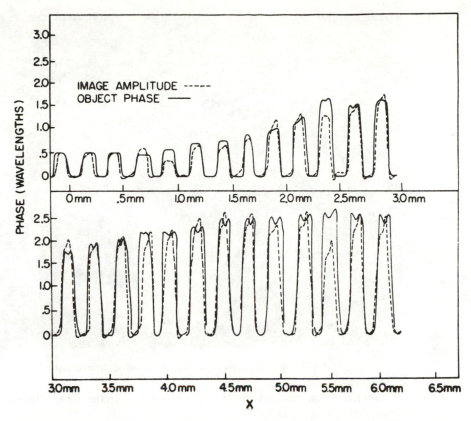

Fig. 35.31 Comparison of image amplitude with object phase for a set of phase steps of gradually increasing height (from Ref. 13).

approximately 20 line pairs/mm with this coherent imaging process.

The range of linearity for a bleached photographic step wedge target is shown in Fig. 35.31. The agreement between this method and an independent interferometric measurement of the object phase height is excellent. Since the bleached phase relief image has a two- to three-wave phase height limitation due to the photographic process,[17] this differentiation-integration method measured all the phase height that existed in this particular object.

35.4.3.2   *Method 2: Electro-Optic System Technique*

The electro-optic system technique[14] involves two steps. In the first step, a Mach-Zehnder interferometer is utilized to change the phase of the object of interest to a phase modulation of a time-varying sinusoidal signal. The output of the above system is then used as an input to a digital phase detection system, which essentially converts the phase modulation of the sinusoidal waveform into a signal proportional to the phase. Two phase detectors are necessary to remove the phase redundancy of the system. The first phase detector records the phase over a range of 2π and the second detector only measures the phase changes that are integral multiples of 2π, thus behaving essentially as a fringe counter in a Michelson interferometer.

In Fig. 35.32, we show a phase contrast map of the phase portion of the

Fig. 35.32 Area scan of biological specimen (from Ref. 14).

Drosophila larva from a 4-μm-thick cross section. In this figure, adjacent phase levels differ by $\lambda/16$ in OPD.

There is essentially no phase redundancy limitation in this technique and, therefore, it is very useful for complex transmitting objects that naturally occur in nature such as biological specimens. The phase is determined by a fringe contour technique, which is not limited by the dynamic range of the object.

35.5   CONCLUSIONS

In this chapter, we have reviewed various techniques using spatial filtering that were introduced to visualize phase objects. The methods include the small phase approximation of Zernike as well as linear and nonlinear methods. Techniques for extending linearity and removing redundancy of phase objects were also discussed.

35.6   APPENDIX: IMAGING WITH AN OBLIQUE ILLUMINATION DOUBLE-SIDEBAND PHASE CONTRAST SYSTEM

To analyze this one-dimensional system, we consider a phase image of the form $\exp[i\phi(x)]$ which for small phase variations is written as

$$\exp[i\phi(x)] \cong 1 + i\phi(x) \text{ for } \phi(x) \ll 1 . \tag{35.48}$$

This is the usual Zernike approximation. If this image is viewed with a one-dimensional oblique illumination microscope having two sources,[4] then the image intensities from the two symmetrically displaced source points add incoherently. The two sources shift the spectra in the transform plane. Each point image (or dc) behaves as a coherent reference beam for its spectrum. For one source, the amplitude distribution in the transform plane from Eq. (35.48) and the equivalent filter is

$$f_1(x) \cong \delta(x) + i\tilde{\phi}(x)\,\text{rect}(x - x_c|x_0) \,, \tag{35.49}$$

where $x_c = x_0 - \epsilon$, and 2ϵ is the diameter of the point source image in the transform plane or lens aperture. In Eq. (35.49), the Fourier spectrum of the phase object is centered on axis and the mask is written in terms of the equivalent filter function. For the other symmetrically displaced source, the amplitude distribution in the transform plane in terms of the equivalent filter is

$$f_2(x) \cong \delta(x) + i\tilde{\phi}(x)\,\text{rect}(x + x_c|x_0) \,. \tag{35.50}$$

The image intensity resulting from the incoherent point sources (whether realized sequentially or simultaneously) after phase shifting the delta function by a $\pi/2$ filter is given by

$$I(\alpha) = |\tilde{f}_1(\alpha)|^2 + |\tilde{f}_2(\alpha)|^2 \,. \tag{35.51}$$

Substituting Eqs. (35.49) and (35.50) into (35.51) yields

$$I(\alpha) = 2K^2 + 2K\,\text{Re}\left[\int_{-2x_0}^{0}\tilde{\phi}(x)\exp(2\pi i\alpha x)\,dx + \int_{0}^{2x_0}\tilde{\phi}(x)\exp(2\pi i\alpha x)\,dx\right]$$

$$+ \left|\int_{-2x_0}^{0}\tilde{\phi}(x)\exp(2\pi i\alpha x)\,dx\right|^2 + \left|\int_{0}^{2x_0}\tilde{\phi}(x)\exp(2\pi i\alpha x)\,dx\right|^2 , \tag{35.52}$$

where K represents the Fourier transform of the delta function. In Eq. (35.52), the squared terms are small by the small phase approximation and, hence, are ignored, so that we obtain

$$I(\alpha) = 2K^2 + 2K\,\text{Re}\int_{-\infty}^{\infty}\tilde{\phi}(x)\,\text{rect}(x|2x_0)\exp(2\pi i\alpha x)\,dx \,. \tag{35.53}$$

Rewriting Eq. (35.53) gives

$$I(\alpha) = 2K^2 + 8Kx_0\,\text{Re}[\phi(\alpha) \circledast \text{sinc}(4\pi x_0\alpha)] \,, \tag{35.54}$$

which shows that the phase image is convolved with a coherent impulse response of a lens which appears to have a diameter of $2x_0$ rather than x_0. This effective doubling of diameter means that the coherent resolution of the system is double. In addition, the coherent noise is simultaneously averaged due to the different positions of the point source.

REFERENCES

1. F. Zernike, "Phasenkontrastverfarhren bei der mikroskopischen Beobachtung," *Physik Z.* 36(22/23), 848–851 (1935).
2. F. Zernike, "Phasenkontrastverfahren bei der mikroskopischen Beobachtung," *Z. Tech. Phys.* 16(11), 454–457 (1935).
3. D. J. Cronin, J. B. DeVelis, and G. O. Reynolds, "Equivalence of annular source and dynamic coherent phase contrast viewing systems," *Opt. Eng.* 15(3), 276–278 (1976).
4. D. J. Cronin and A. E. Smith, "Dynamic coherent optical system," *Opt. Eng.* 12(2), 50–55 (1973).
5. M. Françon, "Le contraste de phase en optique et en microscopie," *Editions de la Rev. d'Optique Theorique et Instrumental,* Paris, 1–109 (1950).
6. E. L. O'Neill *Introduction to Statistical Optics,* Addison-Wesley Publishing Co., Reading, Mass. (1963).
7. D. Malacara, *Optical Shop Testing,* John Wiley and Sons, Inc., New York (1978).
8. D. Gabor, G. W. Stroke, D. Brumm, A. Funkhouser, and A. Labeyrie, "Reconstruction of phase objects by holography," *Nature* 208(5016), 1159–1162 (1965).
9. R. S. Longhurst, *Geometrical and Physical Optics,* 2nd ed., John Wiley and Sons, Inc., New York (1967).
10. J. B. DeVelis and G. O. Reynolds, *Theory and Applications of Holography,* Addison-Wesley Publishing Co., Reading, Mass. (1967).
11. H. B. Henning, "A new scheme for viewing phase contrast images," *Electro-Optical Systems Design* 6(6), 30–34 (1974).
12. P. F. Mueller, "Optical reconstruction of phase images," in *Processing of Images and Data from Optical Sensors,* W. H. Carter, ed., Proc. SPIE 292, 39–46 (1981).
13. R. A. Sprague and B. J. Thompson, "Quantitative visualization of large variation phase objects," *Appl. Opt.* 11(7), 1469–1479 (1972).
14. G. E. Sommargren and B. J. Thompson, "Linear phase microscopy," *Appl. Opt.* 12(9), 2130–2138 (1973).
15. J. W. Goodman, *Introduction to Fourier Optics,* McGraw-Hill Book Co., New York (1968).
16. P. F. Mueller, D. J. Cronin, and G. O. Reynolds, "Novel image duplication technique utilizing Fourier optics," in *Processing of Images and Data from Optical Sensors,* W. H. Carter, ed., Proc. SPIE 292, 47–58 (1981).
17. H. M. Smith, "Photographic relief images," *J. Opt. Soc. Am.* 58(4), 533–539 (1968).

36 Partially Filled, Synthetic Aperture Imaging Systems: Incoherent Illumination *

36.1 INTRODUCTION

In many optical imaging situations (e.g., space optics or infrared imaging systems), the lens is nearly diffraction limited so that object resolution is limited by the diameter of the primary lens or mirror. Thus, to increase object resolution, either the lens diameter must be increased or the wavelength decreased. Since the wavelength reduction is often impractical, systems designers must increase the lens diameter to see the smaller objects. The weight, size, and cost of an optical system are probably the limiting factors in most applications. Thus, some sort of synthetic aperture optical system becomes a candidate for solving the problem of achieving higher resolution. Various kinds of optical synthetic apertures have been discussed in the literature,[1-3] including interferometry, feedback-controlled optics, imaging with partially filled apertures, and coherent optical aperture synthesis.[1]

Historically, one of the earliest demonstrations of optical aperture synthesis was the Michelson stellar interferometer. This instrument was used by Michelson and Pease in the 1920s to measure the angular diameter of the star Betelgeuse. The interferometer consisted of outrigger mirrors in front of a telescope, which were moved apart until the fringe contrast became zero in the focal plane. By utilizing the inverse Van Cittert-Zernike theorem of coherence theory, this mirror separation may be used to determine the angular diameter of the star. (In fact, Michelson actually used a direct convolution in the image space, so that the fringe contrast is zero when the spacing of the fringes is the same as the size of the image of the star.) This concept was utilized by radio astronomers in the 1950s to measure the angular diameter of stars in the radio universe.[3,4] Since these radio measurements require aperture diameters of miles or more to resolve the biggest radio stars, aperture synthesis with these radio interferometers made such measurements feasible. In the late 1950s and early 1960s, the side-looking radar system was developed at the University of Michigan. This active system used

*Reprinted with permission from "A Review of Partially-Filled, Synthetic-Aperture, Imaging Systems," by G. O. Reynolds, in *Infrared, Adaptive, and Synthetic Aperture Optical Systems,* R. B. Johnson, W. L. Wolfe, J. S. Fender, eds., SPIE Proc. 643, 141–179 (1986). This article is continued in Chapter 37.

heterodyne recording to store the individual radar returns and utilized the motion of the aircraft to build up aperture along the flight direction. Also in the 1950s, Hanbury-Brown and Twiss made measurements with their phase-insensitive intensity interferometer to obtain a one-dimensional correlation function of the source. Finally in 1967, the Woods Hole Summer Study on synthetic aperture optics was held to collate these technologies and develop new optical synthetic aperture concepts for use in future spacecraft applications.

In this chapter, and continuing in Chapter 37, we review the work performed on partially filled synthetic aperture optical imaging systems concentrating primarily on the work performed during the late 1960s and early to mid-1970s.

36.2   NONLINEARITIES OF PARTIALLY FILLED SYNTHETIC APERTURES DUE TO DEGREE OF COHERENCE

It has been shown in the literature that partially filled synthetic apertures exhibit an inherent nonlinearity when the object illumination is partially coherent.[5,6] Partially filled optical synthetic apertures have the property that incoherent sources may be studied interferometrically, in a manner usually reserved for systems utilizing coherent radiation. Apertures consisting of pieces from the same lens or mirror, in general, form a nonlinear system. These systems become linear when used to observe incoherent or self-luminous objects. To indicate this difference in linearity, the general nature of the object illumination must be incorporated in the discussion. This has been done by utilizing coherence theory.[5,6]

Assume that there are two apertures, A_1 and A_2, and that their instantaneous outputs are multiplied and averaged (see Fig. 36.1); i.e., we cross-correlate the outputs of the two apertures. The apertures will scan the object and the variations of the correlator output will yield a map of the object distribution. The system output is[5,6]:

$$P(x) = \text{energy distribution}$$

$$= \int_{-\infty}^{\infty} \int_{-\infty}^{\infty} G_1(x - x_1') \, G_2^*(x - x_2') \, \Gamma(x_1', x_2', 0) \, dx_1' \, dx_2' \,, \qquad (36.1)$$

where $G_1(x)$ and $G_2(x)$ are, respectively, the amplitude impulse responses of apertures A_1 and A_2 at the mean frequency $\bar{\nu}$. Equation (36.1) is the general

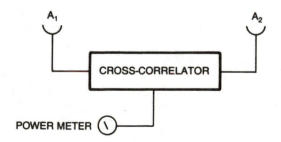

Fig. 36.1 Schematic of synthetic aperture system.

expression for the complex output of the synthetic apertures when the radiation is quasimonochromatic. When the object illumination is coherent, we use Eq. (11.26) in the form

$$\Gamma(x_1', x_2', 0) = V(x_1') V^*(x_2') , \tag{36.2}$$

and Eq. (36.1) reduces to

$$P(x) = \int_{-\infty}^{\infty} G_1(x - x_1') V(x_1') dx_1' \int_{-\infty}^{\infty} G_2^*(x - x_2') V^*(x_2') dx_2' . \tag{36.3}$$

Thus, the mapping of extended coherent objects with the synthetic aperture is nonlinear so that the system does not synthesize (i.e., no cross-correlation). Interpretation of such data would be extremely difficult at best.

When the object illumination is incoherent (or the object is self-luminous, corresponding to the type of sources actually encountered in radio astronomy or optical astronomy), then we use Eq. (11.28) in the form

$$\Gamma(x_1', x_2', 0) = P_{obj}(x_1') \delta(x_1' - x_2') , \tag{36.4}$$

and Eq. (36.1) becomes

$$P(x) = \int_{source} P_{obj}(x_1') G_1(x - x_1') G_2^*(x - x_1') dx_1' . \tag{36.5}$$

Thus, the system is linear in power, having a complex impulse response given by

$$\mathscr{I}(x) = G_1(x) G_2^*(x) , \tag{36.6}$$

due to a point object

$$P_{obj}(x') = \delta(x') . \tag{36.7}$$

Therefore, even though this system is nonlinear for mapping coherent objects, it is definitely linear for the more common case of mapping incoherent sources or objects.

36.3   APERTURE SYNTHESIS WITH INCOHERENT ILLUMINATION

36.3.1   Sequential Aperture Synthesis

Wilczynski demonstrated that a full aperture can be synthesized by a pair of phased apertures being sequentially placed at different separations.[2,3,7] The phasing is done with a lens and the recording is accomplished by sequentially exposing a film in the image plane. The modulation transfer function (MTF) of the system is built up by the sequential exposures, as illustrated in Fig. 36.2 for a one-dimensional case. Wilczynski[7] described a two-dimensional space telescope utilizing these principles. Stroke[3,8] showed that optical processing with a low-frequency attenuating filter can be used to remove the low-frequency bias redun-

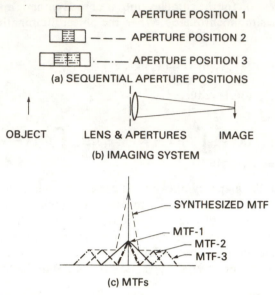

Fig. 36.2 One-dimensional sequential synthetic aperture.

dancy in the MTF caused by these sequential aperture exposures. The processed image clearly showed the increased resolution due to the sequential aperture syntheses.

36.3.2 Simultaneous Aperture Syntheses

36.3.2.1 Mills Cross Aperture

Reynolds et al.[3,9] have shown that optical aperture synthesis in the imaging mode (analogous to radio interferometers) can be achieved by phasing the lens aperture elements as if they were each sections cut from the same optical surface. This phasing creates an interferometric code within each impulse response of the image that can be processed, after detection, to create a transfer function which is a cross-correlation of the various aperture elements; i.e., it has transfer characteristics equal to or better than those of the corresponding full aperture. The phasing must be done to preserve the coherence characteristics between the object and the image on a point-by-point basis; i.e., the path length differences must be less than the coherence length of the radiation to guarantee interference effects within each individual impulse response.

Since it is assumed that incoherent white light is being used to illuminate the object, we see that the restriction of having the apertures made from the same lens or mirror is important. We are utilizing the well-known fact in the lens makers' trade that a lens is designed such that all the rays leaving any particular object point and passing through the imaging element will combine, in phase, at the conjugate point in the image. Obviously, there are an infinite number of geometrical patterns in which we could array the elements.

In Fig. 36.3, the effect of placing a cross aperture in front of a lens in a simple imaging experiment is shown. (We discuss the problem of how to do this with separate aperture elements later.) There is an object plane, a cross lens, and an observing plane, or an image plane. This lens aperture can be described mathematically as a slit in the x direction and a slit in the y direction, i.e.,

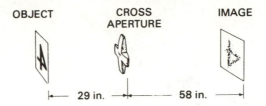

OBJECT   CROSS   IMAGE
                APERTURE

|— 29 in. —|—— 58 in. ——|

Fig. 36.3 Schematic of cross aperture imaging system (Mills cross optical analog) (after Ref. 3).

$$\psi(\alpha,\beta) = \text{rect}(\alpha|a)\,\text{rect}(\beta|b) + \text{rect}(\alpha|b)\,\text{rect}(\beta|a)$$

$$- \text{rect}(\alpha|b)\,\text{rect}(\beta|b) \; ; \tag{36.8}$$

see the Appendix of Sec. 36.5. The amplitude response of this lens is a sinc function in the x direction plus a sinc function in the y direction. Since the lens is properly phased on a point-by-point basis, when this information is recorded on a piece of photographic film, we merely square the sum of the diffraction patterns as shown in Eq. (36.8) to obtain the corresponding normalized intensity impulse response as

$$\mathscr{I}(x,y) = \left|\; \text{sinc}\left(\frac{2\pi ax}{\lambda f}\right) + \text{sinc}\left(\frac{2\pi ay}{\lambda f}\right)\;\right|^2 , \tag{36.9}$$

which is the normalized form of Eq. (36.37) from the Appendix of Sec. 36.5. The Fourier transform of the impulse response describes the transfer function of the system:

$$T(\mu_x,\mu_y) = T(\mu_x|2a) + T(\mu_y|2a) + 2\,\text{rect}(\mu_x|a)\,\text{rect}(\mu_y|a) . \tag{36.10}$$

We see that the transfer function consists of three terms: The first term of the impulse response (a sinc-squared term in the x direction) Fourier transforms into a triangular function in the μ_x direction. The second term, the sinc-squared term in the y direction, gives a triangular function in the μ_y direction. The cross term, or the correlation term, gives rise to the synthesized part of the transfer function that covers a rectangular area in the transform plane; its cutoff frequency is equivalent to that of a square lens of half the length of the cross. These three terms combined in a plane are shown schematically in Fig. 36.4. [Fig. 36.6(b) shows the same result obtained experimentally and is quite dramatic.]

For incoherent illumination, every point in the object plane is independent of every other point, and the imaging system is linear in intensity. Therefore, the image of an extended source is (see Chapters 4 through 6)

$$I_{im}(x,y) = \int I_{obj}(\xi,\eta)\,\mathscr{I}(x-\xi,y-\eta)\,d\xi\,d\eta , \tag{36.11}$$

where $\mathscr{I}(x-\xi,y-\eta)$ is the folded form of the intensity impulse response. The system has a characteristic impulse response as just described, and it has a characteristic transfer function. All object points will add up independently through the superposition integral and give the resulting image described in Eq. (36.11). However, to utilize the synthetic transfer function of the Mills cross, we must eliminate the bias terms—the two terms that run in the perpendicular

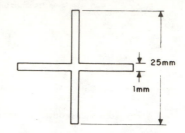

Fig. 36.4 Mills cross incoherent transfer function (after Ref. 3).

Fig. 36.5 Cross aperture imaging system (after Refs. 3 and 9).

directions in transform space. Optical processing methods are utilized to perform this task.

In Fig. 36.5, we show the mask used as the cross aperture in the experimental arrangement of Fig. 36.3. The particular mask used was 25 mm long by 1 mm wide. The experimental results of this cross aperture imaging experiment[3,9] are shown in Figs. 36.6 and 36.7. In the experiments, a cross aperture (Fig. 36.5) over a lens (Fig. 36.3) was used to demonstrate the synthesizing concept. The interference terms in the impulse response are seen as the scalloping effects in Fig. 36.6(a) and the synthesized transfer function (square) is shown in Fig. 36.6(b). The image of the three-bar target collected by the synthetic aperture is shown in Fig. 36.7(a). Notice the image is quite astigmatic and the information in the central region of the target is definitely not resolved.

It was observed that a piece of flat glass placed over one arm of the cross removed the scalloping effect in Fig. 36.6(a) because the phasing is destroyed when the glass thickness exceeds the coherence length of the radiation.

As noted in Fig. 36.6(b), the bias terms (autocorrelation results) are stronger than the synthesized term. This results in an overall raw image (Fig. 36.7) that has low contrast. If the raw image is processed (either optically, digitally, or electronically), the effects of the bias terms are removed and a high-quality high-contrast synthetic aperture image results. To achieve the synthesized transfer function of the cross shown in Fig. 36.6(b), the bias information along the μ_x and μ_y spatial frequency axes should be removed. This can be done through Fourier processing.

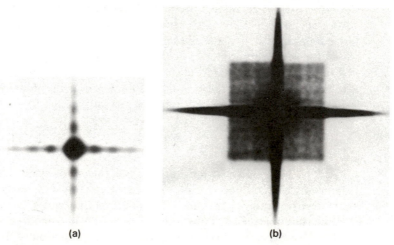

(a)   (b)

Fig. 36.6 (a) Point spread function (impulse response) and (b) transfer function of a Mills cross aperture with correct phasing (after Refs. 3 and 9).

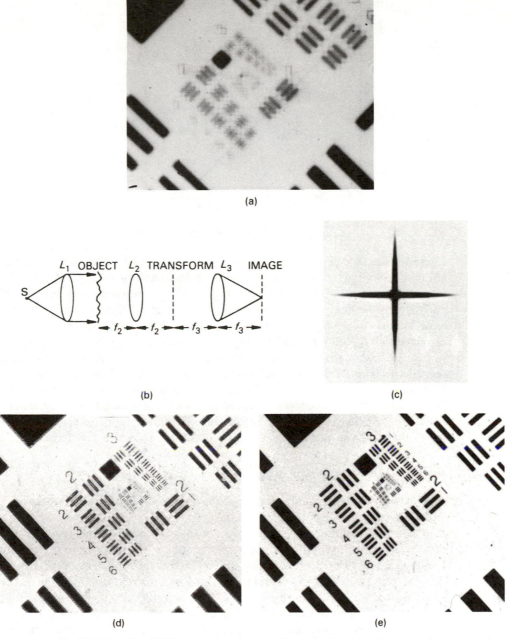

Fig. 36.7 Results of Mills cross experiment: (a) image of the three-bar resolution target taken through the cross aperture (Fig. 36.5) imaging system of Fig. 36.3; (b) conventional coherent Fourier processing system, a modification of that in Fig. 30.1; (c) inverse filter for cross aperture imagery; (d) processed image from cross aperture; and (e) conventional photograph through equivalent full aperture. Resolution observed: 30 line pairs/mm (after Refs. 3 and 9).

In the analog processing method discussed in the literature,[3,9] the coherent filtering system shown schematically in Fig. 36.7(b) was used. A transparency of the collected image of Fig. 36.7(a) is placed in the object plane of (b). The inverse filter shown in Fig. 36.7(c) is placed in the transform plane and aligned with the synthetic transfer function of Fig. 36.6(b). The absorption level in the inverse filter is controlled photographically so that the filtered gray level along the μ_x and μ_y axes in Fig. 36.6(b) equals that of the synthesized transfer function. The

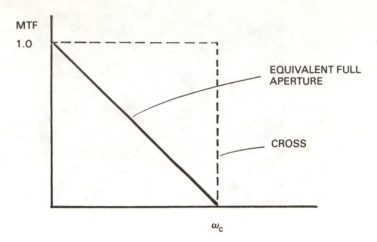

Fig. 36.8 Transfer function of cross aperture.

filtered image is observed in the image plane of Fig. 36.7(b) as shown in (d). The effect of removing the bias image is dramatic since the target is aligned along the axis of greatest synthesis, 45 deg. The three-bar resolution limit observed in Fig. 36.7(d) was 40 line pairs/mm whereas the theoretical limit expected in this experiment was 42 line pairs/mm. For purposes of comparison, the image obtained through the equivalent full aperture is shown in Fig. 36.7(e). The results in Fig. 36.7 agree with the theoretical predictions, showing that the concept is feasible. In comparing Figs. 36.7(e) and (d), we notice that the contrast of the high frequencies (groups 4-1 to 4-6) is decreasing in (e) whereas in the case of the Mills cross picture, the contrast stays constant out to the cutoff frequency; that is, the cross aperture has a flat transfer function. The transfer functions corresponding to the results shown in Figs. 36.7(d) and (e) are shown in Fig. 36.8.

36.3.2.2   *The Covington-Drane and Other Aperture Configurations*

The Covington-Drane aperture configuration shown in Fig. 36.9(a) has also been experimentally verified[3,10] and its synthesized transfer function is shown in (b). This aperture has the interesting property that the outer elements effectively double the system resolution in the synthetic mode over that due to the central aperture alone, as illustrated in Fig. 36.9. The images collected and processed through a one-dimensional Covington-Drane aperture show a doubling of the resolution in the direction of synthesis (vertical bars) and are shown in Figs. 36.9(c) and (d). The vertical bars in Fig. 36.9(d) have a resolution of group 5-4 while the horizontal bars cut off at group 4-4; i.e., a factor of 2 in resolution as expected. Other apertures, for example, the thin annulus, [3,10,11] the minimum redundancy point hex,[2] also called the Golay-6, and the Hex-100 (Ref. 12), have also been studied as aperture configurations for partially filled synthetic aperture imaging systems. The collected and processed images from a nine-segment annular aperture are shown in Figs. 36.9(e) and (f).

36.3.2.3   *Signal-to-Bias Considerations*

One problem inherent with synthetic apertures is the signal-to-bias ratio, which is defined here as the strength of the cross term of the interferometer, or the

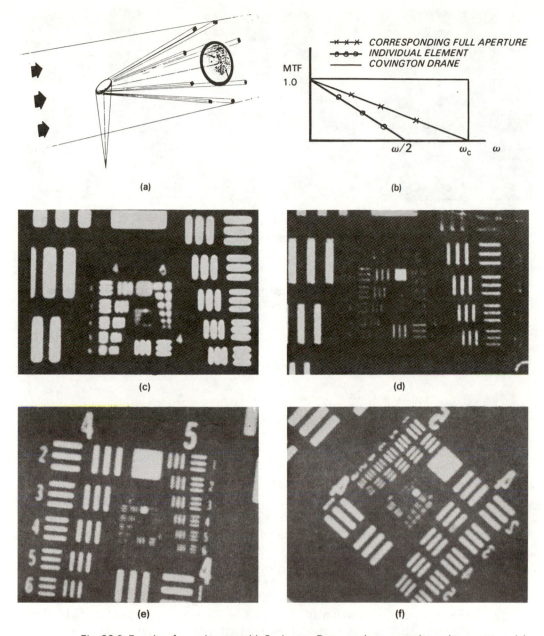

Fig. 36.9 Results of experiments with Covington-Drane and segmented annular apertures: (a) aperture geometry for two-dimensional Covington-Drane system; (b) transfer function of a Covington-Drane aperture; (c) image through and (d) processed image of one-dimensional Covington-Drane aperture; (e) image from a nine-segment annulus; and (f) coherent ground glass image of (e) using coherence reduction techniques to remove speckle [(a)–(d) after Ref. 3].

cross-correlation, relative to the bias or the dc terms. For the Mills cross, the energy ratio of the cross term to the bias terms is proportional to the width-to-length ratio of the cross as seen below.

One must consider the energy contained in the cross term of the Mills cross relative to the bias terms. The cross term must be strong enough to be seen by the

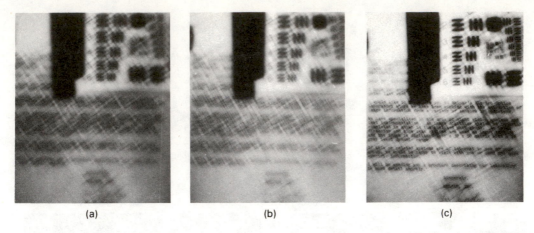

(a)   (b)   (c)

Fig. 36.10 Unprocessed image taken through cross with a length-to-width ratio of (a) 80:1, (b) 40:1, and (c) 20:1.

detector. If we look on axis in transform space, then the energy of the bias terms is proportional to the area of the cross:

$$E_{\text{bias}} = 2la \; ; \tag{36.12}$$

and the energy of the cross term is proportional to the overlap area:

$$E_{\text{cross term}} = 2a^2 \; . \tag{36.13}$$

From the definition of the signal-to-bias ratio given above,

$$\frac{S}{B} = \frac{E_{\text{cross term}}}{E_{\text{bias}}} = a/l = \text{width/length} \; . \tag{36.14}$$

To demonstrate this experimentally, we picked a cross of a given width and varied its length. Figure 36.10(a) is a picture taken through a cross with an 80:1 length-to-width ratio. Since the signal-to-bias ratio is inversely proportional to the length by Eq. (36.14), as the cross is made shorter, a better signal-to-bias ratio should be observed. In Fig. 36.10(b), an image made with a cross having a 40:1 length-to-width ratio is seen, and in (c) the image with a 20:1 ratio is shown. The letters in the 20:1 case are easily readable, showing that the signal-to-bias ratio of the process is a function of the aperture parameters as predicted. In these experiments, the grain noise of the photographic film was made negligible by using high-resolution emulsions.

36.3.2.4   Detector Considerations

Detector Noise: When photographic films are used to record images through partially filled apertures, high-resolution emulsions are necessary to ensure that the interference fringes within the impulse response [scalloping effects in Fig. 36.6(a)] are recorded with adequate modulation. The focal length of the optics can be used to scale the impulse response so that it can be resolved by the film. The parameters of interest for photographic detectors (see Chapter 20) are sensitivity and the ability to record low-contrast imagery. Sensitivity is specified in terms of energy per unit area and must be related to energy per resolution element to make a comparative judgment.

The ability of a photographic detector to record low-contrast imagery is described by the aerial image modulation (AIM) required to obtain an exposure as a function of spatial frequency. It should not be confused with the minimum visibility curve (MV), which describes the minimum detectable modulation on the film as a function of spatial frequency. The AIM and MV curves are related by the film's MTF:

$$\text{AIM} = \frac{\text{MV}}{\text{MTF}} \, . \tag{36.15}$$

The cutoff frequency of a photographic system is usually specified as the point at which the film AIM curve and the system MTF cross when plotted as a function of frequency. The AIM curve includes the criterion for minimum detectable contrast. The focal length of the optical system must be chosen such that the crossover point of the MTF of the unprocessed aperture with the film AIM curve will represent a useful object resolution. Additional information on AIM curves can be found in the literature.[13-15]

Images of various contrast targets were investigated for targets having contrast ratios of 1000:1, 2:1, and 1.6:1. Since the target modulation multiplies the lens MTF to give the signal-to-bias ratio of the aerial image, the requirements on photographic film (detector) are that its AIM curve be less than the modulation of the aerial image at all spatial frequencies.

A typical detector design curve for the Hex-100 array is shown in Fig. 36.11 for observing scenes having contrasts in the image plane of 1000:1, 2:1, and 1.6:1 with 3414 film. At the selected focal length of operation, decreasing the scene contrast to 2:1 results in only a slight degradation in resolution. Below a contrast

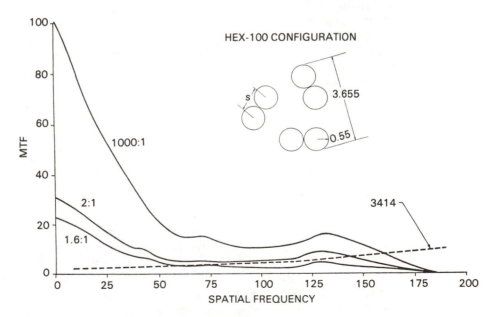

Fig. 36.11 MTF of a Hex-100 array (modified by varying object constrast) and AIM curve of 3414 film.

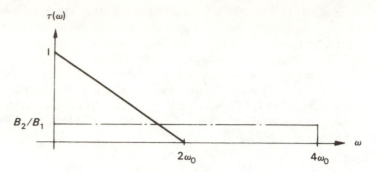

Fig. 36.12 Additive parts of transfer function for one-dimensional optical Covington-Drane aperture (after Ref. 3).

of 2:1, however, the flat portion of the MTF falls below the AIM curve and the resolution degrades rapidly. Thus, if very low contrast scenes are to be resolved with this system, the focal length of the optics must be lengthened considerably. Similar effects occur with electro-optic detectors and must be considered in the design of any system or experiment.

Detector Response (or Exposure Time): In synthetic aperture optics, the trade-offs between (1) total energy collected, compared to a full pupil of the same overall dimension, and (2) information gathered per unit area of the pupil function must be made and evaluated for each application. Of course, detector response (or exposure time) becomes an important parameter in such a trade-off.

Figure 36.12 illustrates the different values of the transfer functions of the cross term and the bias term at the various frequencies for a one-dimensional Covington-Drane aperture. In defining a signal-to-bias criterion, we utilized the fact that the synthetic transfer function is low in energy relative to the bias term at zero frequency.

Recall that the signal-to-bias ratio at zero frequency (defined as the strength of the synthetic term relative to the bias term) is proportional to the areas of the various collecting apertures where B_2 is the area of the cross-correlation term and B_1 is the area of the central aperture. However, notice that at some frequency the two energies are equal. Thus, if the full lens records this spatial frequency, the synthetic aperture will also record it. Based on this crossover point, a very interesting question can be asked: "What exposure times are necessary for a full lens and a synthetic aperture lens to observe the crossover frequency?" (It is easily shown that, relative to zero frequency, the exposure times are proportional to the aperture areas.) To answer this question, an experimental comparison between a cross, a full aperture, a Covington-Drane, and an annular aperture was made. The exposure times necessary to record the crossover frequency at a constant signal-to-noise ratio were compared. A very interesting result was obtained and is shown in Fig. 36.13.

The photograph in Fig. 36.13(b) was taken through the full aperture. The pictures taken with the other three apertures were within one f/stop in speed compared to the full aperture at the high spatial frequencies. However, the advantage of the full aperture is that all the additional energy collected by the full lens effectively acts as a prefog on the film. Since a prefog is equivalent to an f/stop, repeating this experiment and prefogging the three films used to record the synthetic aperture information resulted in systems having essentially equiva-

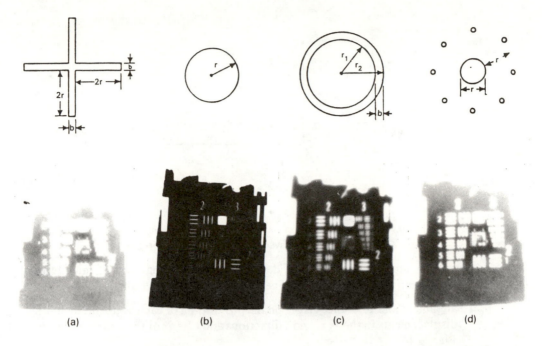

Fig. 36.13 Exposure comparison experiment and aperture used (after Ref. 3).

lent photographic speeds for the four apertures shown in Fig. 36.13. This additional energy level must be large enough to ensure that the unprocessed MTF of the aperture is above the AIM curve of the detector at all spatial frequencies.

36.3.2.5 Alignment Considerations

Computer Model: A computer model was designed to simulate any possible synthetic aperture configuration made up of circular or annular apertures, sector-shaped segments of circular and annular apertures, and multisegmented apertures consisting of combinations of circular and annular apertures.[11,16] The shape of the optical surface (reflecting) can be varied continuously from a sphere to a hyperbola merely by specifying the eccentricity of the desired conic. In addition to the conic surface, the optical surface may be defined as sections of a toroid by specifying the two radii of curvature. The normal optical aberrations were assumed perfect in this model but they were subsequently investigated.[17]

Once the aperture geometry and segment shape have been specified, the computer program calculates and prints out the rms peak-to-peak and mean wavefront deviation from a parabolic reference surface for each piece of the aperture. The system intensity impulse response is obtained by Fourier transforming the aperture data and squaring the result. The optical transfer function (OTF) of the synthetic aperture system is obtained by Fourier transforming the intensity impulse response. The OTF is printed out in two parts: the MTF and the phase of the OTF.

Since the intent of the program was to determine the effect of misalignments and mechanical tolerances on the positions of the various pieces of a multiseg-

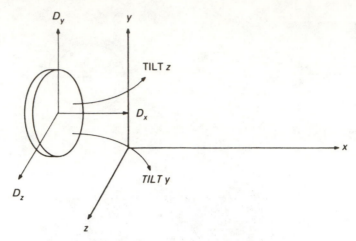

Fig. 36.14 Coordinate system used in computer program.

ment aperture, the program was designed with the capability of moving the various segments independently. The possible motions include:

1. alignment and tolerance errors on the positioning of each segment, including displacements in the x, y, and z directions and tilts about the y and z axes (see Fig. 36.14)

2. a focal length error on each segment due to manufacturing tolerances

3. surface figure error on each piece due to manufacturing tolerances and residual internal stresses in the blank.

Modeling of a One-Dimensional Synthetic Aperture: Figure 36.15 is a drawing of the model synthetic aperture, designed to simulate an experimental mirror system that was used to investigate alignment techniques and verify computer results. This model yields essentially a one-dimensional synthesis in the direction of the outriggers and a direct comparison with the nonsynthesized aperture in the other direction. Figure 36.16 shows the computer results for the perfectly aligned system (with parabolic pieces) showing the MTF in the direction of the outriggers and in the perpendicular (or nonsynthesized) direction. The resolution increase due to synthesis in the direction of the outriggers is easily seen. The result in the perpendicular direction is the same as that for the central aperture alone since there is only synthesis in the y direction.

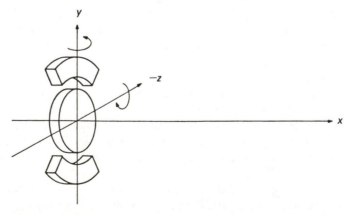

Fig. 36.15 Schematic of model synthetic aperture test mirror (10-in. diameter, f/8 parabola).

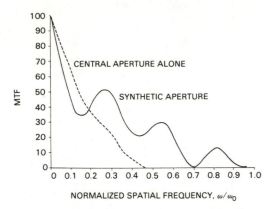

Fig. 36.16 One-dimensional MTFs of the model synthetic aperture.

When one of the outriggers is misaligned through either a tilt or a displacement along the optic axes (defocusing), the transfer function is degraded as shown in Figs. 36.17 and 36.18. The first perturbation was a tilt of the segment about the z axis. For this particular test mirror (10-in. diameter), the overall impulse response (Airy disk) subtends 6.7 microradians (μrad), and the impulse response of the segment in this direction of tilt is ~23 μrad. The tilts introduced in this direction were 1, 2, 3, 7, 9, 13, and 20 μrad, or from one-sixth to three times the maximum aperture Airy disk. The results are shown in Fig. 36.17 for 3, 7, 13, and 20 μrad. Two things should be noted here: At the high frequencies, the MTF decreases gradually with increasing tilt; in the intermediate frequencies, the MTF varies cyclically with increasing tilt. In fact, a 13-μrad tilt about the z axis is not significantly worse in the midfrequencies than a 1-μrad tilt. It should be remembered here that a tilt of a segment on the y axis about the z axis also induces a displacement along the x, or optic, axis.

The next perturbation applied was a displacement of one segment along the x, or optic, axis—focal displacement. It was assumed that since the synthetic aperture impulse response is formed by interference of the impulse responses of the individual segments, a focal displacement of $1/2$ or n wavelengths (path change of 2 or $2n$ waves) would not change the impulse response of the transfer function as long as the displacement was well within the depth of focus of the

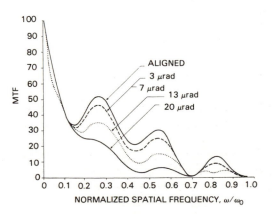

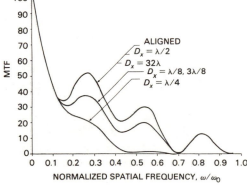

Fig. 36.17 MTF of the test mirror for angular displacement (tilt about the z axis) of one segment.

Fig. 36.18 MTF of test mirror for focal displacement of one segment.

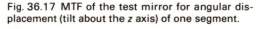

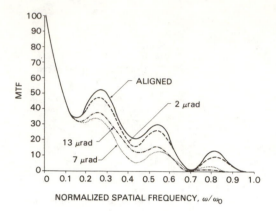

Fig. 36.19 MTF of test mirror for angular displacement about the z axis of one segment and a compensating displacement in the x direction.

overall aperture. Figure 36.18 shows the transfer functions for various displacements along the optic axis. As expected, for displacements of $1/2$ and 32 waves, the transfer function is identical to the unperturbed transfer function. The depth of focus of this mirror is 128λ, if 2λ (f/ number)^2 is used as the depth-of-focus criterion. Thus, the high-frequency portion of the MTF is unaffected by focal displacement (as is evident from the figures) while the midfrequencies have a cyclic variation with a period of $\lambda/2$.

To study further the effect of perturbations on the transfer function, the introduction of a tilt without a displacement in the x direction was measured. This was done in two ways: first, by tilting about the z axis and introducing an appropriate displacement to cancel out the displacement introduced by the tilt, and second, by tilting the segment about the y axis, which does not introduce any mean displacement since the segment is centered on that axis. Because of the geometry of the segments (longer in the z direction), a tilt about the z axis (called tilt-y) has less effect than a tilt about the y axis (called tilt-z). The MTF for tilt-y is shown in Fig. 36.19. As seen in the example plotted, the highest frequency peak gradually drops to zero with increasing tilt, occurring at a faster rate in the direction corresponding to the larger dimension of the segment. The midfrequency MTF drops to a minimum and increases slightly again for increasing tilts. This is more evident in the case of a tilt that is in the direction of the larger dimension of the piece. It is also evident in the other direction, as tilts of both 36 and 48 μrad yield an improved midfrequency MTF over a 24-μrad tilt. A tilt-z of 7 μrad and a tilt-y of 24 μrad both yield an rms wavefront deviation of $\lambda/2$ and cause a minimum in the midfrequency MTF as did the $\lambda/4$ displacement ($\lambda/2$ wavefront deviation) in the focal curves of Fig. 36.18.

An analysis of these results in terms of interference between impulse responses from various segments of the aperture helps to clarify what is happening to the MTF. A single pair of impulse responses from elements that are aligned produces interference fringes whose frequency is a function of the separation of the elements. This fringe frequency contributes to a particular spatial frequency region in the MTF. Another pair of elements with the same separation and hence same fringe frequency will, when aligned, increase the value of the MTF at that particular spatial frequency. A $\lambda/4$ displacement of one piece from one of the pairs (a $\lambda/2$ path change) will shift the fringes from that pair by $\lambda/2$, so that they now add destructively to the fringes from the other pair of elements. The contributions of the two pairs of elements no longer add in the MTF, instead they subtract because of the phase change. This effect can occur whenever the MTF has a redundancy—two or more pairs of elements contributing to the same region of the MTF. It would appear from these results that a nonredundant aperture would be desirable to minimize the effects of misalignments on the transfer function.

In addition to the tilts and displacements discussed above, one segment was displaced along the y axis by 50 μm (0.002 in.) and by 200 μm (0.08 in.). These displacements shift the impulse response in the focal plane, yielding a degraded transfer function. Suitable tilts to realign the impulse responses yielded transfer functions that are identical with the MTF from the aligned mirror. They show that, at least for this size mirror, small radial displacements of the mirror segments can be compensated for by an appropriate tilt.

Experimental Verification of Computer Model: A series of experiments was performed to test the various alignment tolerances predicted by the computer model.[11],[16] The particular apertures studied were: a split f/64 lens with a cross mask; an $\epsilon = 0.7$, f/21 annulus split in half; a full aperture split lens; and a two-dimensional reflective Covington-Drane aperture cut from a 4¼-in. spherical mirror. The apertures were all initially diffraction-limited systems so their experimental performance is readily compared with the model results.

Four different experimental evaluation techniques were used to assess the effects of alignment errors. They were : (1) measurements of the point spread function, i.e., degree of overlapping of the diffraction spots of the individual apertures in the image plane; (2) measurements of the two-dimensional MTF of the synthetic aperture obtained by optically Fourier transforming the impulse response of the overlapped apertures; (3) interferometric examination of the wavefront in the exit pupils; and (4) measuring the modulation in the image plane of the processed image from the synthetic aperture. Of the four methods, the impulse response method was the most accurate and reliable.

The reflective Covington-Drane aperture used in the experiments is shown in Fig. 36.20. A 4¼-in. f/10 mirror was sectioned by separating out a 3-in. center

Fig. 36.20 Synthetic aperture mirror with outriggers.

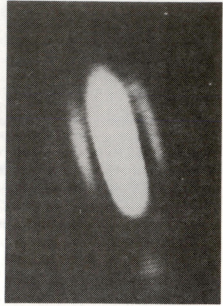

Fig. 36.21 Impulse response of the central aperture.

Fig. 36.22 Impulse response of one of the outriggers.

and forming eight outriggers from the outside section. The outrigger elements were placed on the circumference of a 10-in. circle, making an effective f/4 reflective synthetic aperture. Because of the relatively crude positioning devices used (machine screws on a base plate), the positioning of the elements proved extremely difficult. The elements were positioned for laser light, and the spread function exhibited the expected interference effects; attempts to position the plates with a thermal source were not successful. We conclude that a mount for the mirror elements of a segmented aperture must be carefully engineered with fractions of a wavelength to tolerance the positions of the segments along the axis. Tolerances on the transverse variables and the alignments are basically no more critical than for a refractive device.

In an experiment to investigate some of the alignment tolerances to be expected, the mirror of Fig. 36.20 was set up with only two outriggers. Adjusting screws were provided on the two outriggers to allow positioning of the pieces. An He-Ne laser, expanded through a lens at a distance of 85 ft, was used as a source to photograph the impulse response of the central aperture alone (Fig. 36.21). This is the normal Besinc function that would be expected from a circular aperture. Figure 36.22 shows the impulse response from one of the outriggers alone. In Fig. 36.23, the two outriggers were aligned so that their impulse responses overlap and interfere to produce fringes of high contrast whose frequency is determined by the separation of the outriggers. It is this interference that provides the high resolution of synthetic apertures; the pieces must then be aligned to produce interference. In Fig. 36.24, the central aperture was added. The intensity of the impulse responses of the outriggers is much less than that from the central aperture because of their small size. Their effect is evident here in the fringing which exists across the Besinc from the central aperture in Fig. 36.24. Obtaining interference with the gas laser source was relatively easy because the pieces had to be positioned along the optic axis only to within the coherence

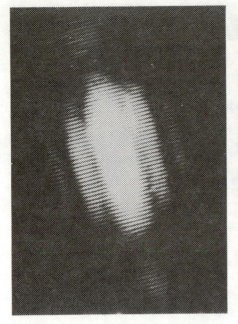

Fig. 36.23 Two overlapping outrigger impulse responses showing the desired interference.

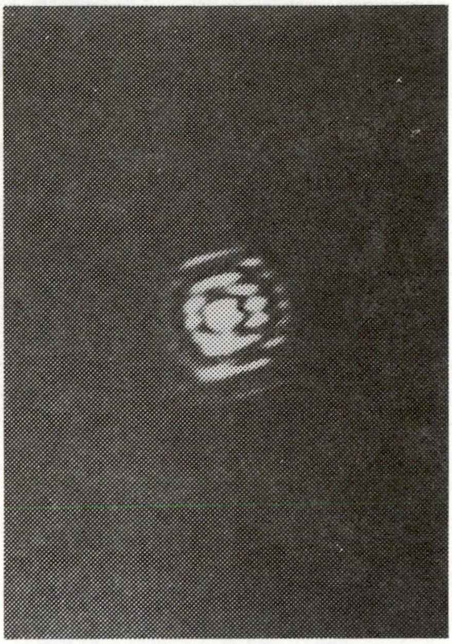

Fig. 36.24 Combined impulse response from two outriggers and the central aperture (aperture synthesis is evident in the fringing seen across the central aperture Besinc).

length of the source, which is many meters. The system was too crude to obtain white-light fringes, for this would mean positioning pieces to within a few microns. This experiment does demonstrate, however, the utility of the impulse response measurement technique and enables us to estimate the experimental tolerances required on a synthetic aperture mirror. The MTF measurements substantiated these data because when the impulse responses of the individual elements do not overlap sufficiently, then the synthesized MTFs degrade rapidly.

The effect of nonoverlapping impulse responses was illustrated in the cross aperture alignment studies. A lens was cut in half and each half mounted in a separate micrometer mount as shown in Fig. 36.25(a). This mount allowed for relative displacement along the optical axis and some tilting of one element. A cross aperture was placed over the mounted elements to create an f/64 Mills cross aperture as illustrated in Fig. 36.25(b).

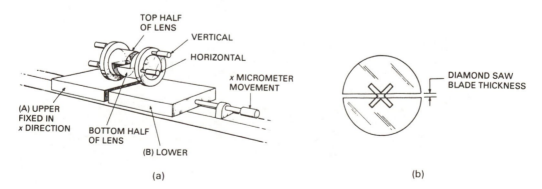

Fig. 36.25 (a) Split lens alignment setup and (b) split lens with a cross aperture.

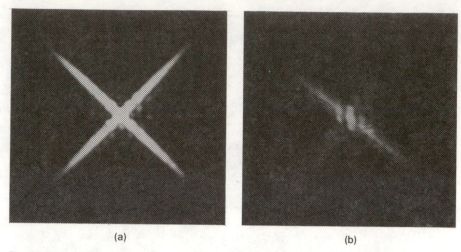

(a)   (b)

Fig. 36.26 (a) Cross point spread function for configuration of Fig. 36.25(b) and (b) point spread function for an angular misalignment of the elements of Fig. 36.25(b).

With the two lens halves closely aligned, the resulting point spread function for the cross aperture is shown in Fig. 36.26(a). Introduction of an angular misalignment in one lens half creates the point spread function shown in (b). The effects of displacement errors are also observed in aperture MTF as seen in Fig. 36.27 where (a) shows the two-dimensional cross MTF at no displacement and (b) shows the MTF when the pieces are misaligned twice the depth of focus. The MTFs from the displaced lens halves are clearly smaller and more distorted compared to the nondisplaced case, illustrating the importance of having the alignment errors small enough so that the impulse responses of the individual elements overlap to create interference.

The alignment data obtained in these experiments agree with those predicted by the computer model.

The extensive experimentation with synthetic apertures indicates that a number of different evaluation methods are useful for measuring and understanding the alignment tolerances. The MTF, point spread function, and wavefront interferogram are all useful for viewing different alignment properties. A

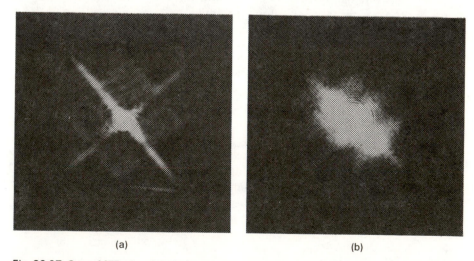

(a)   (b)

Fig. 36.27 Cross MTF when lens halves are (a) aligned as closely as possible and (b) displaced by 2× the depth of focus.

TABLE 36.I

Summary of Alignment Tolerances for Synthetic Apertures

Parameter	Lens Tolerance	Mirror Tolerance	$\epsilon = 0.8$, f/21 Annular Refractive Lens
x relative displacement along the optic axis	$2\lambda(\text{f/number})^2$ (depth of focus)	Coherence length	0.5 mm
Δz Δy^a	0.2 (f/number λ) 0.2 (f/number λ)	0.2 (f/number λ) 0.2 (f/number λ)	20 μm 10 μm
$\Delta\beta$ bending[b] (tilt-y)	$<5\lambda/d_0{}^c$	$<\lambda/4d_0$	0.06 μrad
$\Delta\mu$ twisting[b] (tilt-z)	$<5\lambda/d_0{}^c$	$<\lambda/4d_0$	0.06 μrad

[a]In the annular segments considered, the z dimension was twice that of the y dimension.
[b]$d_0 = 2R$ is the dimension of the outrigger in the dimension of motion.
[c]Empirically determined.

few of the results of these studies are included in Table 36.I. The derivations of the formulas are given in the Appendix, Sec. 36.4.

These tolerances are loose enough in the annular lens case to allow placing two halves of an f/21 annular lens in a well-machined lens holder and simply tightening down carefully. The tolerances in the mirror case with white light are much tighter and require a more involved, but well within the state of the art, alignment procedure.

Jitter Control: Partially filled synthetic apertures require that the interference fringes from the various elements be recorded by the detector. This means that the two outermost elements of the array must be held to a fraction of the angular resolution of a lens of that same diameter so that the fringes remain stable during the recording. Otherwise, the fringes will not be recorded and the recorded data cannot be processed to obtain the synthesized high-resolution image. Therefore, the jitter control specification is much worse than the alignment tolerances given in Table 36.I because it has an angular tolerance of (λ/nD) where D is the distance between the outermost segments and n is a number between 4 and 8 that must be empirically determined.

36.3.2.6 *Synthetic Aperture Camera*

Description of the Camera: A synthetic aperture camera,[11] consisting of a segmented annulus as the aperture, was built and tested to demonstrate the principles of partially filled synthetic aperture optics. The selected lens was an f/21 cemented achromat, corrected for spherical aberration and shown to be diffraction limited within 0.5 deg off axis. An identical full aperture lens was used as a control camera for use in the experimental evaluation. The lens was cored and cut along a diameter to create the segmented annulus. Figure 36.28 is a schematic of the mount for the segmented annulus. A planed surface provides a seat for the elements; a quartz spacer, the width of the saw cut, fixes their separation; and final position adjustments are made by machine screws in the internal plug and in the cell wall. When finally adjusted, the elements are held firmly without strain by an O ring and retainer. Figures 36.29(a) and (b) show, respectively, the spread functions of the well-adjusted segmented annular aper-

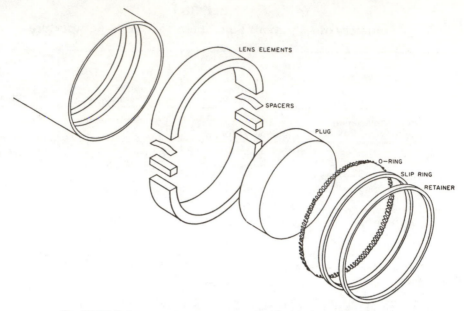

Fig. 36.28 Schematic of mechanical assembly for segmented annulus.

ture and the full aperture of the control camera. The two-dimensional MTF of this segmented annulus is shown in Fig. 36.29(c) with a double folium caused by the cut.

The segmented aperture camera and control camera were gang-mounted on a single tripod (Fig. 36.30) with the intent of taking simultaneous photographs. Although not bore-sighted, the mount allows for that possibility. The camera backs are standard with focal plane shutters. Since atmospheric turbulence is critical in any evaluation, neutral density filters are supplied to give the control camera the same effective aperture as the synthetic aperture. With a double release cable and the same exposure times, the two cameras are operating under identical conditions except for the vagaries of turbulence. The tripod is reinforced to reduce vibration caused by shutter movement. The cameras are focused independently since their focal lengths are slightly different. In critical tests, a microscope was used in focusing but it was found that viewfinder adjustments were adequate.

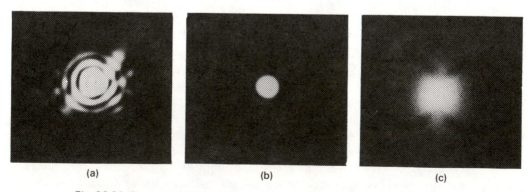

(a)   (b)   (c)

Fig. 36.29 Spread functions (impulse responses) of (a) $\epsilon = 0.8$, f/21 segmented annular aperture, (b) f/21 control camera lens, and (c) photographic representation of transfer function of segmented annulus. (Note lowering of the response in a double folium with an axis in the cut direction).

Fig. 36.30 The segmented annulus and control camera installed and gang-mounted on a tripod.

Camera Alignment Tolerances: The alignment tolerances of an f/21 segmented annulus were investigated with the MTF computer program.

Figures 36.31(a) and (b) show the results of relative defocusing of an $\epsilon = 0.8$, f/21 aperture giving the elements a separation of 0.2 and 0.4 mμ, respectively. The MTF results are essentially equivalent to those of a full annulus having an $\epsilon = 0.8$ as would be expected since the defocusing is less than that allowed in Table 36.I. The phase shift perpendicular to the cut of the lens can be understood as a consequence of the fact that the aberration introduced by a relative defocusing of elements is an odd function in that direction. For small displacements, this phase shift is the most important effect. Similar results on the segmented annulus were obtained for the other degrees of freedom, verifying that the results of Table 36.I can be achieved. Laboratory measurements made on images of three-bar targets also verified the alignment tolerances listed in Table 36.I for the segmented annulus.

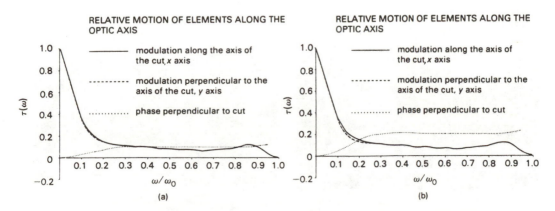

Fig. 36.31 MTF with lens elements separated longitudinally by (a) 0.2 mm and (b) 0.4 mm.

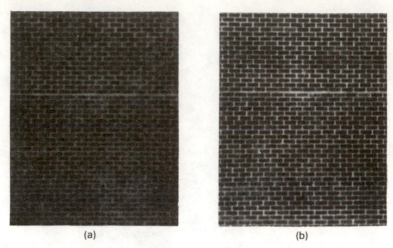

Fig. 36.32 Field condition photographs at 330 m: (a) control aperture and (b) synthetic aperture.

Camera Tests: The camera was tested on a variety of targets both in the laboratory and in the field. The critical tests are of resolution targets at known distances and under minimum turbulence conditions since in the field turbulence is generally too great for the system to approach its theoretical resolution limit. Figures 36.32(a) and (b) show images of a brick wall at 330 m with incoherent illumination (sunlight). This photograph gives some evidence of the greater resolution capability of the synthetic aperture camera. Although the shot was taken over a large parking lot, little evidence of turbulence degradation is seen in the photographs. The scale of reduction at this distance is about $470\times$ (the width of the mortar between the bricks represents a distance of about 0.04 mm at the film plane. Laboratory tests of low-contrast targets and of targets illuminated

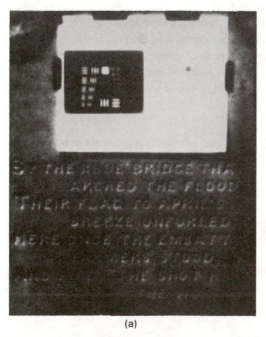
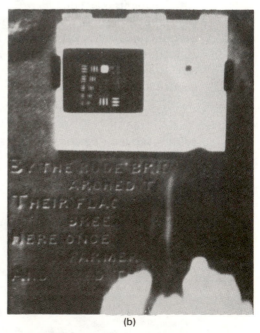

Fig. 36.33 Continuous-tone (a) synthetic and (b) control aperture photographs (turbulence limited).

with low spatial coherence laser light give similar results. Figures 36.33(a) and (b) are taken in normal daylight at 150 m (about 215×) with an exposure time of 0.1 s. The diffraction limit for these photographs at this distance is group 2-3 whereas both lenses are turbulence limited at group 1-5 in these photographs.

Turbulence was discovered as the limiting factor by examining the aerial image with a microscope. Visually, the image could be seen to be moving although the eye could resolve to the diffraction limit. The difference in image contrast between the two apertures is most apparent in the black-on-white resolution target.

The segmented annulus camera also shows the lower contrast expected from an annular aperture. It also demonstrates the slightly greater resolution on limited fields, such as the three-bar target, caused by the narrowing of the impulse response by the annulus.

In summary, the annular segments were assembled with sufficient precision such that there are no apparent effects of the segmentation. The aperture was taken apart and realigned several times during the testing and a procedure developed so that a technician watching a diffraction image can bring it into alignment in a few minutes.

36.3.2.7   *Real-Time Processing*

One method for removing the bias information is for one of the apertures, A_1 in Fig. 36.34, to be modulated in time (e.g., with a phase switch). Then the impulse response becomes

$$\mathcal{I}(y) = \left| 2d\,B_1 \exp\left(\frac{2\pi i \xi y}{\lambda f}\right) \text{sinc}\left(\frac{2\pi dy}{\lambda f}\right) \exp(i\omega t) \right.$$

$$\left. + 2B_2 \cos\left(\frac{2\pi dy}{\lambda f}\right) \exp\left(\frac{2\pi i \beta y}{\lambda f}\right) \right|^2 , \tag{36.16}$$

where B_1 and B_2 are the energies passing through A_1 and A_2, respectively. If $\xi_1 + \beta = 2d$, the cross term is an amplitude-modulated sinusoid:

$$16\,dB_1 B_2 \text{ sinc}\left(\frac{8\pi dy}{\lambda f}\right) \cos\omega t . \tag{36.17}$$

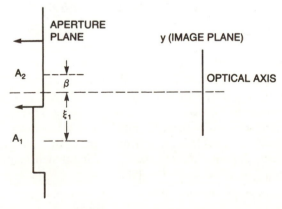

Fig. 36.34 Schematic of compound aperture (one-dimensional Covington-Drane), where A_1 is a clear aperture of width $2d$ and A_2 consists of two delta functioning of separation $2d$.

If the demodulation is realized by a nonintegrating scanning system, such as an image dissector coupled with a synchronous phase and envelope detector, then the cross analog can be realized by this modulation technique. The disadvantage of this method is the resolution limitation imposed by the image dissector. Larger arrays of detectors can increase this resolution.

36.3.2.8   Active Control of the Elements

While the subject of activity controlling the individual elements is not considered here, it should be mentioned that the active optics segmented mirror experiment at Perkin-Elmer[18,19] and adaptive optics technology[20] are both developing areas of technologies that are directly applicable to the future needs of a partially filled synthetic aperture system.

36.3.2.9   Jamieson's Spatial Spectral Interferometer

The basic concept of Jamieson's spatial spectral interferometer[21–23] is that the number of subapertures in a partially filled synthetic aperture (which uses geometry and autocorrelation to create the synthesized transfer function) can be reduced theoretically by using spectral information and image processing techniques to fill in the missing portions of the MTF. This concept uses the fact that each spectral component is imaged independently by the synthetic aperture array and then the point spread function is integrated at the mean wavelength with the spectral distribution of each source point.

A separate measurement of the spectrum of each source point is made simultaneously with an array of Michelson-type interferometers. The multiple wavelength image data are combined with the spectral data from the output of the interferometer in a processing algorithm to create the high-resolution image from the spectrally synthesized MTF.

To date it has been shown that the fringe patterns for the various colors are encoded in the impulse response.[21] However, the processing algorithm for recovering the high-resolution image from the encoded image remains to be developed and shown experimentally.

36.4   APPENDIX: DERIVATIONS OF THE ALIGNMENT TOLERANCES LISTED IN TABLE 36.I FOR SMALL SEGMENT DISLOCATIONS[a]

Any displacement of a segment of an imaging device, lens or mirror, has the effect of causing the exit pupil wavefront to be something other than parts of a perfect sphere.[18,19,24] In this sense, relative position or orientation shifts of the elements of a segmented aperture pupil produce aberrations, and the methods of evaluating aberration tolerances can be used to find tolerances on segment location and orientation. Figure 36.35 illustrates the five significant degrees of freedom of a segment of the annular aperture—two in rotation and three in translation. The third rotation about the optic axis is of minor influence since all segments are part of a surface of rotational symmetry about the axis.

---

[a]The authors are indebted to A. E. Smith for this development.

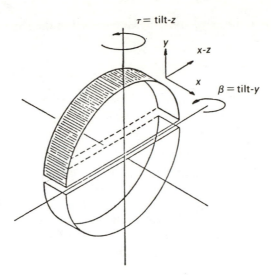

Fig. 36.35 Schematic illustrating the degrees of freedom for a split annulus aperture.

Marechal[25] has shown that for a full aperture and weak aberrations the rms deviation from the reference sphere must not exceed $\lambda/14$ if the irradiance in the center of the diffraction image is to be within 0.8 of its unaberrated value. Since this criterion is generally accepted and simple to apply, it will be used to determine tolerance limits on segment displacement.

To translate the $\lambda/14$ criterion on the rms wavefront into a tolerance on the displacement of aperture segments, first consider refractive elements and a location error parallel to the optic axis. A lens split along a diameter and having the two sections displaced slightly with respect to each other, parallel to the optic axis, is shown in Fig. 36.36. This constitutes an effective defocusing of each element, relative to the original position. To evaluate the tolerance on focusing error, the wavefront in the exit pupil is determined and required to have no more than $\lambda/14$ rms deviation from the reference sphere in the exit pupil. It is recognized that this criterion limits its interest to the irradiance at the center of the diffraction image.

The wavefront sections leaving the neighborhood of the exit pupil are in phase on axis but deformed relative to each other. The axial optical path length has not changed as a result of the element translation but the foci of each section have moved a distance equal to the displacement of the lens elements. The reference

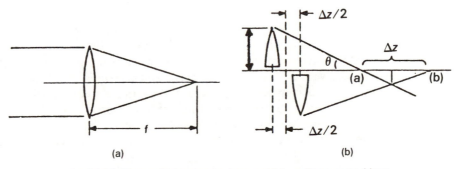

Fig. 36.36 Focus of (a) a regular lens and (b) a split separated lens.

sphere in the exit pupil can be described in the paraxial approximation by

$$\Phi_{ref} = \frac{r^2}{2f} ,$$
(36.18)

where r is a coordinate in the plane of the exit pupil and f is the focal length of the elements. The aberrated wavefront is given by

$$\Phi_{ab} = \frac{r^2}{2\left(f \pm \frac{\Delta x}{2}\right)} ,$$
(36.19)

where the positive and negative signs apply, respectively, to the two halves, each segment having been moved $\Delta x/2$ out of alignment. The deviation of the wavefront from the reference sphere is given by

$$\Delta\Phi = \frac{r^2}{2f}\left[1 - \frac{1}{\left(1 + \frac{\Delta x}{2f}\right)}\right] , \quad 0 < \xi < \pi$$

$$= \frac{r^2}{2f}\left[1 - \frac{1}{\left(1 - \frac{\Delta x}{2f}\right)}\right] , \quad \pi < \xi < 2\pi ,$$
(36.20)

where ξ is the angle in the (y,z) plane. The resulting rms deviation of the aberrated wavefront is

$$\Delta\Phi_{rms} = \frac{\Delta x}{4\sqrt{3}} \frac{R^2}{f^2}\left[1 + \frac{3}{2}\left(\frac{\Delta x}{2f}\right)^2 + ...\right] ,$$
(36.21)

where R is the radius of the segment. The series does not converge but the problem is not seriously violated by neglecting terms of higher order. Requiring $\Delta\Phi_{rms} < \lambda/14$ gives an allowable tolerance on the separation between the lens elements of

$$\Delta x \leq \frac{2\sqrt{3}}{7}\lambda\left(\frac{f}{R}\right)^2 \simeq \frac{\lambda}{2}\left(\frac{f}{R}\right)^2 .$$
(36.22)

This is recognizable as the focal tolerance.[26] Thus, lens elements can suffer relative axial displacements within the range of the focal tolerance without seriously affecting the quality of the diffraction image. For an f/20 system illuminated with 0.5-μm radiation, this displacement is about 400 μm.

The result is completely different when reflective elements are considered. Any motion of a mirror element not only distorts the wave relative to the reference sphere but changes the optical path length between object and image positions. The deviation from the reference sphere for elements separated by a distance Δx is

$$\Delta\Phi = \Delta x + \frac{r^4}{8f^3}\left[1 - \frac{2}{\left(1 - \frac{\Delta x}{2f}\right)^2}\right], \quad 0 < \xi < \pi,$$

(36.23)

$$\Delta\Phi = -\Delta x + \frac{r^4}{8f^3}\left[1 - \frac{1}{\left(1 - \frac{\Delta x}{2f}\right)^2}\right], \quad \pi < \xi < 2\pi.$$

The linear term in Δx is so dominant that $\Delta\Phi_{rms}$ is—neglecting the higher order terms—just Δx. Consequently, the tolerance on the displacement of mirror elements is the same as the allowed rms wavefront deformation, $\lambda/14$, and to a first approximation independent of the relative aperture. Thus, the coherence length of the radiation must be greater than Δx if the impulse responses of the two mirror elements are to interfere.

Positional errors in x and y produce similar effects and are the same in reflective and refractive systems in that the diffraction image and the vertex of the wavefront are moved to the side. The deviation from the reference sphere for an element located $\pm\Delta z/2$ to the side is, to a first approximation,

$$\Delta\Phi = \pm\frac{\Delta z r \cos\xi}{2f}.$$

(36.24)

Repeating the earlier calculations for the aperture of two segments gives

$$\Delta\Phi_{rms} = \frac{\Delta z}{4}\frac{R}{f}.$$

(36.25)

The resulting tolerance on the position of the elements in z or y must be

$$\Delta z \leq \frac{2}{7}\frac{\lambda f}{R},$$

(36.26)

which, for an f/21 system in 0.5-μm radiation, amounts to about 6 μm.

In the neighborhood of the geometrical focus, lines of equal irradiance form a cylindrical structure.[26] The tolerances for longitudinal position of a refractive element and for lateral position of either type indicate that the displaced diffraction image must stay within the 80% isophote. For the reflective element, the longitudinal tolerance is more restrictive since the optical path length to a given field position depends on the segment and, hence, interference effects in the composite diffraction image are set by the coherence length of the radiation.

General statements can be made about the relative sensitivity of the system to segment rotation. Rotations of a refractive element about an axis through or close to a nodal point have no first-order effects on the wavefront in the exit pupil, with the consequence that a refractive aperture has large tolerance limits for either bending or twisting of the effective aperture. However, for segments off axis away from a nodal point the tolerances are similar to those of a mirror.

In reflective systems, a rotation of the segment produces a rotated wavefront and rotations of bending or twisting are similar in their effect. The rms wavefront deviation when a reflective element is rotated through an angle $\Delta\tau$ is

$$\Delta\tau_{rms} = R\,\Delta\tau ,\tag{36.27}$$

with the result that the tolerance on angular position of a mirror segment is

$$\Delta\tau \lesssim \pm \frac{\lambda}{14R} .\tag{36.28}$$

Since the peak-to-peak deviation is greater than the rms wavefront deviation, Eq. (36.28) may be relaxed to[b]

$$\Delta\tau < \pm \lambda/8R .\tag{36.29}$$

In conclusion, it can be seen that segments of a refractive aperture need not be portions of a single original lens and that positioning and orientation of refractive elements are well within normal engineering specifications, but that reflective elements must be positioned as in an interferometer. Similar conclusions for reflective systems have been discussed in the literature.[2,3]

For refractor elements, in the paraxial region we experimentally determined that the angular alignment tolerances are a factor of 5 to 10 less severe than for the case of reflector elements, which behave as interferometer elements. Therefore, the refractive segmented annular camera described in Sec. 36.3.2.6 was relatively easy to align.

36.5   APPENDIX: FORMULATION OF THE OPTICAL SYNTHETIC APERTURE ANALYSIS

Partially filled synthetic apertures were developed originally for use in radio astronomy where the frequencies were low enough to make it possible to mix and cross correlate the electromagnetic fields collected by the separate apertures. At optical frequencies, only intensities are measured and it may appear at first that this fact obviates the use of synthetic apertures at optical frequencies. In fact, such systems have been demonstrated with visible light and some typical results were given in this chapter. The analysis in the main body is heuristic and accordingly some of the approximations were omitted. In this Appendix, a somewhat more complete formulation is provided to bring out the physical significance of the approximations.

Consider an imaging system whose aperture function may be represented as the sum of two functions $A_1(\alpha,\beta)$ and $A_2(\alpha,\beta)$, i.e.,

$$A(\alpha,\beta) = A_1(\alpha,\beta) + A_2(\alpha,\beta) .\tag{36.30}$$

Because of the linearity of the wave equation, the amplitude impulse response will be the sum of the corresponding impulse responses, i.e.,

[b]Note that $\lambda/2R$ is the angular resolution of the segment caused by diffraction. Hence, $\lambda/8R$ describes a tolerance wherein the diffraction spot (impulse responses) in the image plane is approximately held to within their 80% intensity point as specified by the Maréchel criterion.[25] In practice, this criterion could be relaxed to the 50% intensity point without serious degradation of the OTF because some contrast enhancement will be achieved when processing the collected imagery. This relaxed criterion is used in defining the tolerance limits of Table 36.I.

$$\tilde{A}(x,y) = \tilde{A}_1(x,y) + \tilde{A}_2(x,y) \ . \tag{36.31}$$

The resultant intensity impulse response is $\mathscr{I}(\mathbf{x})$ given by

$$\mathscr{I}^2(\mathbf{x}) = |A(\mathbf{x})|^2 = \mathscr{I}_1(\mathbf{x}) + \mathscr{I}_2(\mathbf{x}) + 2\,\mathrm{Re}\,[<A_1(\mathbf{x})\,A_2(\mathbf{x})>] \ . \tag{36.32}$$

Physically, $\mathscr{I}_1(\mathbf{x})$ and $\mathscr{I}_2(\mathbf{x})$ are the images (impulse responses) that would be obtained if $A_2(\alpha,\beta)$ and $A_1(\alpha,\beta)$, respectively, were blocked off. Thus, the effects of the synthetic aperture could be obtained by sequentially making three exposures and subtracting them point by point.

However, in the experiments reviewed in this chapter, a different approach was used. Typically, the geometry of the two apertures was chosen in such a way that a single exposure results in an image from which the synthetic aperture image may be extracted by spatial filtering. We illustrate such a treatment by examining the Mills cross in some detail.

Consider a lens that is masked by a cross. The aperture function $A(\alpha,\beta)$ is described by

$$A(\alpha,\beta) = \mathrm{rect}(\alpha|a)\,\mathrm{rect}(\beta|b) + \mathrm{rect}(\alpha|b)\,\mathrm{rect}(\beta|a)$$

$$- \mathrm{rect}(\alpha|b)\,\mathrm{rect}(\beta|b) \ . \tag{36.33}$$

The corresponding amplitude impulse response is

$$\tilde{A}(\mathbf{x}) = 4ab\,S(a|x)\,S(b|y) + 4ab\,S(b|x)\,S(a|y) - 4b^2S(b|x)\,S(b|y) \ , \tag{36.34}$$

where

$$S(a|x) = \mathrm{sinc}\,\frac{2\pi ax}{\lambda f} \ . \tag{36.35}$$

If $a \gg b$, then $\tilde{A}$ may be approximated as

$$\tilde{A}(x) = 4ab[\,S(a|x) + S(a|y)] \ . \tag{36.36}$$

The resultant intensity is then given by

$$\mathscr{I}(x) = 16a^2b^2[\,S(a|x) + S(a|y)]^2 \ , \tag{36.37}$$

which when normalized is Eq. (36.9) of this chapter. While the approximations of Eq. (36.37) are valid and the results obtained with them are in excellent agreement with experiments, one can see evidence in the Fourier transform plane of the existence of the terms that are omitted as a result of those approximations.

REFERENCES[c]

1. J. W. Goodman, "Synthetic Aperture Optics," *Progress in Optics,* E. Wolf, ed., Vol. 7, pp. 3–48, North Holland Publishing Co., Amsterdam (1969).

2. *Synthetic Aperture Optics,* Vols. 1 and 2, Woods Hole Summer Study, National Academy of Sciences, National Research Council Advisory Comittee to the AFSC (Aug. 1967). Available from the Defense Documentation Center as documents AD680806 (Vol. 1) and AD680797 (Vol. 2).

3. *Symposium on Synthetic Aperture Optics,* January 3, 1970, Martha W. Stockton, ed., Optical Sciences Center Technical Report #58, University of Arizona, Tucson (1970); see also G. O. Reynolds, "Techniques for aperture synthesis and reconstruction," same publication.

4. M. Ryle, "A new radio interferometer and its application for the observation of weak radio stars," *Proc. Roy Soc. London* A211, 351–375 (1952).

5. C. J. Drane and G. B. Parrent, Jr., "On the mapping of extended sources with nonlinear correlation antennas," *IRE Trans.* AP-10(2), 126–130 (1962).

6. M. J. Beran and G. B. Parrent, Jr., *The Theory of Partial Coherence,* Prentice Hall, Englewood Cliffs, N.J. (1964).

7. J. S. Wilczynski, "Double-objective telescope with common focus for astronomical observations from space," *J. Opt. Soc. Am.* 57(4), 579A (1967); see also "Double-beam telescope with common focus for astronomical observations II," *J. Opt. Soc. Am.* 57(11), 1415A (1967).

8. G. W. Stroke, "Image deblurring and aperture synthesis using *a posteriori* processing by Fourier-transform holography," *Optica Acta* 16(4), 401–422 (1969).

9. G. O. Reynolds and D. J. Cronin, "Imaging and optical synthetic apertures (Mills cross analog)," *J. Opt. Soc. Am.* 60(5), 634–640 (1970).

10. D. E. Yansen, G. O. Reynolds, and D. J. Cronin, "Optical synthetic aperture analog of two radio interferometers," *Optica Acta* 18(3), 167–180 (1971).

11. G. O. Reynolds and A. E. Smith, "Evaluation of segmented annular aperture camera," *J. Opt. Soc. Am.* 65(10), 1176A (1975).

12. B. Golay, "Point arrays having compact nonredundant autocorrelations," *J. Opt. Soc. Am.* 60(2), 272–273 (1971).

13. F. Scott, "Three-bar target modulation detectability," *Phot. Sci. Engr.* 10(1), 49–52 (1966).

14. R. Williams, "A re-examination of resolution prediction from lens MTF's and emulsion thresholds," *Phot. Sci. Engr.* 13(5), 252–261 (1969).

15. J. B. DeVelis, Y. M. Hong, and G. O. Reynolds, "Review of noise reduction techniques in coherent optical processing systems," in *Coherent Optical Processing,* H. J. Caulfield, ed., Proc. SPIE 52, 55–81 (1975); reprinted here as Chapters 20 and 21.

16. G. O. Reynolds, D. E. Yansen, D. J. Cronin, and A. E. Smith, "Assessment of alignment tolerances in optical synthetic apertures," *J. Opt. Soc. Am.* 62(5), 743A (1972).

17. R. B. Hooker, "The effects of aberration in synthetic aperture systems," PhD Dissertation, University of Arizona (1974).

18. D. A. Markle, in *Synthetic Aperture Optics,* Vol. 2, Woods Hole Summer Study, National Academy of Sciences, National Research Council Advisory Comittee to the AFSC (Aug. 1967). Available from the Defense Documentation Center as document AD680797.

19. R. Crane, "Interference phase measurement," *Appl. Opt.* 8(3), 538–542 (1969).

20. J. W. Hardy, "Active optics: a new technology for the control of light," *Proc. IEEE* 66(6), 651–697 (1978).

21. J. A. Jamieson, "Cross coherent processing for spatial spectral interferometry," in *Modern Utilization of Infrared Technology IV,* I. J. Spiro, ed., Proc. SPIE 156, 49–61 (1978).

---

[c]In addition to those colleagues referenced, the following deserve special mention for their contributions, including photographs, to various portions of the work reported here: D. J. Cronin, J. P. Fallon, D. Grey, A. Ho, P. F. Kellen, R. Remillard, D. A. Servaes, A. Silvestri, E. Sklar, A. E. Smith, O. E. Toler, A. Walters, D. E. Yansen, and J. L. Zuckerman.

22. J. A. Jamieson, "Spatial spectral interferometry," in *Proc. Joint Strategic Sciences Mtg.*, San Diego, California, September 13–15, 1978, AIAA.
23. J. A. Jamieson, "Imaging apparatus including spatial-spectral interferometer," U.S. Patent #4,136,954 (Jan. 30, 1979).
24. H. J. Robertson and G. T. Volpe, "Evaluation of a real-time figure control system for a spaceborne telescope mirror," *J. Opt. Soc. Am.* 62(11), 1339A (1972).
25. A. Maréchal, "Etude des effets combines de la diffraction et des aberrations geometriques sur l'image d'un point lumineux," *Rev. d'Optique* 26(9), 257–277 (1947).
26. M. Born and E. Wolf, *Principles of Optics,* 2nd ed., Pergamon Press, New York (1974).

37 Partially Filled, Synthetic Aperture Imaging Systems: Coherent Illumination*

37.1 APERTURE SYNTHESIS WITH COHERENT ILLUMINATION

This chapter continues the discussion started in Chapter 36 on synthetic aperture imaging systems, but the emphasis here is on coherent rather than incoherent illumination.

37.1.1 Partially Filled Apertures

In Eq. (36.3), we showed that partially filled apertures when illuminated with coherent light do not synthesize because of the nonlinear nature of coherent imaging systems. This lack of synthesis was verified experimentally with laser illumination.[1] The experiments were carried out for the partially filled Mills cross geometry but the results are equally applicable to other geometries as well. To achieve partially filled aperture synthesis with coherent illumination, the coherence of the radiation must be destroyed. Three conditions were found that satisfy this requirement:

1. when a focused point is scanned over the object

2. when a moving diffuser is placed in the coherent illuminating beam

3. when an object has a roughness greater than the coherence length of the illumination.

Figure 37.1 shows the results of imaging with a cross aperture for the case of laser light transilluminating the object. Figure 37.1(a) shows the image obtained with collimated laser light; (b) shows the image obtained when a stationary ground glass was placed between the laser beam and the object; and (c) shows the image obtained when a moving ground glass was inserted between the illuminating beam and the object. As anticipated, Fig. 37.1(a) shows the beam transmitted through the cross arms to be much stronger than the components diffracted by

*This chapter is a continuation of the article reprinted with permission in Chapter 36: "A Review of Partially-Filled, Synthetic-Aperture, Imaging Systems," by G. O. Reynolds, in *Infrared, Adaptive, and Synthetic Aperture Optical Systems*, R. B. Johnson, W. L. Wolfe, J. S. Fender, eds., SPIE Proc. 643, 141–179 (1986).

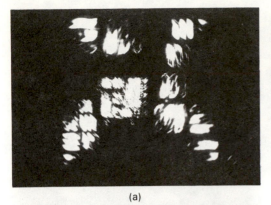

(a)

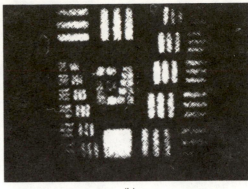

(b)

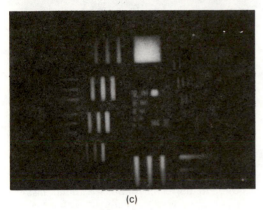

(c)

Fig. 37.1 Transilluminated objects: (a) collimated laser illumination, (b) laser light through stationary ground glass, and (c) laser light through moving ground glass. (Courtesy of D. E. Yansen and A. M. Silvestri.)

the transparency, causing the strong lines in the direction of the cross arms. Use of the stationary ground glass scatters the light, so that the aperture collects light from all sections of the transparency, resulting in Fig. 37.1(b), which exhibits more even illumination. Finally, the moving ground glass destroys spatial coherence properties and thus simulates incoherent illumination. Figure 37.1(c) shows a marked improvement over (a) and (b).

Spatial filtering was attempted on some of the pictures to improve image quality. The transilluminated object of Fig. 37.1(c) with moving ground glass was filtered and the results are shown in Fig. 37.2(a); here there is some noticeable image enhancement. On the other hand, the transilluminated object with no ground glass [Fig. 37.1(a)] was also filtered, and the result is shown in Fig. 37.2(b). Figure 37.2(c) shows the results of filtering the image formed with ground glass in the beam [Fig. 37.1(b)]. In both cases [Figs. 37.2(b) and (c)], neither the filtered image nor the unfiltered image represents a good reproduction of the original object. This demonstrates the nonsynthesis of partially filled synthetic apertures with laser illumination. The results of the laser illumination experiments are summarized in Table 37.I.

37.1.2 Homodyned Aperture Synthesis with Laser Illumination

37.1.2.1 Increased Resolution by Aperture Synthesis

An optical synthetic aperture with laser illumination (analogous to side-looking radar) requires a local oscillator to store the amplitude and phase of the diffracted field as a modulation on a carrier. A method in which the local oscillator

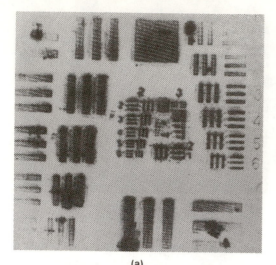

(a)

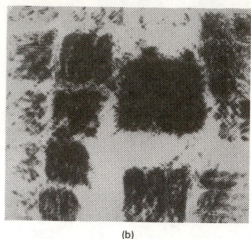

(b)

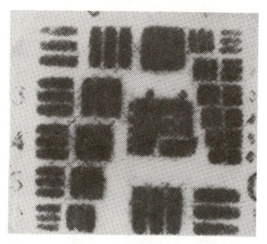

(c)

Fig. 37.2 Filtered images of transilluminated objects: (a) object and moving ground glass of Fig. 37.1(c), (b) object and no ground glass of Fig. 37.1(a), and (c) object and stationary ground glass of Fig. 37.1(b). (Courtesy of P. F. Kellen, D. E. Yansen, and J. L. Zuckerman.)

is replaced by an apertured interferometer achieves this goal. Referring to Fig. 37.3, a specular object fully illuminated by a coherent field (laser light) is shown. The angular spectrum of light from the object exists at the ξ plane. A grating is placed at the ξ plane and, for simplicity, a fiducial mark is placed at the edge of the grating.

Part of the angular spectrum of the object $\psi(\xi)$ is then passed through the

TABLE 37.1
Results of Laser Illumination Experiments with the Cross Aperture

Type of Illumination	$\lambda(\text{Å})$	Coherence in the Image
Collimated transillumination	6328	Highly coherent
Stationary diffuser transillumination	6328	Highly coherent
Reflection off diffuse objects	6328	Moderately coherent
Reflection off rough surfaces	5145 and 4400	Partially coherent
Moving diffuser transillumination	6328	Effectively incoherent
Scanning point source	6328	Effectively incoherent

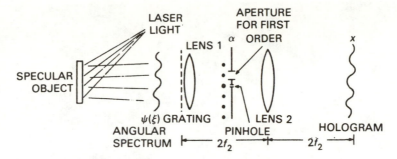

Fig. 37.3 Schematic for aperture synthesis with coherent light. (Courtesy of P. F. Kellen, D. E. Yansen, and J. L. Zuckerman.)

grating and lens L_1. A spatial filter is placed in the Fourier plane of lens L_1. This spatial filter consists of a small hole on axis, which acts as the local oscillator, and a larger hole passing the entire first order. (Note that rather than mixing the diffracted field with a plane wave reference as in normal holography, the reference source is derived from the same field with this interferometric technique.) A hologram is formed in the **x** plane.

That part of the angular spectrum of the object which is passed by the system is stored on the hologram. In addition, the image of a fiducial mark on the grating will also appear on the hologram. The hologram formation system is then displaced in x to record another portion of the angular spectrum of the object. Again, the hologram has on it an image of the fiducial mark.

If the two holograms are placed side by side, aligned by overlapping the fiducial marks to within a fraction of a fringe shift, and illuminated simultaneously by a larger laser beam as shown in Fig. 37.4, the reconstruction in the side order image is the sum of the two angular spectra stored on the two holograms. If the two pieces of the angular spectrum of the object are contiguous, we expect to double the object resolution in the direction of synthesis.

Preliminary experimental results bear out this contention as shown in Figs. 37.5(a) and (b). Figure 37.5(a) shows a reconstructed hologram made with a homodyned grating interferometer. The vertical three-bar resolution in this image is around group 1-6 and the horizontal three-bar resolution is around group 2-6. Figure 37.5(b) shows the reconstructed image obtained when three such holograms are taken sequentially, placed side by side, and reconstructed. Theoretically, we would expect to triple the resolution in the direction of synthesis. In Fig. 37.5(b), we can observe group 3-4 or 3-5 in the vertical direction and group 2-6 in the horizontal direction. This corresponds to a resolution increase

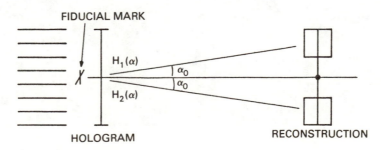

Fig. 37.4 Synthesized hologram reconstruction system (Courtesy of P. F. Kellen, D. E. Yansen, and J. L. Zuckerman.).

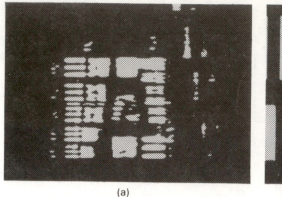
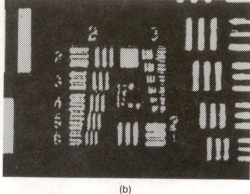

(a)   (b)

Fig. 37.5 Reconstructed images from (a) a single homodyned hologram and (b) three homo-dyned holograms. (Courtesy of P. F. Kellen, D. E. Yansen, and J. L. Zuckerman.)

by a factor of 2.7 or 2.8 in the direction of synthesis, which is close to the theoretical prediction. This result can obviously be extended to a larger number of holograms. Similar results are predicted in the literature.[2,3]

37.1.2.2   Introduction of Turbulent Media[4]

In the laser synthetic aperture experiments discussed above, in which the reference field was created by a self-beating or homodyning technique (a grating over the aperture), a very interesting phenomenon was observed. When a time exposure was used to record the hologram, the noise produced by the random media was averaged out and the quality of the photograph of the object improved. This means that the time-averaged homodyned synthetic aperture is insensitive to random medium fluctuations. This effect could be significant in laser illuminated photography.

When a time-varying medium was introduced in front of the grating in the system shown in Fig. 37.3, the random phase caused the diffraction orders in the **x** plane to move in synchronization. The effect of self-synchronization of the hologram is to wash out noise, but it does not reduce the interference effects. Thus, when the hologram is reconstructed, the pictures are of improved quality when the random medium is present. This is shown in Fig. 37.6, where (a) is a

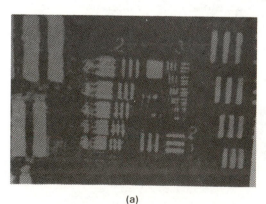
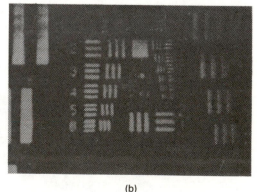

(a)   (b)

Fig. 37.6 A reconstructed hologram (a) with no turbulence and (b) with turbulence. (Courtesy of P. F. Kellen, D. E. Yansen, and J. L. Zuckerman.)

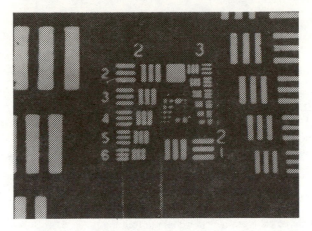

Fig. 37.7 An incoherent photograph made through an imaging system comparable to the one in Fig. 37.3. (Courtesy of P. F. Kellen, D. E. Yansen, and J. L. Zuckerman.)

picture of the reconstructed image resulting from a hologram made in the system of Fig. 37.3 with no turbulence and (b) represents the reconstructed image resulting from a hologram made in the presence of a random medium (turbid air above a hot plate). Note that the three-bar resolutions of Figs. 37.6(a) and (b) are comparable; however, the speckle noise of Fig. 37.6(b) is considerably less. Figure 37.7 is an incoherent photograph made through a comparable imaging system when the hot plate was placed in front of the lens. Notice that the three-bar resolution is less than that observed in both Figs. 37.6(a) and (b).

These experiments demonstrate a very significant effect—the image quality is improved and resolution is not degraded when homodyned coherent images are taken through a random medium. The same effect can also be expected for small field incoherent photography such as astronomy. This effect is further discussed in the literature.[4]

37.1.3   Dynamic Coherent Optical System

The dynamic coherent optical system[5,6] (DCOS) was developed to process optically the photographs made in the annular synthetic aperture camera discussed in Chapter 36, Sec. 36.3.2.6. It has certain distinct advantages over any other coherent imaging system in that it has twice the normal coherent bandwidth of the diffraction-limited aperture and is completely free of the usual coherent noise and edge ringing. Although it has the bandwidth of the incoherent aperture, its transfer function does not taper off at intermediate- and high-spatial frequencies. Its disadvantages as a spatial filtering method arise from the fact that a moving source or rotating prism causes the spatial frequency plane to rotate, which makes filtering for anything but the dc term extremely difficult.

The applications to which the system has been successfully applied are noise reduction in a coherent system,[5] bandwidth doubling of a coherent system,[5] Zernike imaging of a phase target,[5,6] and dark field imaging.[6] Its principal applications may be in noise reduction, discussed in Chapter 38, and in phase microscopy, as discussed in Chapter 35, rather than in the spatial filtering for which it was conceived. It appears to be an ideal system for photoreproduction by projection printing with laser light (see Chapter 38).

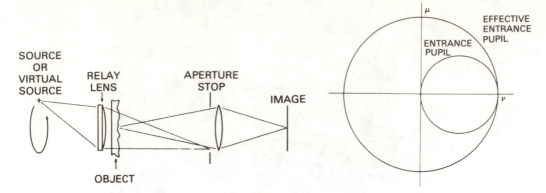

Fig. 37.8 The dynamic coherent optical system (a rotating prism causes the object spectrum to rotate in the plane of the aperture stop) (after Ref. 5).

Fig. 37.9 Superposition of the aperture stop and the spatial frequency plane showing the effective aperture stop (after Ref. 5).

The optical system illustrated in Fig. 37.8 shows the principal components. A coherent source at the focal point of a condensing lens illuminates the system. The prism, or other deviating element, deviates the beam so that the spectrum of the object falls with its central image or dc term close to the edge of the pupil in the Fourier plane (see Fig. 37.9). An alternate system using a rotating light source instead of the virtual rotation produced by the prism has been constructed for a coherent stereoscope.[5] As the prism rotates about the optic axis, the dc term follows a trajectory at the periphery of the physical pupil. The image is recorded as a photograph and is the time average of the spatial frequency components present in the pupil. Since the dc term is present at all times, it is the strongest term and behaves as a reference beam to preserve the phases of the two sides of the angular spectrum of the object, which pass through the optical aperture at different times when the prism is in its θ and $\theta + \pi$ positions.[6] The system transfer function is shown in Fig. 37.10 where it is compared with the coherent, square, and circular pupils of the same diameter.

Figure 37.11 shows a typical set of photographs reproduced through the dynamic coherent system. Figure 37.11(a) exhibits the typical tapering off of the transfer function with incoherent illumination at high spatial frequencies causing

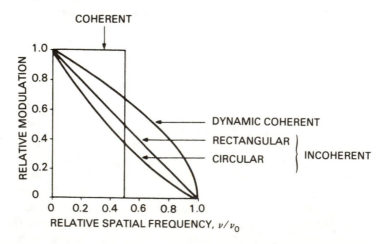

Fig. 37.10 The modulation transfer function (MTF) of the dynamic coherent system with other coherent and incoherent transfer functions for comparison (after Ref. 5).

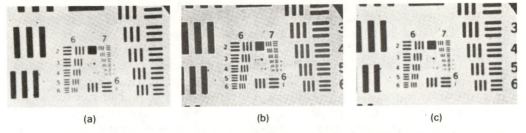

(a)   (b)   (c)

Fig. 37.11 For comparison, photomicrographs of images reproduced in the same optical imaging system with (a) incoherent illumination, (b) coherent illumination, and (c) dynamic coherent illumination (after Ref. 5).

a loss in contrast. Figure 37.11(c) is the dynamic coherent reproduction with the dc term rotating close to the pupil edge. Two characteristics should be specifically noted in this photograph: the freedom from coherent noise and the increase in contrast, which gives an apparent increase in resolution over the incoherent system. The coherent image through the system is shown in Fig. 37.11(b). The lenses and setup were identical in cases (a) and (c) and the photographic processing was the same.

The resolution doubling capability of the system[6] is shown in the dark field mode in Fig. 37.12. In this case, the prism angle was chosen such that the image of the laser (zero frequency) passed just outside the aperture of the imaging lens. When the laser was stationary, the dark field phase image [Fig. 37.12(a)] was observed. Its resolution cutoff is between groups 6-3 and 6-4. The results of the dynamic coherent phase imaging mode are seen in Fig. 37.12(b) where a resolu-

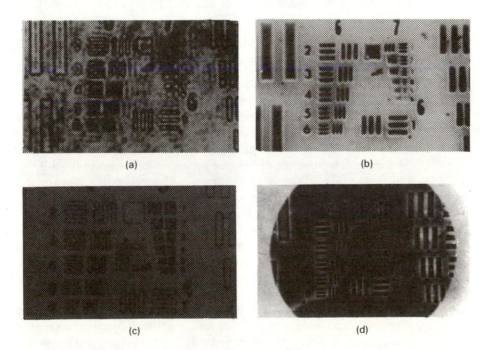

(a)   (b)

(c)   (d)

Fig. 37.12 Images obtained with the DCOS: (a) Coherent image of a target without a prism in the dark field mode, (b) dynamic coherent phase image obtained in the dark field mode, (c) dark field image of a resolution target that was produced by stopping down the entrance pupil of the imaging lens to just exclude the annular trajectory of the source image, (d) phase image reproduced by the dynamic coherent system with a $\lambda/4$ phase annulus over the dc trajectory (after Ref. 6).

tion cutoff between groups 7-3 to 7-4 was observed. Figure 37.12(b) shows a resolution doubling from (a), illustrating the coherent aperture synthesis that occurs with the DCOS.

The continuous-tone capability of the system achieved by rotating the point image onto a $\lambda/4$ phase annulus placed at the edge of the limiting aperture is demonstrated in Figs. 37.12(c) and (d) where the contrast and dark field configurations were experimentally compared. This demonstration indicates that the dark field configuration of (c) highlights the edges, whereas the phase contrast configuration of (d) displays a continuous intensity variation of the phase object.

37.2   MEASUREMENT OF $\Gamma_{12}(0)$ WITH AN OPTICAL SYNTHETIC APERTURE

Since the principle of partially filled optical synthetic apertures is analogous to the measurements made in radio interferometry, the distribution of radiation in the lens aperture (cross) of Fig. 36.3 from Chapter 36 (van Cittert-Zernike theorem) should be the mutual intensity function arising from the distant object.[7,8] When the object is a distant source (for example, a star), then this mutual intensity function should be observed in the transform plane of Fig. 36.7(b) when the image of the source is recorded in Fig. 36.3. This theory was described in Chapters 10 and 11 and examples of the mutual intensity functions arising from various sized circular and rectangular incoherent apertures are shown in Fig. 37.13 when measured with a cross aperture. Obviously, if the source were asymmetric, then an interferometer measurement in the transform plane of Fig. 36.7 would show the phase of $\Gamma_{12}(0)$ as a fringe shift.

37.3   SUPER-RESOLVING PUPIL FUNCTIONS

It was observed in Chapter 36, Fig. 36.32, that the resolution obtained with the narrow annulus appeared better than that obtained with the equivalent full aperture. Further laboratory tests[9] with three-bar targets and narrow annular apertures ($\epsilon = 0.9$) indicated that the resolution cutoff was improved by one or two groups (15 to 30%). This improvement is due to the well-known high-frequency enhancement of annular apertures as shown in the annular transfer functions of Fig. 37.14. Also, as shown in Chapter 36, the thin annulus causes a narrowing of the central lobe of the impulse by 65 to 70%.

If we consider the aerial image of the resolution target convolved with the spread function of the annulus, targets with a narrow spatial extent that do not have a strong contribution from the higher order side lobes of the annulus will be resolved by the annulus even when they are not resolved by a full aperture of the same diameter. This rather simple concept of a narrow central beam to the spread function and a target or object of narrow spatial extent leads one to consider the idea of super-resolution. It follows that a super-resolving pupil that resolves a three-bar target will not necessarily resolve a fifteen-bar target except at the ends, and that resolution of two points may become ambiguous when a third point is introduced into the object field.

Pupil functions having very narrow impulse responses were proposed and analyzed by Toraldo di Francia[10] and further investigated by Frieden.[11] These

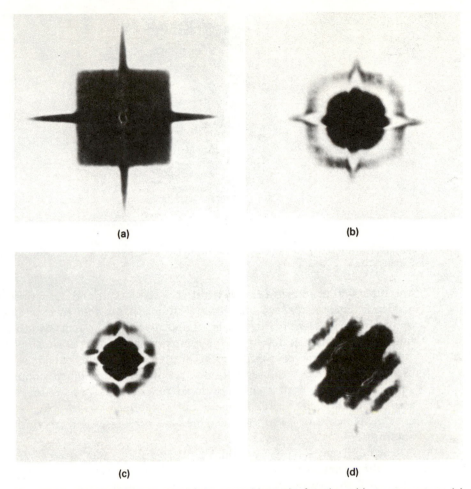

Fig. 37.13 Analog measurements of the mutual intensity function with cross aperture: (a) filtered cross MTF, (b) $\Gamma_{12}(0)$ for a 50-μm pinhole, (c) $\Gamma_{12}(0)$ for a 78-μm pinhole, and (d) $\Gamma_{12}(0)$ for a 30-$\times$80-μm rectangular aperture (after Ref. 7).

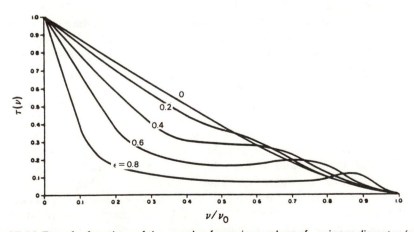

Fig. 37.14 Transfer functions of the annulus for various values of ϵ = inner diameter/outer diameter. (Courtesy of A. E. Smith and D. E. Yansen.)

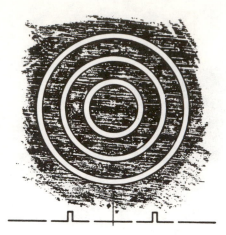

Fig. 37.15 The three-element super-resolving pupil. (Courtesy of A. E. Smith and D. E. Yansen.)

apertures were not experimentally realized by the authors and, thus, remained of academic interest. Yansen and Smith[12] found it possible to experimentally approximate super-resolving pupils for optical imaging systems by utilizing the fabrication techniques developed for complex spatial filter fabrication.[13]

In Ref. 12, a series of annular rings with alternating phase, called the *Toraldo di Francia aperture*,[10] showed considerable promise of practical consequences since it produced both a smaller spread function than the normal circular aperture of the same overall diameter and a significant increase in focal tolerance. This increased focal tolerance has also been discussed for thin annular apertures.[14]

The pupil function fabricated and tested by Yansen and Smith consisted of a set of three annular rings as shown in Fig. 37.15. The radii of the rings and the value of ϵ, the ratio of inner to outer diameters, were chosen to modulate the values of their individual impulse responses. The phase step of π on the middle ring gave its impulse a negative sign with respect to the other two.

The spread function, or intensity impulse response, for the three-element pupil is shown in Fig. 37.16(a), and for reference it is contrasted with the corresponding full aperture (an aperture of the same diameter as the outer ring) and a narrow annulus of the same diameter. Figure 37.16(b) gives the amplitude impulse response of the same pupils. The effect of the narrow annulus is to reduce the diameter of the central core while increasing the energy in the side lobes. The three-element super-resolving aperture maintains the narrow width of the annular central intensity, but it suppresses the irradiance in the first side lobe to a magnitude approximating the first side lobe of the full aperture. It does this at the cost of a tremendous increase in the second side lobe—up to about 1.5 times the central beam intensity. Figure 37.17 shows the experimental spread functions for a full aperture and a three-element Toraldo aperture of the same diameter illustrating that the experimental results are in qualitative agreement with the theoretical predictions of Fig. 37.16. Such apertures could have significant implications for high-resolution small field imaging systems (telescopes) and scanning systems such as microdensitometers and high-resolution laser scanning systems.

These experimental results could be extended by introducing more elements into the annular array, which would have the effect of maintaining the relatively

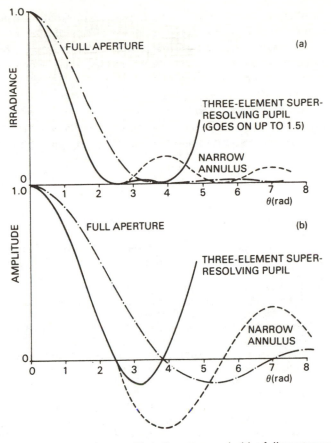

Fig. 37.16 The Toraldo di Francia spread function compared with a full aperture and a narrow annulus (normalized). (Courtesy of A. E. Smith and D. E. Yansen.)

narrow central beam, further broadening the area around the central beam, and increasing the outer side lobes to higher values as predicted by Frieden.[11]

37.4   CONCLUSIONS

In Chapters 36 and 37, we reviewed the concept of partially filled, synthetic aperture optics and showed that it is a viable method for aperture synthesis with

Fig. 37.17 Spread functions for (a) full aperture and (b) equivalent three-element aperture. (Courtesy of A. E. Smith and D. E. Yansen.)

incoherent illumination (including laser illumination when either its spatial or temporal coherence is destroyed). The effects of filtering to recover the synthesized image from the bias image were demonstrated. The effects of alignment tolerances where the apertures are made from segments of lenses and mirrors were analyzed and the results shown to agree with experiments. Alternate methods for aperture synthesis with laser illumination were also briefly considered as was a method for measuring the mutual intensity function of a primary incoherent source with a cross aperture. Finally, an experimental demonstration of increased resolution with a super-resolving pupil was shown.

REFERENCES

1. D. E. Yansen, G. O. Reynolds, and A. M. Silvestri, "Imaging properties of optical synthetic apertures under laser illumination," *J. Opt. Soc. Am.* 61(5), 657–658A (1971).
2. J. W. Goodman, "Synthetic Aperture Optics," *Progress in Optics,* E. Wolf, ed., Vol. 7, pp. 3–48, North Holland Publishing Co., Amsterdam (1969).
3. J. W. Goodman and R. W. Lawrence, "Digital image formation from electronically detected holograms," *Appl. Phys. Lett.* 11(3), 77–79 (1967).
4. G. O. Reynolds, D. E. Yansen, and J. L. Zuckerman, "Time varying random media compensation with holography," in *Developments in Holography,* B. J. Thompson, J. B. DeVelis, eds., Proc. SPIE 25, 183–190 (1971).
5. D. J. Cronin and A. E. Smith, "Dynamic coherent optical system," *Opt. Eng.* 12(2), 50–55 (1973).
6. D. J. Cronin, J. B. DeVelis, and G. O. Reynolds, "Equivalence of annular source and dynamic coherent phase contrast viewing systems," *Opt. Eng.* 15(3), 276–278 (1976).
7. *Symposium on Synthetic Aperture Optics,* January 3, 1970, Martha W. Stockton, ed., Optical Sciences Center Technical Report #58, University of Arizona, Tucson (1970); see also G. O. Reynolds, "Techniques for aperture synthesis and reconstruction," same publication.
8. G. O. Reynolds and D. J. Cronin, "Analog measurement of mutual intensity function using phased arrays," *J. Opt. Soc. Am.* 59(4), 475A (1969).
9. G. O. Reynolds, "Evaluation of segmented annular aperture camera," *J. Opt. Soc. Am.* 65(10), 1176A (1975).
10. G. Toraldo di Francia, "Super-gain antennas and resolving power," *Nuovo Cimento* Supplement 9(3), 426–438 (1952).
11. B. R. Frieden, "On arbitrarily perfect imagery with a finite aperture," *Optica Acta* 16(6), 795–807 (1969).
12. A. E. Smith and D. E. Yansen, "Experimental study of super-resolving pupil function," *J. Opt. Soc. Am.* 61(5), 688A (1971).
13. A. Maréchal and P. Croce, "Un filtre de frequences spatiales pour l'amelioration du contraste des images optiques," *Compt. Rend.* 237, 607–609 (1953).
14. C. Varamit and G. Indebetouw, "Imaging properties of defocussed partitioned pupils," *J. Opt. Soc. Am.* A-2(6), 799–803 (1985).

38   Parametric Design of a Conceptual High-Resolution Optical Lithographic Printer *

38.1   INTRODUCTION

As an example illustrating many of the concepts discussed and developed in this book, we will present a conceptual design for a high-resolution optical lithographic printer having a 0.5-μm linewidth capability. This design entails performing a trade-off analysis of optical system parameters such as wavelength, exposure time, resolution, magnification, coherence, and field of view. The specifications and the trade-off analysis strongly influence the design and choice of hardware for use as the source, condenser, lens, and photoresist (detector), as well as the focal and alignment subsystems. The design is accomplished by considering the system effects discussed elsewhere in this book. These include effects due to coherence nonlinearity (Chapters 11, 17, and 18), a dynamic coherent condenser (Chapters 35 and 37), coherent noise averaging (Chapters 21 and 35), three-beam interferometry (Chapter 22), recording media effects (Chapter 20), and spatial filtering (Chapters 29, 30, 31, and 37). The consideration of these various parameters leads to a design that appears capable of achieving the 0.5-μm linewidth goal. Recent results published in the literature indicate that this goal will be achieved in the integrated circuit industry in the next few years.[1-3] (Note: Since the publication of the referenced article, this resolution goal has been achieved.)

38.2   BACKGROUND

There are currently two concepts in optical projection lithography being utilized by industry. The first concept consists of an all-reflective optical projection system[4] that images a narrow arc of the mask onto the wafer. Simultaneous scanning of the object and image planes (in register) past the stationary optical system creates a two-dimensional image of the mask on the wafer. This system, which is designed to maintain a constant optical path length between the object

*Excerpted with permission from "A Concept for a High-Resolution Optical Lithographic System for Producing One-Half Micron Line Widths," by George O. Reynolds, in *Optical Microlithography V,* H. L. Stover, ed., Proc. SPIE 633, 228–238 (1986).

and image planes, is subject to inaccuracies caused by small variations in wafer thickness and flatness variations in the wafer, chuck, mask, and mirrors. These mechanical problems, coupled with problems producing accurate, small line-width 1:1 masks, cleanliness problems, and diffraction problems associated with imaging these narrow lines, have limited the 1:1 printers to production linewidths of 1.25 μm and larger. Linewidths less than 2 μm are achieved in these systems by introducing short wavelength ultraviolet (UV) sources and are ultimately limited by mechanical run-out problems.

The other concept utilized in optical lithography is the step-and-repeat system.[5–8] This system utilizes a reduction ratio in the production optics (commonly 10:1) and a precise mechanical stage, controlled by a laser interferometer, to replicate the mask onto the wafer. This system relieves the mask fabrication limitation of the 1:1 system and removes the run-out problem (accumulative mechanical error). Since the field of view of this optical system is smaller than the corresponding 1:1 system, smaller linewidths can be accurately reproduced. This system has demonstrated 1- to 2-μm linewidth resolution in a 10:1 production system[5] and linewidths as low as 0.75 μm have been demonstrated.[7] In essence, the 10:1 reduction system reproduces smaller linewidths by relaxing the mechanical error within the lens field of view and utilizing the lens demagnification to reduce the deleterious diffraction (coherence) effects.

To push the capability of optical lithography to 0.5-μm linewidths or less, a new approach is needed. One concept for achieving this limit is to introduce deep UV wavelengths into a step-and-repeat 10:1 reduction system. Since current 10:1 reduction systems are refractive, this higher resolution capability cannot be achieved merely by changing the source wavelength as it was in the 1:1 system.[9] Instead, an entirely new optical design is needed. Improvements have been made in this direction.[10]

Preliminary calculations indicate that it is possible to reach a 0.5-μm linewidth in lithography by using a purely optical system. This optical approach,[11] which combines the best features of the 1:1 and the 10:1 systems, would consist of a deep UV source (2200 to 2500 Å), an optical system tuned to the source wavelength, a 10:1 stepper to avoid the nonlinearities due to the coherence of the radiation, and a photoresist capable of responding to the UV radiation. Focus and alignment systems better than those available in production systems today would also be necessary.

38.3   PROPOSED SYSTEM AND CRITICAL ISSUES

Figure 38.1 shows a block diagram of the proposed system.[11] It entails a deep UV illumination source and a condenser assembly for illuminating the circuit design mask. The optical projection system images the mask at a precise reduction ratio (10:1 was selected for design purposes) onto a photoresist coated on an IC wafer. The mechanical stage, electronic controls, focus control, and alignment subsystems are all essential components of the final system.

In this system the very deep UV source (2000 Å) lowers the linewidth limit directly with wavelength, a reflective optical system removes chromatic aberrations associated with refractive systems, and the 10:1 reduction ratio removes the nonlinear coherence effects characteristic of the 1:1 systems and reduces the limit of linearity to 0.5 μm.

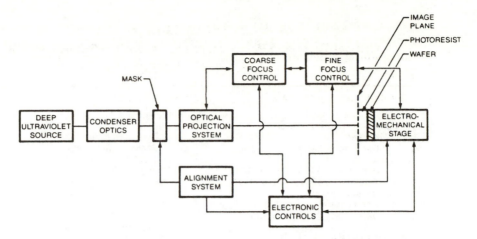

Fig. 38.1 Block diagram of proposed submicron optical projection printer (modified from Ref. 11).

The desired specifications and needs for this proposed system are:

1. a *reflective* optical system having 0.5-μm distortion-free (±250 Å) resolution over a 15-mm field of view[a]

2. a 10×10 array of 1-cm² exposures to fill a 4-in. wafer

3. a 6-in. wafer capability should be designed into the mechanical subsystem to allow for the future growth projected by the industry[b]

4. a UV source wavelength between 2000 and 2500 Å [c]

5. a photoresist having adequate sensitivity at the source wavelength

6. a focal tolerance of about 500 Å (1/20 of a micron)

7. an alignment requirement of 500 Å

8. a very precise mechanical stage, electronic controls, and a very clean environment.

The critical technical issues associated with these various subsystems are listed in Table 38.I and discussed here.

Source: The source is critical from the point of view that enough energy is needed at the UV wavelength of interest to expose the photoresist. Some of the laser sources and conventional sources that might be used are shown in Table 38.I. The conventional sources do not have the energy output of lasers. Laser sources can provide the energy needed at the photoresist. However, other problems with lasers, such as coherence effects, could require the development of a high-power deep UV source.

Reduction Ratio: The reduction ratio of the optics need not be 10:1; a 4:1 or 5:1 reduction ratio may be adequate. Canon has a system that now operates at a 4:1 reduction ratio.[8] Coherence arguments indicate that the 10:1 ratio is better.[12] The optics should be nearly linear down to the linewidth limit. The linearity

---

[a]This is an attractive optical design goal since it is wavelength independent.

[b]One advantage of the stepper is its ability to correct for run-out distortion (accumulated mechanical scan error) before each exposure.

[c]There are two possible sources to consider: an excimer laser and a gas discharge lamp.

TABLE 38.I
Critical Technical Parameters and Issues for Subsystems in
0.5-μm Optical Lithography

SOURCE AND CONDENSER	PHOTORESIST
Type (lamp or laser)	UV sensitivity
Condenser configuration	Speed
Kohler	Wavelength limit
Dynamic coherent	PMMA; 180 to 250 nm
Laser sources	PMIPIC; 180 to 330 nm
ArF; laser 193 nm, 10 W, 90 pps	Multiple reflections
ArF; laser 249 nm, 18 W, 100 pps	Multilayer configuration
Thermal sources	Tune thickness to lens NA
Xenon mercury arc; 205 to 255 nm	FOCUS (z)
Deuterium; 205 to 255 nm	Sensitivity, 500 Å
	Three-beam interferometer
REDUCTION RATIO	ALIGNMENT (x, y)
10:1	Sensitivity, 500 Å
Limit of linearity <0.5 μm	Image processing, correlation
OPTICS	LINEWIDTH MEASUREMENTS
All reflective	0.5-μm accuracy needed
MTF/resolution limit	SEMS more than adequate in research
	Production techniques must be developed

requirement is driven by coherence considerations. We are trying to avoid the nonlinear effects inherent in partially coherent imaging systems as previously discussed in Chapters 11, 17, and 18. With 10:1 optics, this appears feasible.

Optics: Many parameters in the system could be relaxed if an all-reflective optical system can be achieved. Its modulation transfer function (MTF) has to have 60% contrast at the fundamental spatial frequency corresponding to the 0.5-μm linewidth, because of the high clipping nature of the photoresist.

Photoresist: In selecting a photoresist with UV sensitivity, attention must be given to its speed, wavelength, and sensitivity. The UV photoresists work in the 200- to 250-nm range. They behave like thin films and create multiple reflections that look like circuit lines, or little bridges, which must be eliminated. This can be done by using thin layer or tri-layer resists; i.e., the thickness of the resist must be tuned to the numerical aperture (NA) of the lens. Since the lens will have a very fine depth of focus, a thin resist will be needed.

Focus: A 0.5-μm depth of focus needs a focal accuracy of 500 Å. That accuracy is beyond the state-of-the-art of automatic focus measuring devices. Three-beam interferometers should be able to measure focus to this accuracy. We describe this device later in this chapter.

Alignment: The alignment system also has to have 500-Å accuracy. Image processing and correlation mechanisms have shown this potential in the laboratory.

Measurements: How can the 0.5-μm linewidths produced by this system be measured? In the research phase, the scanning electron microscopes are more than adequate. In the eventual production environment, this is not a trivial problem; i.e., with 0.5-μm lines over a 4- or 6-in. wafer, how can they be rapidly and accurately measured? Is visual assessment with a high-power microscope adequate? There are many unanswered questions in this area that are beyond the scope of this book.

These issues will now be addressed separately to illustrate a system that meets these desired specifications.

38.4   OPTICAL SUBSYSTEM CONSIDERATIONS

Due to its importance, the optics is the first subsystem to be analyzed. If the lens cannot be designed and fabricated, then no matter how good the mechanical subsystems are, the central part of the concept is missing. Currently, the 10:1 systems are refractive and they are optimized for the G line (436 nm) and the H lines (404 nm) of the mercury spectrum. They have a linear linewidth limit of ~1 μm. A lens tuned for the I line (365 nm) of mercury has been developed.[13] There has been a significant improvement in the space-bandwidth product of these 10:1 lenses in the last decade. A factor of nearly 2 in resolution at the same image field size has been achieved through good lens design.[10]

The linewidth performance of the 1:1 reflective systems has been reduced by lowering the wavelength. Limitations in linewidth have been attributed to the UV sources that are available, diffraction effects in the one-to-one design, and the run-out problem. Most thermal sources have limited energy in the UV spectrum. However, 1:1 reflective systems have nearly doubled resolution or halved their linewidth by lowering the wavelength and recoating the optics.[9] Both technologies are now approaching 1- to 1.25-μm linewidths in production systems. To date, the wavelength limit has also been driven by available photoresists. Photoresist manufacturers have been improving their photoresist sensitivities in the UV region.[14-16]

In the concept being proposed here, the reflective optical design will remove the wavelength restriction (chromatic aberration) of the system. A 10:1 system will reduce the linearity limit that is inherent from diffraction in the 1:1 system. The larger lines on the mask will reduce cleanliness problems in the production environment. The UV source gives an improvement in resolution directly with the wavelength; thus, by going from the blue to the deep UV, a factor of 2 in wavelength can be realized. Theoretically, it appears feasible to achieve slightly less than the desired 0.5-μm linewidth at a 60% contrast. A photoresist sensitivity matched to the source wavelength is also essential.

In this system, we are considering imaging objects only a few times larger than the wavelength. Therefore, we need high-resolution optical systems. Such systems inherently exhibit nonlinearity effects due to partial coherence.[12] The problem with a partially coherent imaging system is that the MTF becomes a function of the object being imaged. Thus, every time the object changes, the MTF changes. It is difficult to design a lens having a given performance if the performance is going to change as a function of the object that is presented to the lens. Thus, designing a 0.5-μm linewidth optical system requires avoiding or at least controlling these coherence effects.

In optical lithography, some degree of nonlinearity due to partial coherence has been utilized to achieve higher image contrast.[9,10] This works because the

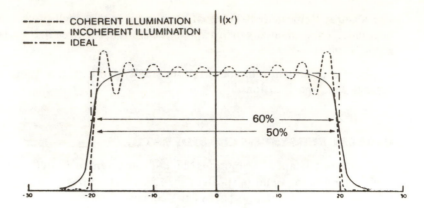

Fig. 38.2 Theoretical intensity distribution in the coherent and incoherent image of a slit (from Chapter 14) showing line narrowing with coherence at the 50 and 60% threshold points.

binary masks are imaged onto the photoresist. Photoresists are forgiving of this nonlinearity for two reasons:

1. The targets are binary, i.e., the circuit lines are either on or off. There is a tremendous advantage in designing an optical system if only on-off targets (as opposed to gray scale) have to be considered.

2. The threshold of the photoresist is very high (60%); i.e., photoresists have a very steep photographic gamma.

An additional advantage of using photoresists is that the 60% contrast criterion helps with linewidth narrowing as discussed in Chapter 15 and seen in Fig. 38.2. If the edges of a bar function are considered, then the 50% crossing point should define the ideal width from a linear system approach. When a 60% crossing point is utilized, some linewidth narrowing occurs because the edge shrinks. The high-contrast clipping action of the photoresist therefore helps make narrower linewidths. Ten and 15% linewidth narrowing should be achievable with a 60% clipping level. This phenomenon is currently helping the industry achieve narrow linewidths. If low-contrast targets were used, the linewidths would broaden.

The coherence nomograph of Chapter 18 was used to evaluate the optics of various lithographic systems. The results are shown in Table 38.II. The optimistic ($N=4$) and pessimistic ($N=10$) linearity limits were determined for the lens of Ref. 9 at blue and UV wavelengths. The linewidths were calculated at 0.5-μm wavelengths and scaled for the blue and the UV. The linewidths obtained were 3.6 μm (optimistic) and 9 μm (pessimistic) for the 0.454-μm wavelength of blue light. The measured linewidths are narrower than predicted by the calculation. This shows that the photoresists with the binary targets are not adversely affected by the partial coherence effects. Experimentally, the 1:1 imaging systems produce narrower linewidths than expected when partially coherent illumination is used.

The 10:1 systems agree quite well with the predicted linewidths. This means that the 10:1 systems are not as coherence limited as the 1:1 systems. Coherence is a diffraction phenomenon and the large lines of the 10:1 system will not diffract as much as the small lines used in the masks for the 1:1 system. The measured values, in each case, are near the optimistic values for reduction systems. This gives us confidence that the proposed No. 1 and proposed No. 2 systems will yield the experimentally predicted linewidths.

TABLE 38.II
Linewidth Linearity Limits for Various Photolithographic Lenses

Lens Configuration Existing	Wavelength (nm)	NA$_{obj}$	Linewidth Linearity Limit[a]		Experimentally Observed Linewidth (μm)
			Optimistic (μm)	Pessimistic (μm)	
1:1 (Ref. 9)	454	0.33	3.6	9	2 to 3
1:1 (Ref. 9)	248	0.33	1.95	4.9	1 to 1.25
1:1 (Ref. 17)	454	0.19	5.0	12.5	2 to 3
10:1[b] (Ref. 10)	436	0.28	0.94	2.4	1.18
10:1 (Ref. 10)	436	0.42	0.81	2.0	0.8
5:1 (Ref. 6)	405	0.35	0.86	2.2	0.88
10:1 (Ref. 7)	436	0.36	0.77	1.9	0.75
New Proposed No. 1 (10:1)	200	0.50	0.2	0.5	TBD[c]
Proposed No. 2 (10:1)	200	0.28	0.23	0.6	TBD

[a]Estimated from coherence nomograph (Chapter 18).
[b]Used in current 10:1 DSW system (Ref. 10).
[c]To be determined.

There may be some unknown photoresist limitations for deep UV lithography but 0.5-μm linewidths should be achievable. Encouraging experimental results have recently demonstrated that 0.5-μm linewidths can be recorded in photoresists with deep UV wavelengths.[18,19]

38.5   EXPOSURE SUBSYSTEM CONSIDERATIONS

In determining the source and resist requirements necessary to create a viable system using our optical concept, the following assumptions were made:

1. a positive resist would be used

2. a throughput of 60 wafer levels/h is necessary

3. 4-in.-diam wafers would be used

4. the field of view of the optical system is 10 mm × 10 mm

5. 100 exposures/wafer level, each with an exposure time of 0.25 to 0.50 s or less, is needed.

Available photoresists sensitive to the deep UV have a sensitivity of ~100 mJ/cm^2 (Refs. 16 and 20). With these resists, a power of 400 mW/cm^2 is needed at the resist plane to create an exposure in 0.25 s. A study of available sources indicated that current UV incoherent sources are not adequate to meet these needs. Therefore, sources richer in the deep UV by a factor of 10 to 50, e.g., cadmium arcs with a visible filter, need to be developed to satisfy the system assumptions stated above. However, a variety of UV laser sources exist that easily meet these requirements. Some of these sources are listed in Table 38.III. With laser sources, a dynamic coherent optical condenser system[22,23] can be used to increase the intermediate MTF values, simultaneously average coherent noise, and achieve the resolution capability of a normal incoherent system. Figure 38.3 shows the basic dynamic coherent optical system and Fig. 38.4 shows its MTF characteristics compared to the normal incoherent MTFs. Notice that

TABLE 38.III
Characteristics of UV Laser Sources*

Laser	Wavelength (nm)	Power	Type
Krypton	312 to 327	4, 5 W	Continuous wave
Ne-Ar	318	350 mW	Continuous wave
Excimer NF$_3$	248	2.4 W	120 mJ/pulse; 20 pulses/s
Argon	231	3.0 W	5 mJ/pulse; 600 pulses/s
Krypton	219	1.2 W	2 mJ/pulse; 600 pulses/s
ArF	193	10.0 W	111 mJ/pulse; 90 pulses/s

*From Ref. 21.

the filtered dynamic coherent MTF has the same bandwidth as the circular MTF of the lens. This is because, with the spinning prism,[d] the laser source at any instant seems to pass one side of the image spectrum through the imaging. One-half of the rotation period later, it passes the other side of the spectrum. Each single sideband has twice the coherent bandwidth so that their sum is the incoherent bandwidth. The dynamic coherent (laser) condenser system gives the full bandwidth image of an incoherent system with a transfer function having higher contrast than the incoherent system, as shown in Fig. 38.4 (Refs. 22 and 23 and Chapter 34). The modulation value of a given spatial frequency with a dynamic coherent condenser is proportional to the amount of time that the frequency component is in the aperture. Since the dc spot is always on the edge of the lens aperture as the prism rotates, its relative MTF value is the circumference of the lens. The higher spatial frequencies transverse lesser arc lengths. The calculated value of arc length versus frequency gives the MTF of the system when normalized by the circumference (characteristic of the zero-frequency component)[22] of the lens.

Thus, in summary, this dynamic coherent laser condenser design could be used advantageously in UV photolithography to create higher midrange values of MTF at the 50 to 60% contrast level. This condenser would also relieve the strain on the optical design, utilize the high power of available UV lasers, remove coherent noise artifacts, and render the exposure time much less than the anticipated 0.25 s. For these reasons, we chose it for our proposed system.

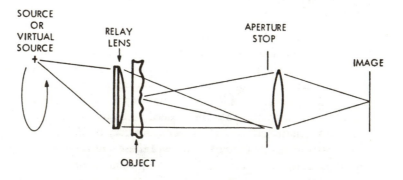

Fig. 38.3 The dynamic coherent optical system (after Ref. 22).

---

[d]The virtual source in Fig. 38.3 can be created with a rotating prism or mirror.

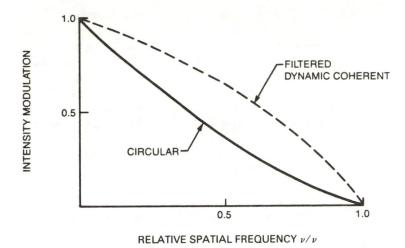

Fig. 38.4 The MTF of the dynamic coherent system and the incoherent transfer function of the same diameter lens (after Ref. 22).

38.6   OPTICAL LENS DESIGN CONSIDERATIONS

Refractive Designs: A refractive system capable of meeting the required specifications (Table 38.II) was uncovered in a preliminary study[24] showing that the concept is feasible. It is possible to meet these specifications by modifying a current design or by generating a new design. A comparison of this design with the S-Planar (a fifteen-element refractor) and the Microtropar (a nine-element refractive system) is given in Table 38.IV.

The proposed design has no third-order spherical aberration, coma, astigmatism, or Petzval sum. It does, however, have some uncorrected distortion. A thorough design to the eleventh or thirteenth order would be necessary to meet a distortion specification of 500 or 1000 Å in the image plane.

Reflective Designs: A solution to the lens design problem that met the specifications of the proposed system was *not* uncovered in our initial design study because of the large NA required. However, the dynamic coherent condenser configurations just discussed allow for obscuration and enable the use of lower NAs than in the reflected designs attempted because of the enhanced MTF of the dynamic coherent condenser shown in Fig. 38.4.

TABLE 38.IV

Comparison of Refractor Specifications for Proposed System with S-Planar and Microtropar as Microlithographic Lenses

	Proposed System	S-Planar	Microtropar
Focal length	100.0 mm	50.0 mm	40.3 mm
Field angle	±3.64 deg	±8.25 deg	±3.25 deg
NA	0.40	0.40	0.40
Reduction ratio	10:1	10:1	10:1
Image diameter	14.1 mm	14.5 mm	5.0 mm
Wavelength	0.20 μm	0.436 μm	0.365 μm
Track length	1210.0 mm	600.0 mm	484.0 mm
Resolution	0.5 μm	1.0 μm	0.8 μm

Effect of Condenser Configuration on Lens Design[25]: The enhanced MTF of the dynamic coherent optical system can be used to advantage in lithography where the linewidth limit occurs at 60% contrast or approximately one-third of the system cutoff frequency. As seen in Fig. 38.4, the enhanced MTF of this dynamic coherent system at one-third the cutoff frequency can reduce the NA of the lens in the design by ~30 to 40%. In reflective designs, this effect becomes even more dramatic because the normal MTF of an annular aperture is reduced 20 to 30% below the diffraction-limited MTF due to the presence of a central obscuration.

As discussed by Cronin and Smith,[22] the enhanced MTF of the dynamic coherent system arises because the modulation of a given spatial frequency is proportional to the arc length that a given spatial frequency makes within the entrance aperture of the lens. When this reasoning is applied to a dynamic coherent annular aperture (or reflective optical system with an obscuration), then some of the arc lengths are reduced due to the presence of the obscuration, and a modified MTF is obtained, as illustrated in Fig. 38.5.

Thus, for lithographic applications where linewidth design is tuned to approximately one-third the cutoff frequency to achieve 50 to 60% contrast in the image, a normal imaging lens with the dynamic coherent configuration can be used to great advantage if

$$\frac{\nu_0}{3} < \frac{\nu_0}{2} - \frac{\epsilon_0}{2} \, , \tag{38.1}$$

where ν_0 is the cutoff frequency of the lens. Equation (38.1) demands that the obscuration diameter be kept smaller than $d/3$. In other words, if the hole size in the reflector, ϵ, is chosen so that the one-third cutoff point is less than the notch in the MTF of Fig. 38.5, the enhanced modulation of the dynamic coherent condenser can be realized with a reflective optical system; i.e., the contrast gain of the dynamic coherent condenser is realized if the hole in the Cassegrain design is small enough. Thus, from Fig. 38.4 we see that we can decrease the NA of the optics by 25 to 30% and keep a 60% contrast for the proposed 0.5-μm linewidth system.

This effect of increasing the MTF is just the opposite of normal Cassegrain systems. In a normal Cassegrain system, the NA must be increased to hold a fixed modulation value to compensate for the midrange modulation decreases due to

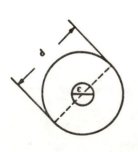
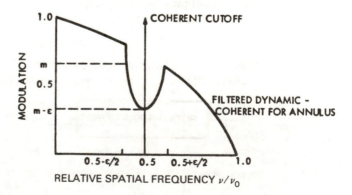

Fig. 38.5 Annular aperture having obscuration of diameter ϵ and a dynamic coherent MTF for the annulus showing the frequency notch due to the obscuration of the annulus (from Ref. 11).

TABLE 38.V
Proposed Reflective Systems

Parameter[a]	Proposed System No. 1	Proposed System No. 2
M	10.0	10.0
NA	0.5	0.28
LW (μm)	<0.5	~0.5
DOF (μm)	±0.32	±1.0
Focal accuracy (μm)	±0.03 to 0.06 (300 to 600 Å)	±0.1 to 0.2 (100 to 200 Å)
Align (μm)	< ±0.05 (500 Å)	±0.5 (500 Å)

[a] Key
M   = Reduction ratio of imaging lens
NA = Numerical aperture of imaging lens
LW = Linewidth capability of system
DOF = Depth of focus of imaging lens in image plane
Align = Alignment accuracy requirement.

the aperture obscuration. This realization should make it possible to consider reflective designs in the proposed lithographic printer and avoid the chromatic aberrations and multiple elements of the refractive designs.

Fiber Optic Illuminator: To avoid mechanical vibrations from the spinning prism or mirror that could cause system problems in the focus and alignment subsystems, we propose utilizing alternative annular source configurations. One such concept utilizes a fiber optic bundle[26] to realize an incoherent annular source which is equivalent to the dynamic coherent configuration.[23,25] The fiber optic bundle will be used to reshape the beam from a laser with a Gaussian intensity distribution to one with an annular intensity distribution. The fiber optic bundle will be bonded to match the laser beam shape and coated to reduce reflections. The individual fiber lengths would be randomized in length to destroy the temporal coherence relationship between the individual fibers. The fibers will then be mounted to an annular fixture of such a dimension as to be imaged to the edge of the high-resolution lithographic lens by the condenser lens.

Summary of Proposed Lens System: These parametric investigations suggest that a deep UV laser source, a dynamic coherent fiber optic condenser system, and a reflective optical system with a small central obscuration should be used in the proposed optical system. The preliminary specifications for two proposed systems of this type are given in Table 38.V. System No. 2 is more attractive because of its lower NA and larger depth of field (DOF). The refractive lens of Table 38.IV could also be used but the laser wavelength tightly controls the lens design and manufacture because of chromatic aberration.

38.7   FOCUSING AND ALIGNMENT CONSIDERATIONS

38.7.1   Focus

The depth of focus requirement of the optical system is given by

$$\text{DOF} \simeq \pm 0.4\lambda/(\text{NA})^2 , \tag{38.2}$$

and the accuracy requirement for an automated focusing device is $\simeq \pm 0.1$ to 0.2 DOF. Current commercial lithographic equipment utilizes various types (pneumatic, intensity profiling, focus wedge, or a fly's-eye lens) of focal sensors[6,7,27] to achieve focal accuracies of ±0.25 μm. The designs proposed here will require focal accuracies between 300 and 2000 Å, depending on the design selected. Equipment currently in use is not good enough to meet these accuracy requirements.

To overcome this limitation, we suggest a three-beam interferometer as an automatic focusing method.[28] In this device, the laser light strikes the beam splitter and part is reflected to the wafer with the remainder going to the reference mirror, which has a central hole. The three reflected beams are incident on the three slits in the entrance aperture of the lens. These three slits pass the two reference beams and the signal beam (the signal beam carries the information regarding the focal shift due to the wafer). These beams combine to form a three-beam interference pattern in the lens focal plane, the fringe amplitude modulation being proportional to the focal shift. Since this device works on the principle of amplitude modulation rather than detection of fringe shifts, it is more accurate than standard interferometers. The theoretical focal accuracy of this device is less than 100 Å, which is finer than that required by the proposed lens system.

The recently described AC interferometric technique[29] is also theoretically capable of measuring focal accuracies to the level required by the proposed system.

38.7.2 Alignment

The alignment requirement for the proposed system is ±500 Å. Accuracies on current production systems are five times larger than this,[7,29] but research systems using laser interferometers or intensity profiling and image processing techniques have reported[30-32] alignment precisions of ±250 Å. Intensity correlation techniques should also be capable of accuracies in this range. Mechanical and temperature controls to hold the alignment to these accuracies will demand state-of-the-art performance since we are approaching fundamental physical limits. These systems would have to be certified for production before the approach could be accepted.

The range of focusing and alignment requirements for the proposed reflective lens systems are summarized in Table 38.V.

38.8 OVERVIEW OF THE PROPOSED SYSTEM

The concept presented here for achieving the 0.5-μm linewidth limit is to introduce deep UV wavelengths into a step-and-repeat 10:1 reduction system. Current 10:1 reduction systems use refractive optics; therefore, this higher resolution capability cannot be achieved merely by changing the source wavelength because the chromatic aberrations would be severe. A new optical design is required for 0.5-μm lithography in a step-and-repeat configuration with a 1-cm die size on a 4-or 5-in. wafer.

Our conclusion for the optical system based on coherence linearity criteria is that this concept should at least reach and possibly exceed the projected linewidth goal of 0.5 μm at throughput rates of forty to sixty 4-in. wafers per hour. A reflective optical system using a Cassegrain or catadioptric configuration to remove chromatic and spherical aberrations coupled to a dynamic coherent excimer laser illuminator is a feasible design. A refractive optical system is also possible but it would be highly chromatic.

Photoresists sensitive to the deep UV have a sensitivity of ~100 mW/cm². This means that 400 mW/cm² is needed at the resist plane to create an exposure in 0.25 s. A variety of UV laser sources exist that easily meet these requirements. The resist thickness must be less than the depth of focus to reduce banding within the resist. With laser sources, a dynamic coherent optical condenser system can be used to increase the intermediate MTF values and simultaneously average the coherent noise. This condenser relieves the strain on the optical design and utilizes the high power of available UV lasers to achieve the required exposure time.

We suggest a three-beam interferometer as an automatic focusing method to maintain focal accuracies of 300 to 2000 Å. This device works on the principle of amplitude modulation rather than the detection of fringe shifts, giving it the required accuracy.

The alignment requirement for the proposed system is ±500 Å. Alignment systems engineered from current research results can be used to meet this alignment specification.

The concept may be adaptable as a modular change to an existing stepper system. In this case, the structural and functional elements of the mechanical subsystems design, including wafer positioning, structural stability (thermal and mechanical), carriage, platen, wafer holding, and work station considerations, could be carried over to this proposed system. The electronic and software subsystem controls for environmental control, wafer handling, reticle handling, and the man/machine interface could also be used with a minimum of change. New controls and procedures would have to be added to support the focus and alignment subsystems in order to operate the system with the operating tolerances suggested by this design.

38.9   CONCLUSIONS

Using several concepts discussed in this book, we have presented a parametric design approach and used it to show that very deep UV optical lithography looks feasible as a method for meeting the 0.5-μm linewidth requirements. However, development is required to achieve this goal. It appears that 0.25- to 0.5-μm linewidths are theoretically possible. However, the practical limit must be determined experimentally.

Resists sensitive in the very deep UV are necessary to meet these goals. Recent research results using UV laser lithography are encouraging in this area.[18,19] A tri-level or thin resist will be needed to reduce banding and to stay within the depth of focus of the optical system.

The dynamic coherent laser condenser yields high energy levels at the resist and it also relieves problems in the design of a reflective optical system. Focal tolerances can be maintained with a three-beam interferometer. Alignment systems currently existing in research must be developed and engineered to support this concept. In addition, this concept may be adaptable as a modular change to a currently existing stepper system.

REFERENCES

1. T. A. Znotins, "Examiner lasers: new vistas in semiconductor processing," *Lasers and Applications* 5(5), 71 (May 1986).
2. V. Pol et al., "Excimer laser-based lithography: a deep ultraviolet wafer stepper," in *Optical Microlithography,* H. L. Stover, ed., Proc. SPIE 633, 6–23 (1986).
3. M. Tipton, V. Marriott, and G. Fuller, "Practical I-line lithography," in *Optical Microlithography,* H. L. Stover, ed., Proc. SPIE 633, 24–31 (1986).
4. M. C. King and E. S. Muraski, "New generation of 1:1 optical projection mask aligners," in *Developments in Semiconductor Microlithography IV,* J. Dey, ed., Proc. SPIE 174, 70–74 (1979).
5. J. Roussel, "Step and repeat wafer imaging," in *Developments in Semiconductor Microlithography III,* R. L. Ruddell et al., eds., Proc. SPIE 135, 30–35 (1978).
6. S. Wittekoek, "Step and repeat wafer imaging," in *Semiconductor Microlithography V,* J. Dey, ed., Proc. SPIE 221, 2–8 (1980).
7. H. E. Mayer, E. W. Loebach, "A new step-by-step aligner for very large scale integration (VLSI) production," in *Semiconductor Microlithography V,* J. Dey, ed., Proc. SPIE 221, 9–18 (1980).
8. R. Hershel, "Optics in the Model 900 projection stepper," in *Semiconductor Microlithography V,* J. Dey, ed., Proc. SPIE 221, 39–43 (1980).
9. J. W. Bossung and E. S. Muraski, "Optical advances in projection photolithography," in *Developments in Semiconductor Microlithography III,* R. L. Ruddell et al., eds., Proc. SPIE 135, 16–23 (1978).
10. J. M. Roussel, "Submicron optical lithography?" in *Semiconductor Microlithography VI,* J. Dey, ed., Proc. SPIE 275, 9–16 (1980).
11. G. O. Reynolds, "Optical lithographic systems," U.S. Patent No. 4,450,358 (issued May 22, 1984).
12. B. J. Thompson, "Image Formation with Partially Coherent Light," *Progress in Optics,* E. Wolf, ed., Vol. 7, pp. 167–230, North Holland Publishing Co., Amsterdam (1969).
13. S. Lee, S. Grillo, and V. Miller, "Submicron optical lithography using an I-line wafer stepper," in *Optical Microlithography IV,* H. L. Stover, ed., Proc. SPIE 538, 17–22 (1985).
14. E. A. Chandross, E. Reichmanis, C. W. Wilkens, Jr., and R. L. Hartless, "Photoresists for deep UV lithography," *Solid State Technol.* 24(8), 81–85 (1981).
15. W. F. Cordes and R. F. Leonard, "Resist materials for high resolution photolithography," in *Semiconductor Microlithography VI,* J. Dey, ed., Proc. SPIE 275, 164–172 (1981).
16. S. Iwamatsu and K. Asanami, "Deep UV projection system," *Solid State Technol.* 23(5), 81–85 (1980).
17. T. W. Novak, "A new VLSI printer," in *Developments in Semiconductor Microlithography,* R. L. Ruddell et al., eds., Proc. SPIE 135, 36–43 (1978).
18. N. Shiotake and S. Yoshida, "Recent advances of optical step-and-repeat system," in *Electron-Beam, X-Ray, and Ion-Beam Techniques for Submicrometer Lithographies IV,* P. D. Blais, ed., Proc. SPIE 537, 168–174 (1985).
19. K. Jain, C. G. Willson, and B. J. Lin, "Ultrafast high resolution contact lithography using excimer lasers," in *Optical Microlithography—Technology for the Mid-1980s,* H. L. Stover, ed., Proc. SPIE 334, 259–262 (1982).
20. R. F. Leonard and W. F. Cordes, "New positive resist designed for use in the mid ultraviolet," in *Optical Microlithography II—Technology for the 1980s,* H. L. Stover, ed., Proc. SPIE 394, 125–133 (1983).
21. M. Lacombat, G. M. Dubrocucq, J. Massin, and M. Brévignon, "Laser projection printing," *Solid State Technol.* 23(8), 115–121 (1980).
22. D. J. Cronin and A. E. Smith, "Dynamic coherent optical system," *Opt. Eng.* 12(2), 50-55 (1973).
23. D. J. Cronin, J. B. DeVelis, and G. O. Reynolds, "Equivalence of annular source and dynamic coherent phase contrast viewing systems," *Opt. Eng.* 15(3), 276–279 (1976).
24. M. R. Hatch, Private Communication.

25. G. O. Reynolds, "Optical lithographic system having a dynamic coherent optical system," U.S. Patent No. 4,498,009 (issued Feb. 5, 1985).
26. D. A. Servaes and G. O. Reynolds, "Fiber optic condenser for an optical imaging system," U.S. Patent No. 4,560,235 (issued Dec. 24, 1985).
27. J. D. Buckley, "Expanding the horizons of optical projection lithography," *Solid State Technol.* 25(5), 77–82 (1982).
28. G. O. Reynolds, "High sensitivity focal sensor for electron beam and high resolution optical lithographic printers," U.S. Patent No. 4,493, 555 (issued Jan. 15, 1985).
29. C. L. Koliopoulos, S. Forbes, and J. Wyant, "Interferometer microscope for surface analyser," *J. Opt. Soc. Am.* 71(12), 1591–1592A (1981).
30. G. L. Resor and A. C. Tobey, "The role of direct step-on-the-wafer in microlithography," *Solid State Technol* 22(8), 101–108 (1979).
31. H. L. Stover, "Stepping into the 80's with die-by-die alignment," *Solid State Technol.* 24(5), pp. 112–120 (1981).
32. *Optical Microlithography II,* H. L. Stover, ed., Proc. SPIE 394 (1983).

Index